AF782171

RELAXIN 2000

RELAXIN 2000

Proceedings of the Third International Conference on Relaxin & Related Peptides
22–27 October 2000, Broome, Australia

Edited by

GEOFFREY W. TREGEAR
Howard Florey Institute,
University of Melbourne, Australia

RICHARD IVELL
Institute for Hormone and Fertility Research,
Hamburg, Germany

ROSS A. BATHGATE
Howard Florey Institute,
University of Melbourne, Australia

JOHN D. WADE
Howard Florey Institute,
University of Melbourne, Australia

KLUWER ACADEMIC PUBLISHERS
DORDRECHT / BOSTON / LONDON

Library of Congress Cataloging-in-Publication Data

International Conference on Relaxin & Related Peptides (3rd : 2000 : Broome, Australia)
Relaxin 2000 : proceedings of the third International Conference on Relaxin & Related Peptides, 2-27 October 2000, Broome, Australia / edited by Geoffrey W. Tregear ... [et al.].
p. cm.
Includes indexes.
ISBN 1-40200-068-5 (alk. paper)
1. Relaxin--Congresses. I. Tregear, Geoffrey W. II. Title.

QP572.R46 I575 2000
573.4'4--dc21

2001038878

ISBN 1-4020-0068-5

Published by Kluwer Academic Publishers,
P.O. Box 17, 3300 AA Dordrecht, The Netherlands.

Sold and distributed in North, Central and South America
by Kluwer Academic Publishers,
101 Philip Drive, Norwell, MA 02061, U.S.A.

In all other countries, sold and distributed
by Kluwer Academic Publishers,
P.O. Box 322, 3300 AH Dordrecht, The Netherlands.

Printed on acid-free paper

All Rights Reserved
© 2001 Kluwer Academic Publishers
No part of the material protected by this copyright notice may be reproduced or utilized in any form or by any means, electronic or mechanical, including photocopying, recording or by any information storage and retrieval system, without written permission from the copyright owner.

Printed in the Netherlands.

Contents

Section 3: Reproductive Physiology of Relaxin II - Uterus and Implantation

Section 4: Reproductive Physiology of Relaxin III – Ovary

Section 7: Physiology of Relaxin II – Male

Section 8: Preparation of Relaxin and Related Peptides

Section 9: Relaxin Bioassays

Section 10: Relaxin-induced Signal Transduction

Section 11: Relaxin-like Factors

Preface

The first international forum on relaxin was a workshop held in Hawaii in 1980. This was followed by the Kroc Foundation conference in Santa Ynes, California in 1981 and then the First International Congress on the Hormone Relaxin in Florence in 1982. After a hiatus of several years, the Second International Congress on the Hormone Relaxin was held in Adelaide in 1994. Since that time, the enormous advances in molecular biology and protein chemistry have led to a stellar growth in relaxin research. Novel sites of action of the hormone have been elucidated, numerous new mammalian primary structures determined, more detailed mechanisms of action postulated, the control of relaxin secretion better understood, non-pregnant roles of the hormone identified, and, importantly, relaxin-related peptides (particularly INSL 3 and 5) identified and possible synergistic actions with relaxin proposed. A raft of novel clinical applications of relaxin has also been explored. Together, these exciting developments made the time ripe for a third international conference. We were therefore delighted to welcome many colleagues, old and new, at *Relaxin 2000*, the first international forum on relaxin and related peptides in the new Millennium. This conference, which we hope will now be held every four to five years, was held at the Cable Beach Inter-Continental Resort at the delightful old pearling town of Broome in the far north-west of Australia from 22-27 October, 2000. It provided all of us with a unique opportunity to learn of the latest advances in the chemistry, biology and clinical applications of relaxin and relaxin-like peptides. It also allowed us to learn from the considerable expertise of the presenters and to develop new directions for future research. The conference was organized by the Howard Florey Institute, University of Melbourne, and supported by our Principal Sponsor, Connetics Inc (USA).

The enormous success of the meeting was testament to many fine contributions made by the participants and we thanks them sincerely. A summary of the scientific presentations has been published in *Trends in Endocrinology and Metabolism* [Ivell, R., 12 (3): 89-91, 2001]. Many individuals were responsible for the fine organization of the conference and in particular, we wish to acknowledge Ms Vivien Agosta of Travel Unlimited who tirelessly managed all aspects of both the resort and travel arrangements. We would also like to take this opportunity to record our sincere appreciation for the outstanding support that has been provided to the organizing committee by many members of both commerce and industry. This has been listed elsewhere in this book. In particular, the contribution of Connetics, Inc. (USA), both financial and scientific, is warmly acknowledged.

We were dismayed to learn of the death after a long illness of our friend and colleague, Fred Greenwood, two months before the conference. We dedicate this book to his memory.

Geoffrey W Tregear, Richard Ivell, Ross A Bathgate and John D Wade

The organizing committee of Relaxin 2000

Back Row, Left to Right: Alastair MacLennan, Roger Summers, John Wade, Ross Bathgate, Geoff Tregear, Richard Ivell, David Sherwood

Front Row, Left to Right: Elaine Unemori, Alastair Summerlee, Laura Parry

Relaxin 2000: 3rd International Conference on Relaxin & Related Peptides

22-27 October, 2000, Cable Beach Inter-Continental Resort, Broome, Western Australia

Chairman

Geoffrey Tregear	Howard Florey Institute, University of Melbourne

Organizing Committee

Geoffrey Tregear (Chair)	Howard Florey Institute, University of Melbourne
John Wade (Secretary)	Howard Florey Institute, University of Melbourne
Roger Summers (Treasurer)	Department of Pharmacology, Monash University
Alastair MacLennan	Department of Obstetrics & Gynaecology, University of Adelaide
Laura Parry	Howard Florey Institute, University of Melbourne

Programme Committee

Richard Ivell (Chair)	Institute for Hormone & Fertility Research, Hamburg
Ross Bathgate	Howard Florey Institute, University of Melbourne
O David Sherwood	Department of Molecular & Integrative Physiology, University of Illinois
Alastair Summerlee	Department of Biomedical Sciences, University of Guelph
Elaine Unemori	Connetics Corp., Palo Alto, California.

Audiovisual and photography

Marc Mathieu

Local Organizers (Howard Florey Institute)

Tania Burazin	Lucy Campbell
Antonia Claasz	Nicola Dawson
Tania Ferraro	Ping Fu
Mary Macris	Marc Mathieu
Sharon McDowell	Eira Parry
Irna Reytomas	Kathryn Smith
Yean-Yeow Tan	Stella Zhao

Sponsors

PRINCIPAL SPONSORS

Howard Florey Institute

Connetics Corporation (USA)

SPONSORS

Advanced Labs
AMRAD
Auspep
Beckman Coulter Australia
Biacore AB
Davies Collison Cave
FH Faulding & Co
Genentech Inc (USA)
Journals of Reproduction and Fertility Ltd
Schering AG (Germany)
The Haworth Press Inc (USA)
Waters Australia

SUPPORTERS

Broome Tourist Bureau
Cable Beach Inter-Continental Resort
Cook & Snell Pty Ltd
Mayflower Worldwide (UK)
Perth Convention Bureau
Travel Unlimited International

Faulding's Young Researcher & Student Travel Award Recipients

Dr Per Kristiansson
Department of Public Health
Uppsala University
Sweden

Dr Cait E MacPhee
Oxford Centre for Molecular Sciences
New Chemistry Laboratory
Oxford
UK

Dr Jacquelien Novak
Department of Obstetrics, Gynecology &
Reproductive Science
University of Pittsburg
USA

Dr Suresh Pillai
Department of Physiology
University of Maryland
USA

Dr Chrishan Samuel
Molecular Medicine Research Institute
California
USA

Dr Ida Vogel
Department of Obstetrics & Gynecology
Aarhus University Hospital
Denmark

Dr Ling Zhao
Laboratory of Genetics & Physiology
NIH/NIDDk/LGP
Maryland
USA

Dr Shuangping Zhao
Department of Molecular & Integrative
Physiology
University of Illinois at Urbana-Champaign
USA

Awards generously sponsored by F H Faulding & Co Limited, GPO Box 1618, Adelaide SA 5001, Australia

In memoriam

Frederick C Greenwood (1927-2000)

Frederick (Fred) C. Greenwood is a legend in the relaxin field. He died of liver cancer on the 8th of August 2000 at the age of 73. Together with his wife Gillian, Fred made a major contribution to relaxin research with many landmark publications over the last 20 years. His early work was with growth hormone. He also developed a method to radioactively label peptides and proteins with iodine (known as the Hunter-Greenwood method) which greatly facilitated the use of radioimmunoassay in endocrine research. His first paper on relaxin with Gillian was on the specificity of radioimmunoassays for relaxin, published in 1979 in the *Journal of Endocrinology*.

Fred was born and educated in England. In 1968, he left the Imperial Cancer Research Fund in London to join the Department of Biochemistry and Biophysics at the University of Hawaii and in 1973 he was appointed director of the Pacific Biomedical Research Center. Together with Gillian, he established the center as a leading international institute for teaching and research and a center of excellence in relaxin research.

We all remember Fred with great affection. He was a great mentor, scientist and friend to many of us in the relaxin field. He was often outspoken and unafraid to voice his convictions but always remained a gentleman. His energy and enthusiasm for the study of relaxin was a driving force in advancing the field. Many of us have very fond memories of one of the very first international meetings on relaxin held in Honolulu in 1980 which was co-hosted by Fred and Gillian.

Our thoughts are particularly with Gill and it is great respect and admiration that we dedicate this volume of the latest in relaxin research to our friend and colleague, Fred Greenwood.

ABBREVIATIONS

ACL	anterior cruciate ligament
BEA	bromoethylamine
cGMP	cyclic guanosine-3',5'-monophosphate
CHF	congestive heart failure
CL	corpus luteum
CX	connexin
E	17β-estradiol
EDRF	endothelium-derived growth factor
EGF	epidermal growth factor
eNOS	endothelial nitric oxide
ER	estrogen receptor
ERPF	effective renal plasma flow
ERVR	effective renal vascular resistance
ET	endothelin
GRF	glomerular filtration rate
hCG	human chorionic gonadotropin
3ß-HSD	3ß-hydroxysteroid dehydrogenase
IGF-1	insulin-like growth factor-1
iNOS	inducible nitric oxide
Ley I-L	Leydig cell insulin-like factor
L-NAME	N^G-nitro-arginine methyl ester
LUSF	lower uterine segment fibroblasts
LV	left ventricular
MALDI-TOF	matrix-assisted laser desorption time-of-flight
MAP	mitogen-activated protein
MDS	microdialysis system
MMP	matrix metalloprotease
nNOS	neuronal nitric oxide
NO	nitric oxide
NOS	nitric oxide synthase
OC	oral contraceptive
P	progesterone
PA	plasminogen activator
PAI	plasminogen activator inhibitor
PBS	phosphate buffered saline
PCK	protein kinase C
PMA	phorbol-12-myristate-13-acetate
RLF	relaxin-like factor
sc	subcutaneous
SVR	systemic vascular resistance
TGFβ	transforming growth factor beta
TFE	trifluoroethanol
TIMP	tissue inhibitor of metalloproteinases
tPA	tissue-type plasminogen activator
uPA	urokinase-type plasma activator
VEGF	vascular endothelial growth factor

SECTION 1

Plenary lecture

Hooked on relaxin: A career-long love affair with a fickle and elusive hormone

Bernard G Steinetz
NYU School of Medicine, Nelson Institute of Environmental Medicine, Tuxedo, NY 10987, USA

At the Second International Congress on the Hormone Relaxin, held in Adelaide, Australia in 1994, Bob Kroc, Roy Greep and I presented an *historical perspective* on relaxin which was published with the conference proceedings [1]. Therefore, rather than go over the same ground again, I would like to review some recent studies that have reinforced my belief that relaxin remains an elusive - yet seductive - hormone for investigation. Your indulgence is then sought for some philosophical observations on relaxin's rightful place in the insulin-like polypeptide family, and some speculations as to what should be investigated as relaxin-related phenomena. However, these observations are best considered within the context of a few sobering realities of where relaxin research stands today.

What is relaxin good for?

One problem that has plagued investigators over the years is the failure to discover practical applications for relaxin in the understanding or treatment of human diseases. As outlined in our presentation at the Second International Congress on Relaxin, the 1950's and 1960's witnessed the attempts of four different pharmaceutical companies to develop relaxin as a drug for human use [1]. Three of these companies actually marketed a relaxin preparation; two for premature labor and threatened abortion and one for cervical ripening. None of these products succeeded. The reasons for failure could largely be ascribed to anemic clinical data, poor choice of route and vehicle of administration and premature fixations on the part of company management (especially marketing!) as to what the hormone should be "good for".

After a long hiatus, during which significant advances continued to be made in basic studies of the chemistry and biology of relaxin, a biotechnologically-oriented pharmaceutical company assembled a team specifically to develop relaxin for cervical ripening at term pregnancy in women. However, in 1994, the disappointing results of clinical trials of relaxin for ripening of the human cervix were presented and discussed at the Adelaide Congress [2,3]. The company then summarily dropped the relaxin project and disbanded the team. Despite those negative results, it is the firm belief of many of us in the field that relaxin does, in fact play an important role in human cervical softening, as this activity has been amply demonstrated in many animal models [e.g., 4 for review].

Failure of the prior studies to have proven clinical efficacy of relaxin for cervical ripening may well have been the result of flawed experimental design rather than lack of

G.W. Tregear et al. (eds.), Relaxin 2000, 3–24.
© 2001 *Kluwer Academic Publishers. Printed in the Netherlands.*

hormonal activity. For example, in the failed clinical studies, human relaxin was administered in a gel by direct application to the cervix. This route was obviously chosen for reasons of expediency, but proof of absorption by demonstration of systemic plasma levels of relaxin was not obtained. In hindsight, a subcutaneous or intramuscular route of administration, together with a clear demonstration of elevation of plasma levels of relaxin would have provided far less ambiguous data. Moreover, as Weiss [5] has pointed out, cervical ripening can occur in the absence of circulating relaxin in women with ovarian agenesis in whom a donor pregnancy is maintained by exogenous progestins. It is unknown if the latter is an autocrine phenomenon involving local production of relaxin by the uterus or even by the cervix, itself. These structures have been shown to produce relaxin in other species such as the guinea pig [6]. However, most importantly, the clinical trial plans ignored some highly relevant basic science data: relaxin's effects upon the cervix can readily be demonstrated in non-pregnant animals [4 for review], thus avoiding confounding actions of endogenous pregnancy hormones, including relaxin, itself. The actions of relaxin on the cervix of nonpregnant women have not been investigated.

Finally, the earlier reports of positive effects of the hormone on the human cervix were conducted with porcine, not human relaxin [4 for review]. To my knowledge, no direct comparisons of the effects of porcine and human relaxin on the cervices of women or non-human primates have ever been made. It is thus conceivable, as Schwabe and Büllesbach [7] point out, that porcine relaxin may be more potent than the human hormone as a cervical "ripener", or be more effective in women with defective relaxin receptors. In any case, relaxin's role in human cervical softening is still moot.

Just before the present meeting in Broome, it was learned that the Connetics Company's Phase III clinical trials of human relaxin for the treatment of systemic sclerosis had failed to show a difference between relaxin and placebo [Erikson and Unemori, this volume]. This was despite encouraging data obtained in Phase II trials [8] and numerous basic studies showing clearcut effects of relaxin on collagen metabolism in human normal dermal and sclerodermal fibroblasts [9-11]. The reader should also be reminded that this was actually the second clinical investigation of relaxin for use in scleroderma, the first being undertaken some forty years ago by Casten and Boucek [see 4 for review]. Those of us who were close to that study were impressed by the dramatic changes in tissue pliability and deep muscle blood flow seen in some of the patients, and also by the fact that improvements were not sustained when, at Bob Kroc's insistence, an extract of porcine ovaries with low relaxin potency was substituted for the high potency porcine relaxin being used in the patients. As in the case of the cervical ripening studies cited above, human relaxin was substituted for porcine relaxin in the recent scleroderma trials, and the method of administration was changed from those employed in the earlier work (i.m. injections in a repository vehicle to constant infusion). Whether these or other factors are responsible for the present clinical failures, only time will tell. But somehow, to those of us who have pursued relaxin research since its infancy, the more things change, the more they seem to stay the same!

In any event, these consistently disappointing clinical trials cannot help but impact negatively on the willingness of granting agencies to fund research projects on relaxin. Both the NIH and private foundations place considerable emphasis on the relevance and potential application of submitted proposals to human medicine. Thus, relaxin's role in human physiology and its potential for human therapeutics remain unproven. Encouraging new results in human subjects are urgently needed! To its credit, the Connetics Company has not given up on relaxin because of the failed scleroderma trial, and has ongoing investigations for possible therapeutic uses in cardiovascular and other disease states. Moreover, as will be seen in a perusal of the present publication, basic research is providing many exciting new leads for potential clinical targets among the remarkable spectrum of biological activities exerted by this chameleon of a hormone.

What is relaxin bad for?

Ironically, the evidence that relaxin can have detrimental effects in human beings is as convincing as the evidence that it has beneficial actions is tenuous. In this age of in vitro fertilization and embryo transfer (IVF/ET), the techniques used to mimic the state of pregnancy in order to achieve implantation have provided ample opportunity to study the effects of relaxin excess on the outcome of human gestation.

Prior studies had clearly shown that excess relaxin can terminate pregnancy in rats [12] and guinea pigs [13], and can induce preterm delivery in cattle [14,15]. Thus, perhaps it is not surprising that the markedly elevated circulating relaxin concentrations, frequently observed after ovarian stimulation with exogenous gonadotropins and/or clomiphene (as used in many IVF/ET protocols), are associated with a high risk for prematurity [16-20]. These data certainly imply that relaxin, itself, (or some factor closely associated with relaxin) exerts a profound effect on human pregnancy. The elevated relaxin levels are not exclusively related to multiple pregnancies. Moreover, the threat of prematurity can be evidenced by preterm labor requiring tocolytics, or by cervical incompetence, requiring placement of a cerclage [5]. (The latter may be the best evidence that relaxin does indeed ripen the human cervix) What is more baffling is the observation that preterm delivery may occur in these women even though the pattern of relaxin secretion is similar to that observed during normal pregnancy; i.e., there is a significant decline in serum relaxin as pregnancy advances. Thus, the supposed consequences of the first trimester elevation in relaxin concentration are not observed until much later in pregnancy, when relaxin levels are much lower.

The above considerations suggest (to me, at least!) that, instead of just falling out in the first trimester, the fetus is normally protected against the dilatory actions of relaxin on the cervix and uterus by some as yet unknown factor(s). These forces apparently can be overridden by markedly elevated relaxin levels. An alternative possibility is that the hypothetical protective factor(s) is not expressed in individuals undergoing exogenous ovarian stimulation, and the elevated relaxin levels are caused by this deficiency. In either case, although there is ample evidence that relaxin, itself inhibits uterine motility [4], studies in laboratory animals show that under certain circumstances it can also induce release of oxytocin from the posterior pituitary [4] and enhance the activity of oxytocin on the pregnant preterm uterus of mice [21]. High doses of relaxin also terminate early pregnancy in rats by an unknown mechanism [12]. It seems possible that the underlying physiological mechanisms may be clarified by studies that focus on the effects of relaxin concentration on progesterone clearance and propagation of oxytocin-driven contractions of the pregnant uterus, and also on a search for the hypothetical relaxin-modulating factor if it is not progesterone.

In a collaborative study with William Swanson of the Cincinnati Zoo (to be reported elsewhere), serum relaxin concentrations were measured by RIA throughout pregnancy in domestic cats following an IVFET procedure which employed ovarian stimulation with exogenous gonadotropins. We wished to determine if the cats showed elevated relaxin levels similar to those seen in women following pregnancy induction by this procedure, and thus might serve as a useful model. There were no significant differences in gestational relaxin concentrations between normally bred cats and those which underwent IVFET. This finding may be related to the fact that cats normally undergo multiple ovulations, whereas women do not. Thus, the amount of luteal tissue induced by gonadotropins in the IVFET procedure and that of normally pregnant cats is probably similar. In women, on the other hand, there is likely to be a much larger amount of luteal tissue due to superovulation following gonadotropin injections, than that after normal ovulation of one or to follicles. The cat would, therefore, not appear to be a useful model for the study of the mechanisms leading to elevated relaxin levels in women undergoing IVFET.

In search of other pathological conditions where relaxin might play a role, we investigated canine hip dysplasia (CHD) in collaboration with George Lust of Cornell University, Christian Schwabe of the Medical University of South Carolina, and Laura Goldsmith of the New Jersey Medical School. CHD is an insidious polygenic disease that affects up to 50% of the larger breeds of dogs. It is similar in many respects to human congenital dislocation of the hip, which afflicts 2-5% of newborn infants. The earliest detectable symptom of CHD is a laxity of the ligaments supporting the hip joint, and if undetected and untreated, inevitably leads to osteoarthritis and loss of joint function in later life. We hypothesized that the hip joint laxity detected early in the lactation period was caused by milk-borne estrogens and relaxin, acting on the hip joint connective tissue of pups genetically predisposed to CHD ("HD+ pups"). ("Genetic predisposition" could mean either elevated levels of estrogens and relaxin in the milk of dysplastic bitches, or early/inappropriate expression of estrogen and/or relaxin receptors in HD+ pups.)

In work to be published elsewhere, relaxin concentrations in dog milk were several orders of magnitude greater than in that of other species, the daily intake by pups being greater than one milligram the first week of suckling. However, no differences were found in estrogen or relaxin concentrations in milk of dysplastic vs. nondysplastic bitches, nor in the sera of the suckling HD+ vs HD- pups. Thus, the hip joint laxity preceding CHD is not caused by selective transmission of elevated milk-borne hormones by dysplastic bitches, or by preferential absorption of these hormones by genetic HD+ pups. The data do not preclude inappropriate expression of estrogen and/or relaxin receptors in joint tissues of HD+ pups, and this avenue needs investigation.

In related experiments, we injected an aromatase inhibitor, CGS 16,949A into newborn HD+ pups daily during the lactation period, to block endogenous estrogen synthesis. Littermate controls received only vehicle. We theorized that if either estrogens or relaxin were responsible for hip joint laxity, CGS 16,949A would either: 1) act directly to prevent laxity by inhibiting estradiol synthesis, or 2) act indirectly by preventing the estrogen "priming" required for relaxin action.

The CGS 16,949A-treated HD+ pups exhibited a significantly reduced hip joint laxity at 8 months as compared with their vehicle-treated littermate controls (Fig. 1). Moreover, the diminished joint laxity was followed by X-ray evidence of a reduced incidence of hip dysplasia at 12 months, the odds-ratio for developing CHD being 8:1 in favor of the vehicle-treated littermates. However, our initial exuberance was tempered by the failure to detect any differences in serum estrogen concentrations between the treated and control pups during the entire treatment period. The estrogens detected in pup serum by RIA thus appear to be derived primarily from the ingested milk, and not from endogenous synthesis. Moreover, the mechanism whereby CGS 16,949A inhibited hip joint laxity remains to be established, and relaxin may have once again led us down a "primrose path"!

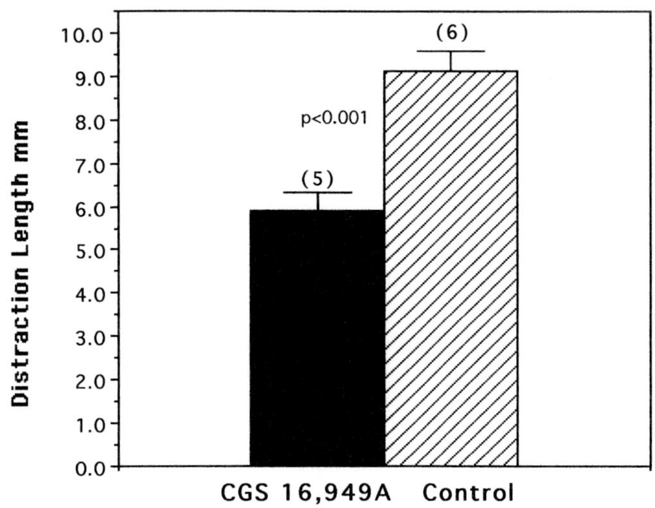

Fig. 1. Distraction lengths of hip joints of genetically dysplasia-predisposed (HD+) pups treated with an aromatase inhibitor, CGS 16,949A (2.5 mg/kg sc daily) or vehicle (control) during the entire lactation period. The distraction lengths were measured using a Smith-Biery distraction apparatus (71) at 8 months of age, and the average of the right and left hips of each pup were used to calculate the mean distractions for each group. Pups from two litters were randomly divided between the "treated" and "control" groups. N=number of pups in each group.

A related retrospective study on human congenital dislocation of the hip (CDH) was conducted in collaboration with Einar Sudmann and his colleagues at the University of Bergen, Norway, and Laura Goldsmith of the New Jersey Medical School. Using cord blood samples from consecutive births in a Norwegian national cohort study of sudden infant crib death (collected at the Rogaland County Hospital, Stavanger, Norway), we determined concentrations of estradiol, estrone, estriol, testosterone, DHEA and relaxin, using specific RIA kits for the steroids, and two methods for relaxin. Twenty five samples of cord blood from known cases of CDH (20 girls and 5 boys) were compared with 49 gender and time of birth-matched normal control samples (39 girls and 11 boys).

Summarizing results to be reported elsewhere, relaxin was only detected in 4 of the CDH cord blood samples and 11 of the controls, and the range of values of the two groups was similar. There were likewise no significant differences in the steroid levels between DH and control samples. Thus, estrogens and relaxin, as measured in cord blood are not associated with the presence or absence of CDH. Similar conclusions were reached by Vogel *et al* [22] and by Forst *et al* [23], although their hypotheses and interpretations of the data differed markedly. Vogel *et al* [22], and we had hypothesized that elevated relaxin levels would induce CDH by initiating joint laxity, whereas Forst *et al* [23] hypothesized that relaxin deficiency would induce CDH by subjecting the fetus to elevated intrauterine pressure, thus causing hip dislocation. It looks like we were all wrong! (Of course, none of these data preclude abnormal hormonal influences on the fetus or inappropriate fetal expression of estrogen and/or relaxin receptors in utero). In any event, another relaxin mystery goes unsolved.

What is relaxin needed for in the female?

Having struck out thus far in efforts to find out what relaxin is good for or bad for, a few relaxinologists, including myself, have turned attention to whether relaxin is needed at all, or whether we are all wasting our time on a vestigial hormone. Let us first consider the consequences of relaxin deficiency. The older literature is replete with data suggesting that relaxin is needed for pregnancy maintenance and parturition in both rats and mice [4 for review]. In pregnant mice, ovariectomized and maintained only on estradiol and progesterone, it was assumed that the resultant impairment of delivery was due to failure of an interpubic ligament to develop in order expand the diameter of the birth canal [4]. Indeed, addition of relaxin to the estrogen/progesterone regimen, restored interpubic ligament growth (the very hallmark of relaxin!), and delivery was once again normal in the ovariectomized mice [4]. These data, obtained using the ancient and venerable techniques of ovariectomy and hormone replacement were seemingly straightforward. However, these older studies did not allow for the possibility of non-ovarian sources of relaxin, or the operation of autocrine or paracrine mechanisms. More recently, new data have been obtained using relaxin-knockout pregnant mice, and relaxin-specific monoclonal antibodies in intact pregnant rats. The results of such studies are somewhat surprising.

Treatment of pregnant rats with monoclonal anti-rat relaxin antibody MCA-1 disrupts pregnancy and prevents normal delivery [24], at least in part because of cervical dystocia [25]. (It is noteworthy that the rat does not normally develop a long interpubic ligament such as that seen in mice and guinea pigs.) Moreover, lactation fails in relaxin deficient rats because the mammary glands and nipples do not develop [26,27]. In knockout relaxin-deficient mice, however, pregnancy and parturition are essentially normal, although no interpubic ligament is formed [28]. As in relaxin deficient rats, lactation fails in relaxin knockout mice, because the mammary glands and nipples fail to develop normally [28].

The above studies show that relaxin is required for normal lactation in rodents, and for parturition in rats. Not only is relaxin not required for parturition in mice, but interpubic ligament formation is also not required for successful delivery (a revelation that may cause Hisaw to roll over in his grave. Incidentally, these elegant experiments in genetically engineered mice actually confirm an ancient report by Crelin [29], who likewise found parturition in mice to be unimpaired when he simply tied their pubic bones together). These findings, however, tell us nothing about relaxin's need (if any) for other physiological functions in females or males, or indeed whether relaxin is needed in other than rodent species.

Serendipity introduced me to Steve Glickman, a noted researcher at the University of California, Berkeley. Glickman and his colleagues have published extensively on the endocrine physiology of the hyena (*Crocuta crocuta*), a species in which the female fetuses are born severely masculinized, due to in utero exposure to excessive concentrations of androgens [30]. Because of this, hyenas have had to undergo some remarkable adaptations in order to reproduce successfully. The clitoris is as large as the penis of the male hyenas, and the vaginal labia are fused to form a pseudo-scrotum. Thus, copulation and also delivery of fetuses must take place through the central urogenital canal of the clitoris (Fig. 2). Upon reading a newspaper article about this work, I decided to contact Steve Glickman; the remarkable changes in the clitoral connective tissues just had to be the work of relaxin! Steve was receptive to a collaborative study, and sent me a collection of blood samples from male and female hyenas in various reproductive stages. Our porcine relaxin antiserum R6 cross-reacted well with hyena relaxin, and we then assayed the samples. Our findings [31] can be summarized as follows: Whereas no relaxin could be detected in serum of males or pubertal or adult nonpregnant females, immunoactive relaxin was found in serum of juvenile females at the exact time of growth of the urogenital meatus. High concentrations of relaxin appeared in the serum of pregnant hyenas in the two weeks preceding parturition. Immunoassays of hyena tissues and serum obtained from uterine and ovarian veins indicated that the placenta was the primary source of relaxin. Further physiological experiments are clearly needed to establish whether relaxin is the hormone responsible for softening and dilation of the clitoris to allow for both insemination and parturition in this strange species.

Fig. 2. Clitoral expansion in a hyena about to give birth to her young. The neonates must pass through the urogenital canal of the clitoris in this remarkable adaptation necessitated by masculinization of all female fetuses in utero. (Reproduced with permission from Frank and Glickman (30).

There are other species where unusual adaptations of the pelvis and the birth canal cry out for study as potential targets for the actions of relaxin. One of these is the pocket gopher (*Geomys bursarius*), which early on was studied by Hisaw because of the remarkable pelvic adaptations that occur at puberty [6 for review]. The pelvis of this burrowing mammal is much too small for the passage of fetuses. With the approach of the first estrus, the pubic bones are gradually resorbed, leaving the pelvis open ventrally, and thus removing this major obstruction to parturition. Neither Hisaw, nor anyone else to my knowledge, has investigated the hormonal influences that bring about this adaptation. One must, however, suspect a role of estrogens and relaxin.

Crelin and Newton [32] were the first to describe the incredible changes in the symphysis pubis of the free tailed bat (*Tadarida brasiliensis cyanocephala*), without which, parturition would be impossible. Adult, nonpregnant females have an interpubic ligament, which grows to an unstretched length of about 2 mm during pregnancy. The ligament is composed of bundles of elastic and collagen fibers. The diameter of the bony birth canal is about 3.4 mm, whereas the diameter of neonatal bats averages 16 mm (Fig. 3). A simple calculation shows that the ligament must therefore increase in length to about 35 mm (about a 15-fold increase) to allow birth of the young, which, incidentally, takes place in about 1.5 minutes [32]. Bats, therefore, would seem likely to depend on estrogens and relaxin for survival of the species, but the endocrinological basis of their pelvic adaptations has not been studied.

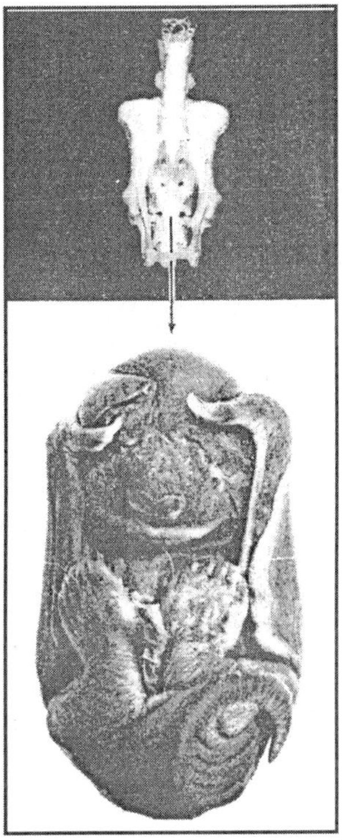

Fig. 3. A newborn free-tailed bat in its folded condition is shown below the bony pelvis of its mother, both at the same magnification (X3). The arrow passes behind the pubic ligament of the mother, indicating the route the newborn took through the birth canal. The ligament, which measures about 2 mm, must increase some 5-fold in length to allow passage of the neonate. From Crelin and Newton (32) with permission.

Circulating relaxin has been measured in several primate species during pregnancy. In the Old-World monkeys, baboons and great apes, the pattern of relaxin secretion is similar to that found in human beings; circulating levels of relaxin are highest in early pregnancy, and then gradually decline over the remainder of gestation. This is in contrast to the pattern observed in most non-primate mammals (e.g., rodents, carnivores, equids), where relaxin concentrations increase with advancing pregnancy, frequently reaching a high peak just prior to parturition [4].

In collaboration with Jim Mahoney at LEMSIP, the NYU Primate Center where I worked for 14 years, we measured serum relaxin levels during pregnancy in a New World Monkey, the marmoset (*Callithrix jacchus*), and found the pattern to be more like that of non-primate species, than that of Old World Monkeys, apes and human beings [33]. This work was confirmed and extended by Almuth Einspanier and her colleagues [34], who also delineated relaxin secretory patterns in the nonconceptive and preconceptive cycles.

We also injected prostaglandin F2α (PGF2α) at midpregnancy in two marmosets to induce luteolysis [33]. The monkeys were maintained for 4 days on the synthetic progestin, 17α-ethyl-19-nortestosterone to prevent fetal loss, but also to allow us to follow the resultant changes in endogenous serum progesterone concentration. PGF2α induced a precipitous decline in serum estradiol-17β, progesterone and relaxin (33). The steroids then rebounded and assumed their normal pattern up to the time of parturition, whereas serum relaxin concentrations remained low for the duration of pregnancy. Thus, elevated circulating relaxin does not appear to be necessary for normal delivery in this species. However, Einspanier *et al* [35] detected relaxin gene expression and relaxin immunoactivity in marmoset uterus and placenta, as well as in the ovarian thecal cells and corpora lutea. Thus, autocrine and paracrine effects of relaxin would not be precluded by functional luteolysis.

Relaxin secretion has also been studied in Old World Monkeys (cynomolgus and rhesus macaques), baboons and chimpanzees [36-42]. All have a circulating relaxin pattern similar to that of the human being during pregnancy, although the absolute concentrations may vary between species. In cynomolgus monkeys and chimpanzees, relaxin has also been detected in the late luteal phase of the cycle and peri-implantation period [38,42], again showing a pattern similar to that seen in women. Ovariectomy or luteectomy in pregnant macaques and baboons induces an immediate fall in serum relaxin (but not progesterone) to undetectable levels, indicating that circulating relaxin is primarily a luteal product [37,40,41]. Fetectomy on day 150 in macaques results in a sharp drop in progesterone, with a much slower decline in relaxin [37]. However, it is difficult to assess the overall importance of relaxin in pregnancy maintenance and parturition in these species, as there is evidence of its local production in the uterus and placenta in macaques and baboons [36,41].

In women, as in the non-human primates, there is abundant evidence that the source of circulating relaxin is the corpus luteum [4 for review]. No circulating relaxin can be detected in women (including those with ovarian agenesis) who achieved pregnancy through IVF-ET or IVF-GIFT and were maintained with exogenous estrogens and progestins [43,44]. Moreover, delivery is normal in such individuals. These findings have led some investigators to conclude that relaxin plays no role in either pregnancy maintenance or parturition. When it is true that ovarian relaxin is not needed for dilatation of the cervix at term [45], there is also important evidence that relaxin can be synthesized locally in the uterus and placenta, where it could play an autocrine or paracrine role [5].

In summary, total relaxin deficiency has only been achieved in knockout mice, and probably also in rats treated with the anti-rat relaxin monoclonal antibody, MCA-1. I say, "probably", because it has not yet been established that MCA-1 can neutralize other than circulating relaxin, and the antibody may not penetrate cells or tissues where the hormone is sequestered, or locally synthesized. Relaxin deficiency in rats impairs parturition, and prevents normal lactation, whereas in mice it primarily affects lactation, In nonhuman and human primates, absence of circulating relaxin has little or no effect on pregnancy maintenance or parturition. However, locally produced relaxin in the placenta or uterus may exert autocrine or paracrine effects, and so it cannot be concluded that it plays no role in these processes. I have also suggested that relaxin is a good bet to play a key role in reproduction of certain other species that have not been adequately studied. These include the hyena, bat and pocket gopher.

We are once again left dangling on the precipice of possible breakthroughs in the understanding of relaxin's role in female reproduction, with the uneasy feeling that if we turn our attention elsewhere, we may miss out on one of the bigger discoveries of our times.

What is relaxin needed for in the male?

The role of relaxin in male reproduction has been far less studied than that in the female, and the evidence is sometimes contradictory and confusing. Many of the studies have been reviewed [4] and can be briefly summarized here. In mammals, relaxin is a product of the prostate gland or seminal vesicles, whereas in birds and sharks it is a product of the testes. I had a pet theory that this dichotomy might be related to the effect of body temperature on relaxin secretion: in mammals, the testes are generally contained in a scrotal sac and maintained below the temperature of the remainder of the body. In birds and sharks, however, the testes are intra-abdominal. My theory held that the testes were the only source of relaxin in primitive male mammals. The evolution of the scrotum with its resultant decrease in testicular temperature adversely affected relaxin synthesis, leading to the secondary adaptation of other glands of the male reproductive tract to assume this function. The armadillo has already punched one hole in this theory: armadillos have abdominal testes, but immunoactive relaxin was detected in armadillo prostates, not testes [4]. (I am still waiting for the weight of the evidence for or against this theory from whales and elephants).

In work conducted with Chris Schwabe of the Medical University of South Carolina, Ian Callard of Boston University and Laura Goldsmith of the New Jersey Medical School, we extracted and affinity purified a relaxin from dogfish shark testes [46]. This material had greater bioactivity (8-16x) than dogfish ovarian relaxin and crossreacted poorly with an antiserum to the latter [46]. Whether male and female shark relaxins are different gene products is a matter of considerable interest, as it has been shown in one mammalian species (*Homo sapiens*) that prostate and ovarian relaxin are identical H2 relaxins [47].

Rather high concentrations of relaxin are found in seminal plasma of human beings and boars [4 for review], and these may be associated with maintenance of sperm motility and fertilizability, although others dispute these findings [4 for review]. In any case, there are no data on the effects of relaxin on sperm behavior in vivo, and no real evidence of its importance in male reproduction. In order to address these issues, collaborative work was begun with Jim Gelsleichter of the Mote Marine laboratory in Sarasota, FL, and Chris Schwabe of the Medical University of South Carolina, using the bonnethead shark as a model. Sharks diverged from the vertebrate stock some 350 million years ago, and many extant species have existed for 200 million years or more. Thus, sharks may provide important insights into the evolution of relaxin in the male, its transition from a testicular to an accessory gland hormone, and its importance in male reproduction.

Bonnethead sharks were captured and bled during the reproductive season, and the serum was assayed for relaxin using our rabbit antiporcine relaxin antiserum R6, 125I-human relaxin-2 as radioligand and dogfish shark relaxin (synthesized by Chris Schwabe) as standards. Both the male and female sharks exhibited a pattern of relaxin secretion that was closely related to the reproductive season. (Relaxin has been found in the circulation of males in only one mammalian species, the boar, and even in that case relaxin levels were not related to reproductive events [4]). In the male sharks, serum relaxin increased during the time of increased spermatogenesis (August-September), and remained elevated at the time of mating (October). No such pattern was seen in immature male sharks. Relaxin was also detected by immunostaining in both the testes and seminal vesicles. These small sharks lend themselves to physiological studies, and we plan to investigate the role of seminal relaxin on sperm transport through the female reproductive tract and fertilization *in vivo* and *ex vivo*. We hope that sharks may provide a key to new understanding of the role of relaxin in males, as they are the only species thus far studied in which the pattern of relaxin secretion is closely related to the events of the reproductive cycle. Thus, while it is much too early to reach conclusions about this work, our hopes are once again rekindled that this different approach will shed new light on the physiology of relaxin in the male.

Relaxin and cancer.

It has been known for years that relaxin stimulates mammary gland growth in mammals [4 for review], and relaxin gene expression has been demonstrated in human, guinea pig and rat breast cells [4]. However a possible role of relaxin in mammary carcinogenesis has received little attention. There is one ancient and obscure unconfirmed report that an impure relaxin containing extract of sow ovaries enhanced the tumor-promoting actions of subcutaneously implanted estrone pellets in an equally obscure strain of hooded rats [48].

Mario Bigazzi and his colleagues in Florence, Italy, who previously published extensively on effects of relaxin on the mouse mammary gland [4 for review], have provided data suggesting that relaxin may play a role in cancer prevention. Thus, relaxin was found to inhibit growth and promote differentiation of human MCF-7 breast cancer cells in vitro, very possibly by activating the l-arginine-nitric oxide (NO) pathway [49-51]. However, when MCF-7 cells were transplanted to nude mice, relaxin treatment did not significantly influence tumor growth or cell proliferation, although it did promote differentiation of the tumor cells towards myoepithelial-like and epithelial-like cells [52].

Breast cancer is the leading cancer cause of death among women, and it is well established that early pregnancy and childbirth dramatically reduces the susceptibility of women to this disease. Similarly, a prior pregnancy protects rats against breast cancer induction with n-methyl-n-nitrosourea (MNU) [53-56]. Because of the proven importance of relaxin for development of the mammary glands during pregnancy in rats

(4, for review), we postulated that these relaxin-induced changes in breast cells confer resistance to the action of carcinogens.

Sherwood's group at the University of Illinois has developed a monoclonal antibody (MCA1) specific for rat relaxin and has published numerous papers on its effects on the physiology of relaxin during pregnancy in rats [4 for review]. Among the activities noted were a marked inhibition of mammary gland development [25,26]. Thus, the MCA1 antibody seemed to offer an ideal way of investigating the possible protective role of relaxin against MNU-induce breast cancer.

Together with Dave Sherwood and Maarten Bosland (of NYU School of Medicine), we investigated the effects of MCA1 on susceptibility of rats to breast cancer. We hypothesized that treatment with MCA1 during pregnancy would negate the protective actions of relaxin on the mammary glands during pregnancy, and the treated rats would then be similar to virgin rats in their response to MNU. Accordingly, groups of rats were mated at age 50 days, and half were treated with MCA1 from days 12-20 of gestation. Controls received non-immune mouse gamma globulin. All rats were delivered at term by cesarean section because MCA1 neutralization of endogenous relaxin prevents normal parturition [24,57]. The previously pregnant rats (MCA1-treated and control) and a group of age-matched virgin rats were then injected with MNU on day 100, and the animals observed for tumors until day 240. Most of the virgin rats developed tumors as expected, but the previously pregnant MCA1-treated and control rats showed equal protection from the carcinogen. Thus prior pregnancy protected rats from subsequent breast cancer induction by MNU, in terms of percentage of tumour bearing rats (20% vs 65%) and number of tumours per rat (0.33 ± 0.1 vs 1.7 ± 0.4). There were no differences in the types of tumours (primary adenocarcinomas) or in tumour volume. However, neutralization of circulating relaxin did not negate the protective action of pregnancy. The data shed no light on a possible autocrine anticancer effect of relaxin synthesized by the mammary glands themselves, and further work is needed to clarify that issue.

What are the non-reproductive functions of relaxin?

The emphasis of this treatise has been put on the role of relaxin in reproductive processes. The older work on non-reproductive actions of relaxin is thoroughly reviewed by Sherwood [4]. However, recent work in many laboratories has demonstrated relaxin receptors in many non-reproductive tissues, including various parts of the brain and the heart atrium. Moreover, physiological and biochemical studies have provided new evidence that relaxin does indeed influence these organs in important ways. As these will be addressed in many of the chapters of this book, suffice it to say that, in the past, we may have been seduced into pursuing relaxin's many reproductive actions, while not paying sufficient attention to other equally important non-reproductive functions. There may also be lessons to be learned from the study of relaxin's effects on connective tissue during pregnancy. Recent evidence reported by Lubahn *et al.* and Galey *et al.* [this volume] revealed relaxin binding sites in anterior cruciate and palmar oblique ligaments. Such evidence suggests there may be more subtle but widespread effects of relaxin on ligamentous connective tissues.

The perennially elusive cow relaxin

This treatise would be incomplete without a paragraph or three about ruminant relaxin. In their original studies, Mike Fields and his associates reported what appeared to be a legitimate, bioactive relaxin-like principle in extracts of cow corpora lutea [58]. This relaxin-like substance(s) seemed to consist of covalently linked two chain molecules but also single chain peptides with differing molecular weights. All of these cross-reacted with an antiporcine relaxin antiserum. This gemisch eventually defied further purification. Subsequent concerted efforts to isolate this relaxin-like substance, the relaxin mRNA or to identify the relaxin gene itself in cow corpora lutea, have likewise uniformly failed [59].

Taking a totally different tack, and buoyed by the knowledge that human, dog and cow's milk contained immunoactive relaxin [52-54] we set about to isolate and affinity-purify cow relaxin from a milk product. Our intention was to send the material to Erika Büllesbach and Chris Schwabe at the Medical University of South Carolina for final purification and sequencing. We investigated fresh skim milk, evaporated skim milk and defatted, powdered milk and found each of these products to contain easily identified relaxin immunoactivity, using three different antiporcine relaxin antisera, including R6 (important because R6 only reacts with relaxins with intact disulfide linkages and thus would not "see" a single chain variant). After using classical relaxin isolation procedures, the extract was further purified by immunoaffinity chromatography on an R6 IgG/Affigel hydrazide column (Fig. 4). This material, which gave a single peak in the area anticipated for 6000 MW relaxin was then sent to Schwabe and Büllesbach for final purification and sequencing. We had finally succeeded in purifying the ephemeral cow relaxin! Or had we?? If the reader is still awake and alert, he or she may have already guessed that the relaxin-like material might once again vanish upon further purification, thus extinguishing any hope of determining its sequence. And that is exactly what happened. Thus, the enigma of cow relaxin persists at the time of this writing.

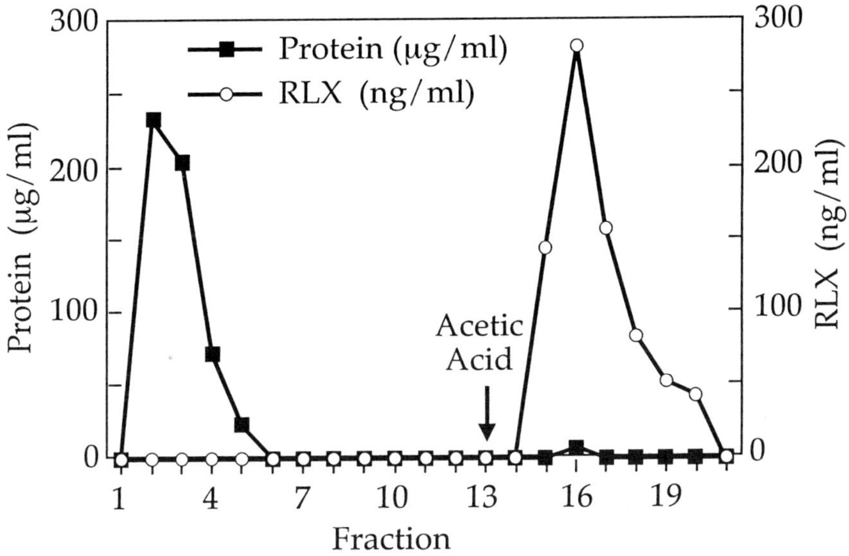

Fig. 4. Immunoaffinity purification of an extract of fat-free powdered cow's milk on an R6 IgG/Affigel column. The powdered fat-free milk was subjected to classical relaxin isolation steps including acid-acetone extraction, gel filtration on Sephadex G50 and carboxymethyl cellulose (CMC) ion exchange chromatography (see 4 for relaxin isolation methods). A 3 ml immunoaffinity column was prepared after protein A isolation of IgG from 1 ml of the R6 antiserum and coupling to the Affigel according to the manufacturer's instructions (Affigel HZ Immunoaffinity Kit, BioRad #153-6060). The fractions were eluted sequentially with phosphate buffered saline (PBS)/0.5M NaCl, PBS (pH7), and 0.1 M acetic acid, started at fraction 13. Radioimmunoassay of the fractions revealed an excellent separation of cow relaxin (peak at fraction 15) from non-relaxin proteins (fractions 1-6).

Philosophy

Evolution of Relaxin and Insulin

Traditionally, relaxin has been classified as an "Insulin-like growth factor", together with IGF-1 and -2. I strongly object — (and I hope that other relaxinologists will join me) — to relegating relaxin to a subservient role to insulin. Gillian Bryant-Greenwood rightly refers to this group of molecules as "Relaxins", because they are structurally so diverse and act upon so many different and seemingly unrelated targets (See Table 1). Insulin only deserves to be the lead molecule in this list by virtue of its earlier discovery. In terms of biological evolution, relaxin molecules likely came first, simply because they are so much more mutable than insulin and the IGFs, which are notably conserved in chemical structure. Of greater importance, Chris Schwabe makes a strong case that relaxins and insulins (and by inference, IGFs, bombyxins, etc.) arose quite independently of each other [7,60,61], and that it is erroneous to think conceptually of relaxin as an insulin mutation. (Readers unfamiliar with Schwabe's thought-provoking "Genetic Potential Hypothesis" should put these references [7,60,61] on their "must read" list. They go a long way towards making sense of protein evolution — or non-evolution, as he suggests). A final point to consider is that we tend to name hormones for their biological source (e.g., insulin) or for their biological activity. Thus relaxin was named for its effect on the guinea pig pubic symphysis. It could just as easily have been named "atriotocin", or how about "NTH" (Nipple Trophic Hormone) had these activities been discovered first.

Table 1. Some Relaxin Targets

Skin	Bone
Uterine myometrium	Cartilage
Uterine endometrium	Ileum
Uterine cervix	Ligaments
Mammary glands	Nipples
Heart atrium	Veins
Central nervous system	Adipocytes
Spermatozoa	

Insulin and IGFs are more than 500 million years old and can be found in non-vertebrate species [62,63]. Similarly, relaxins are more than 500 million years old and are also found among invertebrates and protochordates [7,60,61,64]. In addition, there are other structurally related polypeptides among the vertebrates and invertebrates (e.g., bombyxins and prothoracicotrophic hormone [PTTH]). Polypeptide hormones exert their actions primarily through binding to and activating receptors. Receptors and binding proteins for insulin, IGFs and relaxins are also found among non-vertebrate species [65-67].

In view of the above, I propose that the name of this group of polypeptide hormones be changed to the "Insulaxin Family", a name which more accurately describes these substances. In fact, this relationship has been cemented by Büllesbach and Schwabe's synthesis of a polypeptide with the receptor domain of insulin on the A chain and the receptor domain of relaxin on the B chain [68]. This substance can function as either an insulin or a relaxin. Moreover, the insect hormone, bombyxin was likewise converted to a functional relaxin by appropriate amino acid substitutions in its B chain [68]. This type of experiment underscores the obvious messages that: 1. The basic insulaxin molecule has been extremely important in evolution; 2. Mutations leading to amino acid substitutions in either the A or B chains can have a profound effect on biological activity. Not so obvious, but equally important, proteins with the ability to bind these molecules and translate such binding into hormonal actions have been available (or provided by random mutations) during this time period. The point at which a binding protein becomes a proper receptor is nebulous, but there are examples of proteins that bind both insulin and relaxin, without having the specificity to distinguish between their presumed receptor binding domains. Thus, insulin, like relaxin, enlarges the uterus of dogfish sharks, whereas relaxin, like insulin lowers blood sugar in this species [69]. The proposed "Insulaxin Family " is shown in Table 2. The relationships between these various factors and their receptors have only begun to be studied. It is anticipated that, when the receptor binding domains of each are known, that some surprising relationships may emerge.

Table 2. Insulaxin family

Relaxins	Insulins
Relaxin-Like Factors (RLFs, INSL-3)	IGFs
PTTH (Prothoracicotropic Hormone)	MIPs (Mollusc Insulin- related Peptides)
Bombyxins	
LIPs (Locust Insulin-related Peptides)	

The ultimate use of relaxin in human medicine

The following suggestion is based on both basic experiments and a highly relevant clinical observation. Relaxin should be tested for its ability to reverse the aging process. In concert with estrogens, relaxin "depolymerizes" the ground substance of connective tissue, inhibits collagen synthesis and activates collagenase, causing a change from a water-poor, colloid rich to a water rich, colloid poor state. This can be readily seen in the changes in the connective tissue of the mouse pubic symphysis, where proteoglycan and collagen are degraded over time following treatment with estradiol-17β and relaxin (Fig 5). As the estradiol disaggregates the glycosaminoglycan polymers, water is imbibed and the relaxin-activated collagenases can access the collagen crosslinks, making the tissue softer and more pliable. This is clearly a transformation to a more embryonic state- a reversal of the aging process.

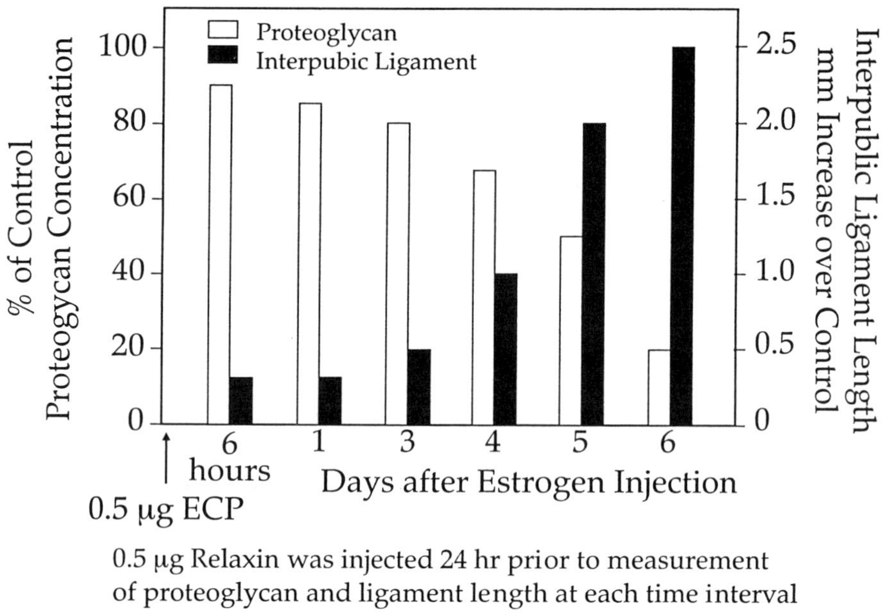

Fig. 5. Transformation of cartilage to "ligamentous" connective tissue: a model of reversal of the aging process by the interaction of estradiol-17β and relaxin? Immature female mice were injected with 5 µg estradiol-17β sc on day 0, following which groups of 10 mice each were killed at the time intervals shown on the graph. Pubic joints were removed, fixed and stained with Safranin O for proteoglycan and the amount of proteoglycan on each section estimated by a scanning technique. Additional groups of 10 similarly estrogen treated mice were then injected s.c. with 0.5 µg relaxin at the times shown on the graph and then killed 24 hrs later for measurement of the interpubic ligaments. The graph thus plots the response to relaxin (ligament length) in relation to the proteoglycan concentration at the time of relaxin injection. We interpret these data to mean that the estrogen induces the chondrocytes to degrade the proteoglycan in the matrix, "deprotecting" the collagen fibers from the collagenases activated by the relaxin.

Similarly, a human subject was immersed up to his armpits in pregnant sow ovaries and relaxin over a period of 30+ years; this exposure has obviously retarded the aging process! (Fig 6).

I rest my case.

Fig. 6. Sequential photographs of Robert L. Kroc taken over the years showing the obvious anti-aging effects of being immersed in the relaxin field for so long a time period

Epilogue

Great strides have been made in relaxin research, and many exciting discoveries will come to light in the proceedings of this, the 3rd International Conference on Relaxin and Related Peptides. Relaxin is earning a rightful place in vertebrate physiology, not just as a reproductive hormone, but as a hormone involved in general maintenance and perhaps, "fine tuning" of the organism, with targets in the central nervous and cardiovascular systems, as well as the better known targets in the integument, bone and cartilage. In this respect, relaxin complements and interacts with the other hormones that regulate metabolic processes: thyroid hormones, growth hormones and insulin. However, fundamental questions still remain concerning its role and importance in human physiology.

Once again I can think of no better way to end this treatise than with the timeless pronouncement of Mario Bigazzi: "(We are) trapped in this fascinating world of relaxin, where all the possible problems and difficulties seem to gather to make researchers crazy and results, always more interesting!"

Acknowledgements

The author gives special thanks to his many collaborators, without whose expertise, advice, discussions and friendship, this work (and these crazy ideas!) would not have been possible: Maarten Bosland, Erika Büllesbach, Jim Gelsleichter, Steve Glickman, Laura Goldsmith, George Lust, Jim Mahoney, Chris Schwabe, Dave Sherwood, Einar Sudmann, Bill Swanson and Gerson Weiss. And very special thanks to Bob Kroc who baited the hook in 1954 and has watched me wriggle ever since!

References

1. Steinetz, B.G., Kroc, R.L. and Greep, R.O., In MacLennan, A.H., Tregear, G.W., Bryant-Greenwood, G.D. (eds.) Progress in Relaxin Research, Global Publications, Singapore, 1995, p. 3.
2. Brennand, J.E., Calder, A.A., Leitch, C.R., Greer, I.A., Chou, M.M. and MacKenzie, I.Z., In MacLennan, A.H., Tregear, G.W., Bryant-Greenwood, G.D. (eds.) Progress in Relaxin Research, Global Publications, Singapore, 1995, p. 380.
3. Bell, R., Permezel, M., MacLennan, A., Hughes, C., Healy, D., Brennecke, S., Trudinger, B. and Mallon, K., In MacLennan, A.H., Tregear, G.W., Bryant-Greenwood, G.D. (eds.) Progress in Relaxin Research, Global Publications, Singapore, 1995, p. 350.
4. Sherwood, O.D., In Knobil, E. and Neill, J.D. (eds.) The Physiology of Reproduction, second edition, Raven Press, 1994, p 861.
5. Weiss, G., Clin. Obstet. Gynecol., 38 (1995) 293.
6. Hisaw, F.L. and Zarrow, M.X., Vit. Horm., 8 (1950) 151.
7. Schwabe, C. and Büllesbach, E.E., FASEB J., 8 (1994) 1152.
8. Seibold, J.R., Korn, J.H., Simms, R., Clements, P.J., MPH, Moreland, L.W., Mayes, M.D., Furst, D.E., Rothfield, N., Steen, V., Weisman, M., Collier, D., Wigley, F.M., Merkel, P.A., Csuka, M.E., Hsu, V., Rocco, S., Erickson, M., Hannigan, J., Harkonen, W.S. and Sanders, M.E., Ann. Intern. Med., 132 (2000) 871.
9. Unemori, E.N. and Amento, E.P., J. Biol. Chem., 265 (1990) 10681.
10. Unemori, E.N., Bauer, E.A. and Amento, E.P., J. Invest. Dermatol., 99 (1992) 337.
11. Unemori, E.N., Beck, L.S., Lee, W.P., Xu, Y., Siegel, M., Keller, G., Liggitt, H.D., Bauer, E.A. and Amento, E.P., J. Invest. Dermatol., 101 (1993) 280.
12. Steinetz, B., Butler, M., Sawyer, M., O'Byrne, E. and Giannina, T., Contraception, 14 (1976) 487.
13. Deanesly, R. and Zarrow, M.X., Endocrinology, 73 (1963) 522.
14. Musah, A.I., Schwabe, C., Willham, R.L. and Anderson, L.L., Endocrinology, 118 (1986) 1476.
15. Bagna, B., Schwabe, C. and Anderson, L.L., J. Reprod. Fert., 91 (1991) 605.
16. Bell, R.J., Sutton, B., Eddie, L.W., Healy, D.L., Johnston, P.D. and Tregear, G.W., Fertil. Steril., 52 (1989) 85.
17. Hanning, R.V. Jr., Canick, J.A., Goldsmith, L.T., Shahinian, K.A., Erinakes, N.J. and Weiss, G., Am. J. Obstet. Gynecol., 174 (1996) 227.
18. Mushayandebvu, T.I., Goldsmith, L.T., Von Hagen, S. Santoro, N., Thurston, D. and Weiss, G., Obstet. Gynecol., 92 (1998) 17.
19. Weiss, G., Goldsmith, L.T., Sachdev, R., Von Hagen, S. and Lederer, K., Obstets. Gynecol., 82 (1993) 821.

20. Kristiansson, P., Svardsudd, K., von Schoultz, B. and Wramsby, H., Hum. Reprod., 11 (1996) 2036.
21. Steinetz, B.G., Beach, V.L. and Kroc, R.L., Endocrinology, 61 (1957) 271.
22. Vogel, I., Andersson, J.E. and Uldbjerg, N. J., Pediatr. Orthop., 18 (1998) 535.
23. Forst, J., Forst, C., Forst, R. and Heller, K.D., Arch. Orthop. Trauma Surg., 116 (1997) 209.
24. Hwang. J-J., Shanks, R.D. and Sherwood, O.D., Endocrinology, 125 (1989) 260.
25. Hwang, J-J. and Sherwood, O.D., Endocrinology, 123 (1988) 2486.
26. Hwang, J-J., Lee, A.B., Fields, P.A., Haab, L.M., Mojonnier, L.E. and Sherwood, O.D., Endocrinology, 129 (1991) 3034.
27. Kuenzi, M.J. and Sherwood, O.D., Endocrinology, 131 (1992) 1841.
28. Zhao, L., Roche, P.J., Gunnerson, J.M., Hammond, V.E., Tregear, G.W., Wintour, E.M. and Beck, F., Endocrinology, 140 (1999) 445.
29. Crelin, E.S., Proc. Soc. Exper. Biol. Med., 86 (1954) 22.
30. Frank, L.G. and Glickman, S.E., J. Zool. Lond., 234 (1994) 659.
31. Steinetz, B.G., Randolph, C., Weldele, M., Frank, L.G., Licht, P. and Glickman, S.E., Biol. Reprod., 56 (1997) 1301.
32. Crelin, E.S. and Newton, E.V., The Anatomical Record, 164 (1969) 349.
33. Steinetz, B.G., Randolph, C. and Mahoney, C.J., Biol. Reprod., 53 (1995) 834.
34. Einspanier, A., Nubbemeyer, R., Schlote, S., Schumacher, M., Ivell, R., Fuhrmann, K. and Marten, A., Biol. Reprod., 61 (1999) 512.
35. Einspanier, A., Zarreh-Hoshyari-Khah, M.R., Balvers, M., Kerr, L., Fuhrmann, K. and Ivell, R., Human Reprod., 12 (1997) 1325.
36. Ghosh, D., Stewart, D.R., Nayak, N.R., Lasley, B.L., Overstreet, J.W., Hendrickx, A.G. and Sengupta, J., Hum. Reprod., 12 (1997) 914.
37. Nixon, W.E., Reid, R., Abou-Hozaifa, B.M., Williams, R.F., Steinetz, B.G. and Hodgen, G.D., In Greenwald, G.S. and Terranova, P.F. (eds.) Factors Regulating Ovarian Function, Raven Press, 1983, p 427.
38. Stewart, D.R., Stouffer, R., Overstreet, J.W., Hendrickx, A. and Lasley, B.L., Endocrinology, 132 (1993) 6.
39. Weiss, G., Steinetz, B.G., Dierschke, D.J. and Fritz, G., Biol. Reprod. 24 (1980) 565.
40. Castracane, V.D., D'Eletto, R. and Weiss, G. In Greenwald, G.S. and Terranova, P.F. (eds.) Factors Regulating Ovarian Function, Raven Press, 1983, p. 415.
41. Castracane, V.D., Lessing, J., Brenner, S. and Weiss, G., J. Clin. Endocrinol. Metab., 60 (1985) 133.
42. Steinetz, B.G., Randolph, C. and Mahoney, C.J., Endocrinology, 130 (1992) 3601.
43. Johnson, M.R., Abdalla, H., Allman, A.C., Wren, M.E., Kirkland, A. and Lightman, S.L., Fertil. Steril., 56 (1991) 59.
44. Petersen, L.K., Larsen, E. and Hansen, A., Eur. J. Obstet. Gynecol. Reprod. Biol., 65 (1996) 255.
45. Eddie, L.W., Cameron, I.T., Leeton, J.F., Healy, D.L. and Renou, P., Lancet, 336 (1990) 243.
46. Steinetz, B.G., Schwabe, C., Callard, I.P. and Goldsmith, L.T., J. Androl., 19 (1998) 110.
47. Winslow, J., Bourell, J., Weiss, G., Reed, B., Stults, J. and Goldsmith, L.T., Endocrinology, 130 (1992) 2660.
48. Cutts, J.H., Canadian Cancer Conf., 6 (1994) 50.
49. Bani, D., Bull. Cancer, 84 (1997) 179.
50. Bani, D., Masini, E., Bello, M.G., Bigazzi, M. and Sacchi, T.B., Cancer Res., 55 (1995) 5272.

51. Sacchi, T.B., Bani, D., Brandi, M.L., Falchetti, A. and Bigazzi, M., Int. J. Cancer, 57 (1994) 129.
52. Bani, D., Flagiello, D., Poupon, M.F., Nistri, S., Poirson-Bichat, F., Bigazzi, M., and Bani Sacchi, T., Virchows Arch., 35 (1999) 509.
53. Russo, J., Calaf, G., Sohl, N. et al. Ann. N.Y. Acad. Sci., 698 (1993) 1.
54. Moon, R.C., Int. J. Cancer, 4 (1969) 312.
55 Thorndarson, G., Jin, E., Guzman, R., Swanson, S.M., Nandi, S., Talamantes, F., Carcinogenesis, 6 (1995) 2847.
56. Grubbs, C.J., Hill, D.L., McDonough, K.C. and Peckham, J.G., J.N.C.I., 71 (1983) 625.
57. Lao Guico-Lamm, M. and Sherwood, O.D., Endocrinology, 123 (1988) 2479.
58. Fields, M.J., Roberts, R. and Castro-Hernandez, A. In Bryant-Greenwood, G.D., Niall, H.D. and Greenwood, F.C. (eds.) Relaxin, New York:Elsevier/North-Holland, 1981, p. 119.
59. Hartung, S., Kondo, B., Abend, N., Hunt, N., Rust, W., Balvers, M., Bryant-Greenwood, G. and Ivell, R., In MacLennan, A.H., Tregear, G.W. and Bryant-Greenwood, G.D., (eds.) Progress in Relaxin Research, Global Publications Services, Singapore, 1995, p. 439.
60. Schwabe, C. and Warr, G., Perspectives Biol. Med., 27 (1984) 465.
61. Schwabe, C. and Büllesbach, E.E., Relaxin and the fine structure of proteins. Springer-Verlag, Berlin Heidelburg, 1998, p. 175.
62. McCrory, J.E. and Sherwood, N.M., Cell. Biol., 16 (1997) 939.
63. Reinecke, M. and Collet, C., Int. Rev. Cytol., 183 (1998) 1.
64. Georges, D. and Schwabe, C., FASEB J., 13 (1999) 1269.
65. Hegyesi, H., Kovacs, P., Falus, A. and Csaba, G., Cell. Biol. Internat., 22 (1998) 493.
66. Hwa, V., Youngman, O.H. and Rosenfeld, R.G., Endocrine Rev., 20 (1999) 761.
67. Kohidai, L., Schiess, N. and Csaba, G., Comp. Biochem. Physiol. Pharmacol. Toxicol. Endocrinol., 126 (2000) 1.
68. Büllesbach, E.E., Steinetz, B.G. and Schwabe, C., Biochemistry, 35 (1996) 9754.
69. Koob, T.J., Laffan, J.J. and Callard, I.P., Biol. Reprod., 31 (1984) 231.
70. Steinetz, B.G., Büllesbach, E.E., Goldsmith, L.T., Schwabe, C. and Lust, G., Biol. Reprod., 54 (1996) 1252.
71. Smith, G.K., Biery, D.N. and Gregor, T.P., J.A.V.M.A., 196 (1990) 59.
72. Crelin, E.S., Trans. N.Y. Acad. Sciences. Ser. II, 31 (1969) 1049.
73. O'Byrne, E.M., Ann. N.Y. Acad. Sci., 380 (1982) 217.

SECTION 2

Reproductive physiology of relaxin I: Perimpartum period

Effects of relaxin on the cervix and the vagina in the rat

O David Sherwood[1,2] and Shuangping Zhao[1]
[1]Department of Molecular and Integrative Physiology and [2]College of Medicine, University of Illinois at Urbana-Champaign, 401 South Goodwin Avenue, Urbana, IL 61801, USA

Introduction

Relaxin and progesterone are secreted by the corpora lutea throughout the second half of 23 - day rat pregnancy [1]. Estrogen secretion by developing follicles is important during this period because relaxin's effects on the reproductive tract are estrogen dependent [1].

Two models were developed to determine the physiological roles of relaxin during rat pregnancy. With the *MCA1-treated pregnant rat model*, a monoclonal antibody for rat relaxin designated MCA1 is administered intravenously daily from day 12 through day 22 to neutralize the biological actions of rat relaxin [2, 3]. With the *ovariectomized pregnant rat model* [4], rats are ovariectomized on day 9 and then given replacement therapy with progesterone and estrogen in subcutaneous (sc) Silastic implants. Relaxin is given sc by miniature osmotic pump.

Studies with both models demonstrated that the duration of delivery is prolonged several fold and the incidence of live births is reduced about 50 percent in relaxin-deficient rats [3,4]. Interestingly, the effects of relaxin on birth are more pronounced in rats than in mice. Six of eight mice that lacked a functional relaxin gene delivered their young as rapidly as did controls [5]. We think relaxin's effects on the cervix are more important in rats than in mice. The rat cervix is potentially more difficult for fetuses to pass through than is the mouse cervix. In the rat, two lumens extend through the uterine horns and the cervix. In the mouse, fetuses pass through one cervical lumen (Fig. 1).

G.W. Tregear et al. (eds.), Relaxin 2000, 27–34.
© 2001 *Kluwer Academic Publishers. Printed in the Netherlands.*

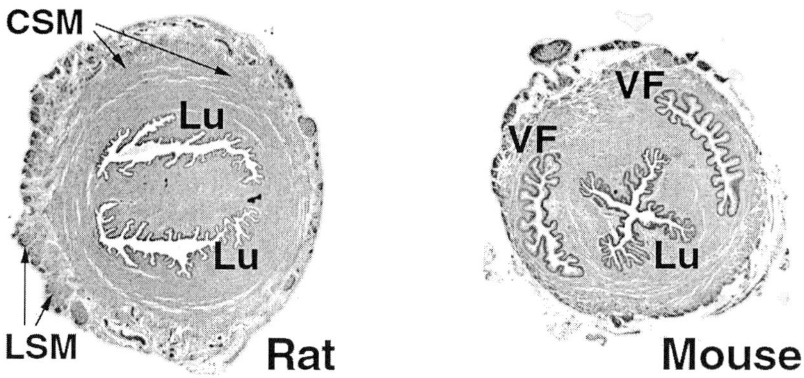

Fig. 1. Cervices obtained at 1400 h on day 22 of pregnancy from a Sprague Dawley rat and at delivery from a BALB/c mouse. CSM, circular smooth muscle; LSM, longitudinal smooth muscle; Lu, lumen; VF, vaginal fornix.

The wet weight of the rat cervix is only about three-fold that of the mouse, and the circumference of the cervical lumina do not differ in the two species at term (Fig. 2). The small cervix is likely a greater impediment at birth in rats than in mice because the pup weight in rats is nearly five times greater than that in mice (Fig. 2).

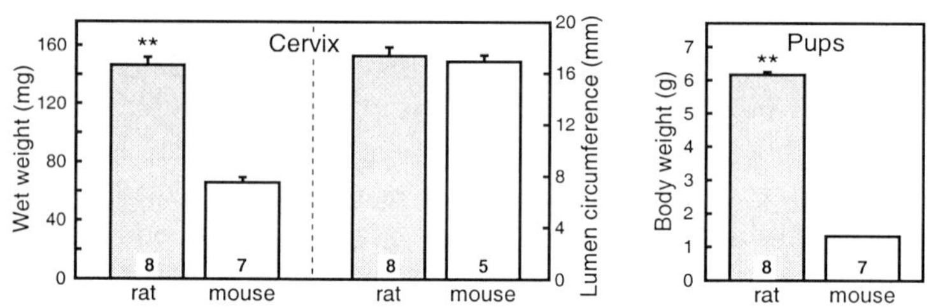

*Fig. 2. Mean (+ SEM) cervical wet weight, lumen circumference, and pup weight at term in rats and mice. Lumen circumference was determined by morphometric analysis of photomicrographs. Data in rats compiled from [4,6-12]. (**, $P < 0.01$). In this and subsequent figures the number of animals is indicated at the base of each column.*

Relaxin has dramatic effects on the rat cervix [4, 6-12]. From day 9 to day 22 of pregnancy the wet weight of the cervix in untreated control rats increases about three fold [4,13], but in relaxin-deficient rats it is only about 55% that controls (Fig. 3A). Relaxin markedly increases the extensibility of the cervix during pregnancy (Fig. 3B).

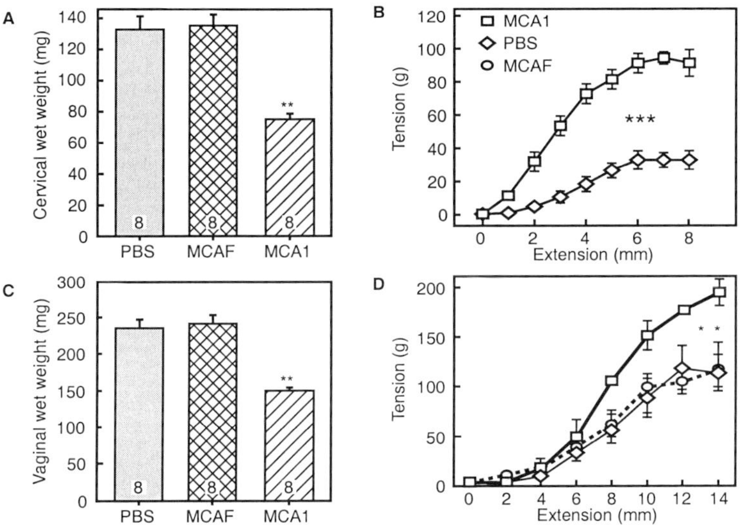

*Fig. 3. Mean (+ SEM) cervical (A) and vaginal (C) wet weights and cervical (B) and vaginal (D) extensibility in MCA1-, MCAF- (monoclonal antibody for fluorescein) and PBS- (phosphate buffered saline) treated rats on day 22 of pregnancy. Significant difference from controls (**, P < 0.01; ***, P < 0.001). From [6,8,10] with permission.*

Relaxin also promotes growth and extensibility of the rat vagina during pregnancy (Fig. 3C, D). This chapter contains an overview of studies conducted to better understand mechanisms associated with relaxin's effects on the rat cervix and vagina.

Results and Discussion

Consider first the cervix (Fig. 1). The stroma adjacent to the epithelial cells that line the lumina consists predominantly of the extracellular matrix components collagen, elastin, proteoglycans and the glycosaminoglycan hyaluronic acid. The stroma also contains cells such as fibroblasts, mast cells, leukocytes and blood vessels. Toward the periphery there are circular smooth muscle cells and longitudinal smooth muscle cells.

Light microscopic analysis of cervices removed on day 22 of pregnancy from MCA1-treated and control rats revealed that relaxin alters several histological characteristics of the cervix. Within the stroma, relaxin reduces the density and organization of collagen fiber bundles, reduces the length of elastin fibers and increases the cross-sectional area of arteriole lumina [14]. The separation of collagen fibers likely permits them to be readily pulled past one another as the fetus enters the cervix during labor. The reduction in length of elastin fibers may contribute to increased cervical extensibility. The enlargement of arteries may contribute to the increase in cervical water content that occurs during the second half of pregnancy [1]. Relaxin also alters the biochemical composition of the rat cervical stroma. Relaxin increases the proteoglycan (dermatan sulfate, heparan sulfate), hyaluronic acid and DNA content [4, 13]. It has been postulated that dermatan sulfate and hyaluronic acid accumulate in the interstices between collagen fibrils and thereby contribute to the dispersion of the collagen [15,16].

Relaxin increases the accumulation of new cells in the cervix (Fig. 4A, C). Within the stroma, relaxin increases new cells associated with arterioles (Fig. 4B) and also isolated cells that are likely largely fibroblasts. Relaxin has little, if any, effect on smooth muscle cell proliferation [9]. The increase in cells associated with cervical arterioles may contribute to the relaxin-dependent increase in cross-sectional area of arteriole lumens [14]. The marked increase in epithelial cells accommodates the more than 50% greater circumference of the cervical lumen in control rats than in relaxin-deficient rats [9].

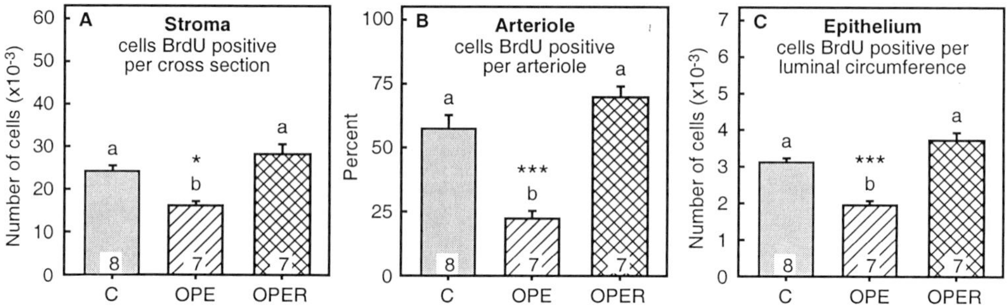

Fig. 4. Influence of relaxin on the accumulation of new stromal cells (A), cells associated with arterioles (B) and epithelial cells (C) in the rat cervix during the second half of pregnancy. The ovariectomized pregnant rat model was used. On day 9 of pregnancy, 78-day-old Sprague-Dawley rats were ovariectomized (0) and fitted sc with Silastic implants that provided blood levels of progesterone (P) and estrogen (E) similar to those that occur during pregnancy. The thymidine analog 5-bromo-2'-deoxyuridine (BrdU) and porcine relaxin (R) [17] were administered sc continuously from day 9 through day 22 via osmotic pump. All values are means + SEM. Values with different superscript letters differ (, $P < 0.05$; ***, $P < 0.001$). From [9] with permission.*

Relaxin may influence the activity of cervical epithelial cells. On day 22 of pregnancy, epithelial cells from control rats appeared have a greater abundance of vacuolated epithelial cells than did cervices from MCA1-treated rats [14]. A recent preliminary study provided limited evidence that relaxin may promote the production of mucin in these cells. Morphometric analysis of cervical sections obtained on day 22 of pregnancy with the *ovariectomized pregnant rat model* (9) indicated that the percentage of cross-sectional area of the epithelium occupied by mucin in relaxin-replete group OPER rats was greater than in relaxin-deficient group OPE rats (Fig. 5). Whereas, it appears that relaxin increases mucin production in the *ovariectomized pregnant rat model*, it cannot be concluded that endogenous relaxin increases mucin production during normal rat pregnancy. Use of the far less invasive *MCA1 pregnant rat model* to determine the influence of neutralization of endogenous relaxin on epithelial cell mucin content is needed to provide much stronger data from which to draw conclusions.

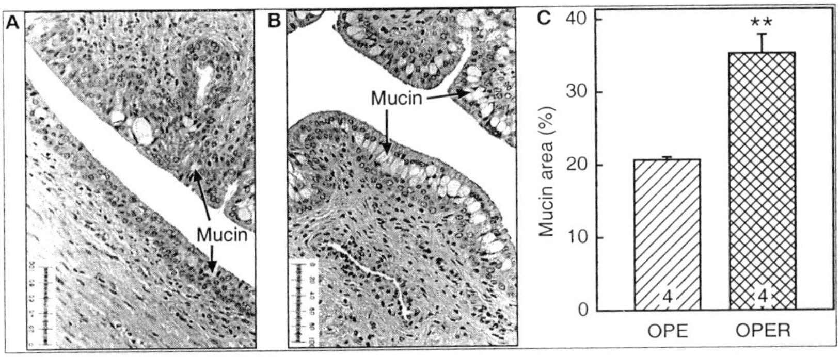

*Fig. 5. Influence of relaxin on mucin content in cervical epithelial cells. Photomicrographs of mucicarmine-stained cross sections of cervices obtained on day 22 of pregnancy from relaxin-deficient group OPE (A) and relaxin-replete group OPER (B) rats. The percentage of the epithelial cell area occupied by mucin is given in C. Values are means + SEM. (**, $P < 0.01$).*

The cells that contain relaxin receptors and trigger the hormones dramatic effects on the physical properties, histological characteristics and biochemical composition of the cervix are not known with certainty. Definitive identification of relaxin's target cells in the cervix and other target tissues awaits the characterization of the relaxin receptor. Nevertheless, studies that used a biotinylated-relaxin probe with an immunohistochemical procedure identified cells that contain specific and saturable relaxin-binding sites (putative relaxin receptors) in the rat cervix [18]. Binding was associated with epithelial cells, circular smooth muscle cells, longitudinal smooth muscle cells and cells associated with blood vessels. We have interpreted the finding that relaxin binding sites are associated with several cell types to indicate that relaxin's effects on the cervix are likely attributable to multiple paracrine interactions involving epithelial cells and also stromal cells within the extracellular matrix.

Molecules that mediate relaxin's diverse effects on the rat cervix remain to be identified. Prostaglandins do not appear to be involved. Employment of an extremely high dose of the cyclooxygenase inhibitor indomethaciin did not attenuate the acute effects of relaxin on growth or extensibility of the cervix in nonpregnant rats [19]. The role of the biological signal molecule nitric oxide (NO) was also examined. During pregnancy, the rat cervix expresses three isoforms of nitric oxide synthase (NOS): the so-called constitutive isoforms, endothelial NOS (eNOS) and neuronal NOS (nNOS), and inducible NOS (iNOS) [20,21]. When NO synthesis was inhibited with the nonselective NOS inhibitor N^G -nitro-L-arginine methyl ester (L-NAME) throughout the last six days of pregnancy, the duration of delivery was prolonged [20]. Additionally, when cervices were removed on day 20 of pregnancy and incubated overnight with L-NAME, they were less extensible than controls [20]. Whereas these findings led to the hypothesis that relaxin's effects on the rat cervix are mediated to a large extent through NO, that does not appear to be the case. When NO synthesis was inhibited in nonpregnant rats by a high dose of L-NAME, relaxin's acute effects on growth of the cervix were attenuated to only a modest degree and relaxin's effects on extensibility of the cervix were not affected [22].

Consider next relaxin's effects on the vagina. Several of relaxin's effects on the vagina are similar to its effects on the cervix. At term, the vagina in relaxin-deficient rats is only about two thirds as large as that in controls [4,8,10] (Fig. 3C), and it is less extensible than in controls (Fig. 3D). Light microscopic analysis supports the view that relaxin likely utilizes similar mechanisms to bring about its effects on the tensile properties of the vagina and cervix. As with the cervix, the collagen density is lower, elastin fiber length is shorter, and arterial lumen cross-sectional area is greater in the vaginas of controls than in the vaginas of relaxin-deficient MCA1-treated rats (Fig. 6).

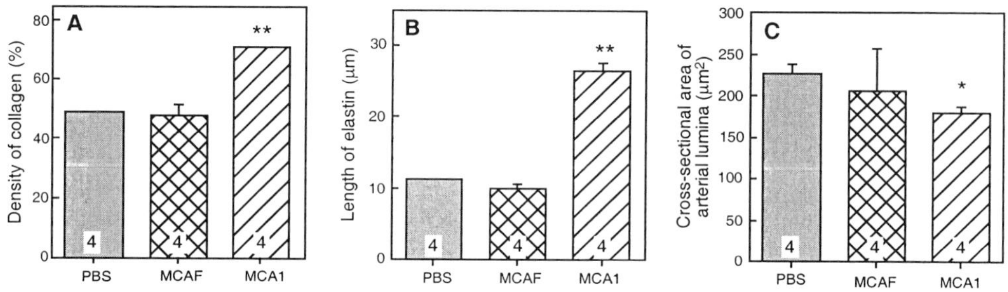

Fig. 6. Mean (+SEM) density of collagen fiber bundles (A), length of elastin fibers (B), and cross-sectional area of arterial lumens (C) in vaginas obtained from PBS-, MCAF-, and MCA1-treated rats on day 22 of pregnancy. Asterisks indicate a significant difference from PBS control (, $P < 0.05$) or both controls (**, $P < 0.01$). From [10] with permission.*

In the vagina, relaxin increases both stromal and epithelial cells [10]. Unlike the cervix, the vagina has a highly stratified epithelium that consists of basal cells and mucus-secreting cells (Fig. 7). Relaxin's effects on the vaginal epithelium are striking. It is readily apparent by visual examination that the height of not only the basal cells but also the mucus-secreting cells is greater in relaxin-replete control rats than in relaxin-deficient MCA1-treated rats (Fig. 7). This finding is attributable, at least in part, to more layers of both basal and mucus-secreting cells in the relaxin-replete animals [10]. Because the number of cell layers per stratum as well as the circumference of the lumen are about 50% greater in controls than in MCA1-treated animals, the total epithelial cells in each stratum per circumference of the vaginal lumen in control rats is more than twice that in MCA1-treated rats [10]. The dramatic relaxin-dependent increase in vaginal epithelial cells that occurs during pregnancy may serve physiological roles that protect both the fetus and the mother. The increase in epithelial cells contributes to the marked increase in the circumference of the vagina, which in turn protects the fetuses by facilitating rapid and safe delivery. Additionally, the increase in both basal and mucus secreting cells may protect the vagina of the mother from injury during delivery. Effects of relaxin on mucin production in the vaginal epithelium, if any, are not known.

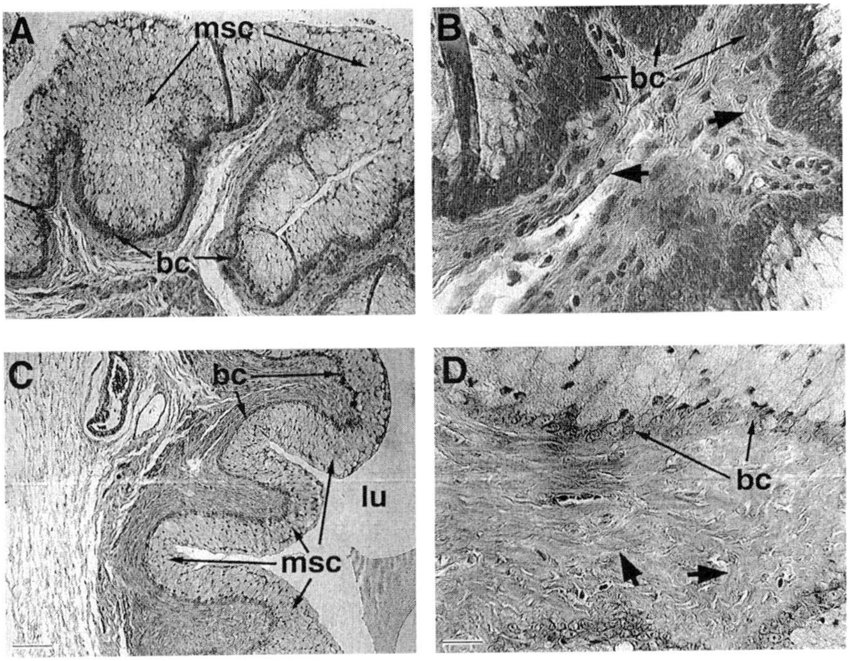

Fig. 7. Photomicrographs of Gomori's trichrome-stained cross-sections of vaginas obtained on day 22 of pregnancy from PBS-treated control (A, B) and MCA1-treated rats (C, D). bc, basal epithelial cells; msc, mucus-secreting epithelial cells; lu, lumen; black arrows, collagen fiber bundles. Bar in C = 100 um. A and C are the same magnification. Bar in D = 25 um. B and D are the same magnification. From [10] with permission.

The mechanisms whereby relaxin brings about growth and softening of the vagina are not known. The findings that relaxin not only has similar effects on the histological characteristics of the cervix and vagina but also binds to similar cell types in the cervix [18] and vagina [8] make it seem likely that relaxin mediates its effects, at least in part, by similar mechanisms in the two portions of the lower reproductive tract.

Whereas progress has been made toward an understanding of the physiological effects of relaxin on the reproductive tract in the pregnant rat, important problems remain to be solved. There is need to characterize the relaxin receptor(s) so that cells that contain relaxin receptors can be identified unequivocally. There is need to identify signal transduction pathways that mediate relaxin's effects so that informed approaches toward the development of pharmacological agents that regulate cervical remodeling can be taken. Whereas it is well established that relaxin's effects on the lower reproductive tract are estrogen dependent, little is known concerning the cellular sites of action or mechanisms whereby estrogen enables relaxin to bring about its dramatic effects. Also unknown are the mechanisms whereby relaxin brings about the accumulation of cells within the cervix and vagina during the second half of pregnancy. There is recent evidence that it does so, at least in part, by inhibiting the rate of apoptosis in both portions of the lower reproductive tract [23]. Finally, it is not known if relaxin has effects on the human cervix that are similar to its effects on the rat cervix. If that is found to be the case, there will be renewed interest in the possibility of employing relaxin to promote growth and/or softening of the cervix at the induction of delivery in women.

Acknowledgments

The authors thank B. Sylavong for supervision of animal care and the School of Life Science Artist Service for preparing the figures.

References

1. Sherwood, O.D., In Knobil, E. and Neill, J.D. (Eds.), The Physiology of Reproduction. 2nd ed., Raven Press, New York, USA, 1994, p. 861.
2. Lao Guico-Lamm, M., Voss, E.W. and Sherwood, O.D., Endocrinology, 123 (1988) 2472.
3. Lao Guico-Lamm, M. and Sherwood, O.D., Endocrinology, 123 (1988) 2479.
4. Burger, L.L. and Sherwood, O.D., Endocrinology, 136 (1995) 4820.
5. Zhao, L., Roche, P.J., Gunnersen, J.M., Hammond, V.E., Tregear, G.W., Wintour, E.M. and Beck, F., Endocrinology, 140 (1999) 445.
6. Hwang, J-J. and Sherwood, O.D., Endocrinology, 123 (1988) 2486.
7. Hwang, J-J., Shanks, R.D. and Sherwood, O.D., Endocrinology, 125 (1989) 260.
8. Zhao, S., Kuenzi, M.J. and Sherwood, O.D., Endocrinology, 137 (1996) 425.
9. Burger, L.L. and Sherwood, O.D., Endocrinology, 139 (1998) 3984.
10. Zhao, S. and Sherwood, O.D., Endocrinology, 139 (1998) 4726.
11. Kuenzi, M.J. and Sherwood, O.D., Endocrinology, 131 (1992) 1841.
12. Zhao, S., Malmgren, C.H., Shanks, R.D. and Sherwood, O.D., Endocrinology, 136 (1995) 1892.
13. Downing, S.J. and Sherwood, O.D., Endocrinology, 118 (1986) 471.
14. Lee, A.B., Hwang, J-J., Haab, L.M., Fields, P.A. and Sherwood, O.D., Endocrinology, 130 (1992) 2386.
15. Kokenyesi, R. and Woessner, J.F. Jr., Biol. Reprod., 42 (1990) 87.
16. Golichowski, A.M., King, S.R. and Mascaro, K., Biochem. J., 192 (1980) 1.
17. Sherwood, O.D. and O'Byrne, E.M., Arch. Biochem. Biophys., 160 (1974) 185.
18. Kuenzi, M.J. and Sherwood, O.D., Endocrinology, 136 (1995) 1367.
19. Sherwood, O.D., Jungheim, E.S., Masferrer, J.L. and Cramer, J.M., Endocrinology, 139 (1998) 867.
20. Buhimschi, I., Ali, M., Jain, V., Chwalisz, K. and Garfield, R.E., Hum. Reprod., 11 (1996) 1755.
21. Ali, M., Buhimschi, I., Chwalisz, K. and Garfield, R.E., Mol. Hum. Reprod., 3 (1997) 995.
22. Sherwood, O.D., Olson, L.M., Zhao, S. and Little, H.R., Endocrinology, 141 (2000) 2458.
23. Zhao, S., Fields, P.A. and Sherwood, O.D., Endocrinology, 142 (2001) 2221.

Relaxin retards apoptosis in the cervix and the vagina during pregnancy in rats

Shuangping Zhao[1] and O David Sherwood[1,2]

[1]*Department of Molecular and Integrative Physiology and* [2]*College of Medicine, University of Illinois at Urbana-Champaign, 407 South Goodwin Avenue, Urbana, IL 61801, USA*

Introduction

During the second half of the 23-day rat pregnancy, the cervix and vagina grow and become more extensible [1-4]. Cervical wet weight increases about three fold [1-3], and vaginal wet weight increases about two fold [3]. Relaxin, which is secreted by the corpora lutea throughout the second half of rat pregnancy [5], plays a major role in promoting this growth of the lower reproductive tract. Relaxin induces more than a 40% increase in the wet weights of the cervix and vagina by term [2-4,6,7]. The relaxin-depended growth of the cervix and vagina is accompanied by increased cell content [3,4,6]. Relaxin increases the number of epithelial cells and thereby contributes to the increase in the circumference of the cervical and vaginal lumina [6-8]. Relaxin also increases the number of stromal cells in the cervix and vagina [6-8]. The increased cellular content of the lower reproductive tract likely contributes to relaxin's vital role in facilitating rapid and safe delivery of the pups [9].

The cellular mechanism(s) that regulates the accumulation of cervical and vaginal cells during the second half of pregnancy is not known. Homeostatic control of cell number results from the dynamic balance between apoptosis and cell proliferation [10]. A striking feature that occurs in nearly all cases of apoptosis is the activation of endonuclease activity, which cleaves cellular DNA between regularly spaced nucleosomal units. The end result is the generation of DNA fragments that can be stained immunohistochemically and localized *in situ* [11].

The first hypothesis tested in this study is that the rate of apoptosis in cervical and vaginal cells declines during the second half of rat pregnancy. Finding that it does, we tested the hypothesis that relaxin inhibits apoptosis by determining if the rate of apoptosis in the cervix and vagina increases following the withdrawal of circulating endogenous relaxin.

Results and Discussion

In order to determine if there is a decline in the rate of apoptosis in the lower reproductive tract during the second half of rat pregnancy, cervices and vaginas were collected from primiparous Sprague-Dawley-derived rats on each of days 5, 10, 15, 18 and 21 of pregnancy and days 1 and 2 postpartum.

Cervical and vaginal wet weights did not differ on days 5 and 10, but they increased progressively between day 10 and day 23 of pregnancy (also designated day 1

G.W. Tregear et al. (eds.), Relaxin 2000, 35–40.
© 2001 *Kluwer Academic Publishers. Printed in the Netherlands.*

postpartum). Wet weight of both portions of the reproductive tract decreased dramatically by day 2 postpartum (Fig. 1).

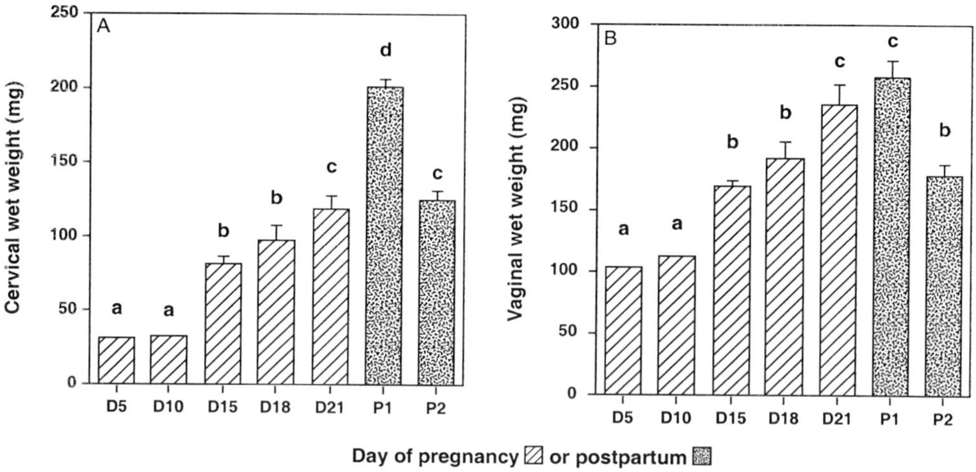

Fig. 1. Mean (± SEM) cervical (A) and vaginal (B) wet weights during rat pregnancy and early postpartum. n=4. Different superscript letters indicate a significant difference (P<0.05).

Tissues were fixed for 24 h in 10% neutral buffered formalin, dehydrated in an ascending series of ethanol, cleared with xylene, embedded in paraffin and sectioned (5 μm). Terminal deoxynucleotidyl transferase-mediated dUTP nick end-labeling (TUNEL) [11] in conjunction with morphological analysis was employed to determine the rate at which cells were undergoing apoptosis. The percentage of TUNEL-labeled cells (labeling index = LI) was determined by counting TUNEL-labeled cells divided by total cells in the same field and then multiplying by 100. Data were obtained from 4 rats/group, 3 sections (approximately 20 μm apart)/rat, and at least 500 cells per section. Thus at least 6000 cells were analyzed per group for the epithelium and 6000 cells were analyzed per group for the stroma.

The rate of apoptosis was reduced during the second half of pregnancy in both the cervix and the vagina. The LI of TUNEL-labeled cells in the cervical epithelium (Fig. 2A) was approximately 2% on days 5 and 10, declined to approximately 0.5% on days 15-21 and then rose dramatically to about 18% by the second day following delivery. The LI in the cervical stroma (Fig. 2B) was generally lower than in the epithelium, but the change in its rate of occurrence followed a similar temporal pattern. The cervical stroma LI was significantly lower on day 21 than earlier in pregnancy, and it increased dramatically during the first two days post partum.

The rates of apoptosis in the vagina followed a temporal pattern that similar to the cervix (Fig. 2C,D). The LI in the vaginal epithelium and stroma were lower on days 18 and 21 than during days 5 or 10 of pregnancy or days 1 and 2 postpartum.

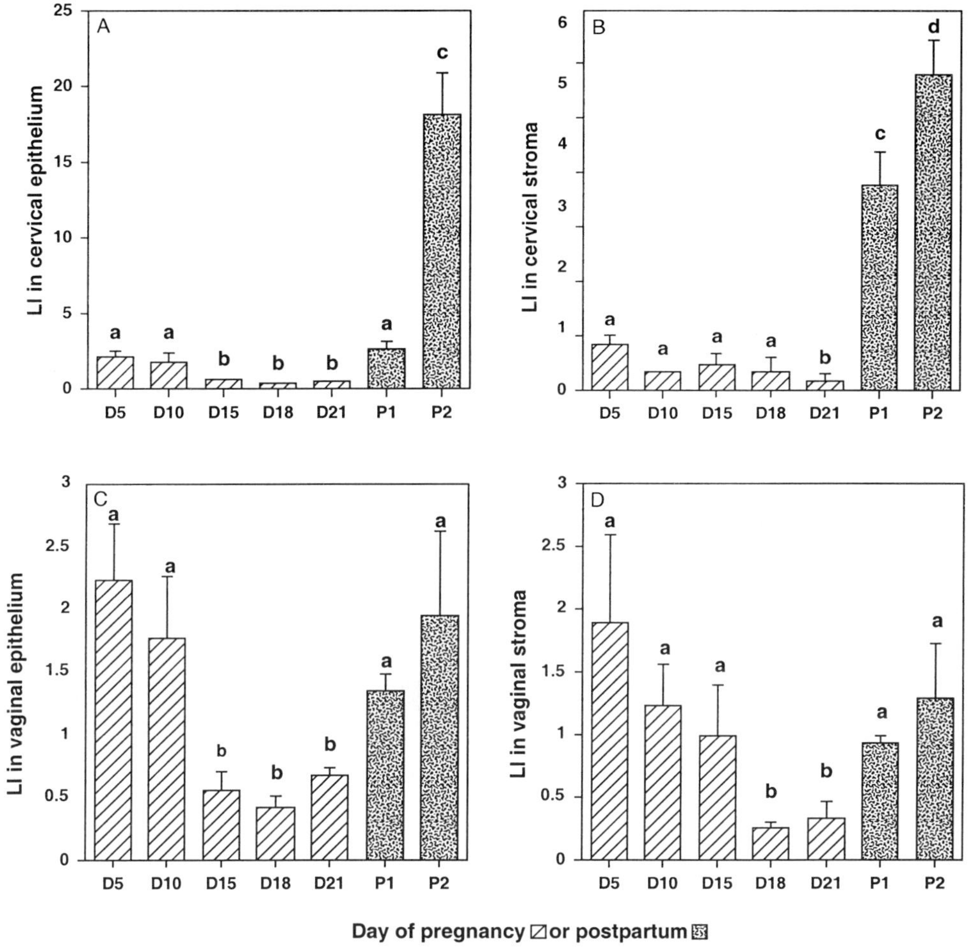

Fig. 2. Mean (± SEM) labeling index of TUNEL-labeled cells in the cervical epithelium (A,), cervical stroma (B), vaginal epithelium (C) and vaginal stroma (D) during pregnancy and early postpartum. n=4. Different superscript letters indicate a significant difference (P<0.05).

One function of apoptosis is to maintain steady cell numbers in tissues [10]. However, when tissues grow, as is the case of the lower reproductive tract during pregnancy, a mechanism that can contribute to an increase in cell numbers is a reduction in the rate at which cells undergo apoptosis. This report indicates that during the first half of pregnancy, when the lower reproductive tract is not growing extensively, the rate of apoptosis is moderate and steady. However, during the second half of pregnancy, when cervical and vaginal growth increase progressively, the rate of apoptosis is reduced several fold. Also consistent with the view that apoptosis plays a role in regulating cell numbers in the lower reproductive tract is the observation that the rate of apoptosis is extremely high after parturition, when the size of the cervix and vagina decrease markedly.

The observation that the rates of apoptosis are relatively low during the second half of pregnancy when relaxin is secreted into the blood provides indirect evidence that relaxin may inhibit the rate of apoptosis. Accordingly, a second experiment was conducted to examine the direct effect of relaxin on apoptosis in the lower reproductive tract. Endogenous relaxin was immunoneutralized with a monoclonal antibody for rat relaxin designated MCA1 [9] on days 19-21 of pregnancy. Withdrawal of relaxin caused an increase in the rate of apoptosis in both the cervix and the vagina (P<0.05). In the cervix, the LI of TUNEL-labeled cells in the epithelium was greater in MCA1-treated rats than in controls from 6h to 48h after the initiation of MCA1 treatment (Fig. 3A). Maximal LI occurred 12h and 24h after treatment. The LI in the cervical stroma was

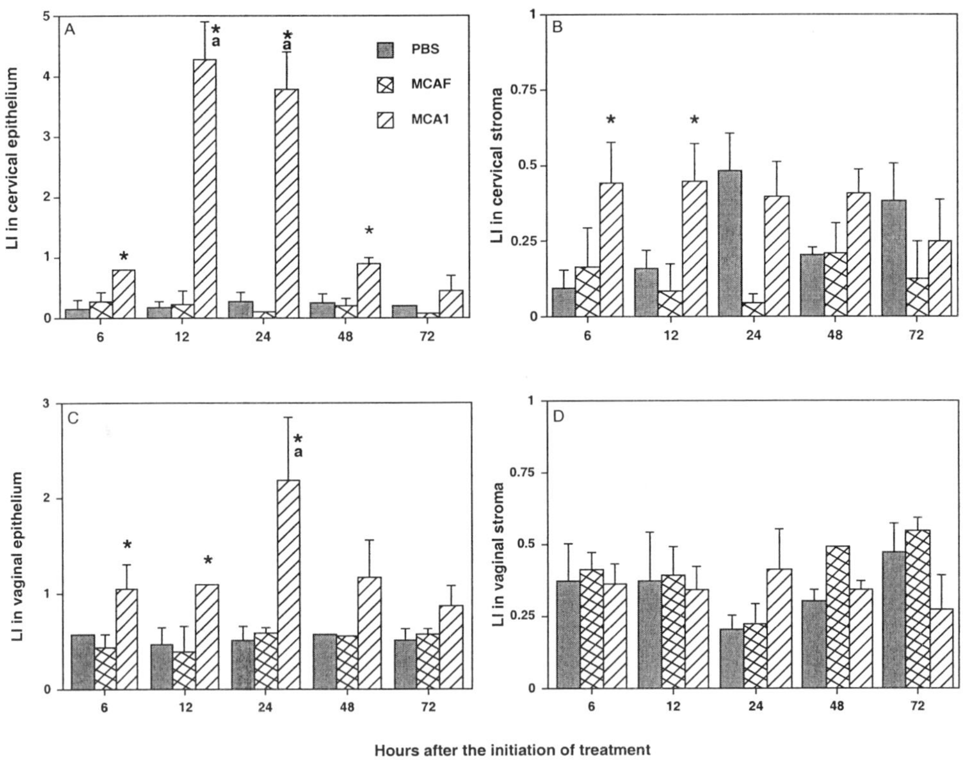

Fig 3. Mean (± SEM) labeling index of TUNEL-labeled cells in the cervical epithelium (A), cervical stroma (B), vaginal epithelium (C) and vaginal stroma (D) after PBS (vehicle control), MCAF (monoclonal antibody for fluorescein, monoclonal antibody control) and MCA1 treatment. n=4. An asterisk indicates a significant difference from controls (P<0.05). The superscript letter indicates a significant difference (P<0.01) from other MCA1-treated groups that have no superscript letter.

greater than in controls 6h and 12h following the neutralization of endogenous relaxin (Fig. 3B). Neutralization of relaxin had less dramatic effects in the vagina. The LI in vaginal epithelial cells was greater from 6h through 24h following initiation of neutralization of endogenous relaxin. Maximal LI occurred 24h after treatment (Fig. 3C). Treatment did not influence the LI in the vaginal stroma (Fig. 3D).

The finding that both cervical and vaginal wet weights were lower than controls following neutralization of relaxin is also consistent with the view that relaxin promotes growth of the lower reproductive tract, at least in part, by inhibiting apoptosis (Fig. 4). Cervical and vaginal wet weights were significantly lower than those in PBS and MCAF controls 48h and 72 h, and 72 h, respectively, following initiation of MCA1 treatment.

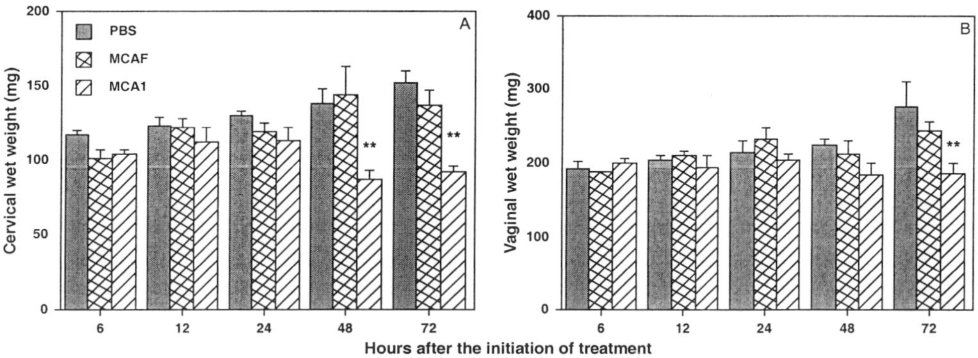

Fig. 4. Mean (± SEM) cervical (A) and vaginal (B) wet weights after PBS, MCAF, or MCA1 treatment. n=4. Two asterisks indicate a significant difference from controls ($P<0.01$).

The findings in this report may provide insight concerning a mechanism whereby relaxin effects other rat tissues and/or target tissues in other species. Specifically, relaxin-induced growth of the rat nipple [12] and pig vagina, cervix, uterus and mammary glands [13,14] during pregnancy may be attributable, at least in part, to a reduction in the rate of apoptosis.

This study provides evidence that relaxin may act as a cell survival factor during the second half of pregnancy. Relaxin is not the only hormone that regulates growth of the cervix and vagina in rats. Studies demonstrated that estrogen alone and/or with relaxin promotes growth of the cervix [15] and vagina [16] in nonpregnant rats. Estrogen has also been demonstrated to inhibit apoptosis in the rat vagina [17]. Therefore, it seems likely that estrogen not only contributes to relaxin's role as a cell survival factor but also, in and of itself, is a cell survival factor in these tissues during rat pregnancy.

The possibility that relaxin also increases the rate of cell proliferation in the cervix and vagina during rat pregnancy cannot be ruled out. Bagnell and coworkers provided evidence that relaxin can promote cell proliferation in cultured porcine follicular granulosa cells and thecal cells [18,19]. The hypothesis that relaxin both inhibits the rate of apoptosis and increases the rate of cell proliferation in the rat cervix and vagina during pregnancy is attractive because it provides complimentary mechanisms to account for the dramatic growth of the lower reproductive tract that occurs during the second half of pregnancy.

In summary, this study provides evidence that the rate of apoptosis in the cervix and vagina declines during the second half of rat pregnancy and that the reduction in the rate of apoptosis is due, at least in part, to elevated serum relaxin levels. Therefore, it is concluded that relaxin-induced growth of the cervix and the vagina is likely attributable, at least in part, to inhibition of apoptosis.

Acknowledgments

The authors thank Ms. L. Hung for assistance with tissue sectioning and cell counting, B. Sylavong for supervision of animal care, and the School of life Science Artist Services for preparing figures.

References

1. Harkness, M.L.R. and Harkness, R.D., J. Physiol., 148 (1959) 524.
2. Hwang, J-J., Shanks, R.D. and Sherwood, O.D., Endocrinology, 125 (1989) 260.
3. Burger, L.L. and Sherwood, O.D., Endocrinology, 136 (1995) 4820.
4. Zhao, S. and Sherwood, O.D., Endocrinology, 139 (1998) 4726.
5. Sherwood, O.D., In Knobil, E. and Neill J.D. (Eds.) The Physiology of Reproduction. 2nd ed., Raven Press, New York, USA, 1994, p.861.
6. Burger, L.L. and Sherwood, O.D., Endocrinology, 139 (1998) 3984.
7. Zhao, S., Kuenzi M.J. and Sherwood, O.D., Endocrinology, 137 (1996) 425.
8. Lee, A.B., Hwang, J-J., Haab, L.M., Fields, P.A. and Sherwood, O.D., Endocrinology, 130 (1992) 2386.
9. Lao Guico-Lamm, M. and Sherwood, O.D., Endocrinology, 123 (1988) 2479.
10. Wyllie, A.H., Cancer Metastasis Rev., 11 (1992) 95.
11. Gavrieli, Y., Sherman, Y. and Bensasson, S.A., J. Cell Biol., 119 (1992) 493.
12. Kuenzi, M.J. and Sherwood, O.D., Endocrinology, 131 (1992) 1841.
13. Hurley, W.L., Doane, R.M., O'Day-Bowman, M.B., Winn, R.J., Mojonnier, L.E. and Sherwood, O.D., Endocrinology, 128 (1991) 1285.
14. Min, G., Hartzog, M.G., Jennings, R.L., Winn, R.J. and Sherwood, O.D., Endocrinology, 138 (1997) 560.
15. Kroc, R.L., Steinetz, B.G. and Beach, V.L., Ann. N.Y. Acad. Sci., 75 (1959) 942.
16. Berman, J.R., McCarthy, M.M. and Kyprianou, N., Urology, 51 (1998) 650.
17. Sato, T., Fukazawa, Y., Kojima, H., Enari, M., Iguchi, T. and Ohta, Y., Anat. Rec., 248 (1997) 76.
18. Zhang, Q. and Bagnell, C.A., Endocrinology, 132 (1993) 1643.
19. Zhang, Q. and Bagnell, C.A., Endocrine J., (UK) 2 (1994) 349.

Synthesis and function of relaxin in human endometrium

Gerson Weiss[1], Smita Palejwala[1], Linda Tseng[2] and Laura T Goldsmith[1]

[1]Department of Obstetrics, Gynecology & Women's Health, New Jersey Medical School,Newark, New Jersey, 07103, USA; and [2]Department of Obstetrics & Gynecology, State University of New York Health Sciences Center at Stony Brook, Stony Brook, New York 11794

Introduction

Circulating relaxin is not required for the establishment, maintenance or successful delivery of human pregnancy. Local synthesis in human endometrium is therefore necessary to suggest a role for relaxin in human endometrial function. Previous data demonstrate that relaxin is present in the endometrium during the late luteal phase. Immunohistochemical studies of endometrium from patients at different stages of the menstrual cycle demonstrated localization of relaxin to glandular epithelial cells during the luteal phase using antibodies to the native hormone and to the human relaxin C-peptide [1]. The presence of relaxin has also been detected by immunohistochemistry in secretory endometrium from a) women having natural cycles after day 3 postovulation, b) anovulatory premenopausal women taking progestins, and c) postmenopausal women undergoing estrogen-progesterone replacement therapy, but not in normal proliferative phase or early secretory endometrium [2]. However, localization may be due to sequestration of circulating relaxin or C-peptide. No data have provided evidence for the synthesis of relaxin in human endometrium.

Results of studies by Hisaw and colleagues were the first to suggest a role for relaxin in endometrial growth [3,4]. Studying ovariectomized rhesus monkeys, Hisaw administered various doses of estrogen, progesterone and porcine relaxin containing ovarian extracts and showed that only in the presence of estrogen did relaxin containing extracts have any influence upon endometrial architecture. Maximum thickening and maturation of endometrial architecture were induced by relaxin containing extracts in the presence of both estrogen and progesterone. A pronounced growth of the endometrium above that produced by the steroids alone was produced by relaxin containing extracts. The effects seem to be a response of the endometrium to relaxin as an enhancement of the alterations produced by the combination of the steroids. Subsequent studies in the rhesus monkey suggest that relaxin induces marked proliferation of the endothelial cells of the endometrial blood vessels and dilatation of the superficial endometrial blood vessels [4]. In addition, results from these *in vivo* studies suggest that relaxin stimulates an intensified differentiation of the endometrial stromal cells into predecidual cells [3]. Relaxin appears to be necessary to maintain endothelial proliferation and vascular dilatation in the endometrium of the rhesus monkey. Animals in which administration of relaxin containing extracts was discontinued, but steroids were continued, showed reduced endothelial proliferation and dilatation [3].

G.W. Tregear et al. (eds.), Relaxin 2000, 41–45.
© 2001 *Kluwer Academic Publishers. Printed in the Netherlands.*

Results from various studies have provided evidence that demonstrates that relaxin is a potent regulator of differentiated function of human endometrial cells *in vitro*. Tseng and colleagues, during extensive and detailed studies of the effects of relaxin in human endometrial differentiation, have found that relaxin appears to be a significant modulator regulating differentiation of the human endometrium [5-8]. These studies have shown that relaxin stimulates cAMP accumulation and the production of several secretory products, including prolactin, insulin-like growth factor, and insulin-like growth factor binding protein-1, in progestin-primed endometrial stromal cells [9-13]. Prolactin and IGFPB-1 are considered to be the major secretory proteins of decidual cells [14] and the induction of expression of these secretory proteins has been widely used as a biochemical marker of decidualization of endometrial stromal cells *in vitro* [15]. IGFBP-1 is the major protein secreted from endometrial stromal cells during hormone stimulation. IGFBP-1 plays an essential role in regulating the mitogenic activity during growth and differentiation phases of endometrial stromal cell decidualization [16]. Detailed studies of regulation of IGFBP-1 gene promoter activity in endometrial stromal cells demonstrate that relaxin, not progestin, is the major inducer of IGFBP-1 gene transcription [17]. Thus, relaxin appears to be a more powerful factor in human endometrial decidualization than progesterone. Studies have also shown that relaxin significantly increases total cellular protein and inhibits progestin-induced DNA synthesis [5], as well as dramatically altering the ultrastructure of progestin treated endometrial stromal cells [14].

Thus, ample data which demonstrate definitive effects of relaxin upon human endometrial cells *in vitro* and suggest effects upon primate endometrium have been provided. However, a physiological role for relaxin in human endometrium has not yet been demonstrated and, since normal human pregnancy can proceed in the absence of circulating relaxin, demonstration of local synthesis is required for such a role to be proposed. Our studies were designed to provide 1) conclusive evidence for human endometrial synthesis of relaxin and 2) evidence for a potential physiological role of relaxin in women.

Results and Discussion

Expression of relaxin mRNA was examined in normal human endometrial tissue from 10 regularly menstruating women, 22 to 36 years old, at two time points: 1) 11-12 days after the start of menses (proliferative) and 2) 8-9 days after an LH surge (secretory) in an ovulatory cycle. In addition, expression of relaxin mRNA was studied in primary cultures of endometrial stromal and glandular epithelial cells, from six women taken during secretory or proliferative phases. Preparation and characterization of the endometrial cell cultures have been described in detail previously [7].

Determination of human relaxin specific mRNA was performed using a nested RT-PCR method. Total RNA was reverse transcribed into first strand cDNA using cloned MuLV reverse transcriptase and random hexamers. The resulting cDNA was used as template to amplify (by PCR) a 306 base pair DNA segment corresponding to amino acids 31-132 in the C peptide of the prorelaxin protein. The primer pair used, designed to span the exon 1-intron-exon 2 boundary, was a 19 nucleotide forward primer with the sequence

5' TCT CTG AGC CAG GAA GAT G 3'

corresponding to amino acids 31-36; and a 22 nucleotide reverse primer with the sequence

5' CTT AGG CTT GGA TAC TCA TTC T 3'

complimentary to the 3' end of the relaxin mRNA amino acids 126-132. RNA from human corpus luteum taken at term pregnancy was used as a positive control. RNA from human lung was used as a negative control. 2% of the resulting DNA product was further amplified using the following pair of nested exon-intron boundary spanning primers: An 18 nucleotide forward primer with the sequence

5' CCA GTG GCA GAA ATT GTG 3'

corresponding to amino acids 43-48, and a 22 nucleotide reverse primer with the sequence

5' GAT TCC AGT CTT CTC TTT GAA G 3'

corresponding to amino acids 96-102. This nested PCR reaction yielded a 181 base pair DNA product (corresponding to amino acids 43-102). Primer pairs were selected using the Wisconsin Package version 9.1 (Genetics Computer Group, Madison, WI) software using stringent specifications for the intron/exon junction spanning region of human prorelaxin cDNA. The pairs were selected based on their GC content and melting and annealing temperatures. Additionally, since not intended to distinguish between the H1 & H2 transcripts, the primer pair used for the first PCR reaction was one which would amplify both H1 & H2 relaxin cDNA and the sequence of the primer pair used for the nested reaction corresponded to H2 cDNA. To normalize for RNA variability, an internal standard PCR reaction for 18 S RNA was performed. The DNA products from both sets of reactions were resolved by agarose gel electrophoresis. Gels were stained with GelStar nucleic acid stain (FMC Bioproducts, Rockland, ME) and photographed under UV light. The nucleotide sequences of purified PCR products from endometrial and corpus luteum samples were authenticated by determination of the precise nucleotide sequence.

Relaxin specific mRNA was detected in endometrial tissue from all 10 subjects and in all cultured epithelial and stromal endometrial cells examined. Relaxin specific mRNA was detectable in proliferative phase endometrium from all 10 subjects and in 7 of the corresponding secretory phase endometrial tissues. Relaxin mRNA expression was consistently higher in proliferative than in secretory phase endometrium in all 10 subjects. We have previously shown that decidua from aluteal baboons contain higher relaxin concentrations than decidua from luteal intact baboons [18]. These data suggest that there is feedback between circulating relaxin levels and local production. In women, circulating relaxin levels are higher in the late secretory phase. Thus, it is possible that lower levels of endometrial relaxin mRNA in the secretory phase are due to higher circulating levels. An additional hypothesis is that greater relaxin levels are required during the proliferative phase to support the rapid thickening of the endometrium.

Since results from previous *in vivo* studies in rhesus monkeys suggest a role for relaxin in endometrial endothelial cell proliferation, we examined the effects of relaxin upon vascular endothelial growth factor (VEGF) expression by human endometrial stromal and glandular epithelial cells maintained *in vitro*. In addition, since a role for relaxin in implantation might involve reorganization of endometrial connective tissue, matrix metalloproteinase expression was determined. Primary cultures of human endometrial cells were prepared as previously described [7]. Replicate dishes were incubated without or with 50 – 100 ng/ml pure porcine relaxin (NIH, 3000 GPU/mg) or progestin for 16 to 48 hours. Conditioned medium samples were collected and assessed for VEGF content using a human VEGF specific ELISA (R&D Systems, Minneapolis, MN) which detects VEGF165 and VEGF121. Interassay and intraassay variations were 5.0 (n=5 assays) and 4.1 (n= 10 observations) % respectively and the precision of the assay was 94 - 98%. Expression of procollagenase (proMMP-1) and prostromelysin (proMMP-3) were determined by Western Blot analysis using specific, well characterized, primary antibodies and methods described previously [19].

Cultures of both stromal and glandular epithelial cells expressed ample quantities of VEGF and both cell types responded to relaxin. In glandular epithelial cell cultures, relaxin caused a significant inhibition of VEGF expression to 62 ± 7.1% of control (M±SEM, n=4 experiments, p = 0.01). In stromal cell cultures, relaxin stimulated VEGF expression to 220±59% of control (M±SEM, n=2 experiments). In contrast to the effect of relaxin, progestin inhibited VEGF expression in stromal cultures to 37+13 % of control (n=3 experiments). These data indicate that relaxin regulates secreted VEGF isoforms differentially in endometrial glandular epithelial and stromal cells.

Relaxin inhibited expression of proMMP-1. Studies of glandular epithelial cells demonstrated that inhibition was more marked in cells from the secretory phase (12.5±4% of control, M±SEM, n=3 experiments, p=0.0002) than in cells from the proliferative phase (55±9 % of control, n=2 experiments). In stromal cells from the secretory phase, an inhibition to 67±8% of control (n=3 experiments, p=0.058) was observed. In contrast, relaxin barely affected proMMP-3 expression in either epithelial or stromal cell cultures. In glandular epithelial cells, proMMP-3 expression was 87±4% and 73±13% of control in proliferative and secretory phases, respectively. In stromal cells from the secretory phase, levels of proMMP-3 were 91±5% of control.

Relaxin regulation of VEGF expression suggests a novel mechanism by which relaxin can regulate endometrial vascularization and in concert with the dramatic inhibitory effect of relaxin upon MMP-1 during the secretory phase, suggests a role for relaxin in implantation.

In summary: 1) Relaxin mRNA is present in human endometrium. There are greater endometrial message levels in the proliferative phase than in the secretory phase; 2) Relaxin is produced by human endometrium; 3) Relaxin stimulates VEGF secretion from endometrial stromal cells, but inhibits VEGF secretion from endometrial glandular epithelial cells, and 4) proMMP-1 is markedly inhibited by relaxin in endometrial glandular epithelial cells from the secretory phase, and relaxin has no effect on proMMP-3. These findings allow for the novel hypotheses that relaxin is a regulator of endometrial remodelling throughout the menstrual cycle, with lower secretory message levels due to feedback control from circulating relaxin, and in conjunction with other regulatory agents, relaxin may be involved in the endometrial secretory phase remodelling which supports implantation.

Acknowledgements

This work was supported by in part by NIH Grants HD 22338 (GW, LTG) and HD 19247 (LT).

References

1. Bryant-Greenwood, G.D., Rutanen, E-M, Partanen, S., Coelho, T.K. and Yamamoto, S.Y., Mol. Cell. Endocrinol., 95 (1993) 23.
2. Yki-Jarvinen, H., Wahlstrom, T. and Seppala, M., Acta Obstet. Gynecol. Scand., 64 (1985) 663.
3. Dallenbach-Hellweg, G., Dawson, A.B. and Hisaw, F.L., Am. J. Anat., 119 (1966) 61.
4. Hisaw, F.L., Hisaw, F.L., Jr. and Dawson A.B., Endocrinology, 81(1967) 375.
5. Tseng, L. and Lane, B., In MacLennan A.H., Tregear, G.W. and Bryant-Greenwood, G.D. (eds.) Progress in Relaxin Research, World Scientific Publishing Co. Ltd., Singapore, 1995, p.325.
6. Liu, H-C. and Tseng, L., Endocrinology, 104 (1979) 1674.
7. Zhu, H.H., Huang, J.R.,Mazella, J., Rosenberg, M. and Tseng, L., J. Clin. Endocrinol. Metab., 71(1990) 889.
8. Tseng, L. and Zhu, H.H J., Soc. Gynecol. Invest., 5 (1998) 149.
9. Chen, G. A., Huang, J. R. and Tseng, L., Biol. of Reprod., 39 (1988) 519.
10. Huang, J.R., Tseng, L., Bischof, P. and Janne, O.A., Endocrinology, 121 (1987) 2011.
11. Bell, S. C., Jackson, J. A., Ashmore, J., Zhu, H. H. and Tseng, L., J. Clin. Endocrinol. Metab., 72 (1991) 1014.
12. Tseng, L., Gao, J. G., Chen, R., Zhu, H. H., Mazella, J. and Powell, D. R., Biol. Reprod., 47 (1992) 441.
13. Gao, J. G., Mazella, J. and Tseng, L., Mol. Cell. Endocrinol., 104 (1994) 39.
14. Lane, B., Oxberry, W., Mazella, J. and Tseng, L., Hum. Reprod., 9 (1994) 259.
15. Irwin, J. C., Utian, W. H. and Eckert, R. L., Endocrinology, 129 (1991) 2385.
16. Frost, R. A., Mazella, J. and Tseng, L., Biol. Reprod,. 49 (1993) 104.
17. Gao, J. G., Mazella, J. and Tseng, L., Mol. Cell. Endocrinol., 104 (1994) 39.
18. Castracane, V.D., Lessing, J., Brenner, S. and Weiss, G., J. Clin. Endocrinol. Metab., 60 (1985) 133.
19. Palejwala, S., Stein, D. and Wojtczuk, A., Weiss, G. and Goldsmith, L., T., Endocrinology, 139 (1998) 1208.

Analysis of relaxin deficient mice

Ling Zhao[1,3], Peter J Roche[2], Jenny M Gunnerson[1], Chrishan Samuel[1], Vicki E Hammond[1], Felix Beck[1], E Marelyn Wintour[1] and Geoffrey W Tregear[1]

[1]Howard Florey Institute, University of Melbourne, Victoria 3010, Australia; [2]Department of Medical Laboratory Science, RMIT University, Victoria 3001, Australia; and [3]Present address: Laboratory of Genetics and Physiology, NIDDK, NIH, Bethesda, MD 20892, USA

Introduction

The corpus luteum of pregnancy is the major source of circulating relaxin in mice and rats [1]. Both messenger RNA and peptide have also been detected in the uterus, brain and prostate gland [2-4]. Binding sites for relaxin include uterus, cervix, mammary gland, nipple, pubic symphysis, cardiac atrium and brain [5-8]. Much of our knowledge of the biological actions of relaxin has been derived from studies in the rat. Removal of the source of circulating relaxin by bilateral ovariectomy established that relaxin in combination with estrogen is required for normal parturition to occur [9]. Similar results were obtained by Sherwood and coworkers in a series of experiments using passive immunization with monoclonal antibodies specific for rat relaxin to neutralize endogenous activity [10]. Underdeveloped mammary glands and nipple with resultant poor lactation performance [11, 12] and effects on cervical softening [13], vaginal growth [14], and water consumption [15] were also observed.

To further understand the physiological role of relaxin in the mouse, we examined the effects of inactivation of the relaxin gene using the technique of homologous recombination in embryonic stem cells [16, 17].

Results and Discussion

The mouse relaxin gene is a single copy gene consisting of two exons encoding a B chain, a C (connecting) peptide and an A chain [18]. Synthetic and comparative studies have shown that the A and B chain on their own are not bioactive [3, 4]. In addition, the A-chain N-terminal α-helix is required for the full biological activity of the intact hormone. Hence, a targeting construct was designed to delete the N-terminal 17 amino acids from the 25 amino acids that comprise the A-chain and the C-terminal 90 amino acids of the 103 residues that make up the C-peptide. An exogenous gene fragment (neomycin phosphotransferase gene) replaced the deleted section of the gene. Thus, the mutated gene consists of exon one and a few base pairs of exon II. Theoretically, this mutated gene possibly encode for a protein that contains a signal peptide, a B-chain, a very short part of the N-terminus of the C-peptide and a small fragment of the A-chain C-terminus. As most of the A-chain is missing and the shortened C-peptide is presumably unable to fold the prohormone to a correct conformation, it is assumed that

G.W. Tregear et al. (eds.), Relaxin 2000, 47–51.
© 2001 *Kluwer Academic Publishers. Printed in the Netherlands.*

this mutation results in a deletion of the bioactive relaxin *in vivo*. An illustration of the relaxin gene structure is shown in Fig. 1. The DNA fragment from the EcoRI to BglII sites was deleted and replaced by the neomycin transferase gene.

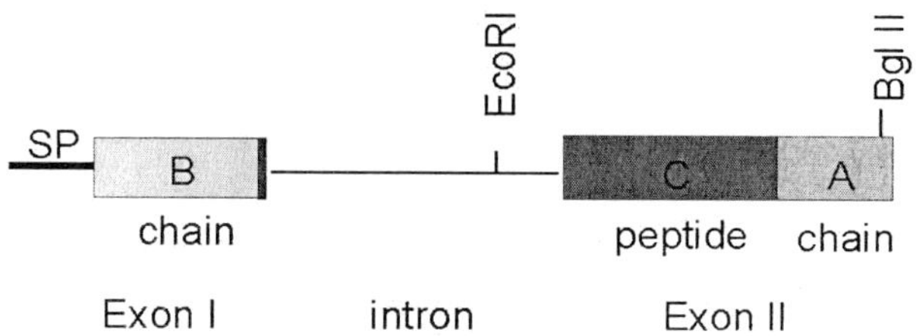

Fig. 1. Schematic diagram of the structure of the mouse relaxin gene. The relaxin gene contains two exons and one intron. Exon I encodes the signal peptide (SP), the B-chain and a small portion of the connecting peptide (C-peptide); Exon II encodes 90% of the C-peptide and the A-chain. The region between EcoRI and BglII has been deleted in the mutated allele of the relaxin gene. This deletion covers 90% of C-peptide and 70% of the A-chain.

Phenotypic analysis of offspring from relaxin heterozygote intercrosses (mixed 129sv × C57BL/6J background) demonstrated that relaxin null mutant mice could survive to adulthood [16]. The genotype ratio of the offspring of heterozygote (rlx +/-) cross mating was 1:2:1 (wild-type:heterozygote:homozygote). This indicates that relaxin is not essential for foetal development. Both male and female relaxin null mutant (rlx -/-) mice are fertile. The knockout female can carry their pregnancy to term. Mean gestation length was 19.7 ± 0.26 days for rlx -/- mothers (n = 17) compared with 19.5 ± 0.20 days for rlx +/+ animals (n = 9). The delivery time (from the first pup to the last) was observed for 8 rlx -/- mice and for 7 age-matched controls. For rlx +/+ animals the average time was 1.7 ± 0.3 h. Six of the mutant animals delivered their young in 2.0±0.3 h but two were unable to deliver normally. One had a protracted labour lasting for 15 h while the other had to be killed because she was in obvious distress having given birth to two dead pups some 12 h earlier. Ten further pups were found dead *in utero* upon post mortem [16].

The gestation length and delivery time did not differ greatly between wild-type and knockout animals although the pubic symphysis in the knockout mice did not relax (Table 1). However, the relaxin null mutant mothers were unable to suckle their young due to the impaired nipple development during pregnancy which suggests that the main role of relaxin in the mouse is in mammary gland and nipple development during pregnancy. Although, the nipple development was arrested in the relaxin knockout animals due to the lack of relaxin, milk secretion was confirmed by the treatment of oxytocin in these null mice [16].

Table 1. Length (mm) of pubic symphesis of virgin (wt) and full term gestation mice.

VIRGIN (n=10)	RLX +/+ (n=5)	RLX +/- (n=6)	RLX -/- (n=11)
1.17 ± 0.65	5.37 ± 0.64*	3.67 ± 0.52*	1.65 ± 0.22

* $P < 0.01$ compared with virgin and rlx -/- mice.

The plasma osmolality stayed higher in the knockout mice compared with the wild-type control during pregnancy (Table 2).

Table 2. Osmolality (mosmol/kg water) of day 14.5 –18.5 pregnant mice.

RLX +/+ (n=11)	RLX +/- (n=10)	RLX -/- (n=22)
306 ± 1	311 ± 3	316 ± 2 *

*P< 0.001 compared with wild-type mice. n indicates the animal number in each group.

The presence of myoepithelial cells was demonstrated immunocytochemically in sectioned mammary tissue using mouse monoclonal anti-α-smooth muscle actin antibody (clone 1A4, Sigma, Saint Louis, Mi, USA), which is a marker for myoepithelial cells. It was readily shown that myoepithelial cells were present in large numbers in the mammary glands of both rlx -/- and rlx +/+ mice. There was no difference in the histology of the mammary gland between non-pregnant rlx -/- mice and their wild-type litter mate [17].

At 14.5 days of gestation, under the light microscope, there was slightly less glandular tissue in the rlx -/- animals compared to the rlx +/+ mice while the ducts were somewhat dilated [17] but this did not reach statistical significance at the 0.05 level. By 18.5 days of pregnancy, the development of the mammary glands of rlx +/+ and rlx -/- mice showed a large difference. There was much more glandular tissue and less fat in the rlx +/+ than in the rlx -/- animals and the ducts of the rlx -/- were grossly dilated compared with rlx +/+ mothers [16]. Secretion was observed in the alveoli and ducts of both rlx +/+ and rlx -/- mice. Mammary glands from heterozygotes appeared similar to the wild-type although morphometric studies were not carried out in these animals.

In 2-day post-partum (P2) mammary glands, the ducts in rlx -/- animals remained very dilated. The alveoli were in the early stage of involution compared with the alveoli of rlx +/+ mice. They were often smaller and lined by flattened cells with heterochromatic nuclei compared with the cubical cells and euchromatic nuclei seen in normally lactating glands from wild-type mice. The ducts in rlx -/- animals were even more dilated than those at 18.5 days and were lined by squamous epithelial cells compared to the more cubical cells lining the ducts of wild-type control animals. At high power, in the knockout mice, there are more fat droplets within the epithelial cells rather than in the lumen compared with their counterpart as shown previously [16].

The relaxin knockout mouse was used to assess the effects of relaxin during pregnancy on water, collagen content, growth and morphology of the nipple, vagina, uterus, cervix, pubic symphysis and mammary gland. Relaxin increases the growth of the nipple, cervix, vagina and pubic symphysis during pregnancy. Large increases in water content (20%) in the pubic symphysis occurred in pregnant (day 18.5) rlx +/+ mice but not in rlx -/- animals. This suggests that in the pubic symphysis, relaxin may increase the concentration of a water-retaining extracellular matrix component (hyaluronate). Histological analysis showed that the collagen stain was more condensed in the tissues (vagina, cervix, pubic symphysis, mammary gland and nipple) of rlx -/- than that of rlx +/+ mice. In the wild-type mice, collagen content decreased significantly in the nipple and vagina, but not in other tissues. It suggests that the failure of collagen degradation and the lack of growth in the nipple underlie the inability of the relaxin null mice to feed their young. Vaginal and cervical luminal epithelia, which proliferated markedly in the rlx +/+ pregnant mice, remained relatively atrophic in the rlx -/- mice. As proliferation and differentiation of uterine and vaginal epithelia are thought to be induced by a paracrine stromal factor upon estrogen stimulation, this suggests that relaxin may be this paracrine factor.

Our results suggest that relaxin affects target tissues by either decreasing the collagen content (such as in nipple and vagina), reorganizing collagen fibres (in vagina, cervix, pubic symphysis and nipple) or increasing the water content (such as in the pubic symphysis). Relaxin may exert these roles through its regulation of metalloproteinases, which are known to be involved in nearly all tissue growth and remodelling processes.

These results directly prove the role of relaxin in elongating the pubic symphysis in mice during the second half of pregnancy. The effects of relaxin on the nipple and vagina confirm similar results found in rats treated with anti-relaxin monoclonal antibodies. The inhibitory effects of relaxin on mammary gland development and nipple growth are more profound than that discovered in other species. These studies suggest that the mode of action of relaxin on target tissues during pregnancy in mice is by remodelling the extracellular matrix.

Acknowledgments

This study was supported by an Institute block grant from the National Health and Medical Research Council of Australia (grant reg key no. 983001) to Howard Florey Institute.

References

1. Sherwood, O.D., In Knobil, E. and Neill, J.D. (eds.) The Physiology of Reproduction, Raven Press, New York, 1994, p. 861.
2. Fields, P.A., Lee, A.B., Haab, L.M., Hwang, J.J. and Sherwood, O.D., Endocrinology, 130 (1992), 2985.

3. Osheroff, P.L. and Ho, W.H., J. Biol. Chem., 268 (1993) 15193.
4. Gunnersen, J.M., Crawford, R.J. and Tregear, G.W., Mol. Cell. Endocrinol., 110 (1995) 55.
5. Weiss, T.J. and Bryant-Greenwood, G.D., Biol. Reprod., 27 (1982) 673.
6. Kuenzi, M.J. and Sherwood, O.D., Endocrinology, 136 (1995) 1367.
7. Osheroff, P.L., Cronin, M.J. and Lofgren, J.A., Proc. Natl., Acad. Sci. USA, 89 (1992) 2384.
8. Osheroff, P.L. and Phillips, H.S., Proc. Natl. Acad. Sci. USA, 88 (1991) 6413.
9. Downing, S.J. and Sherwood, O.D., Endocrinology, 116 (1985) 1200.
10. Zhao, S. and Sherwood, O.D., Endocrinology, 139 (1998) 4726.
11. Hwang, J-J., Lee, A.B., Fields, P.A., Haab, L.M., Mojonnier, L.E. and Sherwood, O.D., Endocrinology, 129 (1991) 3034.
12. Kuenzi, M.J. and Sherwood, O.D., Endocrinology 131 (1992) 1841.
13. Lee, A.B., Hwang, J-J., Haab, L.M., Fields, P.A. and Sherwood, O.D., Endocrinology 130 (1992) 2386.
14. Zhao, S., Kuenzi, M.J. and Sherwood, O.D., Endocrinology, 137 (1996) 425.
15. Zhao, S., Malmgren, C.H., Shanks, R.D. and Sherwood, O.D., Endocrinology, 136 (1995) 1892.
16. Zhao, L., Roche, P.J., Gunnersen, J.M., Hammond, V.E., Tregear, G.W., Wintour, E.M. and Beck, F., Endocrinology, 140 (1999) 445.
17. Zhao, L., Samuel, C.S., Tregear, G.W., Beck, F. and Wintour, E.M., Biol. Reprod., 63 (2000) 697.
18. Evans, B.A., John, M., Fowler, K.J., Summers, R.J., Cronk, M., Shine and Tregear, G.W., J. Mol. Endocrinol., 10 (1993) 15.

Relaxin is a key regulatory peptide in the reproductive tract of the pregnant tammar wallaby, *Macropus eugenii*

Laura J Parry
Howard Florey Institute and Department of Zoology, University of Melbourne, Victoria, 3010, Australia

Introduction

The tammar wallaby *Macropus eugenii* is a monovular, seasonally breeding marsupial with a prolonged 9-10 month phase of lactation. A single corpus luteum (CL) is present in one ovary for the length of gestation and a new CL forms in the contralateral ovary after ovulation between one and two days postpartum. In the presence of a pouch young, the conceptus develops to a unilaminar blastocyst of approximately 100 cells and then enters lactational diapause. The new CL also becomes quiescent. Loss or removal of the pouch young terminates lactational diapause and both the quiescent blastocyst and CL resume development, with births expected 26 or 27 days later [1]. In most marsupials, the lifespan of the CL is not extended by the presence of a conceptus and the luteal phase of the oestrous cycle is similar in duration to the length of gestation. The CL reaches maximum size (40-50 mg) during the mid-luteal phase and shows signs of luteolysis the day before birth [2]. The CL is essential for normal parturition in marsupials. Over 90% tammars lutectomized in late gestation fail to deliver live young [3]. However, parturition is not affected by removal of the preovulatory follicle. Other studies show that neither progesterone nor oestrogen are needed for normal parturition to occur [1]. This chapter reviews the evidence which demonstrates that ovarian relaxin is a key regulatory peptide role in the reproductive tract of the tammar wallaby.

Results and Discussion

Synthesis of relaxin in marsupials

The presence of a bioactive relaxin-like peptide was first demonstrated in the late 1960s in extracts of corpora lutea from the late pregnant tammar wallaby and brushtail possum *Trichosurus vulpecula*. Peak bioactivity was measured in the second half of pregnancy, with a marked fall post-partum [4]. Immunoreactive relaxin was then identified in the CL of the pregnant North American opossum *Didelphis virginiana* [5]. Krause [6] also reported that small electron-dense granules were a prominent feature of the luteal cell cytoplasm in the opossum throughout pregnancy but disappear rapidly after birth. Similar granules were described earlier in the CL of other marsupial species but it was always thought that they contained progesterone as elevated plasma concentrations of this steroid were correlated with increased numbers of granules [7]. However, Tyndale-Biscoe [8] demonstrated that peak numbers of secretory granules coincided with peak relaxin-like

G.W. Tregear et al. (eds.), Relaxin 2000, 53–58.
© 2001 *Kluwer Academic Publishers. Printed in the Netherlands.*

bioactivity in the pregnant tammar and suggested that they contained relaxin rather than progesterone.

The first conclusive evidence of relaxin synthesis in marsupials came in 1997 with the isolation of the tammar relaxin cDNA molecule [9]. Both the CL and yolk sac placenta synthesize relaxin during pregnancy. Relaxin gene expression in the CL is highest in early and mid-pregnancy, reduced at term and absent postpartum or in animals during embryonic diapause. There is a rapid increase in relaxin mRNA in the CL approximately 6 days after reactivation of the diapausing blastocyst [9]. At this early stage of pregnancy, the uterus is secretory and mitoses are observed in the CL. These endocrine changes, which are first observed 3 to 4 days post reactivation, are thought to precede reactivation of the blastocyst as increases in metabolism and blastocyst growth are not seen until at least 8 days post reactivation. Evidence suggests that prolactin maintains diapause in the tammar by suppressing CL activity; removal of this inhibitory factor allows the CL to resume synthetic activity [10]. The presence of high concentrations of prolactin binding sites on tammar luteal cells after ovulation suggests that prolactin has direct effects on the CL [11]. It would be interesting to know if the rapid upregulation of relaxin gene expression on Day 6 of gestation is linked to removal of the inhibitory effect of prolactin on the CL.

The yolk sac placenta is a second source of relaxin in the pregnant tammar wallaby. The tammar has a diffuse, epithelio-chorial placenta, which forms when the shell membrane ruptures and the vascular region of the yolk sac interdigitates with the endometrial epithelium on Day 19 of the 26-day gestation. The vascular region initially comprises about 25% of the total placenta surface area but this expands rapidly to over 50-75% by the time of attachment, with a marked increase in vascularity [1]. Analysis of tissues by Northern hybridization confirmed the presence of a 1.0 kB gene transcript in the placenta from Day 19 of gestation and demonstrated that relaxin gene expression is significantly greater in the early stages of placentation compared with the end of pregnancy (Fig. 1).

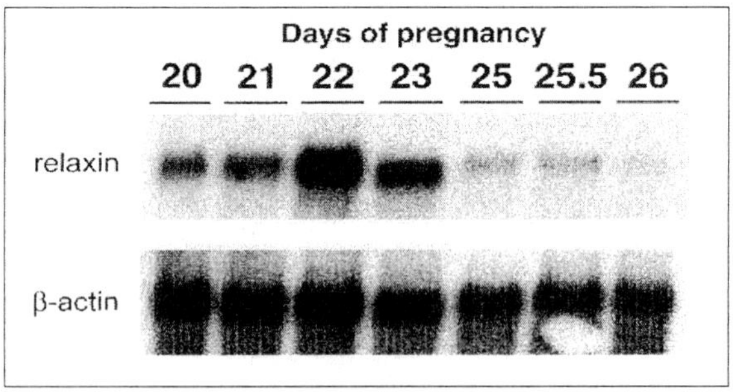

Fig. 1. Relaxin gene expression in the yolk sac placenta (Northern analysis, 20 μg total RNA) at different stages of the 26-day gestation in the tammar wallaby.

The upregulation in relaxin gene expression coincides with the significant increase in the vascular region of the placenta.

Relaxin mRNA and immunoreactive relaxin are both localized specifically to luteal cells in the ovary (Fig. 2). Cytoplasmic, electron-dense granules (200-350 nm in diameter) are the only organelles in the luteal cell that contain immunoreactive relaxin (Fig. 2). There is a progressive increase in the number of electron-dense granules during pregnancy, with a rapid fall in numbers at term and immediately postpartum. These data are consistent with morphological studies in eutherian species. In the tammar, some luteal cells are more abundant with electron-dense granules and rough endoplasmic reticulum whereas others contain few granules but more agranular reticulum. These structural differences do not appear to be linked to particular luteal cell sizes and increased granule numbers are not correlated with larger luteal cells.

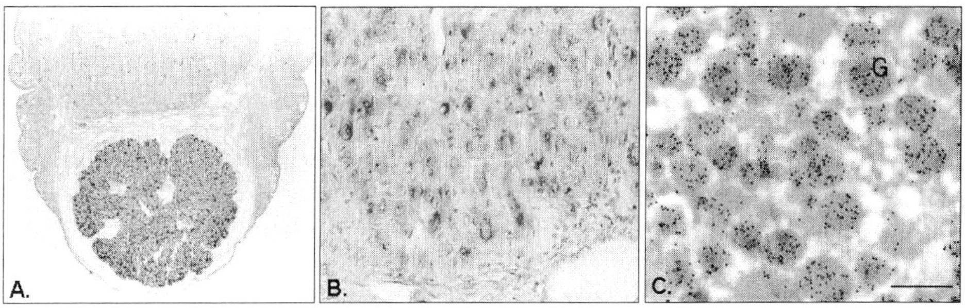

Fig. 2. Localization of (A.) relaxin mRNA and (B.) immunoreactive relaxin in the corpus luteum on Day 23 of gestation. (C.) Immunoreactive gold particles are observed on small, electron-dense granules (G) in the luteal cell cytoplasm.

Analysis of relaxin-labelled secretory granules in the luteal cell cytoplasm throughout gestation demonstrated that relaxin immunoreactivity is low in the tammar CL throughout early and mid pregnancy but increases markedly between Days 21 and 22 of gestation, when the CL is maximum in size and weight. In addition to the increased number of electron-dense granules with advancing pregnancy, the results of this study suggested that the quantity of relaxin packaged in individual secretory granules increases in the late luteal phase. Given the early onset of relaxin gene expression in the tammar CL, it was also thought that preprorelaxin was the predominant form of relaxin stored in the cell cytoplasm and that further enzymatic processing did not occur until later on in gestation [12]. However, using PAGE and Western Blotting with a human H2 relaxin antiserum, it has recently been demonstrated that a large quantity of a 6.0 kD immunoreactive relaxin protein is present in extracts of tammar CL obtained between Days 14 and 21 of pregnancy [R.A.D. Bathgate, *personal communication*]. A quantitative assessment of relaxin protein concentrations in the CL and ovarian vein is now needed to provide information on the release of the hormone into the circulation.

Role of relaxin in the reproductive tract

Marsupials give birth to small, altricial young after a relatively short gestation. For example, the tammar wallaby produces a single fetus (birth weight 350-400 mg) approximately 26 days after reactivation of the diapausing blastocyst. Unlike eutherians, female marsupials have two anatomically separate uteri, which open into an anterior vaginal expansion via separate cervices. Macropodid marsupials (e.g. tammar wallaby) are monovular species so only one uterus will be gravid and contain the single fetus in each pregnancy cycle whereas the contralateral uterus remains non-gravid. Effectively, morphological changes specific to the gravid uterus, which may be influenced by the fetus, can be isolated from the non-gravid uterus. At parturition, the fetus moves from the gravid uterus through the cervix into the median vagina and reaches the urogenital opening via the urogenital sinus. Despite the differences in neonate size and reproductive tract anatomy, marsupials share with eutherians many similar endocrine mechanisms involved in the regulation of the physiology of parturition [1].

The function of the cervix is particularly important in marsupials because it remains tightly closed throughout pregnancy to prevent premature loss of the relatively small marsupial fetus. However, at the end of pregnancy it has to soften and dilate to enable the fetus to pass into the birth canal. The tensile properties of tammar cervices are markedly altered close to term; the cervix of the gravid uterus becomes highly extensible on the day of expected births [D.G. Porter, *personal communication*]. In early and mid-pregnancy, the cervix of the gravid uterus consists of dense collagenous tissue surrounding the lumen. The collagen fibers are tightly-packed together in a parallel organization, with little amorphous ground substance between them. There are several deep invaginations which protrude into the stromal tissue from the lumen. At the end of pregnancy, these invaginations become deeper with many side-branches that may serve to increase the surface area of the lumen. The stromal tissue of these invaginations is less dense compared with the cervix of earlier stages of pregnancy; the collagen fibers appear more loosely organized, with numerous spaces between them and fewer fibroblasts. The cervical epithelium also proliferates due to an elongation of the epithelial cells.

Surprisingly, there are no clear differences in the histological appearance between the cervices of the gravid and non-gravid uteri even in the late stages of pregnancy. On Day 25 of pregnancy, the collagen fibres in the cervix of the non-gravid uterus are loosely organized, with numerous spaces between them so the stromal tissue in the luminal region appears less dense. Moreover, the invaginations into the stromal tissue are deeper, with numerous branches as seen in the cervix of the gravid uterus. Finally, the epithelium proliferates. Recent studies show that these morphological changes associated with the re-organization of collagen fibres in the cervix are pregnancy-specific in the tammar wallaby. In comparison with late pregnant (Day 25) tammars, both cervices of non-pregnant animals at the equivalent stage of the oestrous cycle still consist of dense, highly organized collagen fibres. The epithelium does not thicken and the cells remain cuboidal. The only apparent change is an increase in the diameter of the cervical lumen and the appearance of small, wide clefts protruding into the stromal tissue.

These data suggest that a systemic, but pregnancy-specific, endocrine factor acts on both the gravid and non-gravid cervix and stimulates a re-organization of the collagen fibres in late gestation. The strongest evidence to date that relaxin may be this factor comes from preliminary autoradiography studies in which frozen sections of the cervix from each uterus were incubated with ^{33}P- labelled tammar relaxin. Relaxin binding sites appear to be localized to the stromal tissue surrounding the lumen of both cervices (Fig. 3). This binding is displaced with purified tammar relaxin.

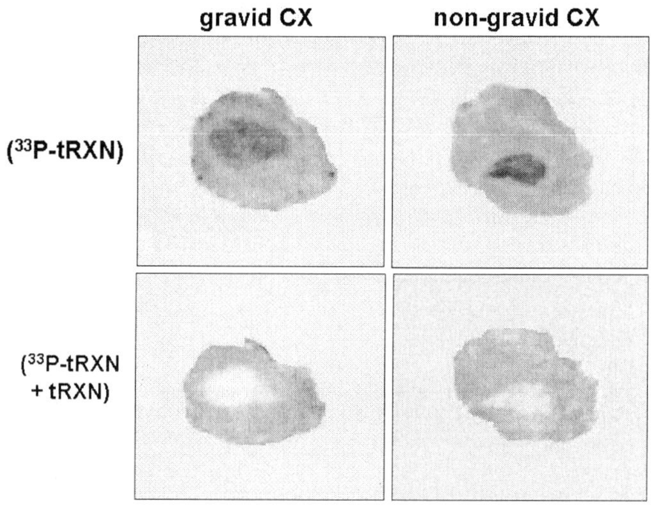

Fig. 3. Autoradiographic localization of relaxin binding sites in the gravid and non-gravid cervix (CX) of the pregnant tammar wallaby on Day 26 of gestation, using ^{33}P labelled-tammar relaxin (tRXN).

In summary, this study has shown that the tammar wallaby synthesizes a true relaxin molecule in the context of a physiology which is present in most mammalian species investigated. The temporal pattern of relaxin gene expression in the tammar CL conforms relatively closely to that of eutherians and implies a conservation of essential functions, particularly in regard to reproduction. The marsupial model also provides unique opportunities to investigate the potential influence of the feto-placental unit in mediating either relaxin synthesis or relaxin receptor numbers. Data show that relaxin binding sites are present and collagen re-organization occurs in the cervix of both uteri, which implies that the fetus does not play a major regulatory role in this aspect of reproductive tract physiology.

Acknowledgements

The author thanks Marilyn Renfree and Richard Ivell for their encouragement and input to this research, and Andrew Siebel for performing the autoradiography study. The author is also grateful to the Australian Research Council (ARC) and the Clive & Vera Ramaciotti Foundation for funding. She is currently a recipient of an ARC QEII Research Fellowship.

References

1. Renfree, M.B., In Lamming, G.E. (eds.) Marshall's Physiology of Reproduction (vol.3, 4th edn.), Chapman & Hall, London, 1994, p. 677.
2. Renfree, M.B., Green, S.W. and Young, I.R., J. Reprod. Fert., 57 (1979) 131.
3. Harder, J.D., Hinds, L.A., Horn, C.A. and Tyndale-Biscoe, C.H., J. Reprod. Fert., 75 (1985) 449.
4. Tyndale-Biscoe, C.H., J. Reprod. Fert., 19 (1969) 1191.
5. Krause, W.J., Zool. Anz. 230 (1993) 249.
6. Krause, W.J. and Sherman, D.M., Ann. Anat., 174 (1992) 341.
7. Gemmell, R.T., Aus. J. Zool., 27 (1979) 501.
8. Tyndale-Biscoe, C.H., In Bryant-Greenwood, G.D., Niall, H. and Greenwood, F.C. (eds.), Relaxin, Elsevier, Amsterdam, 1981, p. 225.
9. Parry, L.J., Rust, W. and Ivell, R., Biol. Reprod., 57 (1997) 119.
10. Hearn, J.P., J. Reprod. Fert., 39 (1974) 235.
11. Stewart, F. and Tyndale-Biscoe, C.H., J. Endocrinol., 92 (1982) 63.
12. Parry, L.J., Clarke, J.M. and Renfree, M.B., Cell Tissue Res., 290 (1997) 615.

Characterization of preprorelaxin in a marsupial, the tammar wallaby *Macropus eugenii*

Laura J Parry[1] and Richard Ivell[2]
[1]*Howard Florey Institute & Department of Zoology, University of Melbourne, Parkville, Victoria, 3010, Australia;* [2]*Institute for Hormone & Fertility Research, University of Hamburg, 22529 Hamburg, Germany*

Introduction

Recent molecular evidence points to a divergence of marsupials, together with monotremes, from the mammalian stock approximately 130 million years ago. A comparison of the primary structure of hormone molecules from such diverging groups of animals is useful to highlight conserved motifs which may be important for the function of the molecule e.g. receptor binding sites and enzyme cleavage sites. The aim of this study was to assess the effects of divergence and geographical isolation on the relaxin gene in one Australian marsupial, the tammar wallaby.

Results and Discussion

The nucleotide and derived amino acid sequences of tammar preprorelaxin were established by combined RT-PCR and 3'- and 5'-RACE methods, using RNA from the corpus luteum (CL). The full-length tammar relaxin preprohormone is 579 bp and encodes a peptide of 192 amino acids (aa). Comparison of the known cleavage sites between the B-, C- and A-domains in other relaxin molecules suggests that tammar preprorelaxin comprises a signal peptide of 26 aa, B-domain of 31 aa, C-domain of 111 aa and A-domain of 24 aa. The N-terminus of the protein-coding region has all the characteristics of a signal peptide that is probably cleaved after the alanine residue at position 26. In all species, the greatest homology at the amino acid level is in the B-chain sequence. The region between residues 9 and 28, including the C-G-R-X-X-X-R motif (the likely receptor binding domain) is highly conserved compared with other species (Fig. 1), supporting the hypothesis that this region is particularly important for biological activity. The cysteine residues in both the B-and A-domains (which are involved in disulphide bond formation) are also conserved across species as is the tetrabasic sequence R-K-K-R at the C-domain/A-domain junction, which corresponds to a consensus cleavage motif for furin. There is also a furin cleavage motif at the B-domain/C-domain junction, indicated by the sequence R-W-K-R, which is unique to tammar preprorelaxin. The percentage homology in the C-domain ranges from 16-27% (Fig. 1) and is dramatically reduced from the 70% reported previously in other species. Two particular regions of relatively high homology include the R-K-K-R sequence at the C-domain/A-domain junction and two peptide motifs, E-E-F (positions 70 to 72) and E-A-V-E (positions 81 to 84). There are also conserved glutamic acid residues at positions 14, 25, 39 and 97 in the

G.W. Tregear et al. (eds.), Relaxin 2000, 59–61.
© 2001 *Kluwer Academic Publishers. Printed in the Netherlands.*

C-domain; one of these, Glu_{14}, corresponds to the glutamic acid residue in the pig, mouse, rat and human relaxins which is at the splice site of the single intron. The only other invariant amino acids in the C-domain are leucine residues at positions 44, 51 and 57 and a serine residue at position 19. Comparison of the tammar relaxin A-domain with other species reveals a maximum of 41% homology with shark, rat and H2 relaxin A-peptides, particularly in the 15 amino acids of the N-terminal end (Fig. 1). There is a serine residue replacing the absolutely conserved arginine in all other species examined. Contrary to previous suggestions, this residue may not be essential for the activity of this peptide as tammar relaxin binds to the rat relaxin receptor with high affinity and is bioactive in the human THP1-cell cAMP bioassay (Bathgate et al., this volume). Also of interest are the Ile-Val residues at positions 6 and 7 in the A-domain; although these are not present in any other known relaxin sequence, they are both found in the same position in the porcine insulin A-chain.

```
tammar MLARL-MFAHLLGAWLLL-SFFPGELGA  QRIDDKPMKLCGREFVRAVIFTCGGSRW-KRI      S--MIQRSLQPEQEPEQMFVSRVN-DVEEGKLDPEVIRGWK

skate                                RPMWEERSRLCGRDLIRAFIYLCGGTRWTRLPNFGNYPIM
shark                               QSLSNAESGIKLCGREFIRAIIFACGGSRW
dogfish                             QSFKNAEPGIKLCGREFIRAVIYTCGGSRW
rat      MSSRL-LLQ-LLGFWLFL-SQPCRA  RVSEEWMDQVIQVCGRGYARAWIEVCGAS-VGRLAL     SQEE-PAPLARQAEAEVVPS-FINKDAEPFDMTLKCLPNLS
hamsterMSCK-FVLQ-LLGFWLLL-SQPCRA  RVTKEWLDEVIHVCGREYVRAILDICAAT-VGLEPLRRRRMTE-----------EAV-SSFIKEDAEPFD-T---MPNLS
g pig                                 GFLDKVIKVCGRDLVRIKIDICGKILLGDMTTG     QEK-QRILGSGQSAEIMPSS-INKEVDSLNMLESIAN-LP
pig      MP-RL-FS-YLLGVWLLL-SQLPREIPG  QSTND-FIKACGRELVRLWVEICGSVSWGRTAL     SLEE-PQ-LETGPPAETMPSS-ITKDAEILKMMLEFVPNLP
equine                               QKPDD-VIKACGRELARLRIEICGSLSWKKTVL     RLEE-PG-LEVGQPVEIVSSS-ISKDAEALNTKLGLNSNLP
human1 MP-RLFLF-HLLEFCLLL-NQFSRAVAA  KWKDD-VIKLCGRELVRAQIAICGMSTWSKRSL     SQEDAPQ-TPRPV-AEIVPS-FINKDTETIIIMLEFIANLP
human2 MP-RLFFF-HLLGVCLLL-NQFSRAVA  DSWMEE-VIKLCGRELVRAQIAICGMSTWSKRSL     SQEDAPQ-TPRPV-AEIVPS-FINKDTETINMMSEFVANLP

tammar   EGQIPLLN-RGWDLLMNILN-LHGHTEDSTPTKEEFMLTSDESK-EAVENNKKEIKATSPLEQYDFMLMTHPRKKR  DRSSPIVDYCCHVSCTKNDIAKLC.

Skate                                                                                  EEKMGFAKKCCAIGCSTEDFRMVC.
Shark                                                                                  ATSPAMSIKCCIYGCTKKDISVLC.
Dogfish                                                                                EGSPGMSSKCCTYGCTRKDISILC.
Rat      EERKAALSEGRAPFPE--LQQHAPALSDSVVSLEGFKKTFHNQLGEA-EDGG------PP-ELKYLGSDAQSRKKR  QSGALLSEQCCHIGCTRRSIAKLC.
Hamster  EKPKTALPEGHPSLPE--QQQYVPVSSDSVGSLDDFKKSFHATQGEA-EDS-------SLPELKSLYLDTLSRKKR  YTSIYMSHQCCFRGCSRRSLTAAC.
g pig    EELRAMLPEKQPSSPQ--LQQYVPALKNSNVAVKELNKIIRGRQEEA-EDN-------SHSLLKDFNLNIYSPKKR  QLDMTVSEKCCQVGCTRRFIANSC.
Pig      QELKATLSERQPSLRE--LQQSA--SKDSNLNFEEFKKIILNRQNEA-EDK-------SLLELKNLGLDKHSRKKRLF--RMTLSEKCCQVGCIRKDIARLC.
Equine   KEQKATLSERQPSWRE-LLQQPA--LKDSNLNLEEFEETIQKTQSEV-EDD-------SLSELKNLGLDKHSRKKR  --MIQLSHKCCYWGCTRKELARQC.
Human1   PELKAALSERQPSLPE--LQQYVPALKDSNLSFEEFKKLIRNRQSEAADSN--------PSELKYLGLDTHSQKKR  RPYVALFEKCCLIGCTKRSLAKYC.
Human2   QELKLTLSEMQPALPQ--LQQHVPVLKDSSLLFEEFKKLIRNRQSEAADS--------SPSELKYLGLDTHSRKKR  QLYSALANKCCHVGCTKRSLARFC.
```

Fig. 1. Comparison of the deduced amino acid sequence of tammar relaxin with other relaxins of different species. Regions of homology are indicated in the bold shading; underlined amino acids in the tammar relaxin sequence differ from conserved amino acids in other species.

Tammar relaxin (B and A peptides only) shows most similarity at the amino acid level with shark, dogfish and both H1 and H2 human relaxins, with sequence homologies of 47%, 49%, 47% and 47% respectively, whereas there is only 25% and 34% homology compared with hamster and rat relaxin.

As in most eutherian species, the tammar wallaby appears to have just one relaxin gene. Southern hybridization of EcoR1-, XbaI- and HindIII-digested genomic DNA with a tammar relaxin-specific probe indicated single specific bands [1], confirming the presence of a single relaxin gene in the tammar genome.

On the basis of amino acid homology, we report that the marsupial preprorelaxin is more similar to elasmobranch and human preprorelaxin compared with other eutherian species despite the evolutionary distance between them. We have also identified conserved protein motifs evidently essential for biological function. The prorelaxin C-peptide is minimally conserved between marsupials and eutherian species. The most common conserved residues are the glutamic acids and leucines. It is thought that the connecting peptide directs the folding of the prorelaxin molecule thereby ensuring correct disulphide bond formation. Our data imply that the majority of amino acids in the C-domain may be functionally redundant. This notion is supported by the observation that a synthetic human relaxin in which the C-domain is reduced to only 13 amino acids retains correct folding and receptor-binding properties [2].

Acknowledgments

The authors thank the Alexander von Humboldt Foundation and the Deutsche Forschungsgemeinschaft for their generous support for this project.

References

1. Parry, L.J., Rust, W. and Ivell, R., Biol. Reprod., 57 (1997) 119.
2. Vandlen, R., Winslow, J., Moffat, B. and Rinderknecht, E., In MacLennan, A., Tregear, G. and Bryant-Greenwood, G.D. (eds.), Progress in Relaxin Research, World Scientific Publishers, Singapore, 1995, p. 59.

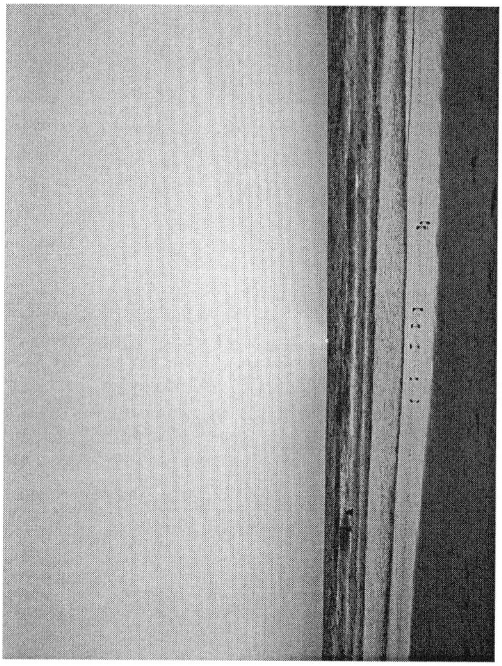

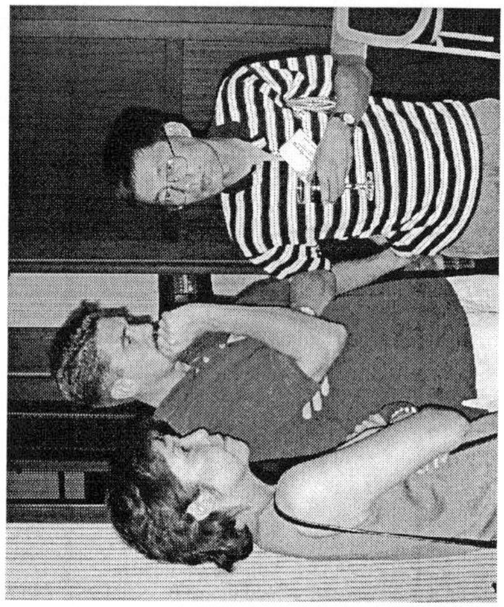

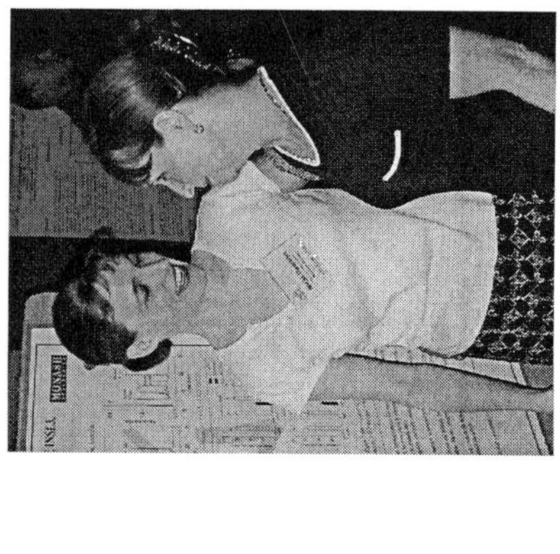

SECTION 3

Reproductive physiology of relaxin II: Uterus & Implantation

Relaxin induces specific alterations in gene expression in the human endometrium

Elaine N Unemori, Martyn Lewis, Beverly H Grove, Usha Deshpande
Connetics Corporation, Palo Alto, California, 94303, USA

Introduction

Tight, coordinate regulation of numerous genes is necessary for the endometrium's sole function, to provide a bed compatible with embryonic implantation and the maintenance of a successful pregnancy. *In vitro*, relaxin has been shown to regulate the expression of a number of proteins believed to be critical for, or contribute to, endometrial receptivity. The list of genes upregulated by relaxin in the endometrial stroma include insulin-like growth factor binding protein-1 [1], prolactin [2], and HOXA-10 [3]. Because relaxin is present in the circulation and in the endometrium itself during the peri-implantation period, it may play a major role in determining endometrial receptivity.

Structural and functional alterations in the endometrial vasculature contribute to endometrial receptivity. Angiogenesis, or the growth of new blood vessels from pre-existing ones, vasodilation, and vascular permeability are features that have been correlated with the period of time during which embryonic implantation occurs, also called the "window of receptivity." In this chapter, data will be presented demonstrating the stimulation of two genes, vascular endothelial growth factor (VEGF) and cyclooxygenase (COX)-2, in normal human endometrial stromal (NHE) cells by recombinant human gene 2 relaxin (rH2 relaxin). The induction of both genes by rH2 relaxin *in vitro* and their coincident expression during the midluteal phase of the menstrual cycle when naturally occurring relaxin circulates, suggests that relaxin may influence the physiology of endometrial blood vessels during the peri-implantation period.

VEGF is an endometrial cell mitogen, vascular permeability, and vasodilatory factor that is expressed in cyclic pattern in the endometrium. VEGF immunoreactivity and mRNA expression are found in the glandular epithelium and stromal cells during the proliferative and secretory phases of the menstrual cycle [4]. Staining patterns are intense in epithelial cells and more diffuse in the stroma. NHE cells respond to rH2 relaxin by induction of VEGF mRNA and protein [5]. Because the endometrium is a tissue that uniquely undergoes shedding and then regeneration and revascularization, the controlled expression of VEGF and other angiogenic factors is of significant interest. Furthermore, because angiogenesis is absolutely required for implantation and early embryonic development [6], the regulation of VEGF and other factors during the peri-implantation period must be highly evolved.

Cyclooxygenase, which exists as two isoforms, COX-1 and COX-2, is the rate-limiting enzyme in the biosynthesis of prostaglandins. In the murine endometrium, both

G.W. Tregear et al. (eds.), Relaxin 2000, 65–72.
© 2001 *Kluwer Academic Publishers. Printed in the Netherlands.*

COX-1 and COX-2 are expressed, but only COX-2 expression coincides with the window of implantation. Targeted disruption of the COX-2 gene caused a failure of embryonic implantation, demonstrated by the inability of wild type blastocysts to implant into the uteri of pseudopregnant COX-2-deficient female mice [7]. COX-2 is also present in the human endometrium in the midluteal phase of the menstrual cycle [8]. Receptors for prostaglandins are also localized in the endometrial epithelium and stroma in a site-specific manner, which has suggested their importance in implantation and decidualization in the mouse [9]. In this chapter, we describe the regulation of both VEGF and COX-2 expression by rH2 relaxin *in vitro* in NHE cells.

Results and Discussion

VEGF protein is constitutively produced by NHE cells in culture, and the addition of rH2 relaxin causes a dose-dependent increase in its secretion by these cells (Fig. 1).

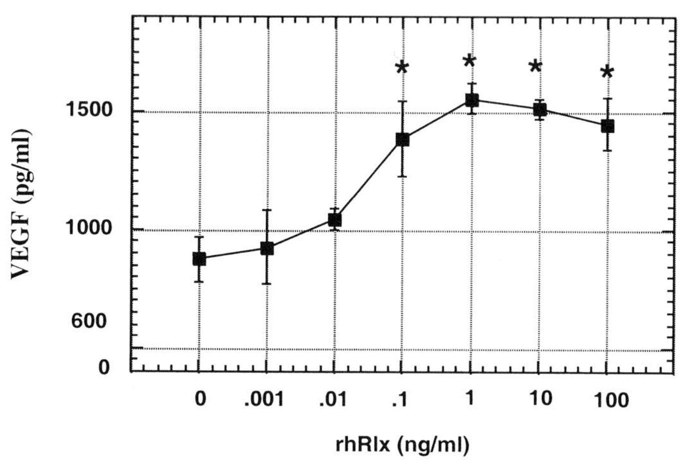

*Fig. 1. VEGF protein expression by NHE cells following 24h stimulation with rH2 relaxin (0.01-100 ng/mL). Cultures were treated in triplicate for each rH2 relaxin dose and assayed for VEGF accumulation in conditioned media in duplicate by ELISA. Points = mean ± SD; *p < 0.05 compared to untreated control by Student-Newman-Keuls test.*

The increase in VEGF expression also occurs at the level of VEGF transcipts, with an induction apparent within 30 min of rH2 relaxin addition to NHE cells, as shown by RT-PCR (Fig. 2). Transcripts of both the 121 amino acid (aa) and 165 aa isoforms are increased in response to rH2 relaxin treatment. To determine whether this upregulation was transcriptional, the 2331 bp promoter region 5' of the translation start site of the VEGF gene was isolated (Fig. 3).

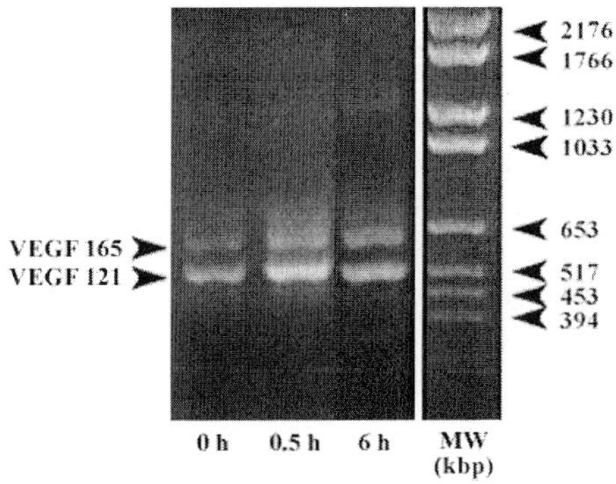

Fig. 2. Semi-quantitative RT-PCR analysis of transcripts of VEGF induced in NHE cells following rH2 relaxin treatment for 0, 0.5, and 6 hours. PCR products, representing the 165 aa and 121 aa isoforms of VEGF are rapidly upregulated following the addition of rH2 relaxin.

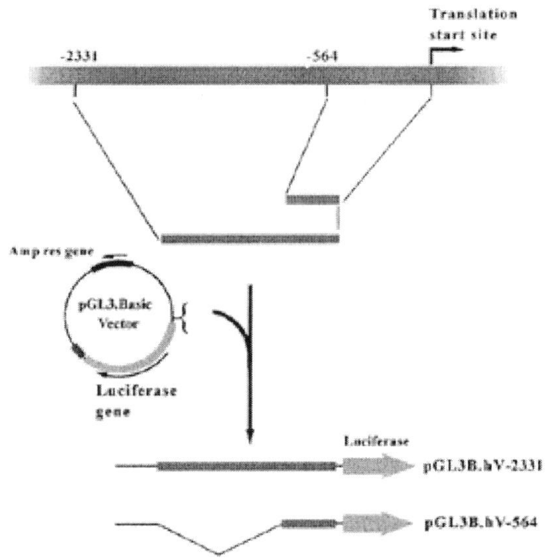

Fig. 3. Two fragments spanning –1 to –2331 and –1 to –564 relative to the translation start site were isolated (kindly provided by Dr. Brian Seed, Harvard University). Both fragments were subcloned into the Bgl II and Hind III restriction digest sites of the Firefly luciferase reporter vector pGL3.Basic (Promega), generating constructs pGL3B.hV-2331 and pGL3B.hV-564, respectively.

In addition, to begin determination of what sequences within the promoter transduced this effect, a 564 bp subfragment was also isolated. Both fragments were cloned into a firefly luciferase reporter vector pGL3.Basic (Promega), generating constructs pGL3B.hV-2331 and pGL3B.hV-564, respectively. The reporter constructs were then transfected into NHE cells, along with a Renilla luciferase reporter vector which provided an internal control for transfection efficiency. Promoter activation was assayed by measuring luciferase accumulation within transfected cells 24 h following exposure to rH2 relaxin (0 or 10 ng/mL). rH2 relaxin stimulated VEGF promoter-driven firefly luciferase expression by two-fold through both promoter fragments ($p<0.05$) (Fig. 4). The level of luciferase expression in untreated cells was higher when the construct containing the larger insert was transfected, compared to expression with the construct containing the 564 bp insert, indicating the presence of additional sequences regulating baseline levels of stimulation in the larger fragment.

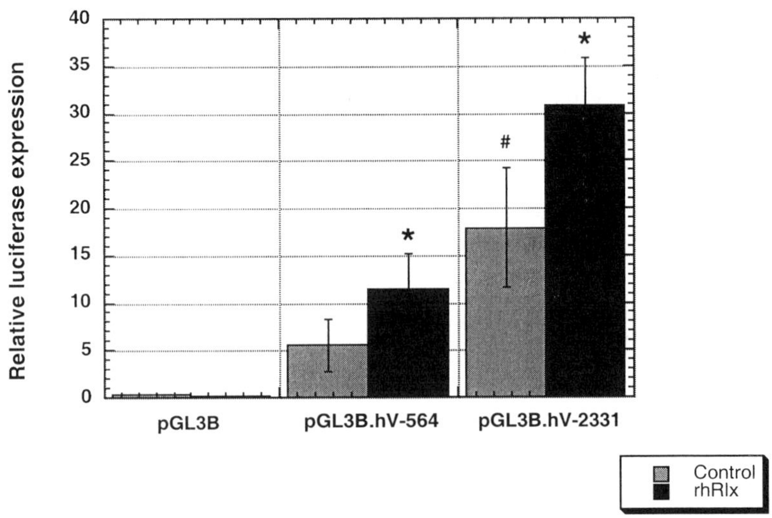

*Figure 4. rH2 relaxin stimulation of luciferase expression through VEGF promoter sequences. NHE cells were transfected with pGL3B, or two constructs containing VEGF promoter inserts. Firefly luciferase accumulation within 24 h was measured following treatment of transfected cells with rH2 relaxin (10ng/mL) or with no treatment. Firefly luciferase expression was normalized to expression of Renilla luciferase to account for transfection efficiency. RH2 relaxin, compared to no treatment, stimulated a significant increase in firefly luciferase expression. Bars = means ± SD; * $p < 0.05$ vs untreated plasmid control; # $p<0.05$ vs pGL3B.hV564 control, by Student-Newman-Keuls test; n = four experiments.*

To compare the upregulation of the endogenous VEGF gene by rH2 relaxin, protein secreted into the conditioned media by transfected cells was quantified by ELISA (Fig. 5). The rH2 relaxin-mediated increase in VEGF protein secretion was markedly higher than two-fold, suggesting that rH2 relaxin regulation of the endogenous VEGF gene involved levels in addition to transcriptional activation of the gene.

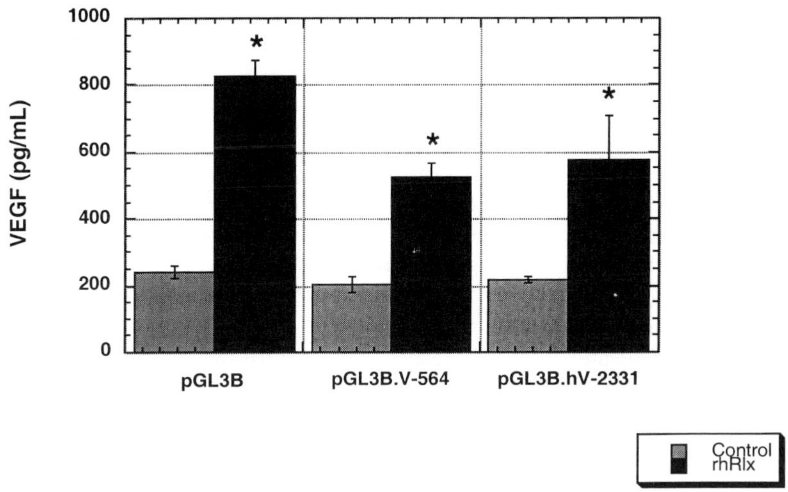

*Fig. 5. Endogenous VEGF protein secreted by transfected NHE cells described in Fig. 4. VEGF accumulated in media conditioned for 24 h following no treatment or treatment with rH2 relaxin (10 ng/mL) was quantified by ELISA. Bars = mean ± SD; *p < 0.05 by Student-Newman-Kuels test, n = 4 experiments.*

There are several known consensus transcriptional control elements within the VEGF promoter (Fig. 6). Estrogen is believed to transcriptionally stimulate the VEGF promoter via a variant of an estrogen response element [10]. There is also a hypoxia-inducible factor response elements (HIE) within the promoter, and it is presumably responsible for the response of the gene to hypoxia [11]. Because the proportional increase in luciferase expression by rH2 relaxin was comparable when driven by either the 2331 bp or 564 bp fragments, rH2 relaxin likely mediates VEGF expression via the Sp-1, AP-2, PEA-3 consensus sequences, or perhaps through a novel sequence contained within the 564 bp fragment.

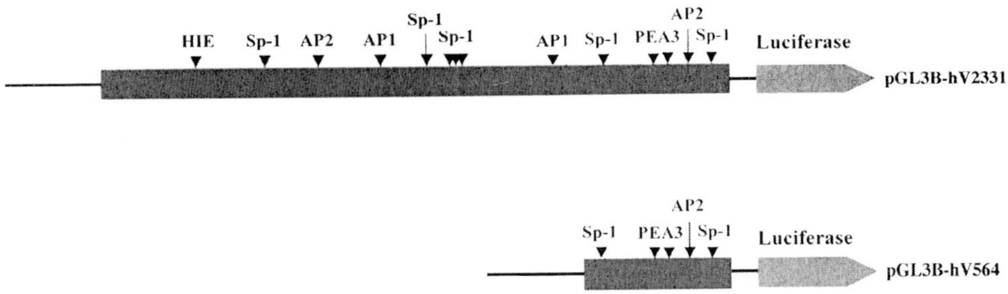

Fig.6. Schematic map of consensus response elements within the VEGF promoter fragments tested.

The expression of the enzyme, COX-2, was also investigated in these cells. Immunocytochemical staining of NHE cells using a polyclonal antibody to COX-2 shows positive staining of a proportion of unstimulated cells, indicating that COX-2 is expressed at a basal level by some resting endometrial stromal cells (Fig. 7). Following 24 h of rH2 relaxin (10 ng/mL) treatment, most cells show intense staining, with much of the stain concentrated peri-nuclearly. This pattern is consistent with the localization of the enzyme in the nuclear envelope [12]. RT-PCR analysis has also demonstrated a rapid upregulation of COX-2 transcripts following exposure of NHE cells to rH2 relaxin (data not shown). The approximately 2kb sequence of DNA 5' of the COX-2 gene contains regulatory elements, which include AP-1, Sp-1, NF-κB, PEA-3 and a cAMP response element. This promoter region is being evaluated for rH2 relaxin responsiveness.

Recently, Connetics has initiated a clinical study to examine the effect of rH2 relaxin on the endometrium of women undergoing controlled ovarian hyperstimulation for the purpose of egg donation in an in vitro fertilization program. This population of women is hormonally stimulated in a manner very similar to women undergoing IVF. Since infertile women in IVF programs may benefit from rH2 relaxin therapy, egg donors are appropriate surrogate subjects for examination of rH2 relaxin's efects on the endometrium. At the present time, 10 of the 24 subjects have been enrolled in this placebo-controlled study. The effects of administration of three rH2 relaxin doses (2.5, 7.5, 25 μg/kg/day) for 14 days on endometrial paramteres associated with implantation success, including endometrial blood flow, endometrial thickness, ultrasonographic pattern, and histology, are being assessed. The expression of biochemical markers, including VEGF and COX-2, is also being evaluated.

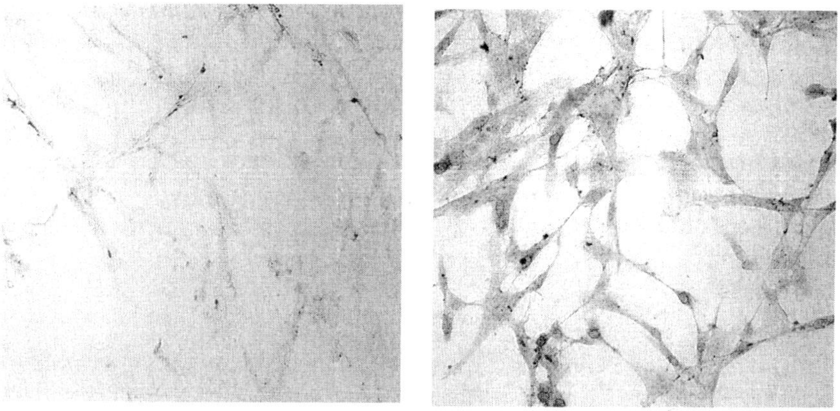

Fig. 7. Immunocytochemical staining of COX-2 in NHE cells without (right panel) and following (left panel) addition of rH2 relaxin (10 ng/mL). Cells were methanol-fixed and stained with a polyclonal antibody to human COX-2 (Santa Cruz Biotechnology, CA).

In summary, relaxin stimulates the expression of a constellation of biochemical markers *in vitro* that are believed to contribute to implantation and decidualization. VEGF and COX-2 are two genes whose expression is modulated by rH2 relaxin *in vitro,* and whose expression *in vivo* may be important in regulating vascular changes that prepare the endometrium for implantation. Angiogenesis and enhanced blood flow, localized endometrial vascular permeability, and uterine edema are all believed to facilitate the penetration phase of the implantation process and ensuing decidualization. Thus, relaxin may play a central role in coordinating gene expression during implantation and early pregnancy.

REFERENCES

1. Tseng, L., Gao, J.-G., Chen, R., Zhu, H. H., Mazella, J. and Powell, D. R., Biol. Reprod., 47 (1992) 441.
2. Huang, J. R., Tseng, L., Bischof, P. and Janne, O. A., Endocrinology, 121 (1987) 2011.
3. Gui, Y., Zhang, J., Yuan, L. and Lessey, B. A., Mol. Hum. Reprod., 5 (1999) 866.
4. Shifren, J. L., Tseng, J. F., Zaloudek, C. J., Ryan, I. P., Meng, Y. G., Ferrara, N., Jaffe, R. B. and Taylor, R. N., J. Clin. Endocrinol. Metab., 81 (1996) 3112.
5. Unemori, E. N., Erikson, M. E., Rocco, S. E., Sutherland, K. M., Parsell, D. A., Mak, J. and Grove, B. H., Hum. Reprod., 14 (1999) 800.
6. Klauber, N., Rohan, R. M., Flynn, E. and D'Amato, R. J., Nature Med., 3 (1997) 443.
7. Lim, H., Paria, B. C., Das, S. K., Dinchuk, J. E., Langenbach, R., Trzaskos, J. M. and Dey, S. K., Cell, 91 (1997) 197.
8. Marions, L. and Danielsson, K. G., Mol. Hum. Reprod., 5 (1999) 961.

9. Yang, Z. M., Das, S. K., Wang, J., Sugimoto, Y., Ichikawa, A. and Dey, S. K., Biol. Reprod., 56 (1997) 368.
10. Mueller, M. D., Vigne, J. L., Minchenko, A., Lebovic, D. I., Leitman, D. C. and Taylor, R. N., Proc. Natl. Acad. Sci. U.S.A., 97 (2000) 10972.
11. Popovici, R. M., Irwin, J. C., Giaccia, A. J. and Giudice, L. C., J. Clin. Endocrinol. Metab., 84 (1999) 2245.
12. Morita, K., Schindler, M., Regier, M. K., Otto, J. C., Hori, T., DeWitt, D. L. and Smith, W. L., Proc. Natl. Acad. Sci. U.S.A., 270 (1995) 10902.

Relaxin is an important factor for uterine differentiation and implantation in the marmoset monkey

Almuth Einspanier
Department of Reproductive Biology, German Primate Center, Göttingen, FRG

Introduction

Relaxin is known to be associated with changes in the uterus and cervix in relation to birth, particularly in terms of a softening and widening of the birth canal and at least in some species a suppression of spontaneous contractions [1,2]. Such findings are generally observed in animal models, like the rat and pig, which show a marked relaxin physiology with maximum relaxin levels during the last trimester of pregnancy. For primates, however, this classic role of relaxin is open to question.

Unlike rodents and the pig, most primates show highest serum relaxin concentrations in the first trimester of pregnancy [eg. 3-6]. Furthermore, the measured peripheral relaxin levels are much lower in primates, attaining only a few percent of those found in non-primate species. These low levels and this early peak level strongly suggest a more paracrine physiology for relaxin in the ovary and uterus, possibly associated with ovulation and the endometrial changes required for implantation.

With its different sites of local biosynthesis, contrasting patterns of secretion, and variable peripheral concentrations depending on the species, relaxin remains a confusing hormone. On the basis of much of the primate data, it seems logical to regard relaxin more like a broad-range growth factor involved in tissue remodelling and reproduction in general than as a specific hormone of late pregnancy. Nevertheless, earlier experiments using porcine relaxin were indeed able to induce cervical ripening in women during labor [7-10], although these experiments were not reproducible when using recombinant human (rH2) relaxin [11-12]. It is to be hoped that the reason the rH2 relaxin failed to show a significant effect on the cervix in these studies will be soon understood and that a better application paradigm can be developed.

Thus, much of our knowledge about relaxin derives from studies on rodents and pigs [1], and yet the situation in humans appears to differ markedly from these other species. There is therefore a considerable need for intensive studies of relaxin physiology in an appropriate primate model with relevance to human reproductive health and disease.

The marmoset monkey

New World Monkeys and in particular the common marmoset monkey (*Callithrix jacchus*) are being increasingly used for research on reproductive physiology. The similar cycle length, similar uterine/ ovarian morphology and patterns of hormone expression to these in the human make the marmoset ideally suited for biomedical

G.W. Tregear et al. (eds.), Relaxin 2000, 73–82.
© 2001 *Kluwer Academic Publishers. Printed in the Netherlands.*

research [13]. The marmoset is a small primate (around 400g), can adapt well in captivity, and is easy to handle and breed. Puberty occurs at 8 months and full sexual maturity is reached at 18-24 months [14], considerably earlier than in Old World monkeys. The Callitrichidae are multiovulatory primates, are non-seasonal, and usually give birth to twins or triplets after around 145 days of pregnancy [e.g. 13, 14,]. Since these primates do not have a post-partum amenorrhoea as in other monkeys, there is a relatively short inter-birth interval. All these characteristics lead to a high fecundity in captivity.

An experimental advantage of the marmoset in the context of reproductive studies is its sensitivity to the luteolytic action of prostaglandin. The ovarian cycle can be precisely controlled by a single application of a synthetic prostaglandin $F_{2\alpha}$ ($PGF_{2\alpha}$) given during the mid luteal phase [15]. This induces immediate luteolysis and hence the onset of a new wave of preovulatory follicle development. There are also some disadvantages to the marmoset. Because of its small size, frequent and large blood samples cannot be taken. Also, as known for other primates, there are some metabolic differences between marmosets and man [13]. In general, however, the marmoset monkey offers a more convincing model for the human than most laboratory species like rats and mice.

Patterns of relaxin expression and secretion in the marmoset monkey

The primary structure of relaxin, derived from its cDNA-sequence, is now known for the human, chimpanzee, rhesus monkey, gorilla, orangutan, baboon and marmoset monkey [16-22]. While all anthropoids have two relaxin-genes, only in the human do both genes appear to be expressed. The marmoset monkey and all other primates appear to have only a single relaxin encoding gene.

In women and non-human primates, the main source of relaxin during the cycle and early pregnancy appears to be the corpus luteum (CL) [e.g. 23-25]. In the cycle, relaxin expression increases with the age of the CL, such that peripheral concentrations of relaxin are high during the mid- to late luteal phase and continue to increase if pregnancy occurs. If there is no conception, relaxin levels fall concomitant with the resulting luteal regression. Similar patterns of peripheral relaxin have been detected and described in several different primates [eg. 5, 6, 26-32].

While highest levels of relaxin are detected in pregnancy, relaxin can also be measured at low levels in the circulation of the human [33], chimpanzee [30], rhesus monkey [6], baboon [28] and marmoset [32]. Like the human, the marmoset has a cycle length of around 28 days, with approximately 10 days follicular phase and 18 days luteal phase. The hormone profiles are very similar to those in the human, though the relaxin concentrations are all somewhat higher in the marmoset. Relaxin concentrations are low during the follicular phase and increase after ovulation [32], as in the other primates (including humans). Additionally, there may be a small rise in relaxin shortly before ovulation [32].

In a recent experiment we analysed the behaviour of peripheral progesterone and relaxin levels in the marmoset following $PGF_{2\alpha}$ application during the mid/late luteal phase (n=12 animals). $PGF_{2\alpha}$ was applied at a concentration of 8 μg/400 g per animal on day 12 of the luteal phase [15]. Blood samples (0.3 ml) were collected for four days starting at the day of $PGF_{2\alpha}$ application and analysed for progesterone [34] and relaxin [32]. Progesterone showed an immediate decline and within 1 day levels had declined to between 5 and 20 ng/ml. In contrast, relaxin remained elevated for a further day, and only subsequently began to decrease (Fig.1). The reason for this discrepant pattern of release is still not fully understood.

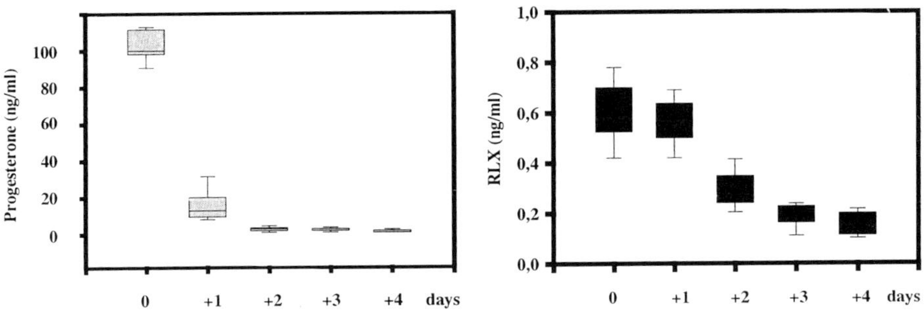

Fig. 1. Progesterone and relaxin (RLX) concentrations following $PGF_{2\alpha}$ application in the mid-luteal phase. day 0: day of $PGF_{2\alpha}$ application; N= 12 monkeys, ± SEM

Although the CL appears to be the principal source of relaxin during the cycle, in the marmoset there is also a low level expression of relaxin in the proliferative and secretory endometrium during the cycle [22]. This local uterine expression appears to increase further if conception occurs, and may extend also to placental and fetal tissues. The first relaxin profiles in peripheral blood of pregnant marmoset monkeys were described by Steinetz and colleagues [31], who observed maximal relaxin levels in the second trimester of pregnancy, somewhat later than in the human. However, there is a significant difference between the conceptive and non-conceptive cycles at around the time of implantation [32], similar to what has been described for the human [35, 36] and other primates [6]. Therefore, relaxin, like chorionic gonadotrophin (CG), is a useful marker of early pregnancy also in the marmoset monkey.

What is the major source of relaxin in the marmoset if pregnancy proceeds? In two spontaneously aborted pregnancies (weeks 9 and 16), while there was an immediate decline in blood concentrations of estradiol and progesterone, relaxin only declined some days later [31]. In contrast, a luteolytic application of $PGF_{2\alpha}$ at the 9th week of pregnancy, at a level insufficient to cause abortion, induced an immediate decline of all three hormones, which is different to the situation in the cycle (Fig. 1). In this study gestagen was substituted to protect the fetus [31], making it difficult to distinguish the precise role and relevance of relaxin at this stage of pregnancy. The results of this study and the physiological decline of relaxin at around 16-18 weeks after conception point to the conclusion that relaxin at this stage of pregnancy is still mostly a product of the CL, and that relaxin is not required for maintenance of pregnancy or parturition [31]. This hypothesis of Steinetz and colleagues [31] that the CL graviditatis is the major source of relaxin production could be partly supported by our own experiments, where we have been able to detect relaxin-immunoreactivity in the CL graviditatis during early, mid and late pregnancy in the common marmoset monkey (Fig. 2). There is strong staining for relaxin, 3ß-hydroxysteroid dehydrogenase and the progesterone receptor in nearly all luteal cells (90%) during early and mid pregnancy. During mid to late pregnancy (>100 days) there is a steady decline in numbers and intensity of positively stained cells for all three antigens. In the endometrium, there is also strong staining for relaxin in early and mid pregnancy in both maternal and fetal cells. Thus, at least in mid pregnancy both the ovary and the uterus appear to be producing relaxin, leaving it to an open question what is the relative contribution of each source for the pregnancy maintenance.

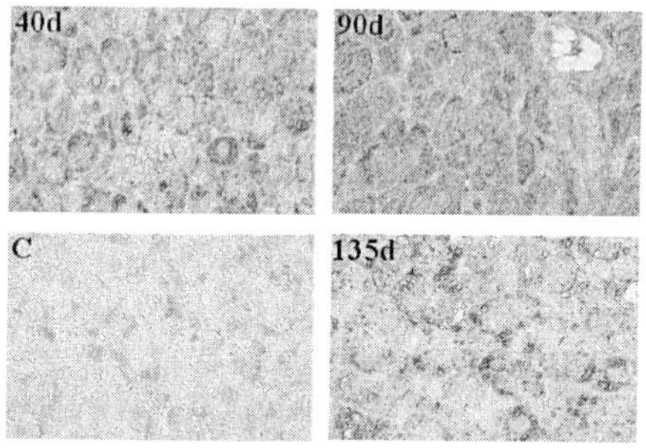

Fig. 2: Immunoreactivity of relaxin (red staining) in the CL of pregnant marmoset monkeys C: control animal (IgG), 40d: day 45, 90d: day 90, 135d: day 135. Immunostaining is localized in the cytoplasm, nuclei are counterstained with haematoxylin; Magnification x 400.

A pathological situation during pregnancy: the cystic mole

Since relaxin appears to be a useful marker for early pregnancy and adequate placental development in the marmoset monkey, a dysfunction of the placenta should be reflected in the peripheral relaxin levels. Although there is no significant correlation between the number of implanted blastocysts and peripheral relaxin levels, resorption of an embryo or pregnancy failure does appear to be reflected either by a decline in peripheral relaxin levels or by a slower increase in relaxin, as has been demonstrated previously [32].

Another form of dysfunctional placenta is represented by a cystic mole, which results normally in an abortion during early to mid pregnancy. This rare clinical picture was found in one marmoset monkey, which aborted at day 66 of pregnancy (Fig. 3). Blood samples were collected throughout gestation 1-2 times per week. The cystic mole was visible by ultrasound around day 37 of pregnancy, shortly after a significant peripheral relaxin decline. Spontaneous abortion then occurred at day 66 of pregnancy. Peripheral relaxin concentrations, which initially had been no different from those in normal pregnancies, began to decline more noticeably around day 50 of pregnancy, while progesterone remained unaffected until abortion itself occurred, when it also declined. Comparing the one abortion in the marmoset described by Steinetz and colleagues [31], which occurred in the 9th week (around 63 days) of pregnancy and where relaxin declined only some days after the abortion, this could have been due to a still functional luteal and placental relaxin production. In the marmoset with the cystic mole that we examined, both ovaries had luteal cysts visible by ultrasound examination, which would indicate a dysfunction of the local ovarian hormone system. Taking all these observations together suggests that peripheral relaxin is probably produced mainly by the CL graviditatis, though the function of this relaxin and its relevance for the maintenance of pregnancy still needs clarification. In particular, the local uterine production of relaxin in maternal and fetal structures suggests it also might be involved in some aspects of fetal/maternal interaction.

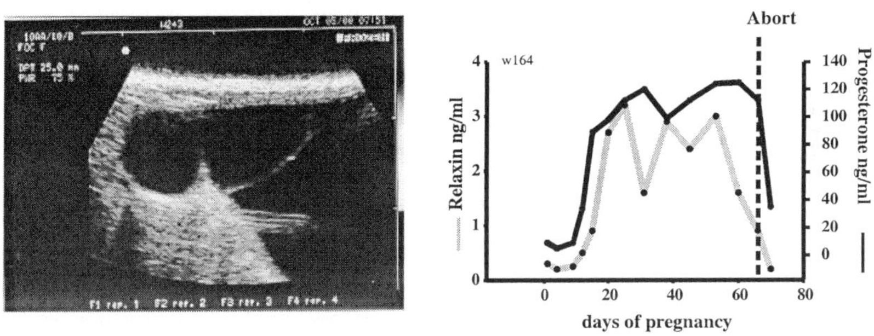

Fig. 3. Cystic mole in a pregnant marmoset monkey. Ultrasound documentation of a fluid filled uterus without any fetal structures (left side). Hormone profiles of progesterone and relaxin during the failed pregnancy. Abortion occurred at day 66 of pregnancy. Progesterone: black lines; relaxin: grey lines

Implantation and early pregnancy

Both, CG and relaxin, hormones increase during early pregnancy, however, the relationship between CG and relaxin is still unclear. Earlier studies in primates which have looked at a possible causal relation between luteinizing hormone (LH)/CG and relaxin levels in peripheral blood have yielded quite contrary results. While some authors have suggested a direct control function for hCG on relaxin production [28, 37-41], other groups refute any such relationship [32, 35, 42, 43]. In humans, multiple pregnancies or gonadotropin-stimulated IVF procedures indicate a positive link between gonadotropins and relaxin concentrations [44-46]. In the marmoset monkey, while there is a parallel increase of both LH/CG and relaxin, no causal interaction between the two can be demonstrated [32]. An exogenous application of hCG at the expected time of implantation failed to induce any significant peripheral relaxin increase in non pregnant animals. A similar conclusion could be drawn from experiments in which there was local application of exogenous hCG to *ex vivo* mid-luteal phase CL-tissue using a microdialysis system (see chapter by Verhagen and Einspanier in this volume). Thus both *in vivo* as well as *ex vivo* techniques suggest more an indirect action between LH/hCG and relaxin than any direct control of relaxin by CG in the marmoset monkey.

An involvement of relaxin in the implantation process itself and early pregnancy still requires direct confirmation, although a number of indirect lines of evidence exist. The peripheral blood relaxin content is maximal in primates at this time [6, 32, 35], and at least in women the relaxin concentration correlates with the number of implanted embryos [47]. There is also a significant effect on peripheral relaxin levels of early abortion in the primate compared to the normal physiological situation in early pregnancy [6, 32, 35, 48]. Possibly most important here is that in *in vitro* cell culture experiments using human endometrial stroma cells, relaxin is a highly significant inducer of decidualization [49-52]. Both progesterone and relaxin cause a significant intracellular cAMP increase, which is necessary for stromal differentiation from a proliferative to a secretory type. This differentiation process induces a cascade of biochemical actions which enable the blastocyst to implant and invade.

Changes in the endometrium are necessary for preparing an appropriate environment for the implanting embryo and to enable a well coordinated interaction between maternal and fetal tissues. Relaxin is likely to be a major player here, since it is produced locally in maternal and fetal tissues, is able to modulate extracellular matrix [53], and stimulates VEGF expression [54, 55]. It therefore seems plausible that such local decidual or placental relaxin [22, 56, 57] could have a paracrine effect on fetal-maternal interaction. Such effects could also occur at later stages of pregnancy, for example, in influencing genes for MMPs, cytokines, prostaglandin production or TIMPs within fetal membranes [58, 59]. This idea is supported by the indirect identification of putative relaxin receptors in the chorion, amnion, placental villi and cervix of the human [60].

35 years ago the first *in vivo* experiments carried out on rhesus monkeys [61, 62] demonstrated that relaxin, under the influence of steroids, had an angiogenic and differentiating effect on the endometrium, thus providing a foundation for all subsequent experiments. Encouraged by these experiments, and those of Sherwood [1] carried out in the pig and mouse, as well as by our own initial results in the marmoset monkeys, we have carried out a set of *in vivo* experiments aimed at understanding the role of relaxin in and on the marmoset uterus. Adult ovariectomized (ov, 3 months previously) female marmoset monkeys were subcutaneously injected with either progesterone (50 μg/day, days 7-10), estradiol (35 μg/d, days 1-10), porcine relaxin (100 μg/6h, days 7-10) or the combination of estradiol (day 1-10) and relaxin (day 7-10) [63, 64]. Before collection each uterus was examined by ultrasound, and the vagina and nipples were measured.

There was a significant increase in uterine wet weight after treatment with estradiol, relaxin, and their combination compared to the control, but not with progesterone alone. Ultrasound examination confirmed the wet weight measurements, a uterine size increase was only measurable with estradiol, relaxin and their combination. The nipples did not increase in size under any treatment. Vaginal diameter increased from 0.2 to 0.3-0.4 cm under estradiol, relaxin and estradiol/relaxin application.

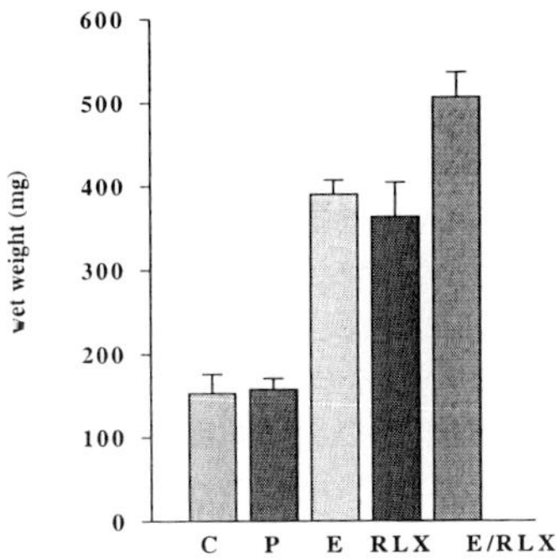

Fig. 4. Wet weights (mg) of each treatment group, C: control, P: progesterone, E: estradiol, RLX: relaxin, E/RLX: estradiol and relaxin. Mean ± SEM (n=4 animals per group)

The control group showed the thinnest endometrium, with least endometrial glands and blood vessels, and lowest levels of VEGF-expression. All hormone treatments caused an increase in the thickness of the endometrium as well as in the number of glands. The typical picture of each hormone treatment was as follows: estradiol induced a loose and edematous stroma with many blood vessels, progesterone showed dense collagen with a highly organized stroma, and relaxin demonstrated a marked increase in blood vessel content and the differentiation of the stroma with active glands. The combined application of estradiol and relaxin resulted in the thickest endometrium with luminal edema, most glands, many blood vessels as well as a high VEGF expression throughout the whole endometrium. In this treatment group collagen fibers are less dense and less organized than the other groups. Results from this study indicate that estradiol, relaxin and their combination promote uterine growth and angiogenic action [62, 63]. Moreover, these results provide evidence that relaxin and estradiol can act synergistically to remodel the endometrium in this species. Further studies are currently underway to understand the role of progesterone in this context and therefore to understand the basis for the endometrial changes at the time of implantation. These are still poorly understood and many open questions remain. A better understanding of what is going on here should provide a valuable background to design better therapeutic approaches to aid implantation and remediate infertility.

Future perspectives

Recent research in the marmoset and other species has drawn the focus of attention away from relaxin as solely a hormone of parturition. There appear to be functions for relaxin during ovulation, in CL maintenance and at implantation. Relaxin seems to share many characteristics with growth factors and cytokines, though while there are similarities, there are also differences which we need to recognize.

Relaxin appears to be a very variable hormone with different species-specific peripheral concentrations, sites of local synthesis, and sites of action. There may also be differences between local or systemic relaxin functions in the different species. Local effects of relaxin could also be masked by the effects of other paracrine-acting agents, preventing clear-cut answers in some experiments.

From the literature and our own results, relaxin appears to be an important factor during implantation in primates. However, this does not necessarily exclude an additional role during parturition, though this still needs to be verified in primate species. The mitogenic and angiogenic actions of relaxin during early pregnancy and implantation in primates definitely need further study, especially in the context of early pregnancy failure. Studies in my own laboratory are currently in progress to see whether there is a supportive effect of relaxin on the pre-pregnant and early pregnant uterus in the marmoset. Such studies also need performing for the human, though it would be wise to await the results in a suitable primate model like the marmoset before embarking on the ethically more problematic situation in the human.

Acknowledgements

I gratefully acknowledge the Deutsche Forschungsgemeinschaft (grant Ei 333/8-1) for their generous support of my research. I should also like to thank the Schering Research Foundation (Berlin, FRG) and Connetics Cooperation (Pablo Alto, USA) for supporting my attendance at this conference. Thanks are also due to Professor David Sherwood for providing the porcine relaxin antibody as well as porcine relaxin, to Dr. Elaine Unemori for providing human recombinant relaxin and Professors Ralf Einspanier and Dieter Schams for the generous gift of VEGF antibodies and primers as well as Dr. Knabe for blood vessel examination. I should also like to acknowledge the help of Professor Dallenbach-Hellweg and Professor Richard Ivell for fruitful and exciting discussions about possible roles of relaxin in primates as well as for guidance in the examination of the marmoset uteri. Thanks are also due to K. Fuhrmann, A. Marten, A. Jurdzinski for excellent technical assistance and Professor Keith Hodges for providing research facilities.

References

1. Sherwood O.D., In: Knobil, E., Neill, J.D. (eds), The Physiology of Reproduction. 2nd ed., Raven Press, New York, USA, 1994, p. 861.

2. Bryant-Greenwood, G.D. and Schwabe, C., Endocr. Rev., 15 (1994) 5.

3. Weiss, G., Steinetz, B.G., Dierschke, D.J. and Fritz, G., Biol. Reprod., 24 (1981) 565.

4. Eddie, L.W., Bell, R.J., Lester, A., Geier, M., Bennett, G. and Johnston, P.D., Lancet, 8494 (1986) 1344.

5. Bell, R.J., Eddie, L.W., Lester, A.R., Wood, E.C., Johnston, P.D. and Niall, H.D., Obstet. Gynecol., 69 (1987) 585.

6. Stewart, D.R., Stouffer, R.L. and Overstreet, J.W., Endocrinology, 312 (1993) 6.

7. MacLennan, A.H., Green, R.C., Bryant-Greenwood, G.D., Greenwood, F.C. and Seamark, R.F., Lancet, 8162 (1980) 220.

8. MacLennan, A.H., Green, R.C., Grant, P. and Nicholson, R., Obstet. Gynecol., 68 (1986) 598.
9. Evans, M.I., Dougan, M.B., Moawad, A.H., Evans, W.J., Bryant-Greenwood, G.D. and Greenwood, F.C., Am. J. Obstet. Gynecol., 147 (1983) 410.
10. Leppert, P.C., In: MacLennan, A.H. Tregear, G.W., Bryant-Greenwood, G.D. (eds), Progress in Relaxin Research, Global Publications, Singapore, 1995, p341.
11. Bell, R.J., Permezel, M., MacLennan, A., Hughes, C., Healy, D. and Brennecke, S., Obstet. Gynecol., 82 (1993) 328.
12. Brennand, J.E., Calder, A.A., Leitch, C.R., Greer, I.A., Chou, M.M. and MacKenzie, I.Z. In: MacLennan, A.H., Tregear, G.W., Bryant-Greenwood, G.D. (eds), Progress in Relaxin Research, Global Publications, Singapore, 1995, p380.
13. Hearn, J.P., In: Hearn, J.P. (ed): Reproduction in New World Primates-New models in Medical Science. Lancaster/Boston/The Hague, MTP Press, 1983, p181.
14. Hearn, J.P., In: Kleinman, D.G. (ed), Biology and Conservation of the Callitrichidae; Washington DC, Smithsonain Institute Press, 1978, p163.
15. Summers, P.M., Wennink, C.J and Hodges, J.K., J. Reprod. Fert., 73 (1985) 133.
16. Hudson, P., Haley, J., John, M., Cronk, M., Crawford R., Haralambidis, J., Tregear, G., Shine, J. and Niall, H., Nature, 301 (1983) 628.
17. Crawford, R.J., Hudson, P., Shine, J., Niall, H.D., Eddy, R.L. and Shows, T.B., EMBO J., 3 (1984) 2341.
18. Crawford, R.J., Hammond, V.E., Roche, P.J., Johnston, P.D. and Tregear, G.W., Journal of Molecular Endocrinology, 3 (1989) 169.
19. Evans, B.A., Fu, P. and Tregear, G.W., Endocrinology, 40 (1994) 385.
20. Evans, B.A., Fu, P. and Tregear, G.W., In: Progress in Relaxin Research; MacLennan, A., Tregear, G.W., Bryant-Greenwood, G.D. (eds), Global Publications Services, Singapore, 1995, p85.
21. Ivell, R., Hunt, N., Khan-Dawood, F. and Dawood, M.Y., Mol. Cell. Endocrinol., 66 (1989) 251.
22. Einspanier, A., Zarreh-Hoshyari-Khah, R., Balvers, M., Fuhrmann, K. and Ivell, R., Hum. Reprod., 12 (1997) 1325.
23. Weiss, G., O´Bryne, E.M. and Steinetz, B.G., Science, 194 (1976) 948.
24. Mathieu, P., Rahier, J. and Thomas, K., Cell Tiss. Res., 219 (1981) 213.
25. Sakbun, V., Ali, S.M., Greenwood, F.C. and Bryant-Greenwood, G.D., J. Clin. Endocrinol. Metab., 70 (1990) 508.
26. O´Bryne, E.M., Carriere, B.T., Sorenson, L., Segaloff, A., Schwabe, C. and Steinetz, B., J. Clin. Endocrinol. Metab., 47 (1978) 1106.
27. Weiss, G., Steinetz, B.G., Dierschke, D.J. and Fritz, G., Biol. Reprod., 24 (1981) 565.
28. Castracane, V.D., D´Eletto, R. and Weiss, G., In: Greenwood, G.S., Terranova, P.F. (eds), Factors Regulating Ovarian Function, Raven Press, New York, 1983, p415.
29. Castracane, V.D., Lessing, J., Brenner, S. and Weiss, G., J. Clin. Endocrinol. Metab., 60 (1985) 133.
30. Steinetz, B.G., Randolph,C. and Mahoney, C.J., Endocrinology, 130 (1992) 3601.
31. Steinetz, B.G., Randolph, C. and Mahoney, C.J., Biol. Reprod., 53 (1995) 834.
32. Einspanier, A., Nubbemeyer, R., Schlote, S., Schuhmacher,M., Ivell, R., Fuhrmann, K. and Marten, A., Biol. Reprod., 61 (1999) 512.
33. Stewart,D.R., Celniker, A.C., Taylor, C.A. Jr, Cragun, J.R., Overstreet, J.W., Lasley, B.L., J. Clin. Endocrinol. Metab., 70 (1990) 1771.
34. Heistermann, M., Tari, S. and Hodges, J.K, J. Reprod. Fertil., 99 (1993) 243.

35. Stewart, D.R., Overstreet, J.W. and Celinker, A.C., J. Clin. Endocrinol., 38 (1993) 379.
36. Stewart, D.R., Nakajima, S.T., Overstreet, J.W., Boyers, S.P. and Lasley, B.L., In: MacLennan A, Tregear GW, Bryant-Greenwood GB (eds.), Progress in Relaxin Reseach, Global Publication Services, Singapore, 1995, p. 214.
37. Quagliarello, J., Goldsmith, L., Steinetz, B.G., Lustig, D.S. and Weiss, G., J. Clin. Endocrinol. Metab., 51 (1980) 74.
38. Nixon, W.E., Reid, R., Abou-Hozaifa, B.M., Williams, R.F., Steinetz, B.G. and Hodges, G.D., In: Greenwald G.S., Terranova P.F. (Eds), Factors Regulatoring Ovarian Function, Raven Press, New York, 1981, p. 421.
39. Ottobre, J.S., Nixon, W.E. and Stouffer, L., Biol. Reprod., 31 (1984) 1000.
40. Ottobre, A.C., Ramsey, K.R. and Ottobre, J.S., J.Reprod. Fertil., 91 (1991) 313.
41. Duffy, D.M., Stouffer, R.L. and Stewart, D.R., Biol. Reprod., 53 (1995) 447.
42. Seki, K., Uesato, T., Tabei, T. and Kato, K., Endocrinol. Jpn., 32 (1985) 741.
43. Gosh, D., Stewart, D.R., Nayak, N.R., Lasley, B.L., Overstreet, J.W., Hendrickx, A.G. and Sengupta, J., Hum. Reprod., 12 (1997) 914.
44. Hanning, R.V., Steinetz, B.G. and Weiss, G., Obstet. Gynecol., 66 (1985) 42.
45. Weiss, G., Goldsmith, L.T., Sachdev, R., Von Hagen, S. and Lederer, K., Obstet. Gynecol., 82 (1993) 821.
46. Johnson, M.R., In: MacLennan A, Tregear GW, Bryant-Greenwood GB (eds), Progress in Relaxin Research, Global Publications, Singapore, 1995, p. 225.
47. Norman, R.J., McLoughlin, J.W., Borthwick, G.M., Yohkaichiya, T., Matthews, C.D., MacLennan, A.H. and de Kretser, D.M., Fertil. Steril., 59 (1993) 130.
48. Witt, B.R., Wolf, G.C., Wainwright, C.J., Johnston, P.D. and Thorneycroft, I.H., Fertil. Steril., 53 (1990) 1029.
49. Chen, G.A., Huang, J.R. and Tseng, L., Biol. Reprod., 39 (1988) 519.
50. Fei, D.T., Gross, M.C., Lofgren, J.L., Mora-Worms, M. and Chen, A.B., Biochem. Biophys. Res. Commun., 170 (1990) 214.
51. Bell, S.C., Jackson, J.A., Ashmore, J., Zhu, H.H. and Tseng, L., J. Clin. Endocrinol. Metab., 72 (1991) 1014.
52. Telgmann, R. and Gellersen, B., Hum. Reprod. Update, 4 (1998) 472.
53. Hwang, J.J., Lin, S.W., Teng, C.H., Ke, F.C. and Lee, M.T., Biol. Reprod., 55 (1996) 1276.
54. Tseng, L., Gao, J.K., Chen, R., Zhu, H.H., Mazella, J. and Powell, D.R., Biol. Reprod., 47 (1992) 441.
55. Unemori, E.N., Erikson, M.E., Rocco, S.E., Sutherland, K.M., Parsell, D.A., Mak, J. and Grove, B.H., Hum. Reprod., 14 (1999) 800.
56. Hansell, D.J., Bryant-Greenwood, G.D. and Greenwood, F.C., J. Clin. Endocrinol. Metab., 72 (1991) 899.
57. Gunnersen, J.M., Fu, P., Roche, P.J., Tregear, G.W., Mol. Cell. Endocrinol., 118 (1996) 85.
58. Bryant-Greenwood, G.D., Placenta, 19 (1998) 1.
59. Millar, L.K., Boesche, M.H., Yamamoto, S.Y., Killeen J., DeBuque, L., Chen R., and Bryant-Greenwood, G., Am. J. Obstet. Gynecol., 179 (1998) 126.
60. Kohsaka, T., Min, G., Lukas, G., Trupin, S., Campbell, E.T. and Sherwood, O.D., Biol. Reprod., 59 (1998) 991.
61. Hisaw, F.L. and Hisaw, F.L., Am. J. Obstet. Gynecol., 89 (1964) 141.
62. Dallenbach-Hellweg, G., Battista, J.V. and Dallenbach, F.D., Am. J. Anat., 119 (1966) 61.
63. Einspanier, A., Fuhrmann, K. and Sherwood, O.D., Biol. Reprod., 60 (1999) Suppl. 1, 182.
64. Husen, B., Binder, C., Hagemann, T. and Einspanier, A., Biol. Reprod., 62 (2000) Suppl. 1, 391.

Relaxin and porcine uterine growth and remodeling

Judy Lenhart[1], Kathleen Ohleth[1], Peter Ryan[2], Janet Wang-Lee[1] and Carol Bagnell[1]

[1]*Department of Animal Sciences, Rutgers University, New Brunswick NJ 08901, USA;*
[2]*Department of Animal and Dairy Sciences, Mississippi State University, Mississippi State, MI 39762, USA*

Introduction

Uterine growth is the result of mitogenic activity of uterine cells, as well as remodeling of the connective tissue framework of the uterus, to facilitate uterine expansion. The stimulation of uterine growth during pregnancy is critical to accommodate growing fetuses. Among the causes of early embryo loss is competition for limited maternal resources, due to insufficient uterine accommodation, which reduces uterine capacity. Relaxin stimulates uterine growth and remodeling in prepubertal [1], pregnant [2] and ovariectomized, steroid-treated animals [3]. While the uterotropic effects of relaxin have been documented, the mechanism of relaxin action in promoting uterine growth is less clear. In our laboratory we are studying the ability of relaxin to stimulate uterine growth, independently of estradiol, using an in vivo, prepubertal gilt model in which the uterotropic actions of relaxin were first reported by Hall *et al.* [1]. Studies from our laboratory indicate that relaxin interacts with the IGF-I system in promoting uterine growth [4]. In addition, our working hypothesis is that relaxin acts as a growth factor in the uterus by increasing expression of elements that enhance remodeling and cell communication such as connective tissue proteases, connexin gap junction proteins and vascular endothelial growth factor (VEGF).

The action of relaxin on tissue remodeling involves changes in a complex cascade of proteolytic enzymes, including those of the matrix metalloprotease (MMP) and serine protease (plasminogen activator, PA) systems [5,6]. Urokinase-type (uPA) and tissue-type (tPA) plasminogen activators are serine proteases that convert plasminogen to plasmin, which in turn can activate some of the MMPs, including MMP-9 [7]. Our studies focus on the gelatinases, MMP-2 and MMP-9, which are the MMP subclass responsible for degrading type IV and V collagens, a major component of basement membranes. The gelatinases are crucial for cell migration, tissue reorganization and angiogenesis [8,9]. The regulation of connective tissue protease activity is complex, occurring at the level of protein synthesis as well as, inhibition of active proteases by specific tissue inhibitors of metalloproteinases (TIMPs) and plasminogen activator inhibitors (PAIs). In addition to connective tissue proteases, enhanced cell-cell communication via gap junctions is thought to be critical in regulating growth and development. In the uterus, the major endometrial gap junction proteins are connexin-26 (CX-26) and –32 (CX-32) while the predominant connexin in

G.W. Tregear et al. (eds.), Relaxin 2000, 83–89.
© 2001 *Kluwer Academic Publishers. Printed in the Netherlands.*

the myometrium is CX-43 [10,11]. Uterine connexin gap junction expression is tissue-specific and regulated by steroid hormones during implantation [reviewed in 12] and at the time of parturition [13]. Thus, we were interested in whether the uterotropic effects of relaxin involve tissue-specific changes in connexin protein expression. Uterine vascularization is key in facilitating growth and implantation and VEGF is an important mediator of these actions. Relaxin stimulates VEGF expression in the rat uterus [14] and in human endometrial stromal cells [15]. However, whether the uterotropic effects of relaxin in the pig involve increased angiogenesis or expression of VEGF protein has not been investigated.

Therefore, in this paper we focus on the impact of relaxin on uterine (a) connective tissue protease activity, focusing on the plasminogen activator and MMP systems (b) connexin gap junction proteins and (c) vascular endothelial growth factor. In these studies prepubertal gilts were treated with porcine relaxin (0.5 mg every 6 h for 54 h) to induce uterine growth [4]. Uterine tissue and flushings, collected 3 h after the last injection, were used for analysis.

Results and Discussion

A. Relaxin and uterine connective tissue protease activity

Given the importance of the plasminogen activator system in reproductive tissue remodeling, we started by focusing on the impact of relaxin on tPA and uPA in the uterus. Following systemic administration of relaxin to prepubertal gilts, secreted uPA activity in uterine flush samples was significantly greater ($P < 0.05$) than in control animals (Fig 1b). This increase in uPA activity in relaxin-treated animals was linked to an increase in immunoreactive uPA protein in the uterine flushes (Fig 1a).

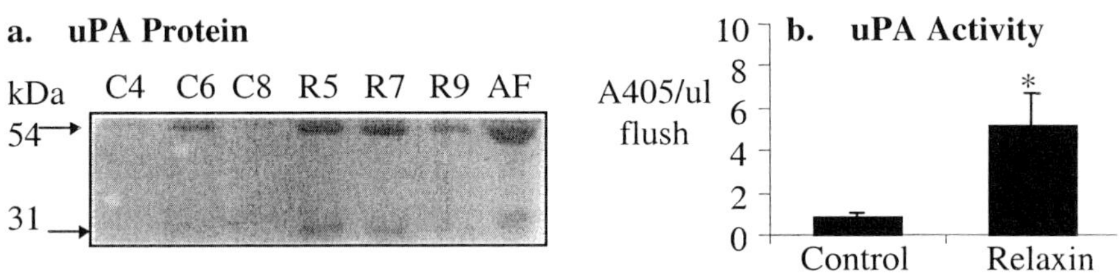

Fig. 1. The effect of relaxin on PA protein and activity in uterine flushes of prepubertal gilts. (a) immunoblot of uPA protein (b) uPA activity. Uterine flushes from control (C; n=3) and relaxin-treated pigs (R; n=3). Amniotic fluid (AF) served as a positive control for uPA protein. Asterisk indicates significant difference (P< 0.05)[16].

However, there was no evidence for a change in uPA activity or uPA protein content of uterine tissue in response to relaxin. Since uPA is important for extracellular matrix degradation, the observation that relaxin increases uterine secretion of uPA, rather than tissue storage, suggests that one way relaxin may promote uterine growth and remodeling is by increasing uPA-mediated extracellular proteolysis. In the uterus, neither cell-associated nor secreted tPA was influenced by relaxin treatment of gilts (data not shown). Likewise, Casslen *et al.* [17] reported that while uPA secretion by explants of proliferative phase human endometrium increased in response to estradiol, there was little or no effect on tPA secretion.

We next examined the effect of relaxin on porcine uterine gelatinases. Both MMP-2 and MMP-9 were present in the porcine uterus regardless of treatment. However, while protein expression was similar in relaxin-treated and control animals, enzyme activity was attenuated by relaxin. In contrast, relaxin significantly increased secretion of MMP-2 and MMP-9 protein into uterine fluid, and MMP-associated enzyme activity in the uterine flushes was significantly elevated following relaxin administration ($P<0.05$; data not shown). Target substrates for these enzymes, specifically type IV and V collagens, have been localized to the pericellular basement membrane and external lamina surrounding differentiated decidual cells [18]. The relaxin-mediated release of MMP-2 and MMP-9 into the uterine lumen, and their subsequent activation may facilitate endometrial tissue reorganization in preparation for implantation and placentation.

Extracellular matrix remodeling is crucial for cell migration and growth, and we hypothesized that relaxin may increase expression of MMP-2 and MMP-9, facilitating type IV collagen basement membrane remodeling in the uterine stroma to accommodate the growing fetuses. However, in the relaxin-treated prepubertal pig, we found that the activity of both MMP-2 and MMP-9 was attenuated in uterine tissue. Thus, we were interested in determining whether the decrease in gelatinase activity was a direct result of relaxin administration, or if relaxin stimulates uterine expression of TIMPs, important regulators of gelatinase activity. Concurrent production and secretion of proteases and their inhibitors allows strict regulation and localization of proteolytic activity. We found that relaxin significantly increased ($P<0.05$) expression of TIMP-1 and TIMP-2 proteins (Fig. 2a) as well as TIMP inhibitor activity (Fig. 2b) in uterine tissue homogenates as well as in uterine flushes (data not shown).

One explanation for the decline in uterine tissue gelatinase activity we observed in response to relaxin administration was this increase in tissue-associated uterine TIMP activity (Fig. 2b). Another possibility is that, after 54 h of exposure to relaxin in vivo, MMP-2/9 released from uterine cells into the extracellular space may have already moved into the uterine flush compartment. We reported a similar pattern of relaxin-induced increases in uPA activity in uterine luminal fluid with no change in tissue-associated uPA [16]. The presence of MMP-2/9 and uPA in the extracellular space is likely an early event in uterine remodeling important for growth. However, in order to regulate growth once tissue remodeling begins, a means of controlling connective tissue protease activity becomes paramount. This includes the generation of inhibitors such as the TIMPs to inactivate MMP-2/9 in situ. In addition, as growth proceeds, the shifting of MMPs and/or TIMPs from the tissue compartment into the uterine lumen over time may serve as another means of regulating uterine remodeling.

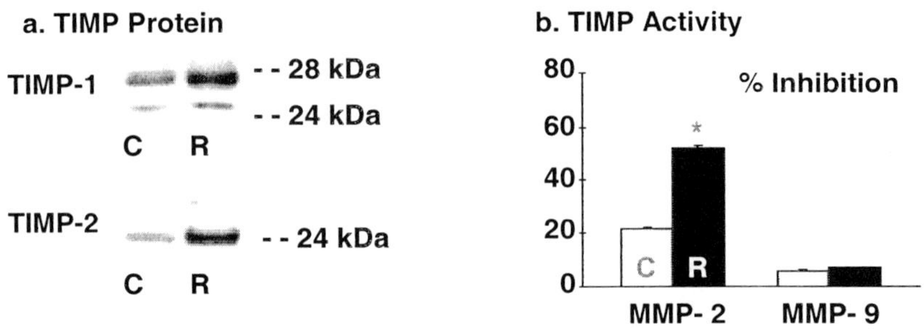

Fig. 2. The effect of relaxin on TIMP protein and activity in uterine tissues from prepubertal gilts. (a) Representative immunoblots of TIMP-1 and TIMP-2 protein in a control (C) and relaxin- treated (R) animal. (b) TIMP Activity. Uterine tissues from control and relaxin-treated gilts (n=3 each) were incubated with hMMP-2 or hMMP-9 (15 ng) and TIMP activity was measured by the cleavage of a fluorescently-labeled peptide substrate after 3 h incubation with uterine homogenates. Data are expressed as % inhibition of MMP substrate degradation. Asterisk indicates significant difference (P<0.05)

B. Relaxin and uterine gap junction expression

The importance of uterine gap junctional communication in facilitating implantation [10,19] and controlling myometrial contractions at the time of parturition is well-documented [20,21]. Furthermore, regulation of CX protein expression during pregnancy has been correlated with changes in estrogen and progesterone levels [21,22]. Gap junctions also play a role in regulating cell proliferation and maintaining tissue homeostasis during tissue growth [23]; activities which are not always steroid dependent. Relaxin raises cAMP levels [24] and activates protein kinase pathways [25], which can affect gap junction communication [26]. Given relaxin's uterotrophic effects [1,3], we investigated whether cell-cell communication was enhanced following relaxin administration. We found that uterine expression of gap junction proteins CX-26, CX-32 and CX-43 was significantly increased in response to relaxin ($p \leq 0.05$; Fig. 3), and that relaxin enhanced expression of these proteins in the absence of endogenous steroids [27].

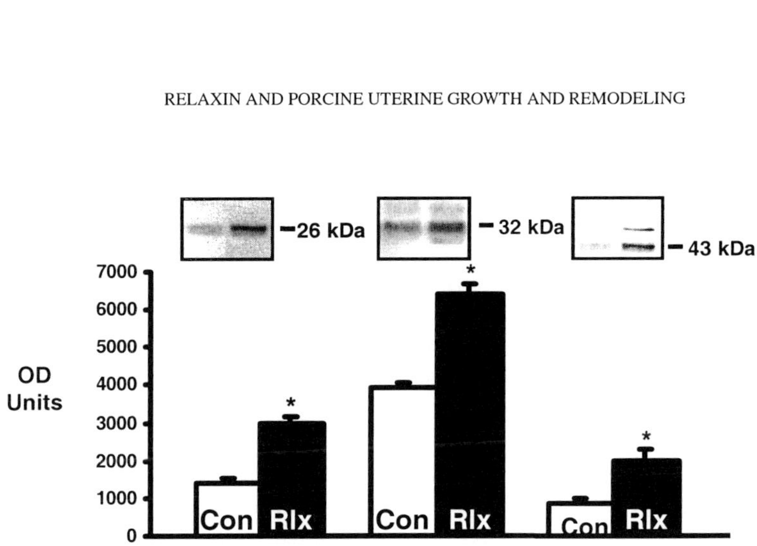

Fig. 3. Effect of relaxin (Rlx) on CX expression in the porcine uterus. Uterine proteins (15 μg) were resolved by SDS-PAGE and CX proteins detected using monoclonal CX-specific antibodies. Representative blots from control and relaxin-treated animals are shown. An asterisk (*) indicates values were significantly different (P<0.05). [Adapted with permission from Lenhart et al.[27].

Additionally, scattered endometrial CX-32 and CX-43 immunopositive cells and punctate CX-26, CX-32 and CX-43 immunofluorescence were observed in the myometrium of both control and relaxin-treated gilts. The intensity of uterine CX immunostaining in response to relaxin could best be described as intermittent between what has been reported following exposure to estrogen or progesterone [11,13,22,28]. This is consistent with the observation that the growth-promoting effects of relaxin in the pig are not dependent on prior exposure to estrogen [1] and are attenuated by progesterone [29].

C. Relaxin and uterine VEGF expression

Uterine growth and remodeling during pregnancy imposes a constant need for additional vasculature. Histological studies have shown that relaxin stimulates uterine vascularization, enhances endothelial cell proliferation, and increases uterine edema [30,31]. Relaxin also increases the expression of VEGF [14,15], an endothelial cell-specific growth factor. VEGF stimulates uterine angiogenesis and vascular permeability during the implantation period and is involved in the processes of decidualization and placentation [32,33]. When we examined VEGF protein in the uterus of prepubertal gilts, we found that expression was significantly greater ($P \leq 0.05$) following relaxin administration. While VEGF protein was localized to the vascular endothelium and myometrium of both control and relaxin-treated gilts with similar intensity, the strong, punctate staining observed in the glandular and luminal epithelium of relaxin-treated animals was minimal or absent in control gilts (data not shown). Thus, we hypothesize that relaxin may be an important local regulator of uterine angiogenesis by increasing VEGF expression and stimulating endothelial cell proliferation. Additionally, relaxin may facilitate uterine angiogenesis by orchestrating cell-cell cross-talk and extracellular matrix remodeling. For example, relaxin significantly increased uterine CX-43 protein expression and CX-43 immunostaining was strongly associated with the endothelial layer of the blood vessels [27]. Likewise, relaxin shifted the balance in expression of connective tissue protease/inhibitor to a more active profile, which would increase connective tissue remodeling and favor vascular migration. Furthermore, while TIMPs are important regulators of MMP activity, they also exhibit growth factor-like activity [34, 35] and can influence angiogenesis [36, 37]; actions that are independent of their inhibitory role. TIMPs are strongly associated with blood vessels in the endometrium [38] and may be responsible for maintaining the integrity of these vessels, regulating angiogenesis and preventing excessive ECM degradation. In conclusion, these studies suggest that relaxin's ability to promote uterine growth and remodeling involve an array of changes directed toward increasing cell-cell communication, connective tissue protease activity and angiogenesis. Together these changes promote the delivery of growth factors and nutrients to uterine tissue and alter the framework of the uterus to promote uterine accommodation and facilitate growth.

Acknowledgments

Supported by USDA 99-35203-7812 and the NJ Agricultural Experiment Station.

References

1. Hall, J.A., Cantley, T.C., Day, B.N. and Anthony, R.V., Biol. Reprod., 42 (1990) 769.
2. Min, G., Hartzog, M.G., Jennings, R.L., Winn, S.R. and Sherwood, O.D., Endocrinology, 138 (1997) 560.
3. Hall, J.A., Cantley, T.C., Galvin, J.M., Day, B.N. and Anthony, R.V. Endocrinology, 130 (1992) 3159.
4. Ohleth, K.M., Lenhart, J.A., Ryan, P.L., Radecki, S.V. and Bagnell, C.A., Endocrinology, 138 (1997) 3652.
5. Koay, E.S.C., Bryant-Greenwood, G.D., Yamamoto, S.Y. and Greenwood, F.C., J. Clin. Endocrin. Metab., 62 (1986) 513.
6. Qin, X., Garibay-Tupas, J., Chua, P.K., Cachola, L. and Bryant-Greenwood, G.D., Biol Reprod., 56 (1997) 800.
7. Ramos-DeSimone, N., Han-Dantona, E., Sipley, J., Nagase, H., French, D.L. and Quiqley, J.P., J. Biol. Chem., 274 (1999) 13066.
8. Hulboy, D.L, Rudolph, L.A., Matrisian, L.M., Mol. Hum. Reprod., 3 (1997) 27.
9. Cross, J.C., Werb, Z. and Fisher, S., Science, 266 (1994) 1508.
10. Winterhager, E., Grummer, R., Jahn, E., Willecke, K. and Traub, O., Dev. Biol., 157 (1993) 399.
11. Saito, T., Oyamada, M., Yamasaki, H., Mori, M. and Kudo, R., Int. J. Cancer, 73 (1997) 479.
12. Grummer, R. and Winterhager, E., Cell Tiss. Res., 293 (1998) 189.

13. Ou, C., Orsino, A. and Lye, S., Endocrinology, 138 (1997) 5398.
14. Cullinan-Bove, K. and Koos, R.D., Endocrinology, 133 (1993) 829.
15. Unemori, E.N., Erikson, M.E., Rocco, S.E., Sutherland, K.M., Parsell, D.A., Mak, J. and Grove, B.H., Hum Reprod., 3 (1999) 800.
16. Wang-Lee, J.L., Lenhart, J.A., Ohleth, K.M., Ryan, P.L. and Bagnell, C.A., J. Reprod. Fertil., 114 (1998) 119.
17. Casslen, B., Andersson, A., Nilsson, I.M. and Astedt, B., Proc Soc. Exp. Biol. Med., 182 (1986) 419.
18. Wewer, U.M., Faber, M., Liotta, L.A. and Albrechtsen, R., Lab. Invest., 53 (1985) 624.
19. Day, W.E., Bowen, J.A., Barhoumi, R., Bazer, F.W. and Burghardt, R.C., Anat. Rec., 251 (1998) 277.
20. Orsino, A., Taylor, C.V. and Lye, S.J., Endocrinology, 137 (1996) 1545.
21. Lye, S.J., Nicholson, B.J., Mascarenhas, M., MacKenzie, L. and Petrocelli, T., Endocrinology, 132 (1993) 2380.
22. Risek, B., Clier, F.G., Phillips, A., Hahn, D.W and Gilula, N.B., J. Cell. Sci., 108 (1995) 1017.
23. Goodenough, D.A, Goliger, J.A and Paul, D.L., Annu. Rev. Biochem., 65 (1996) 475.
24. Sanborn, B.M . and Sherwood,O.D., Endocrin. Res., 8 (1981) 179.
25. Sanborn, B.M., Kuo, H.S, Weisbrodt, N.W and Sherwood, O.D., Endocrinology, 106 (1980) 1210.
26. Cole, W.C. and Garfield, R.E., Am. J. Physiol., 251 (1986) C411.
27. Lenhart, J.A., Ryan, P.L., Ohleth, K.M. and Bagnell, C.A., Biol. Reprod., 61 (1999) 1452.
28. Jahn, E., Classen-Linke, I., Kusche, M., Beier, H.M., Traub, O., Grummer, R. and Winterhager, E., Hum. Reprod., 10 (1995) 2666.
29. Adams, W.C., Hanousek, C.A. and Frieden, E.H., Proc. Soc. Exp. Biol. Med., 191 (1989) 159.
30. Vasilenko, P., Mead, J.P. and Weidmann, J.E., Biol. Reprod. 35 (1986) 987.
31. Pepper, M.S, Spray, D.C, Chanson, M., Montesano, R, Orci, L. and Meda, P., J. Cell Biol., 109 (1989) 3027.
32. Smith, S.K., Hum. Reprod., 4 (1998) 509.
33. Cheung, C.Y., J. Soc. Gynecol. Invest., 4 (1997) 169.
34. Gomez, D.E., Alonso, D.F., Yoshiji, H. and Thorgeirsson, U.P., Eur. J. Cell Biol., 74 (1997) 111.
35. Hayakawa, T., Yamashita, K., Ohuchi, E., Shinagawa, A. J., Cell Sci., 107 (1994) 2372.
36. Johnson, M.D., Kim, H.R.C., Chesler, L., Tsao-Wu, G., Bouck, N. and Polverini, P.J., J. Cell Physiol., 160 (1994) 194.
37. Takigawa, M., Nishida, Y., Suzuki, F., Kishi, J., Yamashita, K. and Hayakawa, T., Biochem. Biophys. Res. Commun., 171 (1990) 1264.
38. Zhang, J. and Salamonsen, L.A., Mol. Hum. Reprod., 3 (1997) 735.

Relaxin and placental dysfunction in the horse

Peter Ryan[1], Wendy Vaala[2], Karen Bennett-Wimbush[3], and Carol Bagnell[4]
[1]*Animal and Dairy Sciences and College of Veterinary Medicine, Mississippi State University, Mississippi State, MS 39762, USA;* [2] *Mid-Atlantic Equine Medical Centre, Ringoes, NJ 08551, USA;* [3]*Ohio State University Agricultural Technical Institute, Wooster, OH 4469, USA; and* [4]*Department of Animal Sciences, Rutgers University, New Brunswick, NJ 08901, USA*

Introduction

The equine industry would benefit from a hormonal indicator that reflects placental and fetal well-being in the horse. Pregnancy loss during the third trimester and death of foals weakened by abnormal periparturient events constitute a large percentage of fetal and neonatal mortality [1]. Due to the expense of breeding contracts and the long gestation of the mare, late-term fetal death represents a major financial loss and time investment for the breeder. Early identification of placental insufficiency would make it possible, in some cases, to sustain the pregnancy through medical intervention. Furthermore, aggressive prepartum therapy affords foals a better chance of survival. For example, when placental function is compromised in mares suffering from fescue toxicosis, timely drug therapy rescues the placenta and improves pregnancy outcome [2,3,4,5]. Likewise, if diagnosed early, placentitis can be treated with systemic antibiotics, progestins (progesterone, 17α-hydroxyprogesterone, 5α-pregnanes), non-steroidal anti-inflammatory drugs or steroids (cortisol, to hasten fetal organ maturation).

Current research is focused on the evaluation of equine placental function in a non-invasive manner. Serum-based hormone assays have been used in the past to determine pregnancy and fetal viability in the horse [6,7,8]. However, they have not been useful in predicting problematic pregnancies in late gestation. To our knowledge, there is no simple hormonal diagnostic aid available to veterinarians to predict problematic pregnancies and deliveries and reduce the incidence of late-term abortions and perinatal deaths in horses. Relaxin, a placental hormone that can be measured in systemic blood in high concentrations [9,10], has the potential to be a useful marker of placental function. Our working hypothesis is that relaxin is a valuable clinical tool for both diagnosing placental insufficiency and monitoring treatment efficacy in mares. To this end, we have conducted a series of retrospective studies examining the relationship between systemic relaxin and pregnancy outcome in the late-term pregnant mare.

Placental Insufficiency and Late-term Abortion

Vaala and Sertich [1] have developed three categories to which mares with atrisk pregnancies can be assigned: 1) mares with histories of problematic pregnancies, deliveries, and/or birth of compromised foals, 2) mares with reproductive abnormalities, premature udder development, systemic illness, and 3) mares with no apparent risk

G.W. Tregear et al. (eds.), Relaxin 2000, 91–99.
© 2001 *Kluwer Academic Publishers. Printed in the Netherlands.*

factors that experience an abnormal periparturient event. Examples of recurrent periparturient problems include premature placental separation, dystocia, placentitis, premature delivery and prolonged gestation. Conditions developing during gestation that increase the risk of an abnormal pregnancy outcome regardless of the mare's past reproductive history include severe maternal illness and debilitation, acute abdominal crisis, uterine discharge, premature lactation, twinning and fescue toxicosis (Table I).

Table I. Factors that contribute to placental insufficiency and late-term abortion in the mare.

Maternal	Placental
Hydrops allantois	Premature placental separation
Oligohydrallantois, oligohydramnious	Small placenta (twins)
Abdominal surgery	Placentitis
Severe endometrial fibrosis	Vasculitis
Purulent vaginal discharge	Congenital anomalies
Severe malnutrition	Umbilical cord complications
Severe maternal illness	Developmental anomalies
Pelvic injuries	Thrombosis infarction
Fescue toxicosis	

However, one of the main factors contributing to placental insufficiency and late term abortion is placentitis. Giles and colleagues [11] reviewed pathology case records of 3,500 aborted fetuses, still-born foals and foals that died within 24 hours of birth to determine cause of death. Their findings indicated that over 60% of cases were associated with some form of placental insufficiency (placentitis, premature placental separation, placental thickening) that contributed to pregnancy loss and perinatal death. Subsequent studies have identified placentitis as the single most common factor contributing to late term abortion and perinatal death of foals [12]. Untreated placentitis can result in premature delivery, neonatal sepsis and peripartum hypoxia (i.e. dummy foal syndrome). Such foals are subject to high mortality and often survive only with expensive intensive care.

Another factor that contributes to placental insufficiency is fescue toxicosis resulting from grazing endophyte-infected tall fescue. Tall fescue is an important forage grass for horses in the United States [13,14]. However, it frequently is infected with an endophyte, *Acremonium coenophialum* [15] now identified as *Neotyphodium coenophialum*, which produces ergot alkaloids that cause adverse effects in pregnant mares [16]. Consequences of fescue toxicosis include placental thickening, premature placental separation, complicated deliveries, prolonged gestation, dystocia, agalactia/hypogalactia (poor milk production) and high foal mortality [3,14]. Pregnant mares that suffer from fescue toxicosis have reduced serum prolactin and progesterone [17]. Foal mortality is attributed to dysmaturity and hypoxia due to poor blood flow to the placenta [18]. In this paper, we will provide evidence that systemic relaxin is abnormal in mares that experience problematic pregnancies associated with placental dysfunction.

Relaxin in the Mare

The placenta is the primary source of relaxin in the horse during pregnancy [9,19] although the ovary may be an important source during the reproductive cycle [20]. The purification of equine relaxin from placental tissue has led to its biochemical characterisation [21] and development of a homologous equine relaxin radioimmunoassay (RIA) [10]. While plasma relaxin is low in non-pregnant mares, during pregnancy relaxin increases from day 80 of gestation, peaks at day 175 and remains elevated (80-100 ng/ml in Standardbred mares) until foaling [22]. However, shortly after expulsion of the placenta, relaxin concentrations decline rapidly and return to pregestational concentrations [10]. Little is known concerning the role of relaxin in equids during pregnancy and parturition. There is some variation among breeds with respect to relaxin profiles during pregnancy [22]; however, the functional significance of these differences in terms of actions of relaxin is unknown.

The importance of relaxin during pregnancy and at delivery has been reported in a number of species. For example, women with symptoms of impending miscarriage in early pregnancy had lower systemic concentrations of relaxin compared to women with normal pregnancies [23,24]. This led to the suggestion that relaxin might be a useful epidemiological tool in predicting pregnancy outcome in women [24]. Also, there is evidence that the aborting dog exhibits depressed serum relaxin prior to pregnancy loss [25]. Relaxin is essential for the live delivery of pups in the rat [26,27] and piglets in the sow [28]. Furthermore, relaxin treatment improves calving in beef and dairy heifers by advancing the dilatation of the cervix and increasing pelvic area, thereby decreasing the incidence of dystocia [29,30,31]. There is some evidence that relaxin secretory pattern is compromised during at risk pregnancies and complicated deliveries in horses. Stewart and co-workers [22] reported that plasma relaxin concentrations were low in three Standardbred mares with abnormal termination of pregnancy and suggested that relaxin might be a useful indicator of placental function. However, little is known about the factors that regulate relaxin synthesis and secretion in the pregnant mare. Understanding these factors will allow us to develop relaxin-related therapeutic protocols for high-risk mares.

Results and Discussion

A. Evidence that relaxin is a useful marker of placental function

In recent years we have collected data that supports this view, and our working hypothesis that relaxin could be used as a true biochemical marker of placental function, fetal well-being and predictor of pregnancy outcome in the mare. To this end, a retrospective analysis of plasma relaxin was performed on mares (light breeds) that were presented to the Perinatology Unit at the New Bolton Centre, University of Pennsylvania, with problematic pregnancies [32]. Factors contributing to problematic pregnancies included placentitis, premature placental separation, hydrops, oligohydrallantois, twinning and mares with pituitary tumours. Blood was collected during the third trimester of pregnancy and plasma relaxin was determined using a homologous equine relaxin RIA [10]. In a population of mares with normal pregnancies and deliveries, systemic relaxin ranged from 45.0-85.0 ng/ml, with a mean weekly value of 63.0 ng/ml during the last 7 weeks of gestation. Plasma relaxin declined markedly in mares with problematic pregnancies. Relaxin concentrations in a mare with placentitis declined during the last 45 days of gestation from 53.0 ng/ml (day 298 of gestation) to a low of 33.0 ng/ml (day 312), with an average concentration of 46.0 ng/ml. Two mares with pregnancy-related fluid problems, one with hydrops (excess fetal fluids) and the second with oligohydrallantois (severe loss of fluids) had variable relaxin titres, with mean values of 33.2 and 42.3 ng/ml, respectively. In the latter mare an ultrasound scan of the pregnancy on day 302 of gestation indicated loss of fetal fluid and severe folding of membranes consistent with oligohydrallantois. The foal from this mare was euthanized at birth due to severe hypoxia and sepsis.

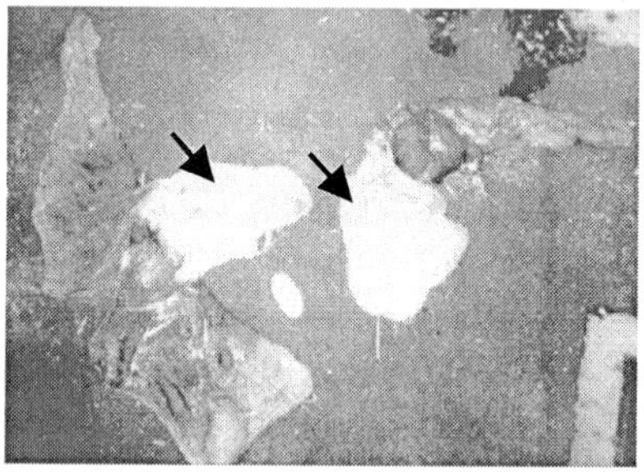

Fig. 1. Gross pathology of placentae from a twin pregnancy in a Thoroughbred mare. Note the large pale avillous areas (arrows) where the allantochorions made contact with each other. The mare delivered small but viable twin foals at term.

In the case of a mare with a twin pregnancy, systemic relaxin declined from a high of 68.9 ng/ml on day 244 to a low of 6.1 ng/ml on day 301 of gestation. The mare delivered both foals (day 332), one of which was dead (mummified) at birth and the second was euthanized at 4 weeks of age. Ultrasonography at 244 days of gestation indicated that the one twin had died. Gross pathology examination of the placenta from the live twin showed large avillous areas of the allantochorion where contact between placentae was made. This reduces the ability of the placentae to support the developing fetuses which is reflected in a corresponding decline in circulating relaxin. Twin pregnancies in mares are regarded as abnormal. The loss of twin fetuses has been attributed to placental insufficiency due to the increased avillous areas of the allantochorions associated with placental contact (Fig. 1), and the combined surface area of the placentae is only slightly larger than that of singletons [33,34]. In all pregnancies reported herein in which systemic relaxin was low, foals were born hypoxic and immature but survived with aggressive medical intervention.

B. Relaxin, fescue toxicosis and placental dysfunction

One of the consequences of fescue toxicosis in the pregnant mare is placental dysfunction which include abnormal placental thickening and premature placental separation [3,14]. Thus, we were interested in determining the effects of fescue toxicosis and fluphenazine (a long acting D_2-dopamine antagonist) treatment on relaxin concentrations in pregnant pony mares [5]. This study is an extension of previous work that examined serum prolactin and the efficacy of fluphenazine as a treatment for fescue toxicosis in pregnant pony mares [4]. Mares grazed on endophyte-infected fescue pasture (80% infected) were treated with either saline (UTRT) or a one-time injection of 25 mg of fluphenazine decanoate (TRT) intramuscularly on day 320 of gestation. Daily blood samples were collected from day 300 of gestation until day 3 postpartum and assayed for relaxin (Fig. 2).

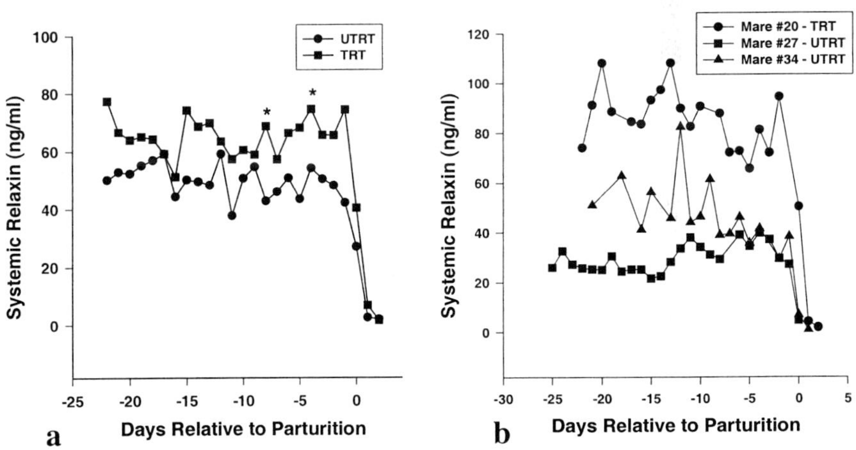

*Fig. 2. (a) Mean daily plasma relaxin concentrations of gravid pony mares grazed on endophyte-infected pasture. Data are shown relative to day of parturition. Mares were given either a one-time injection of saline (UTRT) or fluphenazine decanoate (25 mg i.m.; TRT) on day 320 of gestation. (b) Systemic relaxin profiles of individual mares during the last 20-25 days of gestation are shown. In the absence of fluphenazine treatment, plasma relaxin was markedly depressed in two mares (#27 and #34) that exhibited severe signs of complicated pregnancies consistent with fescue toxicosis. *Values significantly different (p<0.05). Printed with permission from Theriogenology, [5].*

Gestation lengths were shorter (p = 0.07) in the TRT compared with the UTRT mares (330 ± 0.7 and 336.5 ± 3.2 days, respectively). Mean plasma relaxin concentrations in both groups of mares during the week prior to treatment (day 313-319) were variable but not different (UTRT, 53.4 ± 11.3 ng/ml; TRT, 61.4 ± 9.3 ng/ml). In the two weeks prior to delivery, there was a marked difference in relaxin between treatments (Fig 2a); this effect was most dramatic during the last week of gestation (UTRT, 45.7 ± 6.7 ng/ml; TRT, 64.7 ± 6.4 ng/ml). Figure 2b shows representative relaxin profiles for three mares, one receiving fluphenazine and the other two, saline. Three of the six untreated mares and one treated mare showed clinical symptoms of fescue toxicosis (placental thickening, stillbirth, and agalactia). In the week prior to delivery, circulating relaxin in mares with problematic pregnancies (39.9 ± 7.8 ng/ml) regardless of treatment was significantly (p = 0.03) lower than mares with uneventful pregnancies (63.4 ± 5.4; Fig. 3).

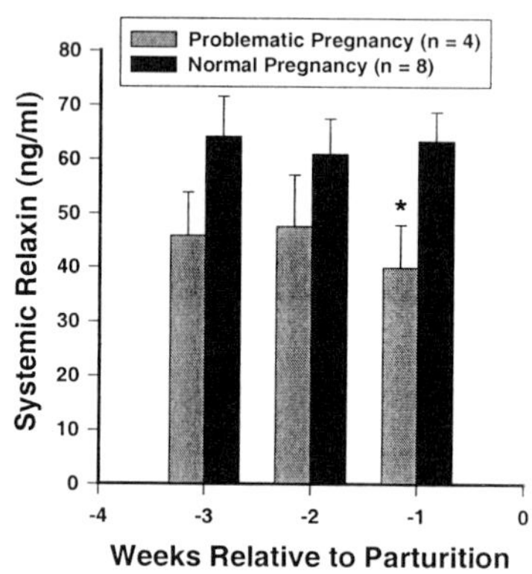

*Fig. 3. Circulating relaxin in gravid pony mares grazed on endophyte-infected pasture experiencing normal or problematic pregnancies associated with fescue toxicosis. Plasma relaxin was markedly lower in mares experiencing pregnancy complications regardless of treatment and significantly so during the week prior to delivery (*p=0.03). Printed with permission from Theriogenology, [5].*

The consequences of fescue toxicosis on pregnancy outcome in mares are well-documented [2,3,17,35]. However, this study was the first to report differences in circulating relaxin concentrations in mares affected by fescue toxicosis [5]. Data presented here show that circulating relaxin in pregnant pony mares exhibiting signs of fescue toxicosis were lower than in mares with uneventful pregnancies and deliveries. Importantly, the drug therapy, fluphenazine decanoate (a decanoate ester of a fluoromethyl phenothiazine), used to treat fescue toxicosis was effective in maintaining systemic relaxin and improving pregnancy outcomes in five of the six mares treated. Taken together, these data suggest that systemic relaxin may offer the veterinarian a potentially useful clinical tool for monitoring treatment efficacy of drugs for placental insufficiency associated with fescue toxicosis.

Significance

Some of the standard reproductive management practices for performance horses have added to the incidence of parturient complications in mares. For example, some horse owners breed mares on the foal heat or mate fillies earlier than three years of age. The incidence of dystocia is estimated to be about 4% in Thoroughbreds, 10% in draft breeds and higher than 10% in the miniature horse [36]. The need for a simple, reliable marker of placental function in the horse is evident. It is our contention that an enzyme-linked immunoabsorbent assay (ELISA) for relaxin could be valuable in assessing at risk pregnancies, and allow time for intervention to sustain the pregnancy and maintain a healthy foal. Currently, efforts in our laboratory are under way to develop an equine relaxin ELISA for this purpose and preliminary data on the production of an immunoreactive recombinant equine relaxin product from Mac-T cells (transfected with the equine relaxin gene) look promising. However, relaxin is not a panacea for all pregnancy problems in the mare, nor will screening for relaxin salvage early pregnancy losses, most of which are thought to be associated with genetic anomalies. Furthermore, whether relaxin could be used therapeutically during pregnancy to ensure normal delivery in the mare has yet to be established, since sufficient quantities of relaxin for this type of study are not available. However, until we can test the feasibility of relaxin as a therapeutic agent, the importance of relaxin to the equine industry is in providing the horse owner/veterinarian with a reliable, non-invasive marker of placental function. Systemic concentrations of relaxin provide a "window" through which the clinician can view the function of the equine placenta. This information, supported by ultrasound data of the placenta, is critical in deciding whether therapeutic intervention is necessary to sustain the pregnancy and ensure delivery of healthy foals.

Acknowledgements

We thank Dr. Patricia Sertich, New Bolton Centre, PA and Dr. David Meirs, Walnridge Farms, NJ for collecting blood samples from mares, Ms. Colleen Boa, Ms. Jackie Agans and Ms. Anita Grochulski for technical assistance. A special thanks to Dr. Dennis Stewart, for the generous gift of relaxin and antiserum. Support provided by New Jersey Agricultural Experiment Station Special Initiative Fund (Equine) 1996-99 and Mississippi Agricultural Experiment Station (MAFES). Journal article #PS9802 of MAFES, Starkville.

References

1. Vaala, W.E. and Sertich, PL. In Vaala, W.E. (ed.) The Veterinary Clinics of North America, Equine Practice, Perinatology, Saunders, Philadelphia, USA, 1994, p. 237.
2. Redmond, L.M., Cross, D.L., Strickland, J.R. and Kennedy, S.W., Amer. J. Vet. Res., 55 (1994) 722.
3. Cross, D.L., Redmond, L.M. and Strickland, J.R., J. Anim. Sci., 73 (1995) 899.
4. Bennet-Wimbush, K. and Loch, W., J. Equine Vet. Sci., 18 (1998) 169-173.
5. Ryan, P.L., Bennet-Wimbush, K., Vaala, W. and Bagnell C.A., Theriogenology, (in press).
6. Terqui, M. and Palmer, E., J. Reprod. Fertil., 27 (1979) 441.
7. Santschi,. E., LeBlanc, M. and Weston, P., J. Reprod. Fert. (suppl.), 44 (1991) 627.
8. Stabenfeldt, G.H., Daels, P.F., Munro, C.J. Kindahl H., Hughes, J.P. and Lasley. B., J. Reprod. Fert. (suppl.), 44 (1991) 37.
9. Stewart, D.R., Stabenfeldt, G.H., Hughes, J.P. and Meagher, D.M., Biol. Reprod., 27 (1982) 17.
10. Stewart, D.R., Endocrinology, 119 (1986) 1100.
11. Giles, R.C., Donahue, J.M., Hong, C.B., Petrites-Murphy, M.B., Poonacha, K.B., Roberts, A.W., Smith, B.J., Tramontin, R.R. and Swerczek, T.W., J.A.V.M.A., 203 (1993) 1170.
12. Hong, C.B., Donhahue, J.M., Giles, R.C., Petrites-Murphy, M.B., Poonacha, K.B., Roberts, A.W., Tramontin, R.R., Smith,, B., Tuttle P.A., Swerczek, T.W., J. Vet. Diagn. Invest., 5 (1993) 560.
13. Hoveland, C.S., Ag. Ecosyst. Environ., 44 (1993) 33.
14. Putnam, M.R., Bransby, D.I., Schumacher, J., Boosinger, T.R., Bush, L., Shelby, R.A., Vaughan, J.T., Ball, D. and Brendemuehl, J.P., Amer. J. Vet. Res., 52 (1991) 2071.
15. Lyons, PC., Plattner, RD. and Bacon, C.W., Science, 232 (1986) 487.
16. Porter, J.K. and Thompson, F.N., J. Anim. Sci., 70 (1992) 1594.
17. McCann, S.J., Caudle, A.B., Thompson, F.N., Steudemann, J.A., Heushner, G.L. and Thompson, Jr. D.L., J. Anim. Sci., 70 (1992) 217.
18. Poppenga, R.H., Mostrom, M.S., Haschek, W.M., Lock, T.F., Buck, W.B., Beasely, V.R., Proc. 27th Am. Assoc. Vet. Lab. Diagn., (1984) 325.
19. Klonisch, T., Ryan P.L., Yamashiro, S. and Porter, D.G., Biol. Reprod., 52 (1995) 1307.
20. Ryan, P.L., Klonisch, T., Yamashiro, S., Renaud, R., Wasnidge, C. and Porter, D.G., J. Reprod. Fert., 110 (1997) 329.
21. Stewart, D.R. and Papkoff, H., Endocrinology, 119 (1986) 1093.
22. Stewart, D.R., Addiego, L.A., Pascoe, D.R., Haluska, G.J. and Pashen, R., Biol. Reprod., 46 (1992) 648.
23. Witt, B.R., Wolf, G.C., Wainwright, C.J., Johnston, P.D. and Thorneycroft, I.H., Fertil. Steril., 53 (1990) 1029.
24. Stewart, D.R., Overstreet, J.W., Celniker, A.C., Hess, D.L., Cragun, J.R., Boyers, S.P. and Lasley, B.L., Clin. Endocrin., 38 (1993) 379.
25. Steinetz, B.G., Bullesbach, E.E., Goldsmith, L.T., Schwabe, C. and Lust, G., Biol. Reprod., 54 (1996) 1252.
26. Downing, S.J. and Sherwood, O.D., Endocrinology, 116 (1985) 1200.

27. Lao-Guico-Lamm, M.L. and Sherwood, O.D., Endocrinology, 123 (1988) 2472.
28. Nara, B.S., Welk, F.A., Rutherford, J.E. and Sherwood, O.D., J. Reprod. Fertil., 66 (1982) 359.
29. Musah, A.I., Schwabe, C., Willham, R.L. and Anderson, L.L., Endocrinology, 118 (1986) 1476.
30. Musah, A.I., Schwabe, C., Wilham R.L. and Anderson, L.L., Anim. Reprod. Sci., 16 (1988) 237.
31. Bagna, B., Schwabe, C. and Anderson, L., J. Reprod. Fertil., 91(1991) 605.
32. Ryan, P.L., Vaala, W. and Bagnell, C., Proceed. 44th Annual Conven. Amer. Assoc. Equine Pract., 44 (1998) 62.
33. Jeffcott, L.B. and Whitwell, K.E., J. Comp. Path., 83 (1973) 91.
34. Whitewell, K.E., Vet. Clin. N. Amer. Large Anim. Prac., 2 (1980) 313.
35. Bossinger, T.R., Brendemeuhl, J.P., Schumacher, J., Bransby, D.I., Kee, D. and Shelby, R.A., Biol. Reprod. (Monograph), 1 (1995) 61.
36. Vandeplassache, M. In McKinnon, A.O. and Voss, J.L. (eds.). Equine Reproduction. Lea and Febiger, Philadelphia, USA, 1993, p. 537.

Intersection of the relaxin and estrogen signalling pathways in the uterus

Robert D Koos and Suresh B Pillai
Department of Physiology and Center for Studies in Reproduction, University of Maryland School of Medicine, 655 West Baltimore St., Baltimore, Maryland 21201, USA

Introduction

The protein hormone relaxin exerts a wide range of effects on the reproductive system, as well as other tissues, including stimulation of growth of the uterus during pregnancy, remodelling of the reproductive tract in preparation for parturition, and development of the mammary glands [1]. The mechanisms by which relaxin exerts these effects, however, remain unclear, in large part because the relaxin receptor has not yet been cloned. Interestingly, many of relaxin's acute effects on the uterus closely resemble those induced by estrogens. For example, relaxin rapidly induces a large increase in uterine wet weight in immature, ovariectomized rats, and also stimulates uterine metabolic processes, regardless of whether or not there is prior estrogen priming [2-5]. Uterine growth is also stimulated in other species, including pigs [6-8] and primates [9,10]. Estrogen and relaxin can induce the expression of many of the same genes, including vascular endothelial growth factor [VEGF; 11,12, Pillai SB et al., this volume], insulin-like growth factor I [IGF-I; 7,13], the uterine transcription factor HOXA-10 [14], and endothelial-type NO synthase [15-17], all of which could play roles in mediating relaxin's uterotrophic effects.

A similarity in the effects of two such structurally diverse hormones as estrogen and relaxin suggests that their signalling pathways may converge at some point in target cells. Estrogen receptors (ERs)*, which are ligand-activated nuclear transcription factors, are one possible point of convergence. This conclusion is based on a large number of reports that ERs, as well as other members of the nuclear hormone receptor family, can be activated by a diverse group of signalling molecules in the absence of ligand [reviewed in 18,19]. This includes various growth factors, neurotransmitters, cyclic nucleotides, and cyclins [20-25]. In view of this, it may be more accurate and instructive to think of ERs as nuclear transcription factors that can be activated in multiple ways, only one of which is through the binding of ligand. Among the reported activators of ERs are insulin [23] and IGF-I [22], members of the same gene family as relaxin [1]. This suggested to us that the estrogen-like effects of relaxin on the uterus might also be mediated through estrogen-independent activation of ERs. We now report studies that support this hypothesis.

* For convenience, we will refer generically to "ERs" or "the ER." It is now known of course that there are at least two distinct forms of ER, referred to as ER-α and ER-β [reviewed in 26]. Their relative roles in estrogen action, however, are still far from clear. Because ER-α appears to be the predominant receptor in the uterus and because evidence is accumulating that ER-β functions mainly as a modulator of ER-α action [27,28], it seems likely that ER-α mediates the interactions reported

G.W. Tregear et al. (eds.), Relaxin 2000, 101–108.
© 2001 *Kluwer Academic Publishers. Printed in the Netherlands.*

here.

Results and Discussion

Inhibition of relaxin-induced uterine edema by an ER antagonist

To determine if relaxin might activate ERs, we first evaluated whether its effects on the uterus would be blocked by the specific ER antagonist ICI 182,780 [29]. This antagonist [30], or the similar but less potent ICI 164,384, has been shown previously to block the estrogen-like effects of other growth factors [20-22]. As shown in Fig. 1, 17β-estradiol and purified porcine relaxin [31] increased uterine weight by 98% and 77% respectively (from 0.26 ± 0.01 mg/g to 0.51 ± 0.04 mg/g and 0.46 ± 0.05 mg/g, respectively; $p < 0.001$) in immature, ovariectomized rats. Pre-treatment with ICI 182,780 completely inhibited the hypertrophy induced by both estradiol and relaxin. Neither the ICI/estradiol nor the ICI/relaxin groups were significantly different from either the control or ICI alone groups.

Uterine histology confirmed that ICI 182,780 inhibits relaxin's induction of endometrial edema [29]. Uteri from animals treated with ICI 182,780 alone resembled those from controls. Treatment of animals with relaxin alone caused interstitial edema in the endometrium within 6 hours. The swollen endometrial compartment was filled with fluid from the luminal epithelium to the myometrium. This accumulation of fluid was completely prevented by pre-treatment with ICI 182,780 [29].

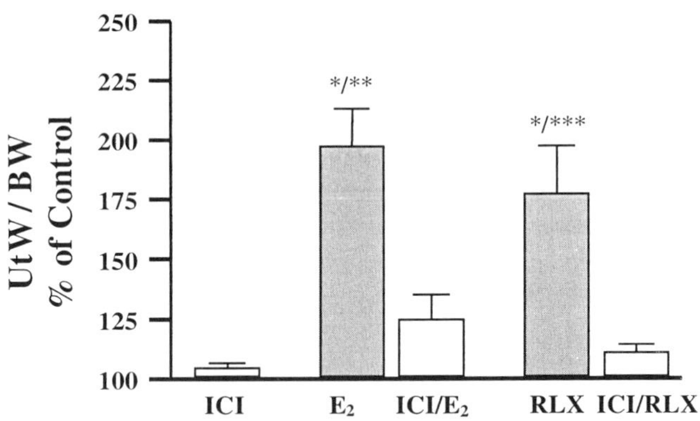

*Fig. 1. The effect of ICI 182,780 treatment (3 μg/g body weight, BW) prior to hormone treatment (17β-estradiol, 5 μg/100 g BW or porcine relaxin (RLX), 85 μg/ 100 g BW) on changes in uterine weight (UtW, mg/g BW). All values are presented as percent of control (Mean ± SEM; n = 5 animals/treatment group). *$p < 0.0005$ vs. control, **$p < 0.001$ vs. ICI/E_2, ***$p < 0.0005$ vs. ICI/RLX. Modified from [29].*

The results of this study demonstrated that the ER antagonist ICI 182,780 almost completely inhibits both estrogen- and relaxin-induced increases in uterine weight in ovariectomized immature rats, strongly suggesting that relaxin's effects involve activation of ERs.

Stimulation of an ERE-luciferase reporter gene plasmid by relaxin

We next evaluated the ability of relaxin to stimulate expression of an ERE-luciferase reporter gene plasmid in transfected MCF-7 breast cancer cells. MCF-7 cells were chosen because they express both ERs (mainly ER-α) and, and respond to relaxin with increased proliferation, indicating that they also express the relaxin receptor [32]. The development of such an *in vitro* model may also facilitate studies on the pathways that link the relaxin receptor and the ER.

ER-positive MCF-7 cells (HTB-22, American Type Culture Collection, Manassas, VA) were plated in 24-well culture dishes in phenol red-free DMEM plus 10% charcoal-stripped FBS at a density of 2.6 x 10^5 cells/well. Twenty-four h after plating, the medium was changed to serum-free, phenol red-free DMEM, and the cells were transfected with the ERE-SV40-luciferase reporter vector [0.2 μg/well; 33], as well as a pSV-β-galactosidase control vector (0.2 μg/well, Promega, Madison, WI) to control for transfection efficiency, using Lipofectamine-Plus transfection reagent (Life Technologies, Gaithersburg, MD). After a further 24 h, the medium was changed again and replicate cultures treated with vehicle (controls), 17β-estradiol (10^{-10} to 10^{-6} M), or relaxin (also 10^{-10} to 10^{-6} M). Twenty-four h after treatment, the cells were lysed and assayed for both luciferase and β-galactosidase activity (Dual-Light assay system, Tropix, Bedford, MA). As shown in Fig.

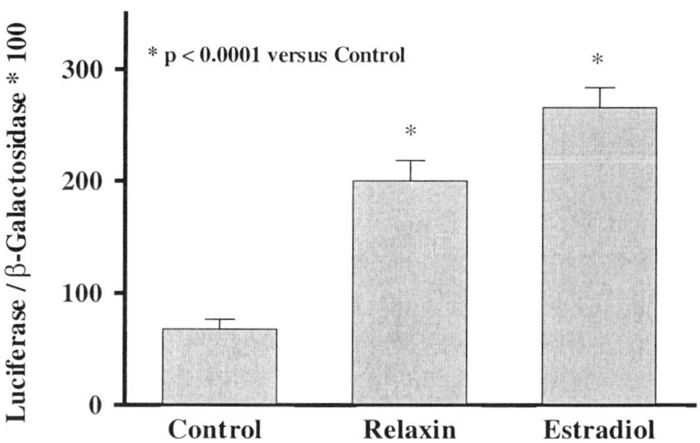

Fig. 2. Stimulation of an ERE-luciferase reporter gene expression by relaxin. Twenty-four hours after transfection, the cells were treated with either vehicle control, porcine relaxin (10^{-6} M) or 17β-estradiol (10^{-6} M) for 24 h. Results are expressed as the mean (± SEM) of luciferase activity as a percent of β-galactosidase activity. N = 24 wells/treatment group.

2, 17β-estradiol and relaxin stimulated similar significant increases in luciferase activity compared to unstimulated controls (approximately 4-fold and 3-fold, respectively). In a second experiment, the time course of the induction by relaxin (10^{-6} M) was examined (at 0, 3, 6, 12, and 24 h). A significant relaxin effect was already evident by 3 h and peaked at 12 h (data not shown). These results demonstrate that relaxin can rapidly activate the expression of a reporter gene linked only to a minimal consensus ERE, again indicating that activation of the relaxin receptor leads to activation of ERs.

These studies provide the first evidence that relaxin can induce gene transcription via activation of the ER. This could explain why the acute responses of the uterus to estrogen and to relaxin are so similar. Thus, relaxin joins a growing list of signalling molecules that have been reported to regulate gene expression via pathways that converge on a common transcription factor: the ER [18-19]. This could involve phosphorylation of the ER itself, an event essential for both estrogen and growth factor-induced ER activation. Alternatively, it may be due to alterations in the activity of co-activators and/or co-repressors of ER activity. Once activated, ERs could increase the expression of downstream mediators of uterine growth, such as IGF-1 [7, 13] and VEGF [11-12]. While the body of evidence supporting the concept of ligand-independent activation of ERs is now quite substantial, Gehm et al. [33] have recently presented data which casts some doubt about the ability of EGF to activate ERs. In contrast to the effects of relaxin observed in this study, EGF had little ability to stimulate luciferase synthesis in MCF-7 cells transfected with the same ERE-SV40-luc plasmid. The similarity in the degree of induction of luciferase by estrogen in that study to that seen here and the much stronger induction by relaxin as compared to EGF indicates that relaxin likely acts via activation of the ER and ERE. It is still possible, however, that there may be other pathways by which non-ligands induce estrogen-like effects. Further studies of this phenomenon are clearly needed.

VEGF is probably responsible for the relaxin-induced increase in uterine edema, as it is for that caused by estrogen. We have demonstrated previously that estrogen stimulates VEGF expression in the rat uterus prior to the onset of edema [11] and that the administration of a specific antibody to VEGF prevents that edema as well as implantation [Rockwell, L.C. et al., unpublished]. Relaxin has likewise been reported to stimulate VEGF expression by human endometrial cells *in vitro* [12] and we have now confirmed that it has a similar effect in the rat uterus [Pillai, S.B. et al., this volume].

The identification of ERs as possible mediators of relaxin's acute effects on the uterus is significant because almost nothing is known about relaxin's mechanisms of action. High affinity cell-surface receptors for relaxin are present on target tissues [8, 34,35], but neither the cDNA nor gene for the receptor has been cloned. In human uterine fibroblasts, relaxin induced an increase in tyrosine phosphorylation of a 220 kDa protein, presumably its receptor [36]. This suggests that the receptor is a tyrosine kinase, as are those for the related factors insulin and IGF-1. It has also been reported that a tyrosine kinase inhibitor blocks relaxin-induced porcine granulosa cell proliferation [37]. Even if its receptor is a tyrosine kinase, relaxin has also been linked to increases in cAMP levels in several studies [reviewed in 36]. Cyclic AMP can also activate ERs [38], apparently by a mechanism different from that involved in EGF activation [39]. Given its diverse effects on the reproductive system, as well as on other tissues, relaxin may act via several pathways. It is also possible that this varies from one cell type to another, and with variations in the overall hormonal milieu, leading to different outcomes in different cell types and at different times in the reproductive sequence.

Relaxin probably functions both as a blood-borne hormone and as a locally-produced autocrine and/or paracrine factor [40]. While the corpus luteum is the principal source of circulating relaxin, considerable evidence indicates that it may also be produced within the uterus and other tissues. For example, relaxin was detected in both glandular and luminal epithelial cells of the human uterus during both the proliferative and secretory

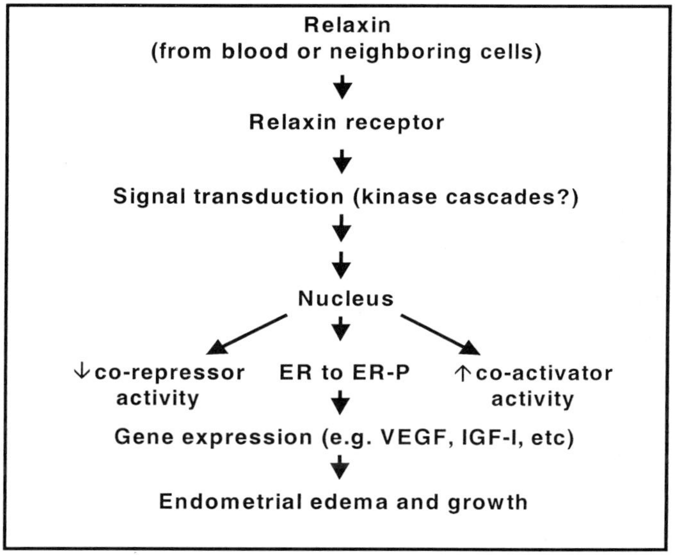

Fig. 3. Possible pathway by which relaxin might activate ERs. Relaxin binding to a specific cell surface receptor may activate a kinase cascade that could lead to phosphorylation of ERs, or to increases/decreases in nuclear receptor co-activators/co-repressors.

phases of the menstrual cycle by immunocytochemistry [41]. Immunoreactive relaxin was also present in endometrial epithelial cells of pregnant rats [42] and of both pregnant and non-pregnant pigs [43]. Relaxin mRNA has likewise been detected by RT-PCR in the uterus of the rat [44], both pregnant and non-pregnant, and the pig [43]. Additional mapping of the patterns of expression of relaxin, and its receptor when identified, in the uterus may provide better insight into its physiological role there relative to that of estrogen.

Relaxin activation of ERs could be involved in the growth of not only the uterus but also other tissues that grow during pregnancy, such as the vagina [45], cervix [46,47] and breast [48]. Estrogen stimulation of uterine growth is believed to be mediated by growth factors such as EGF and IGF-1 [13,18-22]. Relaxin has also been shown to up-regulate IGF-1 expression in the pig uterus [7]; whether this involves ER activation remains to be determined.

It is possible that some of relaxin's acute effects on other tissues may also involve activation of ERs. For example, relaxin, like estradiol [49, 50], increases prolactin levels in both anterior pituitary [51] and endometrial cells [52] as well as IGF-1 levels in porcine granulosa cells [53]. Cervical ripening also involves vasodilation and increased vascular permeability similar to the effects exerted by both estrogen and relaxin on the uterus [54-56]. Outside the reproductive system, both relaxin and estrogen exert effects on the heart and blood vessels [57,58]. It will be necessary to repeat such studies of relaxin's effects in the presence of ICI 182,780 to determine if any of these effects are mediated by ERs.

In conclusion, the present study provides evidence that relaxin activates ERs in the uterus. This should help to clarify past observations and facilitate the elucidation of relaxin's mechanisms of action and its physiological functions, both in and outside the reproductive system. The ability of relaxin to mimic some effects of estrogen would also have to be taken into account in evaluating its use as a drug for the treatment of human disorders and diseases. On the other hand, relaxin might be used, in certain circumstances (i.e., where the use of estrogen itself is undesirable), in lieu of estrogen to elicit selective effects.

Acknowledgments

This work was supported by NIH grants CA45055 and HD36207. We thank Dr. David Sherwood for the generous gifts of porcine relaxin, as well as for his advice and encouragement during these studies. We also thank Dr. Barry D. Gehm and Dr. J. Larry Jameson of Northwestern University Medical School for kindly providing the ERE-SV40-luc reporter plasmid.

References

1. Sherwood, O.D., *In* Knobil, E. and Neill, J.D. (Eds), The Physiology of Reproduction. 2nd ed., Raven Press, New York, USA, 1994, p. 861.
2. Steinetz, B.G., Beach, V.L., Blye, R.P. and Kroc, R.L., Endocrinology, 61 (1957) 287.
3. Vasilenko, P., Frieden, E.H. and Adams, W.C., Proc. Soc. Exp. Biol. Med., 163 (1980) 245.
4. Vasilenko, P., Mead, J.P. and Weidmann, J.E., Biol. Reprod., 35 (1986) 987.
5. Adams, W.C., Hanousek, C.A. and Frieden, E.H., Proc. Soc. Exp. Biol. Med., 191 (1989) 159.
6. Hall, J.A., Cantley, T.C., Day, B.N. and Anthony, R.V., Biol. Reprod., 42 (1990) 769.
7. Ohleth, K.M., Lenhart, J.A., Ryan, P.L., Radecki, S.V. and Bagnell, C.A., Endocrinology, 138 (1997) 3652.
8. Min, G., Hartzog , M.G., Jennings, R.L., Winn, R.J. and Sherwood, O.D., Endocrinology, 138 (1997) 560.
9. Hisaw, F.L.Jr and Hisaw, F.L., Am. J. Obst. Gyn., 89 (1964)141.
10. Hisaw, F.L., Hisaw, F.L.Jr and Dawson, A.B., Endocrinology, 81 (1967) 375.
11. Cullinan-Bove, K. and Koos, R.D., Endocrinology, 133 (1993) 829.
12. Unemori, E.N., Erickson M.E., Rocco, S.E., Sutherland, K.M., Parsell, D.A. Mak, J. and Grove, B.H., Hum. Reprod., 14 (1999) 800.
13. Murphy, L.J. and Ghahary, A., Endo. Rev., 11 (1990) 443.

14. Gui, Y., Zhang, J., Yuan, L. and Lessey, B.A., Mol. Hum. Reprod., 5 (1999) 866.
15. Weiner, C.P., Lizasoain, I., Baylis, S.A., Knowles, R.G., Charles, I.G. and Moncada, S., Proc. Natl. Acad. Sci. USA, 91 (1994) 5212.
16. Bani, D., Baccari, M.C., Nistri, S., Calamai, F., Bigazzi, M. and Sacchi, T.B., Endocrinology, 140 (1999) 4434.
17. Yallampalli, C., Dong, Y.L., Biol. Reprod., 63 (2000) 34.
18. Smith, C.L., Biol. Reprod., 58 (1998) 627.
19. Weigel, N.L. and Zhang, Y., J. Mol. Med., 76 (1998) 469.
20 Ignar-Trowbridge, D.M., Nelson, K.G., Bidwell, M.C., Curtis, S.W., Washburn, T.F., McLachlan, J.A. and Korach, K.S., Proc. Natl. Acad. Sci. USA, 89 (1992) 4658.
21. Ignar-Trowbridge, D.M., Teng, C.T., Ross, K.A., Parker, M.G., Korach, K.S. and McLachlan, J.A., Mol. Endocrinol., 7 (1993) 992.
22. Aronica, S.M. and Katzenellenbogen, B.S., Mol. Endocrinol., 7 (1993) 743.
23. Patrone, C., Ma, Z. Q., Pollio, G., Agrati, P., Parker, M.G. and Maggi, A., Mol. Endocrinol., 10 (1996) 499.
24. Smith, C.L., Conneely, O.M. and O'Malley, B.W., Proc. Natl. Acad. Sci. USA, 90 (1993) 6120.
25. Trowbridge, J.M., Rogatsky, I. and Garabedian M.J., Proc. Natl. Acad. Sci. USA, 94 (1997) 10132.
26. Gustafsson, J.A., J. Endocrinol, 163 (1999) 379.
27. Hall, J.M. and McDonnell, D.P., Endocrinology, 140 (1999) 5566.
28. Weihua, Z., Saji, S., Makinen, S., Cheng, G., Jensen, E.V., Warner, M. and Gustafsson, J.A., Proc. Natl. Acad. Sci. USA, 97 (2000) 5936.
29. Pillai, S.B., Rockwell, L.C., Sherwood, O.D. and Koos, R.D., Endocrinology, 140 (1999) 2426.
30. Wakeling, A.E., Dules, M., Bowler, J., Cancer Res., 51 (1991) 3867.
31. Sherwood O.D. and O'Byrne, E.M., Arch. Biochem. Biophys., 160 (1974) 185.
32. Bigazzi, M., Brandi, M.L., Bani, G., Sacchi, T.B., Cancer, 70 (1992) 639.
33. Gehm, B.D., McAndrews, J.M., Jordan, V.C. and Jameson, J.L., Mol. Cell. Endocrinol., 159 (2000) 53.
34. Mercado-Simmen, R.C., Bryant-Greenwood, G.D. and Greenwood, F.C., Endocrinology, 110 (1982) 220.
35. Tan, Y.Y., Wade, J.D., Tregear, G. W. and Summers, R.J., Brit. J. Pharmacol., 127 (1999) 91.
36. Palejwala, S., Stein, D., Wojtczuk, A., Weiss, G. and Goldsmith, L.T. 1998. Endocrinology, 139 (1998) 1208.
37. Zhang, Q. and Bagnell, C.A. 1993. Biol. Reprod., 48S1 (1993) 493.
38. Ince, B.A., Montano, M.M. and Katzenellenbogen, B.S., Mol. Endocrinol., 8 (1994) 1397.
39. El-Tanani, M.K. and Green, C.D., Mol. Endocrinol., 11 (1997) 928.
40 Bryant-Greenwood, G.D. and Schwabe, C., Endo. Rev., 15 (1994) 5.
41. Bryant-Greenwood, G.D., Rutanen, E.M., Partanen, S., Coelho, T.K. and Yamamoto, S.Y., Mol. Cell. Endocrinol., 1-2 (1993) 23.
42. Fields, P.A., Lee, A.B., Haab, L.M., Hwang, J.J. and Sherwood, O.D., Endocrinology, 130 (1992) 2985.
43. Knox, R.V., Zhang, Z., Day, B.N. and Anthony, R.V., Endocrinology, 135 (1994) 2517.
44. Gunnersen, J.M., Crawford, R.J. and Tregear, G.W., Mol. Cell. Endocrinol., 110 (1995) 55.
45. Zhao, S., Kuenzi, M.J. and Sherwood, O.D., Endocrinology, 137 (1996) 425.
46. Hwang, J.J. and Sherwood, O.D., Endocrinology, 123 (1988) 2486.
47. O'Day, M.B., Winn, R.J., Easter, R.A., Dzuik, P.J. and Sherwood, O.D., Endocrinology, 125 (1989) 3004.

48. Hurley, W.L., Doane, R.M., O'Day-Bowman, M.B., Winn, R.J., Mojonnier, L.E. and Sherwood O.D., Endocrinology, 128 (1991) 1285.
49. Randolph, J.F.Jr, Peegel, H., Ansbacher, R. and Menon, K.M., Am. J. Obstet. Gynecol., 162 (1990) 1109.
50. Yoshikawa, M., Uozumi, T., Kawamoto, K., Arita, K., Ito, A. and Takahashi, S., Endocr. J., 42 (1995) 235.
51. Sortino, M.A., Cronin, M.A. and Wise, P.M., Endocrinology, 124 (1989) 2013.
52. Huang, J.R., Tseng, L., Bischof, P. and Janne, O.A., Endocrinology, 121 (1987) 2011.
53. Ohleth, K.M. and Bagnell, C.A., Biol. Reprod., 53 (1995) 1286.
54. Winn, R.J., O'Day-Bowman, M.B. and Sherwood, O.D., Endocrinology, 133 (1993) 121.
55. Chwalisz, K. and Garfield, R.E., J. Perinat. Med., 26 (1998) 448.
56. Sherwood, O.D., Olson, L.M., Zhao, S. and Little, H.R., Endocrinology, 141 (2000) 2458.
57. Bani, D., Gen. Pharmacol., 28 (1997) 13.
58. Gustafson, J.A., J. Soc. Gynecol. Investig., 7 (2000) S8.

Relaxin up-regulates rat uterine vascular endothelial growth factor mRNA *in vivo*

Suresh B Pillai[1], O David Sherwood[2], and Robert D Koos[1]
[1]*Department of Physiology, University of Maryland School of Medicine, Baltimore, Maryland, 21201, USA;* [2]*Department of Molecular and Integrative Physiology and College of Medicine, University of Illinois at Urbana-Champaign, Urbana, Illinois, 61801, USA.*

Introduction

Relaxin, a member of the insulin-like growth factor family, has long been know to play a key role in numerous species in the alteration of the female reproductive tract prior to parturition [1]. Most of these well-established effects of relaxin have been characterized with regard to relaxin's systemic effects as a circulating factor released from the corpus luteum during pregnancy and prior to parturition [2,3]. Many of relaxin's acute effects, however, may result from the local production and action of relaxin in the reproductive tract. Regardless of the source, relaxin has been found to induce a doubling of uterine wet weight in immature ovariectomized rats within 6 h as well as increases in uterine glycogen content and protein synthesis [4-7]. Furthermore, this increase in uterine wet weight can be blocked by ICI 182, 780, a highly specific estrogen receptor inhibitor [8]. This finding classifies relaxin as a ligand-independent activator of estrogen receptors.

Vascular endothelial growth factor, or VEGF, is a potent inducer of vascular permeability and angiogenesis [9,10] and is believed to play a primary role in the preparation of the uterus for implantation. VEGF induces changes in vascular permeability within the endometrium, contributing to the exudation of serum growth factors and other cellular effectors responsible for extracellular matrix remodelling and subsequent angiogenesis both during implantation [11] or, in the absence of implantation, in cyclic endometrial remodelling [12,13]. Relaxin has been established as a factor which contributes to alterations of uterine vasculature in numerous species, including monkeys [14] and rats [6]. The question then arises: could relaxin's effects on the uterine vasculature be mediated by VEGF.

Production of VEGF is rapidly up-regulated by estradiol in the uterus of the rat [12]. In cell cultures, ICI 182, 780 inhibits activation of a VEGF-specific reporter gene construct [15]. Recently, Unemori *et al* showed that treatment of cultured human uterine endometrial cells with relaxin increased VEGF expression [16]. Although relaxin's effects upon VEGF production have been demonstrated *in vitro,* no studies on relaxin's influence on VEGF production had been conducted *in vivo*. Based upon relaxin's rapid induction of uterine edema via an estrogen receptor-dependent pathway, we hypothesized that this may be occurring, in part, via an increase in uterine VEGF. In this study, we examined the acute effects of relaxin administration on VEGF mRNA expression in the immature female rat uterus following relaxin treatment. The results

G.W. Tregear et al. (eds.), Relaxin 2000, 109–113.
© 2001 *Kluwer Academic Publishers. Printed in the Netherlands.*

demonstrate an increase in VEGF mRNA which mirrors that induced by estradiol, providing further evidence that relaxin's acute effects in target tissues may involve estrogen receptors.

Results and Discussion

Eighteen day-old intact immature female rats were treated with PBS/ethanol (control) 17β-estradiol (5 μg/100 g body weight) or pure porcine relaxin [17] (85 μg/ 100 g body weight). Animals were sacrificed at 3 and 6 h, uteri were dissected out, and total RNA was extracted and reverse transcribed for PCR analysis. PCR primers generated product sizes of 331 base pairs for β-actin (internal control gene), 431 base pairs for $VEGF_{120}$ and 560 base pairs for $VEGF_{164}$. PCR products were separated via polyacrylamide gel electrophoresis, stained with ethidium bromide, and photographed. Densitometric analyses were conducted on photo negatives to assess changes in VEGF mRNA expression.

PCR products for $VEGF_{120}$ and $VEGF_{164}$ were generated from the immature rat uterine mRNA. Treatment with estrogen markedly up-regulated (6-10 fold) the amount of product generated for both splicing forms at both time points compared to controls. This agrees with our earlier observation on the effect of estradiol on VEGF expression in the uterus [12]. Relaxin's effects on $VEGF_{120}$ and $VEGF_{164}$ expression were almost identical, with increases in the 6-8 fold range. The yield of β-actin PCR product was unaffected by either estradiol or relaxin treatment.

In addition to VEGF, the expression of several other relevant genes, including estrogen receptors (ERs) and progesterone receptors, was examined. Neither estrogen nor relaxin caused any changes in levels of ER-α mRNA, yet both markedly down-regulated both ER-β1 and ER-β2 (data not shown). In contrast to the parallel effects of estrogen and relaxin on VEGF and ER-β expression, estrogen strongly up-regulated progesterone receptor mRNA levels while relaxin had no effect.

These data show for the first time the induction of VEGF expression in the uterus of rats *in vivo* by relaxin. Furthermore, the pattern of this induction closely resembles that induced by estrogen, providing more evidence in support of relaxin as a ligand- independent activator of estrogen receptors in the uterus. These findings also provide evidence for VEGF mediating relaxin's uterotropic effects. Although the relaxin receptor has not been characterized beyond binding studies [18], numerous studies concerning the signal transduction pathways stimulated by relaxin have been done. Palejwala *et al* found that in human uterine fibroblasts, relaxin caused an increase in tyrosine phosphorylation, indicating that one possible signal transduction pathway for relaxin involves phosphorylation cascades, a characteristic shared with insulin and IGF-I, two of the other members of the relaxin family [19]. Other studies implicate relaxin in stimulating cAMP-dependent pathways *in vivo* in rat uterine cells [20]. Within the myometrium of intact, estrogen-primed rats, relaxin inhibits oxytocin-induced increases in phosphoinositide turnover through increased cAMP-dependent protein kinase activity [21]. This uterine elevation in cAMP-dependent signal transduction could also lead to activation of ERs [22]. Thus, the pathways induced by relaxin in the uterus could involve several different phosphorylation cascades.

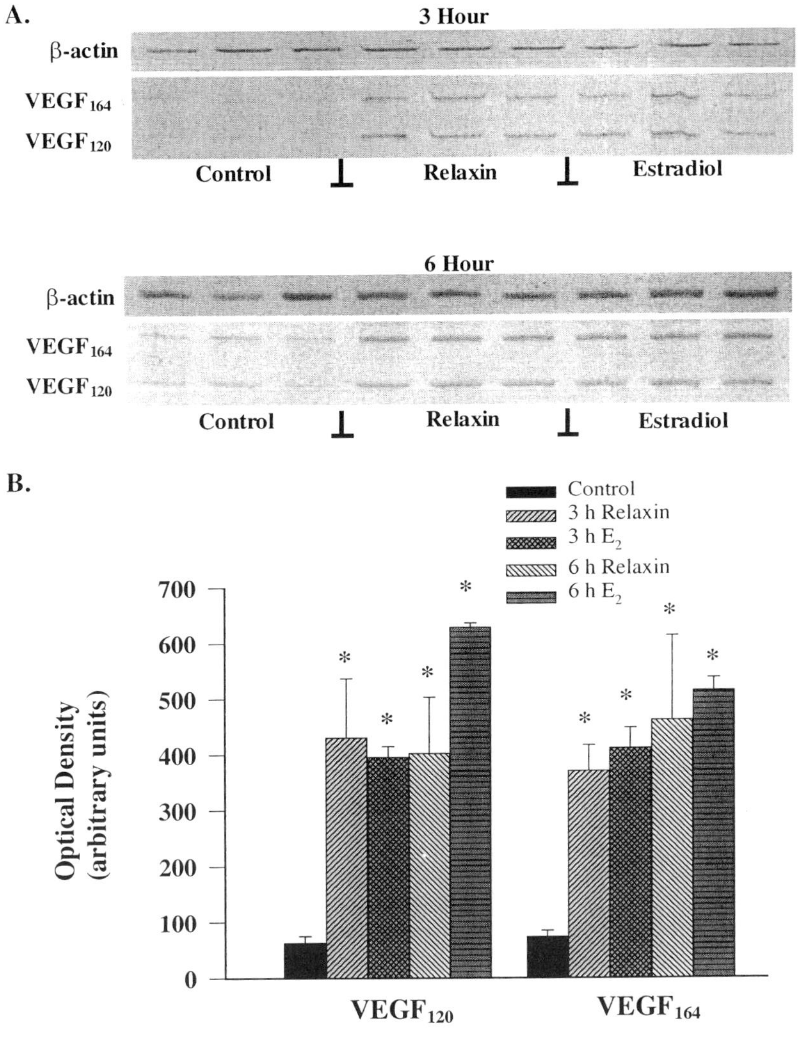

Fig.1. Total RNA was extracted from immature female rat uterus following treatment with 17-β estradiol or pure porcine relaxin at 3 h and 6 h. RNA was then reverse transcribed, and PCR was conducted. ***A.*** *Representative electrophoretic products for β-actin, $VEGF_{120}$, and $VEGF_{164}$ 3 and 6 h following treatment. Images are from the photo negatives of gels.* ***B.*** *Densitometric analysis of products. Data are presented in arbitrary units of optical density; * $p < 0.01$ versus controls for both $VEGF_{120}$ and $VEGF_{164}$.*

Circulating relaxin contributes to its effects within the reproductive tract during pregnancy and parturition, but local production could also be important. RT/PCR analysis shows that relaxin mRNA is detectable in the brain, kidney, uterus, and ovary of both pregnant and non-pregnant rats [23]. Local production of relaxin has been found in the endometrium of the marmoset monkey, specifically within the luminar and glandular epithelium [24]. Furthermore, this epithelial relaxin increases in concentration through the secretory phase of the cycle. This data is supported by studies in humans, where the increase in endometrial relaxin secretion coincides with the period of implantation, suggesting a role for relaxin in that process [25].

In conclusion, these data support our hypothesis that relaxin induces its acute uterotropic effects, specifically uterine edema, through up-regulation of VEGF production. Further studies are required to clarify if relaxin's up-regulation of VEGF occurs via estrogen receptor pathways.

References

1. Sherwood, O.D., In Knobil, E. and Neill, J., (eds) The Physiology of Reproduction. 3 ed., Raven Press, New York, 1994, p. 861.
2. Weiss, G., Science 194 (1976) 948.
3. Goldsmith, L. T., Weiss, G. and Steinetz, B. G., Endocrinol. Metab. Clin. North Am., 24 (1995) 171.
4. Steinetz, B.G., Beach, V.L., Blye, R.P. and Kroc, R.L., Endocrinology, 61 (1957) 287.
5. Vasilenko, P., Frieden, E. H. and Adams, W. C., Proc. Soc. Exp. Biol. Med., 163 (1980) 245.
6. Vasilenko, P., Mead, J. P. and Weidmann, J. E., Biol. Reprod., 35 (1986) 987.
7. Adams, W. C., Hanousek, C. A. and Frieden, E. H., Proc. Soc. Exp. Biol. Med., 191 (1989) 159.
8. Pillai, S. B., Rockwell, L. C., Sherwood, O. D. and Koos, R. D., Endocrinology, 140 (1999) 2426.
9. Dvorak, H. F., Brown, L. F., Detmar, M. and Dvorak, A. M., Am. J. Pathol., 146 (1995) 1029.
10. Shibuya, M., Adv. Cancer Res., 67 (1995) 281.
11. Shweiki, D., Itin, A., Neufeld, G., Gitay-Goren, H. and Keshet, E., J. Clin. Invest., 91 (1993) 2235.
12. Cullinan-Bove, K.and Koos, R. D., Endocrinology, 133 (1993) 829.
13. Hyder, S. M., Stancel, G. M., Chiappetta, C., Murthy, L., Boettger-Tong, H. L. and Makela, S., Cancer Res., 56 (1996) 3954.
14. Hisaw, F. L., Hisaw, F. L. J. and Dawson, A. B., Endocrinology, 81 (1967) 375.
15. Karas, R. H., Gauer, E. A., Bieber, H. E., Baur, W. E. and Mendelsohn, M. E., J. Clin. Invest., 101 (1998) 2851.
16. Unemori, E. N., Erikson, M. E., Rocco, S. E., Sutherland, K. M., Parsell, D. A., Mak, J. and Grove, B. H., Hum. Reprod., 14 (1999) 800.
17. Sherwood, O. D. and O'Byrne, E. M., Arch. Biochem. Biophys., 160 (1974) 185.
18. Tan, Y. Y., Wade, J. D., Tregear, G. W. and Summers, R. J., Br. J. Pharmacol., 127 (1999) 91.
19. Palejwala, S., Stein, D., Wojtczuk, A., Weiss, G. and Goldsmith, L. T., Endocrinology, 139 (1998) 1208.

20. Hughes, S. J., Hollingsworth, M. and Elliott, K. R., J. Reprod. Fertil., 109 (1997) 289.
21. Sanborn, B. M., Qian, A., Ku, C. Y., Wen, Y., Anwer, K., Monga, M. and Singh, S. P., Adv. Exp. Med. Biol., 395 (1995) 469.
22. Smith, C. L., Biol. Reprod., 58 (1998) 627.
23. Gunnersen, J. M., Crawford, R. J., and Tregear, G. W., Mol. Cell Endocrinol., 110 (1995) 55.
24. Einspanier, A., Zarreh, H. K. M., Balvers, M., Kerr, L., Fuhrmann, K., and Ivell, R., Hum. Reprod., 12 (1997) 1325.
25. Yki-Jarvinen, H., Wahlstrom, T., and Seppala, M., Acta Obstet. Gynecol. Scand., 64 (1985) 663.

Comparative aspects of relaxin and relaxin-like factor (Insulin 3) in the placental trophoblast

Thomas Klonisch and Sabine Hombach-Klonisch
Institute of Anatomy and Cell Biology, Martin Luther University, Faculty of Medicine, Halle, Germany

Introduction

Developing as a complex temporary system of membranes surrounding the fetus of viviparous vertebrates, the placenta consists of fetal membranes apposing or fusing with the maternal mucosa. The dominating placenta of most mammals is the chorio-allantoic placenta. Both members of the relaxin family, relaxin itself and the structurally closely related relaxin-like factor (RLF) also known as INSL3 [1, 2], are expressed at the feto-maternal interface of various species including the human [3-5]. Relaxin is regarded a marker of pregnancy in the horse [6], the dog [7], and the cat [8] and has been implicated to play a role in embryo implantation [9-11], placentation [12, 13] and parturition [14, 15]. Given the detection of specific binding sites for relaxin in the uterus/ placenta [14, 16-19] and RLF in the uterus [20, 21], both hormones [22] have both been implicated in paracrine functions within the placenta. Previous work on the placental expression of relaxin mainly in rodents and the human indicated that (a) there appears to be no clear pattern of pregnancy-related physiological factors that might predict the primary source of relaxin in species and (b) relaxin production is not restricted to the trophoblast but can also be expressed by maternal cells (camel [23]; pig [24]; rat [25]; guinea pig [26]; hamster [27]).

Comparative anatomy has revealed a surprising number of placental phenotypes among mammals which include differences in the histological type of materno-fetal interdigitation, the degree of trophoblast invasiveness, the layers of the placental interhemal barriers, maternofetal blood flow interrelationships or placental shape, to name only a few terms to describe placental structure [28]. The histological type of materno-fetal interdigitation and the degree of trophoblast invasiveness comprise, both on a structural and cellular level, physiological processes which are central to our understanding of feto-maternal communication. To elucidate the expression and potential roles of both relaxin and RLF at the feto-maternal interface, we present here our results of a systematic analysis of both hormones of the relaxin family in placental tissues of species with different degrees of trophoblast invasiveness.

Results and Discussion

Species with epitheliochorial placentation lack the invasive trophoblast population found in the placentas of carnivores or higher primates. In the epitheliochorial placentas of the horse and the strepsirrhine primate *Galago crassicaudatus*, the placenta is a

G.W. Tregear et al. (eds.), Relaxin 2000, 115–120.
© 2001 *Kluwer Academic Publishers. Printed in the Netherlands.*

source of relaxin [13, 29, 30]. Being an early product of the trophoblast of the allantochorionic microcotyledon, the villous trophoblast is the sole source of relaxin within the equine placenta. Pseudostratified trophoblast located at the base of microcotyledons are devoid of relaxin [13, 29]. Among lower primates, the villous trophoblast is also a source of relaxin as demonstrated in term placental tissues of the African lorisiform primate *Galago crassicaudatus* and the Malagasy lemur *Variecia variegata rubra* which is plesiomorphic to all lemurs [30]. In contrast, employing RT-PCR analysis with specific primers for Galago-RLF, the placentas of both strepsirrhine species were devoid of RLF amplification products. In the horse placenta, preliminary data seem to indicate that RLF, in contrast to relaxin, appears also not to be expressed in at the feto-maternal interface (Klonisch, unpublished observation). Further work is currently being performed to clarify the RLF expression pattern in the equine placenta.

Although the ruminant syndesmochorial placenta is in large parts an epitheliochorial type of placenta, its unique binucleate trophoblast cells (BNC), accounting for some 20% of the total trophoblast cell population, represent a trophoblast population with limited invasiveness. In the syndesmochorial placenta of domestic ruminants and deer, uninucleate trophoblast cells of the placental villi continuously develop into large binucleate cells [31, 32]. This highly differentiated trophoblast has the ability to migrate within the trophoblast multilayer and eventually fuse with maternal luminal epithelial cells [31, 33]. With the exception of the camel [23], ruminants do not express a relaxin molecule indicating that RLF may serve as a substitute for relaxin in the ruminant placenta. In the placenta of the fallow deer and the goat, as likely in the placentas of most ruminants, RLF was a product of the fetal trophoblast and maternal luminal uterine epithelium. Fetal villous uni- and binucleate trophoblast cells expressed RLF mRNA [4] and protein (Hombach-Klonisch, unpublished observation). Differentiation of the BNC along the migratory pathway resulted in the silencing of RLF gene activity. Migratory BNC located in the apical trophoblast cell layer apposing the maternal luminal epithelium and BNC located at the tip of the interdigitating fetal villus were devoid of RLF transcripts. However, these differentiated BNC had accumulated within their cytoplasmic granules placental lactogen to be released into the maternal circulation upon continuous fusion with maternal epithelial cells to form temporary hybrid trinucleate cells [31]. Thus, RLF may be regarded a new marker of differentiation for early stages of BNC maturation preceeding the expression of the placental lactogene/prolactin gene family, known markers of the mature BNC within the deer placenta [31].

The carnivore placenta exhibits a composite structure in terms of the interhemal membrane barriers. An endotheliochorial zonary placental girdle, where the materno-fetal contact area is elaborate and available for physiological exchange, is bordered by a marginal hematoma representing a hemochorial condition which, in turn, is followed by epitheliochorial paraplacental cupules to both sites of the zonary placenta [34]. In the cat placenta at mid-term, the trophoblast of the zonary girdle lining the fetal lamellar villi expressed relaxin [35]. In contrast, the trophoblast at the tip of the placental villi protruding into the junctional placental zone, where fetal and maternal tissues face each other and intimately intermingle, as well as the hemophagous trophoblast in the marginal hematoma of the placental girdle were devoid of relaxin [35]. Although the bitch also displays an endotheliochorial girdle placenta with the syncytiotrophoblast being separated from endothelial cells of the maternal capillaries by a basal lamina, materno-fetal interdigitation is not lamillar as in the cat placental girdle but labyrinthine. In the dog placenta, both relaxin and RLF expression was studied and revealed differential expression pattern. Relaxin was an early product of the syncytiotrophoblast protruding into the openings of the maternal endometrial glands [36]. The villous cytotrophoblast and the hemophagous cytotrophoblast of the mariginal hematoma were devoid of relaxin [36]. Like in the deer and, as will be reported later, in the human placenta, RLF was a product of both fetal and maternal tissues. For relaxin and RLF, differences in the expression pattern were evident in the early canine placenta and in the hemophagous paraplacental area. In the early canine placenta, the villous cytotrophoblast strongly expressed RLF whereas the syncytiotrophoblast displayed only weak RLF expression at this gestational stage [37]. With pregnancy progressing, however, the syncytiotrophoblast became the major source of RLF within the canine placenta. The relaxin-negative hemophagous cytotrophoblast in the marginal hemochorial zone also expressed RLF. Similar to the BNC differentiation in the syndesmochorial placenta of the deer, differentiation of the cytotrophoblast along the syncytial or hemochorial pathway in the carnivore placenta affects the expression of both members of the relaxin family [36, 37].

Hemochorial placentation in the human is a highly invasive process involving differentiation of the placental cytotrophoblast into two different populations of cells, extravillous and villous trophoblast [38]. Villous cytotrophoblast cells fuse to create the multinucleate syncytiotrophoblast covering the floating chorionic villi, while the extravillous trophoblast starts proceeding along the invasive pathway from the cell columns of anchoring villi. These cells invade the uterine wall and reach the maternal blood vessels as far as the first third of the myometrium [39]. Extensive extravillous trophoblast invasion observed in the first two trimesters of pregnancy results in the attachment of the placenta to the uterus and the establishment of sufficient maternal blood flow into the intervillous space [40]. While the human villous syncytiotrophoblast and extravillous cytotrophoblast have been shown to express both H1 and H2 relaxin [12, 41, 42], syncytium formation of the villous cytotrophoblast resulted in a down-regulation but not a complete loss of RLF expression [5]. Constitutive expression of RLF was observed in the extravillous interstitial and intravascular cytotrophoblast and in trophoblast derived from ectopic tubarian implantation sites and placental tissues of patients with symptomatic pre-eclampsia [5]. Thus, RLF expression in the human placenta is independent of (a) the implantation site, (b) the lack of maternal decidualization at the tubarian implantation site and (c) placental insufficiency with inadequate trophoblast invasion. As in species like the deer and the dog, RLF was also weakly expressed in the maternal decidual cells. Co-expression of human relaxins and RLF in villous and extravillous trophoblast populations [5, 12, 43] might suggest cooperative effects of both hormones. Facilitating the rupture of fetal membranes at term may be just one of the placental functions of human relaxin as H2 relaxin has been demonstrated to induce the release of tissue plasminogen activator, collagenases and matrix-metalloproteinases MMP-1, MMP-3 and MMP-9 in cultured human fetal membranes [14, 15]. Relaxin binding sites have been identified within the human fetal membranes [14, 19] and given the ability of the RLF hormone [22] to weakly bind to relaxin binding sites [20, 21], the human placenta appears to be a target tissue for both relaxin and RLF.

Our analysis on the expression of newly cloned members of the relaxin family in a given placental phenotype of materno-fetal interdigitation indicates that expression of relaxin and RLF is differentially regulated in the resident trophoblast population. Changes in the expression of both members of the relaxin family are the result of trophoblast differentiation. Trophoblast invasiveness, one of many facets of trophoblast differentiation, may correlate with RLF gene activity as there is an apparent lack of RLF expression in the trophoblast of species with epitheliochorial placentation. In the human and carnivore placenta, relaxin and RLF gene activity appear to be affected by factors leading to the formation of a trophoblastic syncytium or phagocytotic cytotrophoblast in the paraplacental and junctional zone of the carnivore placenta, respectively.

Neither relaxin (-/-) mice [44, 45] nor RLF (-/-) mice [46, 47] have been reported to display placental insufficiency. This could provide indirect evidence of the fact that in the hemochorial placenta of the mouse both hormones may functionally complement each other. Double-knockout mice for both relaxin and RLF and, once cloned, for their corresponding receptors will provide valuable new insides into the relaxin-like placental puzzle.

Acknowledgements

The authors would like to thank Christine Froehlich, Elke Bernhard and Elisabeth Schlüter for their excellent technical assistance. We would also like to thank Professor Bernd Fischer for his support. Sabine Hombach-Klonisch was funded by a grant „Wiedereingliederungsstipendium für Frauen“ by the Land Saxony-Anhalt.

References

1. Adham, I.M., Burkhardt, E., Benahmed, M. and Engel, W., J. Biol. Chem., 268 (1993) 26668.
2. Ivell, R., Rev. Reprod., 2 (1997) 133.
3. Tashima, L.S., Hieber, A.D., Greenwood, F.C. and Bryant-Greenwood, G., J. Clin. Endocrin. Metabol., 80 (1995) 707.
4. Hombach-Klonisch, S., Kauffold, J., Rautenberg, T., Steger, K., Tetens, F., Fischer, B. and Klonisch, T. , Mol. Cell. Endocrinol., 159, (2000) 147.
5. Hombach-Klonisch, S., Seeger, S., Tscheudschilsuren, G., Buchmann, J., Huppertz, B., Seliger, G., Fischer, B. and Klonisch, T., Mol. Hum. Reprod., 7 (2001) 349.
6. Stewart, D.R., Addiego, L.A., Pascoe, D.R., Haluska, G.J. and Pashen, R., Biol. Reprod., 46 (1992) 648.
7. Steinetz, B.G., Büllesbach, E.E., Goldsmith, L.T., Schwabe, C. and Lust, G., Biol. Reprod., 54 (1996) 1252.
8. Stewart, D.R. and Stabenfeldt G.H., Biol. Reprod., 32 (1985) 848.

9. Stewart, D.R., Nakajima, S.T., Overstreet, J.W., Boyers, S.P. and Lasley, B.L. In McLennan, A.H., Tregear, G.W. and Bryant-Greenwood, G.D. (Eds.), *Progress in Relaxin Research.* Global Publication Services, Singapore, 1995, p. 214.
10. Stewart, D.R., Stouffer, R., Overstreet, J.W., Hendrickx, A. and Lasley, B.L., Endocrinology, 132 (1993) 6.
11. Einspanier, A., Zarreh-Hoshyari-Khan, M.R., Balvers, M., Kerr, L., Fuhrmann, K. and Ivell, R., Hum. Reprod., 12 (1997) 1325.
12. Sakbun, V., Ali, S.M., Greenwood, F.C. and Bryant-Greenwood G.D., J. Clin. Endocrinol. Met., 70 (1990) 508.
13. Klonisch, T., Mathias, S., Cambridge, G., Hombach-Klonisch, S., Ryan, P.L. and Allen, W.R., Placenta, 18 (1997) 121.
14. Qin, X., Garibay-Tupas, J., Chua, P.K., Cachola, L. and Bryant-Greenwood, G.D., Biol. Reprod., 56 (1997) 800.
15. Qin, X., Chua, P.K., Ohira, R.H. and Bryant-Greenwood, G.D., Biol. Reprod., 56 (1997) 812.
16. Palejwala, S., Stein, D., Wojtczuk, A., Weiss, G. and Goldsmith, L.T., Endocrinology, 139 (1998) 1208.
17. Garibay-Tupas, J.L., Maaskant, R.A., Greenwood, F.C. and Bryant-Greenwood, G.D., J. Endocrinol.,145 (1995) 441.
18. Osheroff, P.L. and King, K.L., Endocrinology, 136 (1995) 4377.
19. Koay, E.S., Bryant-Greenwood, G.D., Yamamoto, S.Y. and Greenwood, F.C., J. Clin. Endocrinol. Metabol., 62 (1986) 513.
20. Büllesbach, E.E., and Schwabe, C., J. Biol. Chem., 274 (1999) 54.
21. Büllesbach, E.E., and Schwabe, C., J. Biol. Chem., 270 (1995) 16011.
22. Büllesbach, E.E., Rhodes, R., Rembiesa, B. and Schwabe, C., Endocrine, 10, (1999) 167.
23. Hombach-Klonisch, S., Abd-Elnaeim, M., Skidmore, J.A., Leiser, R., Fischer, B. and Klonisch, T., Biol. Reprod., 62 (2000) 839.
24. Knox, R.V., Zhang, Z., Day, B.N. and Anthony, R.V., Endocrinology, 135 (1994) 2517.
25. Fields, P.A., Lee, A.B., Haab, L.M., Hwang, J.-J. and Sherwood, O.D., Endocrinology, 130 (1992) 2985.
26. Lee, Y.A., Bryant-Greenwood, G.D., Mandel, M. and Greenwood, F.C., Endocrinology, 130 (1992) 1165.
27. Renegar, R.H., Cobb, A.D. and Leavitt, W.W., Biol. Reprod., 37 (1987) 925.
28. Leiser, R. and Kaufmann, P., Exp. Clin. Endocrinol., 102 (1994) 122.
29. Klonisch, T., Ryan, P.L., Yamashiro, S. and Porter, D.G., Biol. Reprod., 52 (1995) 1307.
30. Klonisch, T., Froehlich, C., Tetens, F., Fischer, B., Hombach-Klonisch, S. Mol. Biol. Evol., 18 (2001) 393.
31. Wooding, F.B.P., Morgan, G. and Adam, C.L. Microsc. Res. Tech., 38 (1997) 88.
32. Lee, C.S., Wooding, F.B.P. and Morgan, G., J. Anat., 187 (1995) 445.
33. Wooding, F.B.P., Hobbs, T., Morgan, G., Heap, R.B. and Flint, A.P.F., J. Reprod. Fertil., 98 (1993) 275.
34. Leiser, R. and Koob, B., J. Exp. Zool., 266 (1993) 642.
35. Klonisch, T., Hombach-Klonisch, S., Froehlich, C., Kauffold, J., Steger, K., Huppertz, B. and Fischer, B., Biol. Reprod., 60 (1999) 305.
36. Klonisch, T., Hombach-Klonisch, S., Froehlich, C., Kauffold, J., Steger, K., Steinetz, B.G. and Fischer, B., Biol. Reprod., 60 (1999) 551.
37. Klonisch, T., Kauffold, J., Steger, K., Bergmann, M., Leiser, R., Fischer, B. and Hombach-Klonisch, S., Biol. Reprod., 64 (2001) 442.

38. Benirschke, K. and Kaufmann, P. (Eds.), Pathology of the human placenta. 3rd ed., Springer-Verlag, New York (1995).
39. Zhou, Y., Fisher, S.J., Janatpour, M., Genbacev, O., Dejana, E., Wheelock, M. and Damsky, C.H. J., Clin. Invest., 99 (1997) 2139.
40. Kaufmann, P., and Castellucci, M., Trophoblast Res., 10 (1997) 21.
41. MacLennan, A.H., Grant, P., Ness, D. and Down, A., Reprod. Fertil. Dev., 3 (1991) 577.
42. Hansell, D.J., Bryant-Greenwood, G.D. and Greenwood, F.C., J. Clin. Endocrinol. Metab., 72 (1991) 899.
43. Bogic, L.V., Mandel, M., and Bryant-Greenwood, G.D., J. Clin. Endocrinol. Metab., 80 (1995) 130.
44. Zhao, L., Samuel, C.S., Tregear, G.W., Beck, F. and Wintour, E.M., Biol. Reprod., 63 (2000) 697.
45. Zhao, L., Roche, P.J., Gunnersen, J.M., Hammond, V.E., Tregear, G.W., Wintour, E.M. and Beck, F., Endocrinology, 140 (1999) 445.
46. Zimmermann, S., Steding, G., Emmen, J.M.A., Brinkmann, A.O., Nayernia, K., Holstein, A.F., Engel, W. and Adham, I.M., Mol. Endocrinol., 13 (1999) 681.
47. Nef, S. and Parada, L.F., Nature Genetics, 22 (1999) 295.

Secretion of matrix metalloproteinase-2, tissue inhibitor of metalloproteinase-1 and human chorionic gonadotropin by cytotrophoblastic cells: Is there a regulation by relaxin?

Pascal Mock, Arielle Meisser, Melvina Araman, Christine Wuillemin, Aldo Campana and Paul Bischof

Department of Obstetrics and Gynaecology, University Hospital and WHO Collaborating Center in Human Reproduction, University of Geneva, 1211 Geneva 14 Switzerland

Introduction

The peptide hormone relaxin is a member of the insulin-related gene family comprising insulin, insulin-like growth factors I and II (IGF I and II), Leydig insulin-like peptide (LEY I-L, RLF, INSL3) and peptides encoded by INSL4 (Early placenta insulin-like), INSL5 and INSL6 genes [1-7]. Whereas the physiological role of relaxin has been well assessed in late pregnancy and parturition [8,9], recent studies in primates and humans have shown that relaxin secreted by the corpus luteum and/or endometrium during the luteal phase may be involved in reproductive processes such as endometrial decidualization [10,11], early pregnancy [12] and even embryo implantation [13].

Human implantation is the result of a still mysterious and not yet well understood process where the endometrium and trophoblast are the two main instruments. It is characterized by apposition and adhesion processes followed by invasion of trophoblastic tissue. This invasive property of trophoblasts is regulated in part by matrix metalloproteinases (MMPs) and their inhibitors known as tissue inhibitors of metalloproteinases (TIMPs). They are synthesized and secreted from both sides of the embryonic-maternal interface [14].

Relaxin has been implicated directly in the up-regulation of matrix metalloproteinase-1 (MMP-1) in human dermal [15] and lung fibroblast [16]. Furthermore, it appears that the prorelaxin H2 gene is expressed in *in vitro* decidualized human endometrial stromal cells [17] confirming its presence in endometrium during the luteal phase as previous studies demonstrated. Based on these two observations, we hypothesized that relaxin present at the embryo-maternal interface may regulate the secretion by cytotrophoblastic cells (CTB) of matrix metalloproteinase-2 (MMP-2) and possibly matrix metalloproteinase-9 (MMP-9) and even one of their inhibitors, the tissue inhibitor of matrix metalloproteinase-1 (TIMP-1) and thus be implicated in the regulation of trophoblast invasion.

To this end, we decided to evaluate the effect of relaxin on the secretion of MMP-2, TIMP-1 and human chorionic gonadotrophin (hCG) in pure *in vitro* cultured first trimester CTB.

G.W. Tregear et al. (eds.), Relaxin 2000, 121–125.
© 2001 *Kluwer Academic Publishers. Printed in the Netherlands.*

Results and Discussion

CTB were isolated and purified according to our previous reports [18]. Briefly, trophoblastic tissue was obtained from legal abortions at 6 to 12 weeks from the last menstrual period, digested with trypsin, separated from blood cells and syncytia on a discontinuous percoll gradient and immunopurified by antibody-coated magnetic particles (anti CD 45, Dyna Beads, Milan Geneva) in order to eliminate all contaminating leucocytes. CTB viability was assessed by Trypan Blue (Sigma) exclusion and cells diluted to 10^6 cells/ml. Synthetic Human gene 2 (H2) relaxin was kindly provided by the Howard Florey Institute, University of Melbourne (Victoria, Australia).

Culture conditions

CTB (100 μl, 10^6 cells/ml, >90 % viability) were incubated in duplicate in the presence or absence of H2 relaxin (1-5000 ng/ml). Incubation was performed in 24-well tissue culture plate (Costar, Cambridge, Mass) under a 5 % CO_2 and 95 % air atmosphere in a humid incubator at 37°C. Culture medium was Dulbecco's modified Eagle's medium (Gibco, Basel, Switzerland) containing 2 mmol/l L-glutamine, 4.2 mmol/l magnesium sulfate, 25 mmol/l Hepes, 1% gentamycin, 1% amphotericin B, 100 ug/ml streptomycin and 100 U/ml penicillin in the absence of serum. Medium was harvested at 24 and 48 hours and the culture stopped at 48 hours. The supernatants were aliquoted and stored at -20°C until assayed. Duplicate wells were run for each treatment condition and the experiments repeated 3 times with different CTB preparations.

Hormone and protein assays

MMP-2 and MMP-9 immunoreactivities (MMP-2i and MMP-9i) were measured in supernatants using our own enzyme immunoassays as described elsewhere [19]. Tissue inhibitor of metalloproteinases-1 (TIMP-1) was measured in supernatants by a commercially available ELISA system (Biotra, Amersham Pharmacia, Biotech). Total hCG (hCG + free β-hCG) was measured in the supernatants by a microparticle enzyme immunoassay with a sensitivity of 1 mIU/ml (Abbott, Abbott Park, IL, USA).

Statistics

Results were expressed as percent of their respective controls per day ((conc. day 1 + conc. day 2)/2) and per 10^6 cells (CTB in the absence of H2 relaxin, mean ± SEM). Statistical analysis was performed by ANOVA using Fisher's test using the Statview program.

Figure 1 illustrates the effect of H2 relaxin on MMP-2i and total HCGi secretions by CTB. When compared to controls, MMP-2i and total HCGi secretions were significantly decreased after incubation with 1 and 100 ng/mL of H2 relaxin (P =

0.0133 to < 0.0001). In contrast, an incubation with a higher concentration of H2 relaxin significantly increased total HCGi and decreased MMP-2 secretions ($P < 0.0001$).

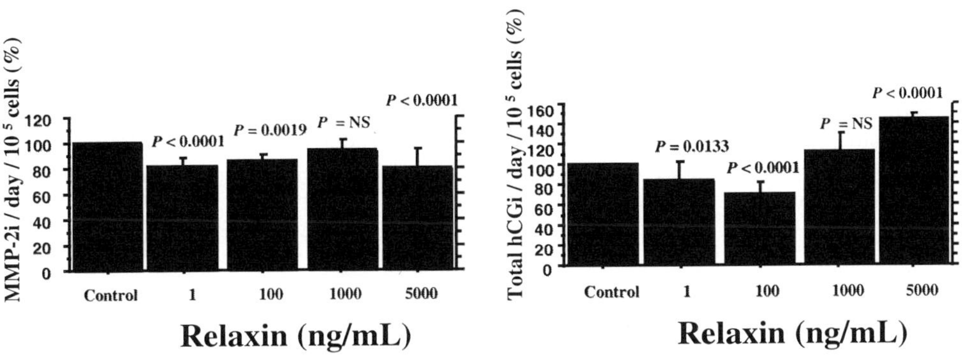

Fig. 1. Effects of H2 relaxin on MMP-2i and total HCGi secretions by first trimester human cytotrophoblast cells.

Figure 2 demonstrates the effect of H2 relaxin on TIMP-1 secretion by CTB. When compared to controls, TIMP-1 secretion was 77±14 % ($P = 0.05$) and 83±27 % (P = NS) after incubation with 1 ng/mL and 100 ng/mL respectively. When we considered only the 48 h period, TIMP-1 secretion decreased significantly to 62±5 % ($P = 0.0328$), 61±11 % ($P = 0.0298$) and 67±28 % ($P = 0.0535$) after incubation with 1, 100 and 5000 ng/mL respectively (data not shown).

These findings indicate that H2 relaxin has an effect on first trimester CTB by regulating trophoblastic factors implicated in implantation. To know whether relaxin improves or inhibits cytotrophoblast invasion remains to be further evaluated. Indeed, our study has shown contrasting results with up- and down-regulation of hCG with high (5000 ng/mL) and low (1 ng/mL) relaxin concentrations respectively. Similarly, whereas relaxin decreases MMP-2i secretion by CTB, indicating an inhibitory effect, TIMP-1 appears also to be decreased which may enhance the process of trophoblast invasion. However, it remains to be elucidated if the overall effect of relaxin on gelatinases is associated with an alteration of its proteolytic activity.

It has been recently demonstrated that relaxin may be implicated in human implantation [13]. Indeed, in patients involved in an IVF program, the authors observed that granulosa cells producing highest levels of relaxin in culture have higher implantation and pregnancy rates. Interestingly, our findings demonstrating a down-regulation of TIMP-1 secretion by CTB may explain the relationship between relaxin and implantation rate. This observation should be considered as another hypothesis for a role of relaxin in implantation as apposed to the ones usually mentioned such as the correlation between low relaxin concentration and out-of-phase endometrial biopsies [20] or the synergic effect with medroxyprogesterone acetate on the *in vitro* endometrial stromal differentiation [21].

More recent studies have assessed the effect of relaxin at the time of implantation on the endometrium, demonstrating that relaxin is implicated in mouse uterine quiescence by an up-regulation of NO synthases [22], that relaxin stimulates the expression of VEGF in normal human endometrial cells [23], induces the expression of glycodelin in human endometrial glandular cells [24] and finally that relaxin increases the expression of HOXA-10, a transcription factor increased during the window of implantation [25]. The present and preliminary studies indicate, for the first time to the best of our knowledge, that relaxin may also play a crucial role in the regulation of well known trophoblastic factors implicated in the maintenance of pregnancy and trophoblast invasion.

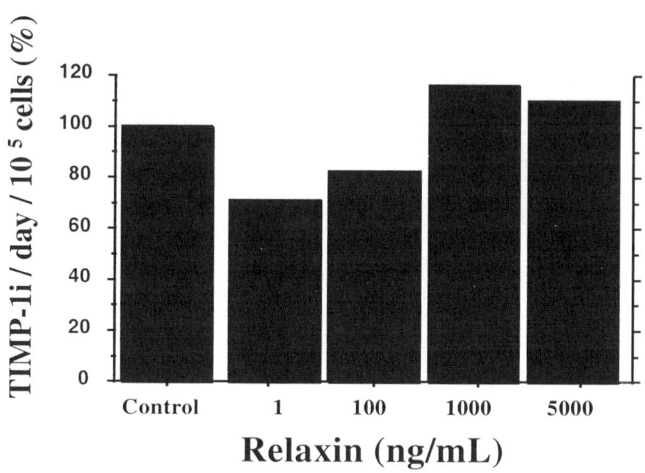

Fig. 2. Effect of H2 relaxin on the TIMP-1i secretion by first trimester human cytotrophoblast cells (n=2).

Acknowledgements

This work was supported by a grant (no. 3200-059471.99/1) from the Swiss National Fund for scientific research to P.M. The authors wish to thank also the Howard Florey Institute (Dr JD Wade), University of Melbourne, for its valuable contribution by providing us synthetic H2 relaxin. We thank also Lydia Delajoux for her excellent technical assistance.

References

1. Le Roith, D.N., Engl. J. Med., 336 (1997) 633.
2. Bryant-Greenwood, G.D. and Schwabe, C., Endocr. Rev., 15 (1994) 5.
3. Bellet, D., Lavaissiere, L., Mock, P., Laurent, A., Sabourin, J.C., Bedossa, P., Le Bouteiller, P., Frydman, R., Troalen, F. and Bidart, J.M., J. Clin. Endocrinol. Metab., 82 (1997) 3169.

4. Conklin, D., Lofton-Day, C.E., Haldeman, B.A., Ching, A., Whitmore, T.E., Lok, S. and Jaspers, S., Genomics, 66 (1999) 50.
5. Kasik, J., Muglia, L., Stephan, D.A. and Menon, R.K., Endocrinology, 141 (2000) 458.
6. Mock, P., Frydman, R., Bellet, D., Diawara, D.A., Lavaissière, L., Troalen, F. and Bidart, J.-M., J. Clin. Endocrinol. Metab., 84 (1999) 2253.
7. Mock, P., Frydman, R., Bellet, D., Chassin, D., Bischof, P., Campana, A., and Bidart, J.-M., J. Clin. Endocrinol. Metab., 85 (2000) 3941.
8. Qin, X., Garibay-Tupas, J., Chua, P.K., Cachola, L. and Bryant-Greewood, G.D., Biol. Reprod., 56 (1997) 800.
9. Qin, X., Chua, P.K., Ohira, R.H. and Bryant-Greewood, G.D., Biol. Reprod., 56 (1997) 812.
10. Tseng, L., Gao, J., Chen, R., Zhu, H., Mazella, J. and Powell, D.R., Biol. Reprod., 47 (1992) 441.
11. Bryant-Greewood, G.D., Rutanen, E.-M., Partanen, S., Coehlo, T.K. and Yamamoto, S.Y., Mol. Cell. Endocrinol., 95 (1993) 23.
12. Einspanier, A., Nubbemeyer, R., Schlote, S., Schumacher, M., Ivell, R., Fuhrmann, K. and Marten, A., Biol. Reprod., 61 (1999) 512.
13. Stewart, D. and VandeVoort, C.A., Hum. Reprod., 14 (1999) 338.
14. Hurskainen, T., Hoyhta, M., Tuuttila, A., Oikarinen, A. and Autio-Harmainen, H.J., Histochem. Cytochem., 44 (1996) 1379.
15. Unemori, E.N. and Amento, E.P., J. Biol. Chem., 265 (1990) 10681.
16. Unemori, E.N., Pickford, L.B., Salles, A.L., Piercy, C.E., Grove, B.H., Erikson, M.E. and Amento, E.P., J. Clin. Invest., 98 (1996) 2739.
17. Popovici, R. M., Kao, L.-C. and Giudice, L.C., Endocrinology, 141 (2000) 3510.
18. Bischof, P., Haenggeli, L. and Campana, A., Hum. Reprod., 10 (1995) 734.
19. Meisser, A., Chardonnens, D., Campana, A., and Bischof, P., Mol. Hum. Reprod., 5 (1999) 252.
20. Stewart, D.R., Cragun, J.R., Boyers, S.P., Overstreet, J.W. and Lasley, B.L., Fertil. Steril., 57 (1992) 453.
21. Zhu, H.H., Huang, J.R., Mazella, J., Rosenberg, M. and Tseng, L., J. Clin. Endocrinol. Metab., 71 (1990) 889.
22. Bani, D., Baccari, M.C., Nistri, S., Calamai, F., Bigazzi, M. and Sacchi, T.B., Endocrinology, 140 (1999) 4434.
23. Unemori, E.N., Erikson, M.E., Rocco, S.E., Sutherland, K.M., Parsell, D.A., Mak, J. and Grove, B.H., Hum. Reprod., 14 (1999) 800.
24. Tseng, L., Zhu, H.H., Mazella, J., Koistinen, H. and Seppälä, M., Mol. Hum. Reprod., 5 (1999) 372.
25. Gui, Y., Zhang, J., Yuang, L. and Lessey, B.A., Mol. Hum. Reprod., 5 (1999) 866.

Molecular remodelling of members of the relaxin family during primate evolution

Thomas Klonisch, Christine Froehlich, Frank Tetens, Bernd Fischer and Sabine Hombach-Klonisch

Department of Anatomy and Cell Biology, Martin Luther University, Faculty of Medicine, Halle, Germany

Introduction

The members of the relaxin family include the two hormones relaxin and the structurally closely related relaxin-like factor (RLF). Relaxin displays high sequence variability within and between taxonomic groups among land animals. Widely regarded as a hormone of the female reproductive tract, relaxin is expressed in the ovary and at the feto-maternal interface [1]. Known as a marker of human postpubertal testicular Leydig cells in various species [2-4], RLF is essential for the development of the Gubernaculum testis. Male RLF -/- mice are infertile due to impaired testicular descent resulting in cryptorchidism [5, 6]. Strepsirrhini, which includes the Afro-Asian loris and the Malagasy lemurs, are assumed to have diverged from other primates about 65 Million years ago (Myr) and have developed independently for over 60 Myr [7]. To study the dynamics in the molecular remodelling of both members of the relaxin family during primate evolution, we have cloned the cDNA sequences of the preprorelaxins and RLF in the African lorisiform Galago crassicaudatus (*Gcr*) and of preprorelaxin in the Malagasy lemur Varecia variegata (*Vva*) which is phylogenetically regarded plesiomorphic to all Malagasy lemurs [8].

Results and Discussion

Identical cDNA coding sequences for the preprorelaxin molecules of *Gcr* and *Vva* were determined by RT- and RACE-PCR. This has provided the first endocrinological evidence for a monophyletic origin of all Strepsirrhini. Alignment of the deduced amino acid sequence with other primate relaxins revealed more than 70% difference in aa sequence of the B-domain of strepsirrhine relaxin. In contrast, the A-domain of strepsirrhine relaxin was much more conserved. The receptor binding motif within the B-domain of strepsirrhine relaxin was unique and contained three arginine residues which might engage in a „three prong“ interaction with the corresponding relaxin receptor [9]. In contrast to preprorelaxin, strepsirrhine RLF was much more conserved among primates suggesting selective evolutionary pressures to be effective on both hormones [10]. Based on the deduced primary aa sequences, maximum likelihood analysis placed strepsirrhine relaxin and RLF in outgroups to other primate sequences indicating independent molecular remodelling of both reproductive hormones during the

G.W. Tregear et al. (eds.), Relaxin 2000, 127–128.
© 2001 *Kluwer Academic Publishers. Printed in the Netherlands.*

evolution of Strepsirrhini. Similar to all non-hominoid primates, Strepsirrhini possessed a single gene copy for relaxin and RLF. RT-PCR analysis, non-radioactive in situ hybridization and immunohistochemistry were employed to study the expression of preprorelaxin and RLF mRNA and protein in reproductive tissues. We found mutually exclusive expression of preprorelaxin in the trophoblast of placental tissues of *Gcr* or *Vva* and of RLF in postpubertal testicular tissue of *Vva*, respectively. This implicates both members of the relaxin family to possess different functional roles during reproduction in Strepsirrhini.

Acknowledgements

The authors would like to thank Elisabeth Schlueter for their excellent technical assistance. We are also greatful to Mrs. Ruempler and her collegues at the Lemur Centre, Cologne Zoo, for providing testicular tissue of Varecia variegata. The authors also thank Professor Bernhard Steinetz for generously providing the R6 relaxin antiserum.

References

1. Sherwood, O.D., In Knobil, E. and Neill, J.D. (Eds.), The Physiology of Reproduction. 2nd ed., Raven Press, New York, USA, 1994, p. 861.
2. Ivell, R., Balvers, M., Domagalski, R., Ungefroren, H., Hunt, N. and Schulze, W., Mol. Hum. Reprod., 3 (1997) 459.
3. Adham, I.M., Burkhardt E., Benahmed M. and Engel W., J. Biol. Chem., 268 (1993) 26668.
4. Hombach-Klonisch, S., Tetens, F., Kauffold, J., Steger, K., Fischer, B., Klonisch, T., Mol. Reprod. Dev., 53 (1999) 135.
5. Zimmermann, S., Steding, G., Emmen, J.M.A., Brinkmann, A.O., Nayernia, K., Holstein, A.F., Engel, W., Adham, I.M., Mol. Endocrinol., 13 (1999) 681.
6. Nef, S., Parada, L.F., Nat. Genet., 22 (1999) 295.
7. Yoder, A.D., Cartmill, M., Ruvolo, M., Vilgalys, R., Proc. Natl. Acad. Sci. USA, 93 (1996) 5122.
8. Yoder, A.D. and Irwine, J.A., Cladistics, 15 (1999) 351.
9. Büllesbach, E.E., Yang S. and Schwabe, C., J. Biol. Chem., 267 (1992) 22957.
10. Büllesbach, E.E., Rhodes R., Rembiesa B. and Schwabe C., Endocrine, 10 (1999) 167.

Relaxin in the hamster

Randall H Renegar
Department of Anatomy and Cell Biology, Brody School of Medicine, East Carolina University, Greenville, NC 27858, USA

Introduction

Following domestication of the hamster its primary usefulness in human society was not that of a childhood pet, but as a laboratory research animal. In recent years rats and mice have become the laboratory animal of choice to the near exclusion of other viable rodent species. The physiology of rats and mice is tacitly considered the norm for rodent species thus overlooking valuable information that is to be gained by considering the diversity of mechanisms utilized in achieving the same physiological goal. Nowhere is this diversity more evident than in the process of eutherian reproduction. More diverse still, and with special relevance to this gathering, are the cellular, molecular, and physiological details of relaxin. The purpose of this presentation is to present a contemporary overview of relaxin expression in the hamster and to highlight areas that are in need of additional investigation. An exhaustive comparison of the many aspects of hamster relaxin with that from other species is not possible in this brief presentation. The author apologizes for omission of data pertinent to this presentation.

Relaxin function in the hamster

Within the rodent species, relaxin is primarily expressed in the ovary of rats and mice, uterine glands of guinea pigs, and placenta of the hamster, and substantial quantities of this circulating hormone are present in each species. Although, the biological effects of relaxin are believed to be primarily exerted in an endocrine manner, the role of paracrine/autocrine actions of this hormone should not be overlooked given the variety of cell types that synthesize this hormone, and the diverse roles that these cells have in reproductive processes. The possibility that a relaxin-like hormone is present in the hamster was first suggested by the work of Khaligh [1] that reported that porcine relaxin inhibited spontaneous contractions of the hamster uterus *in vitro*. Subsequently, O'Byrne *et al.* [2] measured plasma relaxin in the hamster using a porcine radioimmunoassay and demonstrated that cervical softening is correlated positively with circulating concentrations of relaxin during pregnancy. Currently, these are the only data that describe a role for relaxin during pregnancy in the hamster. The observation from this laboratory [3] that relaxin is produced early in gestation by the invading trophoblast giant cells (GTC) suggests that this hormone may have a role in this implantation. A role for relaxin in regulation of extracellular matrix enzymes during rupture of the fetal membranes [4, 5] and follicular development [6] has been suggested. Relaxin may regulate decidual cell response to the invading embryo and/or

G.W. Tregear et al. (eds.), Relaxin 2000, 129–136.
© 2001 *Kluwer Academic Publishers. Printed in the Netherlands.*

differentiation of the ectoplacental cone trophoblast cells in the hamster. These possibilities are consistent with the report by Gui et al. [7] that relaxin stimulates expression of the HOXA-10 transcription factor in human endometrial stroma and the observation that relaxin stimulates the IGF/IGFBP system in the pig uterus [8]. Identification and isolation of the relaxin receptor and development of a reliable method for its localization would certainly facilitate efforts to establish a role for relaxin in implantation in the hamster. Studies of post-receptor binding events and their interaction in order to further characterize the role of relaxin in the implantation process are also necessary. Relaxin secreting GTC of the hamster placenta also synthesize placental lactogen II [9], the major lactogenic hormone circulating in the hamster during the second half of pregnancy [10]. The pattern of haPL-II serum concentrations parallels that of relaxin suggesting that relaxin may have a role in regulating haPL-II. Consistent with this idea is the observation that relaxin stimulated prolactin secretion from dispersed human pituitary cells [11] and endometrial stromal cells [12]. Recently, our laboratory [unpublished data] determined that significant quantities of relaxin are present in serum of Day 15 fetuses. The source of this hormone and its role in fetal development and/or physiology remains to be determined. A previous retrospective multiple regression study in humans determined that fetal heart rate was highly correlated with maternal blood hormone concentrations [13]. This data along with the observation that relaxin specific binding sites are present in 1-day postnatal rats [14] suggest a possible role in fetal cardiovascular function.

Chemistry of hamster relaxin

Hamster relaxin was purified in this laboratory [15] from placentas recovered on Days 14 and 15 of pregnancy using the acid-acetone method of Griss [16]. Acetone precipitated proteins were solubilized and applied to a carboxymethyl cellulose ion-exchange column and bound proteins were eluted with 0.1 and 0.3 M NaCl. Western blot analysis using an antibody (R6) to porcine relaxin detected 16.5, 18.7 and 36.0 kDa relaxin immunoreactive (IR) proteins within the 0.1 M NaCl eluant, and a 5.6 kDa relaxin IR protein within the 0.3 M NaCl eluant. The 5.6 kDa protein was purified to homogeneity by gel filtration (Sephadex G-50), ion exchange HPLC, and C_{18}-HPLC. Reduction of the 5.6 kDa protein prior to electrophoresis resulted in a single band of lower molecular weight suggesting that hamster relaxin consists of two chains of approximately equal molecular weight. Isoelectric point of the 5.6 kDa protein was 7.78 which is considerably lower than that reported for porcine relaxin [17] but similar to that of rat relaxin [18]. The 16.5 and 18.7 kDa IR proteins were co-purified by gel filtration and ion exchange HPLC, and at least 5 isoelectric point variants were observed for these proteins. Recovery of relaxin was very low, 0.34 ug/kg of fresh tissue, compared to that reported for the rat [18] and pig [17], but low yields were not unexpected given that relaxin is not stored in secretory granules in the trophoblast giant cell (see below). Values were similar to those reported for other species with a placental source of relaxin [reviewed in 19]. Amino acid sequence information was initially obtained from the 18.7

kDa protein described above [15], and this information was used to construct a primer to use in 3' RACE methodology [20]. The cDNA fragment that was generated (940 bp) was utilized to screen a Day 16 hamster placental cDNA library (gift from F. Talamantes), and a clone containing the entire coding region was identified [21]. Nucleotide sequence analysis revealed a 531 bp open reading frame for preprorelaxin. The derived amino acid sequence contained a 22 amino acid signal peptide followed by a 155 amino acid prorelaxin sequence with a calculated molecular weight of 17,500. The derived prorelaxin amino acid sequence had 51.8, 42.9, 41.7 and 38.2% homology with rat, human-1, human-2, and pig prorelaxins, respectively. Amino acid sequence data was utilized to synthesize hamster relaxin by F-moc chemistry [unpublished data]. The synthesized molecule was found to be 10-fold less potent than porcine relaxin in the mouse interpubic bioassay, similar to the value previously reported for isolated native hormone [15], and was utilized to develop a specific antibody for RIA and ICC methodologies (see below).

Relaxin, like all members of the insulin family of hormones, is synthesized as a precursor (prohormone) requiring enzymatic activity to yield the mature hormone. Enzymes involved in this process include prohormone convertase (PC) 1 [22], PC2 [23] and furin [24, 25]. PC1 and PC2 are localized within the secretory granules of endocrine and neuroendocrine cells where they cleave protein precursors at dibasic residues [26]. In contrast, furin is localized to the trans Golgi membrane [27] of all tissues and cell types. Proteins processed through this compartment are believed to be cleaved at the consensus motif -Arg^{-4}-X^{-3}-Lys/ Arg^{-2}-$Arg^{-1}\downarrow X^{+1}$ where X represents any amino acid [28, 29]. Human prorelaxin has a dibasic site at the B-chain/C-peptide junction and tetrabasic residues at the C-peptide/A-chain junction. Therefore cleavage by furin and PC1 or PC2 may be required for human relaxin processing. Accordingly, when human type 2 prorelaxin [30] was expressed in cells containing only the constitutive pathway, mature hormone was not secreted. In contrast, cotransfection of human kidney 293 cells (native cell contains only the constitutive pathway) with human type 2 prorelaxin and murine PC2 caused secretion of mature relaxin. The amino acid sequence of pig [31] and rat [32] relaxin contain the consensus furin motif at the C-peptide/A-chain junction suggesting a role for furin in this process, however, the amino acid sequence at the B-chain/C-peptide junction does not conform to the furin motif or contain dibasic residues. Haley *et al.* [31] suggested that porcine relaxin may require an enzyme with chymotrypsin-like specificity at the B-chain/C-peptide cleavage site. The subcellular compartment responsible for this process is not known. Interestingly, hamster prorelaxin contains consensus furin processing sites at the C-terminal end of the A-chain/C-peptide (Arg-Arg-Arg-Arg) and C-peptide/B-chain (Arg-Lys-Lys-Arg) junctions suggesting that complete processing could occur in the absence of other prohormone convertases. This conclusion is consistent with the fact that hamster placental GTCs do not contain secretory vesicles [33] or PC 1 [unpublished data] and therefore appear to secrete proteins via the constitutive pathway.

Circulating concentrations

O'Byrne et al. [2], using a homologous porcine radioimmunoassay, were the first to report circulating concentrations of relaxin in the hamster. Although their assay did not quantitatively measure hamster relaxin, it provided surprisingly accurate data concerning the pattern of circulating hormone in this species. These authors reported that relaxin is first detected on day 8 of pregnancy, and then increases dramatically to day 14 in parallel with development of the chorioallantoic placenta. Concentrations increased 3-fold on Day 15 (peak value of 29 ng/ml) before rapidly falling to low levels on Day 1 postpartum. Recently, our laboratory has developed a homologous hamster relaxin radioimmunoassay using synthetic hamster relaxin. Using this antibody, the pattern of circulating relaxin in sequentially sampled pregnant hamsters was similar to that reported by O'Byrne *et al.* [2], however, peak concentrations on the morning of Day 15 averaged 826 ng/ml which far exceeds blood concentrations of this hormone in any species reported to date [unpublished data]. This assay also detected moderate quantities of relaxin immunoactivity in pooled samples of blood from Day 15 fetuses raising the issue of a role for this hormone in the fetus.

Tissue and cellular source

Following their determination that a molecule with relaxin-like immunoactivity was present in the blood of pregnant hamsters, Steinetz *et al.* [34] reported that the ovary was not a significant source of circulating relaxin in the hamster since concentrations were not reduced following ovariectomy. Extracts of placentae recovered during the latter half of pregnancy contained high levels of relaxin immunoactivity suggesting that circulating hormone may be of placental origin. The cellular source of relaxin in the hamster was determined by immunohistochemistry (IHC) methods in this laboratory [3] initially using an antibody to porcine relaxin (R6; gift from Dr. Steinetz) but more recently an antibody to synthetic hamster relaxin has been employed. Relaxin is first localized on Day 8 of pregnancy (16 day pregnancy) in primary GTCs located at the antimesometrial and lateral regions of the trophoectoderm: cells associated with destruction of the uterine epithelium and invasion of the antimesometrial decidua. As pregnancy progresses (Day 10), relaxin expression expands to include secondary giant trophoblast cells (GTC-2) which are located along the mesometrial front of the invasive trophoblast immediately adjacent to the uterine decidua. These cells differentiate from superficial cells of the ectoplacental cone: the origin of many of the cells making up the fetal portion of the chorioallantoic placenta. On Day 10, relaxin was also expressed in endometrial granulocytes surrounding sheathing arteries in the mesometrial decidua when the R6 antibody to porcine relaxin was used [3], however, recent experiments using our antibody to synthetic hamster relaxin did not confirm localization to these cells. With continued placental development the GTC-2 become localized to the trophospongium or junctional portion of the placenta, and the number of cells expressing relaxin increases dramatically in parallel with placental development to reach a maximum on Day 14. Beginning on Day 15 isolated aggregates of cytotrophoblast cells and relaxin containing GTC-2 appear within the placental labyrinth [3; unpublished data]. It is these cells that may be the source of relaxin in the fetus. Recently, localization of relaxin mRNA using *in situ* hybridization histochemistry (ISHH) methodology has confirmed that the GTC synthesize this hormone [unpublished data]. There was no evidence of relaxin synthesis in the ovary or oviduct using ICC or ISHH.

Recent evaluation of implantation site (Day 6, 8, 10) and placental (Days 12, 14, 15) tissues indicate that concentrations of relaxin mirror changes in circulating levels of the hormone [unpublished data]. Similarly, McCaslin and Renegar [21] reported that the 900 bp transcript for prorelaxin mRNA was first detected in the placenta on Day 8, with levels increasing to a plateau on Days 10, 12, and 14 and then increasing approximately 50% by Day 15. In agreement with previous IHC data, prorelaxin mRNA was not detected in the ovary. A large portion of the increase in implantation site/placental relaxin gene expression is likely due to an increase in the number GTC-2 during placental development. Jones and Renegar [35] determined this to be the major reason for changes in haPL-II expression during this same period. However, the increase in cell number was unable to completely account for changes in mRNA levels, especially between Days 12 and 14, suggesting that pregnancy related factors may control transcription of the relaxin gene by individual cells.

Subcellular source

The corpus luteum is the primary source of relaxin in rats and mice during pregnancy, and epithelium of the late pregnant endometrial glands is the sole source of relaxin in the pregnant guinea pig [reviewed in 19]. Relaxin has been immunolocalized to dense secretory granules contained in each of these cell types [36, 37, 38]. Circulating levels of hormone are believe to be the result of regulated release of stored hormone, and factors that regulate relaxin release have been investigated [39, 40, 41]. GTC of the hamster placenta do not contain the dense secretory granules characteristic of luteal and endometrial gland cells [35]. Although early descriptions indicated that relaxin immunoreactive staining product was localized in granular cytoplasmic inclusions in the perinuclear region [3], a subsequent electron microscope study [35] determined that reaction product was associated with the numerous Golgi bodies characteristic of GTCs. This finding was supported by the fact that the vast majority of tissue relaxin was associated with the secretory granule fraction from corpora lutea but not placenta. These findings raised many interesting questions concerning the biochemistry and cellular biology of relaxin. Burgess and Kelly [42] had previously described two cellular mechanisms for release of synthesized proteins: constitutive and regulated. Proteins leaving the cell by the constitutive pathway are not concentrated and stored in large dense secretory granules, but are secreted soon after synthesis, and secretogogue stimulation is not required. Proteins released by the regulated secretory pathway are concentrated 10-200 times, packaged in membrane-bound cytoplasmic granules for storage in the cell, and released upon stimulation by the appropriate secretogogue. Thus control of relaxin synthesis and secretion by the GTC may be more closely tied to gene expression than that of luteal and glandular epithelial cells. There has been little effort to evaluate factors controlling relaxin gene expression in any cell type. Recent studies in the human have sequenced 5' untranslated region of H1 and H2 relaxin [43], however functional studies of the relaxin promoter are limited [44]. Much remains to be done with regard to relaxin gene expression, especially in animals without an ovarian source for this hormone. To this end our laboratory has recently sequenced the 5' UTR of the hamster relaxin gene [unpublished data] and is currently examining the role of various promoter constructs on reporter gene expression in homologous and heterologous cell types.

Summary

The hamster is the only laboratory rodent identified to date with a fetal (placenta) rather than a maternal source (corpus luteum, uterus) of relaxin. Accordingly, this species, with its short gestation time and relatively high endogenous levels of relaxin during pregnancy, offers a unique opportunity to examine the structure, function and expression of relaxin from a fetal tissue. Relaxin receptor binding and post-receptor signaling events as well as the role of this hormone in tissues of the implanting embryo and developing placenta have received little attention and can readily be examined using the hamster as a model system. Given differences in the origin, cell physiology, and function during pregnancy of tissues secreting relaxin in rodents, the external factors and intracellular mechanisms that control relaxin gene expression are also likely to differ. This work is just beginning and offers the promise of many years of fruitful study.

Acknowledgements

The author acknowledges the excellent technical support of Mr. Charles "Chuck" R. Owens in the work from our laboratory and wishes to thank Dr. Hubert W. Burden for his thoughtful review of this manuscript.

References

1. Khaligh, H.S., J. Endocrinol., 40 (1986) 125.
2. O'Byrne, E.M., Sawyer, W.K., Butler, M.C. and Steinetz, B.G., Endocrinology, 99 (1976) 1333.
3. Renegar, R.H., Cobb, A.D. and Leavitt, W.W., Biol. Reprod., 37 (1987) 925.
4. Koay, E.S.C., Too, C.K.L., Greenwood, F.C. and Bryant-Greenwood, G.D., J. Clin. Endocrinol. Metab., 56 (1983) 1332.
5. Koay, E.S.C., Bryant-Greenwood, G.D., Yamamoto, S.Y. and Greenwood, F.C., J. Clin. Endocrinol. Metab., 62 (1986) 513.

6. Too, C.K.L., Bryant-Greenwood, G.D. and Greenwood, F.C., Endocrinology, 115 (1984) 1043.
7. Gui, Y., Zhang, J., Yuan, L. and Lessey, B.A., Mol. Hum. Reprod., 5 (1999) 866.
8. Ohleth, K.M., Zhang, Q., Lenhart, J.A., Ryan, P.L. and Bagnell, C.A., Steroids, 64 (1999) 634.
9. Renegar, R.H. and Southard, J.N., J. Histochem. Cytochem., 38 (1990) 935.
10. Southard, J.N., Campbell, G.T. and Talamantes, F., Endocrinology, 121 (1987) 900.
11. Sortino, M.A., Cronin, M.J. and Wise, P.M., Endocrinology, 124 (1989) 2013.
12. Telgmann, R. and Gellersen, B., Hum. Reprod. Update, 4 (1998) 472.
13. Johnson, M.R., Jauniaux, E., Ramsay, B., Jurkovic, D. and Meuris, S., Br. .J Obstet. Gynaecol., 101 (1994) 1003.
14. Osheroff, P.L. and Ho, W.H., J. Biol. Chem., 268 (1993) 15193.
15. Renegar, R.H., Owens, C.R. and Chalovich, J.M., Biol. Reprod., 49(1993) 154.
16. Griss, G., Keck, J., Engelhorn, R. and Tuppy, H., Biochim. Biophys. Acta., 140 (1967) 45.
17. Sherwood, O.D. and O'Byrne, E.M., Arch. Biochem. Biophys., 160 (1974) 185.
18. Sherwood, O.D., Endocrinology, 104 (1979) 886.
19. Sherwood, O.D., In Knobil, E. and Neill, J.D. (Eds.), The Physiology of Reproduction. 2nd ed., Raven Press, New York, USA, 1994, p. 585.
20. Frohman, M.A., Dush, M.K. and Martin, G.R., Proc. Natl. Acad. Sci. USA, 85 (1988) 8998.
21. McCaslin, R.B. and Renegar, R.H., Biol. Reprod., 53 (1995) 454.
22. Seidah, N.G., Gaspar, L., Mion, P., Marcinkiewicz, M., Mbikay, M. and Chrétien, M., DNA Cell. Biol., 9 (1990) 415.
23. Smeekens, S.P. and Steiner, D.F., J. Biol. Chem., 265 (1990) 2997.
24. Fuller, R.S., Brake, A.J. and Thorner, J., Science, 246 (1989) 482.
25. Hatsuzawa, K., Hosaka, M., Nakagawa, T., Nagase, M., Shoda, A., Murakami, K. and Nakayama, K., J. Biol. Chem., 265 (1990) 22075.
26. Bailyes, E.M., Bennett, D.L. and Hutton, J.C., Enzyme, 45 (1991) 301.
27. Bresnahan, P.A., Leduc, R., Thomas, L., Thorner, J., Gibson, H.L., Brake, A.J., Barr, P.J. and Thomas, G. J., Cell Biol., 111 (1990) 2851.
28. Watanabe, T., Murakami, K. and Nakayama, K., FEBS Letts., 3203 (1993) 215.
29. Yanagita, M., Hoshino, H., Nakayama, K. and Takeuchi, T., Endocrinology, 1332 (1993) 639.
30. Marriott, D., Gillece-Castero, B. and Gorman, C.M., Mol. Endocrinol., 6 (1992) 1441.
31. Haley, J., Hudson, P., Scanlon, D., John, M., Cronk, M., Shine, J., Tregear, G. and Niall, H., DNA, 1 (1982) 155.
32. Hudson, P., Haley, J., Cronk, M., Shine, J. and Niall, H., Nature, 291 (1981) 127.
33. Johns, T.C. and Renegar, R.H., Am. J. Anat., 189 (1990) 167.
34. Steinetz, B.G., O'Byrne, E.M., Goldsmith, L.T. and Anderson, M.B., Endocrinology, 122 (1988) 795.
35. Jones, M.A. and Renegar, R.H., Biol. Reprod., 50 (1994) 581.
36. Anderson, M.B. and Sherwood, O.D., Endocrinology, 114 (1984) 1124.
37. Anderson, M.B., Vaupel, M.R. and Sherwood, O.D., Biol. Reprod., 31 (1984) 391.
38. Pardo, R.J., Larkin, L.H. and Renegar, R.H., Anat. Rec., 209 (1984) 373.
39. Taylor, M.J. and Clark, C.L., Biol. Reprod., 37 (1987) 377.
40. Taylor, M.J. and Clark, C.L., Endocrinology, 125 (1989) 1389.
41. Taylor, M.J. and Clark, C.L., J. Endocrinol., 135 (1992) 543.

42. Burgess, T.L. and Kelly, R.B., Annu. Rev. Cell Biol., 3 (1987) 243.
43. Garibay-Tupas, J.L., Csiszar, K., Fox, M., Povey, S. and Bryant-Greenwood, G.D., J. Mol. Endocrinol., 23 (1999) 355.
44. Gunnersen, J.M., Roche, P.J., Tregear, G.W. and Crawford, R.J., J. Mol. Endocrinol., 15 (995)153.

SECTION 4

Reproductive physiology of relaxin III: Ovary

Relaxin: a luteotrophic factor in the marmoset corpus luteum

Svenja Verhagen and Almuth Einspanier
Department of Reproductive Biology, German Primate Center, Göttingen, FRG

Introduction

In the marmoset monkey (*Callithrix jacchus*), relaxin is produced within the corpus luteum (CL) [1-3], serving as the main source of the hormone during the cycle and early pregnancy, like in other primates [4-6]. Relaxin expression is high during the mid-luteal phase and increases further in early pregnancy, suggesting a possible role in CL maintenance during this period. Furthermore, the circulating relaxin blood levels of early pregnant animals are significantly higher than from non-conceived monkeys, however, compared to the classical animal like pig and rat [7], primates have much lower peripheral relaxin levels suggesting a para/autocrine action in the ovary and uterus.

Previous studies, in which a possible causal relationship between LH/hCG and relaxin was examined from peripheral blood measurements, showed conflicting results. Whereas some researchers postulate a direct control function for hCG on relaxin production [4, 8, 9], others find no evidence for this [3, 10-12]. Furthermore, the more general issue concerning the local action and regulation of relaxin synthesis and secretion in the primate CL remains unresolved. We have therefore carried out additional studies to examine the local influence of relaxin within the ovary, as well as on the interaction between LH/CG and relaxin, making use of a microdialysis system (MDS). MDS acts like an artificial capillary, which is placed into the intact CL [human: 13; primate: 14-16; pig: 17, 18; bovine: 19]. This is illustrated schematically in Fig. 1. The molecular cut-off of the microdialysis membrane used is approximately 30 kDa, and the transfer rate for small peptide hormones (e.g. relaxin or hCG) is about 1%.

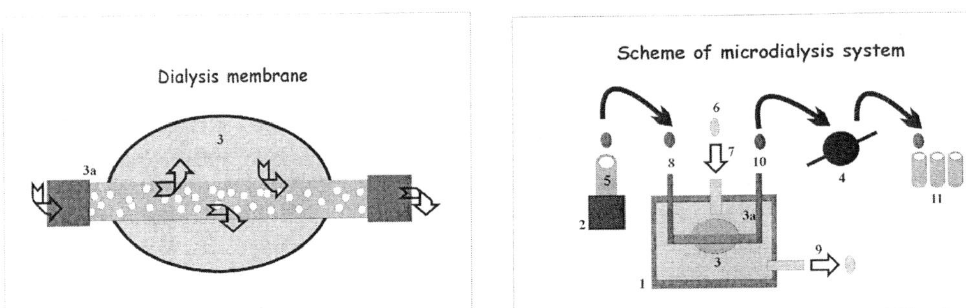

Fig. 1. Schematic representation of the microdialysis system. The microdialysis membrane (left hand side), which is placed into the whole CL-tissue (3) is routed through an incubation chamber (1) filled with warmed M199 medium. The membrane is continuously perfused with ringer solution or ringer plus the effectors of interest (5) using a peristaltic pump (4) and

G.W. Tregear et al. (eds.), Relaxin 2000, 139–144.
© 2001 Kluwer Academic Publishers. Printed in the Netherlands.

fractions collected every 15 min. (11). Influx of media/ringer (6,7) and efflux of media/ringer (9, 10).

Firstly, the advantages of the MDS are their ability to collect hormones being secreted from the luteal cells situated in the neighbourhood of the dialysis membrane, and secondly, that substances of interest can be applied through the microdialysis membrane to induce a local hormonal response, which in turn can be measured in the collected medium.

In contrast to a static culture system, MDS allows immediate changes in hormone responses to be determined and is highly sensitive in terms of speed of response, concentration applied and mode of application for test substances. Moreover, unlike similarly sensitive perfusion systems, MDS maintains the integrity and hence paracrine coherence of the natural tissue. The aim of this study was to examine the local effects of pure porcine relaxin (courtesy Dr. O.D. Sherwood) on steroid production in mid-luteal phase corpus luteum from the marmoset monkey [14,16] using MDS, either alone or in combination with gonadotropins. CL tissue gained by lutectomy from the mid-luteal phase was placed in explant culture, fitted with the MDS, and then continuously perfused with either perfusion medium (standard Ringer solution) only, or supplemented with hCG (100 ng/ml), porcine relaxin (100, 500 ng/ml for 1h) or given sequentially a gonadotropin preapplication (1IU FSH or LH for 30 min) followed by porcine relaxin at two different concentrations (100, 500 ng/ml) for 1h.

Results and Discussion

The basal relaxin and progesterone concentrations of a perfused mid-phase corpus luteum were analysed. The measured concentrations were about 500 pg/ml for progesterone and 80 pg/ml for relaxin, after a perfusion period of 8 h. In the subsequent experiment, hCG as a luteotrophic and possible regulatory factor on luteal relaxin secretion was applied to the mid-luteal phase tissue. The application of 100 IU/ml hCG induced a significant and immediate increase in progesterone response of approximately 150%, whereas hCG had no effect on relaxin secretion (Fig. 2).

After estimating the basal relaxin concentration, the CLs were then perfused with ringer solution for 3h, followed by different concentrations of relaxin (100, 500 ng/ml) for 1h, and then by a further 3h of ringer solution alone. 15 minute fractions were collected and hormone content measured for progesterone, estradiol and estradiol 1-glucuronid (E1G). Consecutive application of 100 and 500 ng/ml relaxin (transfer rate of relaxin is 1% through the membrane, thus equivalent to 0.1 and 0.5 ng/ml) led to a significant and immediate increase in all three hormones analysed (Fig. 3). Relaxin application alone gave a dose-dependent increase of 250% (100 ng/ml relaxin) or 400% (500 ng/ml relaxin) compared to basal values for progesterone, and an increase of 150 % (100ng/ml relaxin) and 300% (500 ng/ml relaxin) for estradiol. For each perfused CL, the specific hormone concentrations in the the dialysate remained elevated during the period of effector application and declined slowly thereafter. In contrast, progesterone and estradiol levels remained elevated for an extended period (15-45 min.)

after application of 500 ng/ml relaxin. Following the period of stimulation, subsequent baseline hormone values were higher than those measured prior to application of the test substance.

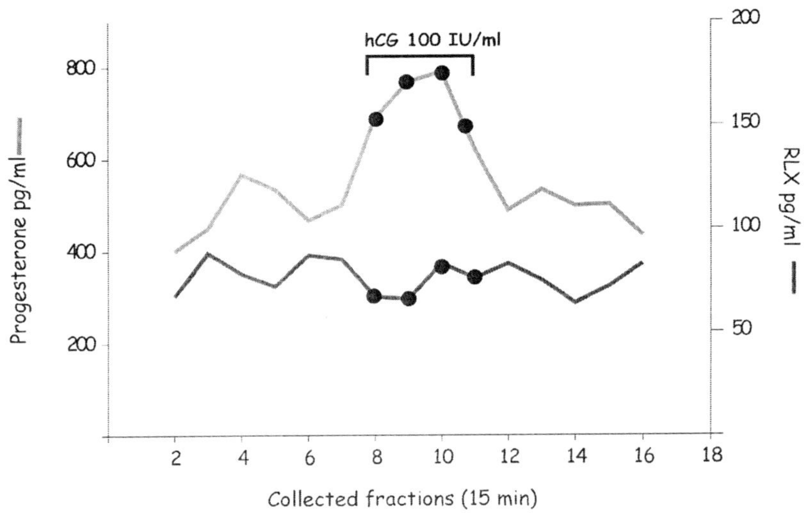

Fig. 2. Progesterone and porcine relaxin (RLX) reponses to application of hCG (100 IU/ml) in microdialysed luteal tissue. Fractions were collected every 15 min. Filled dots represent time period under hCG application (4 fractions, 15-19).

In a second set of experiments, mid-luteal CLs received a low concentration of gonadotropin (1IU hCG or FSH/ml for 30 min.), doses which by themselves do not cause any significant steroid production. This was followed by application of relaxin concentrations similar to those previously described. An application of 100 ng/ml relaxin gave rise to a 400% increase in progesterone when given together with hCG and a 200% increase when given with FSH, compared to basal levels. The progesterone increase was greater after the higher concentration of relaxin (500 ng/ml), giving rise to an increase in progesterone production of 650% with additional hCG, and 950% with additional FSH. Interestingly, estradiol production did not show these marked increases.

Morphological examination of the perfused tissue following an MDS perfusion of 8h showed a close association of the perfusion membrane with the luteal tissue (Fig. 4). The luteal cells juxtaposed to the microdialysis membrane were intact and stained positively for 3ß-hydroxysteroid dehydrogenase (3ß-HSD), one of the essential enzymes in the progesterone synthesis pathway (Fig.4).

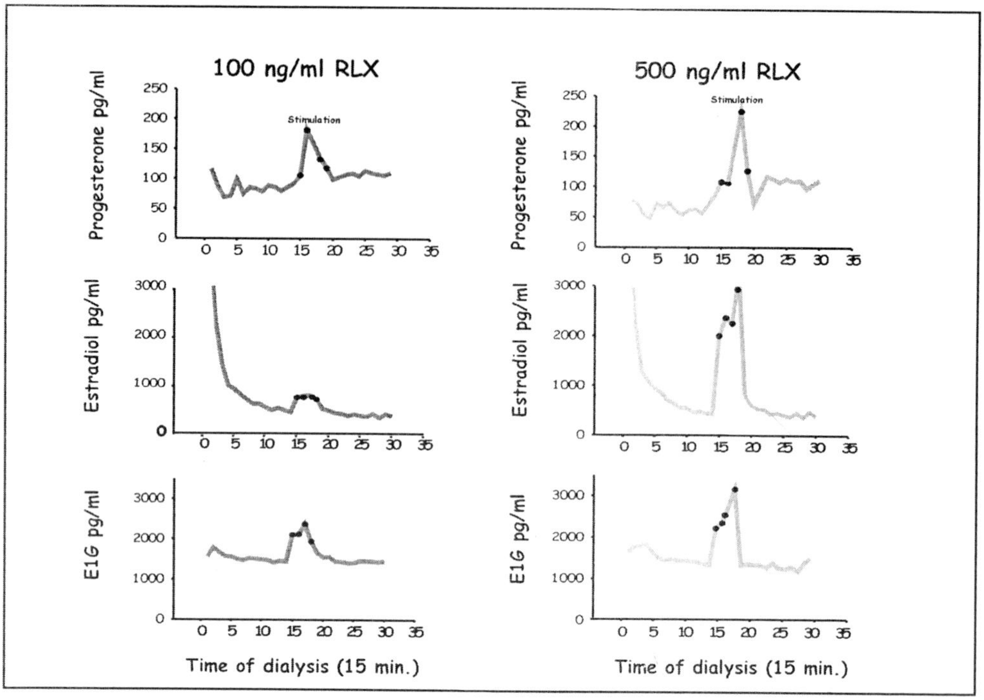

Fig. 3. Dose-dependent effect of porcine relaxin (RLX; 100 and 500 ng/ml). Filled dots represent time period under relaxin application (4 fractions, 15-19)

In general, MDS appears to be a highly appropriate system with which to study dose-dependent effects on luteal tissue, providing very sensitive and immediate responses to changed local situations, thus differing from conventional cell culture systems. The convincing morphological appearance of the perfused marmoset tissue, and the establishment of optimal conditions have been described previously [15, 16]. We have been able to confirm here both the maintenance of tissue integrity following perfusion, as well as the induced activation of the luteal tissue, witnessed by the translocation of mitochondria, the presence of lipid droplets in the perinuclear region, and the vesiculation of the smooth endoplasmatic reticulum in all tisues after the dialysis. These morphological and endocrine changes are typical characteristics of stimulated and actively secreting cells. This is further supported by the presence of 3ß-HSD staining in luteal cells around the microdialysis membrane.

The present study indicated that there was no increase in relaxin production following a relatively high dose of hCG through the microdialysis membrane, although this did cause a considerable increase in progesterone synthesis. In a previous study, we have analysed the peripheral blood levels of relaxin and LH/hCG in the marmoset monkey during early pregnancy. No direct relationship was found between blood levels of relaxin and LH/hCG during the establishment of pregnancy, nor was an increase in peripheral relaxin concentration observed following repeated application of exogenous hCG [3]. Taken together, both the *in vivo* as well as *in vitro* results suggest that there is no direct influence of LH/hCG on the relaxin production by the CL of the marmoset monkey. However, there are indirect mechanisms present, by which hCG could support relaxin secretion.This also implies that the luteotrophic effects of the two hormones (hCG and relaxin) and their maintenance of CL function might involve discrete intracellular systems. An earlier study has also suggested different mechanisms by which hCG and relaxin might be stimulating progesterone production in luteinized human granulosa cells [20], although both are thought to act via the PKA pathway.

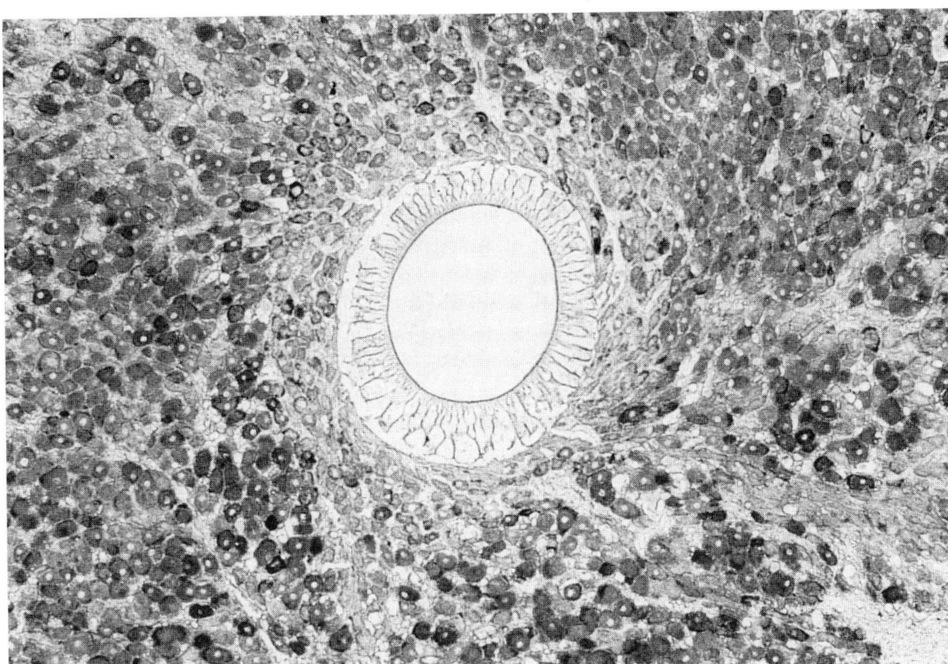

Fig. 4. Marmoset luteal tissue after 8h dialysis stained for 3ß-hydroxysteroid dehydrogenase (red staining). Original magnification x175.

In summary, the data presented suggest a luteotrophic effect of high concentrations of relaxin on luteal cell metabolism, which is further increased by a low level of gonadotropin acting synergistically. Since the main time of increased relaxin production *in vivo* is during the shift from non-pregnant to pregnant status, these results would support the hypothesis that locally produced relaxin is part of the luteotrophic support mechanism and the maternal recognition process in primates.

Acknowledgments

This work was supported by DFG grant Ei 333/8-1. The authors would like to thank Prof. David Sherwood for providing porcine relaxin antibody and purified porcine relaxin, and Prof. Keith Hodges for facilities at the German Primate Center. Special thanks also go to Kerstin Fuhrmann, Alexandra Marten, and Angelika Jurdzinski for excellent technical assistance, and to Karl-Heinz Lamprecht for animal help. This work forms part of the PhD thesis by Svenja Verhagen at the University of Göttingen, FRG.

References

1. Steinetz, B.G., Randolph, C. and Mahoney, C.J., Biol Reprod., 53 (1995) 834.
2. Einspanier, A., Zarreh-Hoshyari-Khah, M.R., Balvers, M., Kerr, L., Fuhrmann, K. and Ivell, R., Hum. Reprod., 12 (1997) 1325
3. Einspanier, A., Nubbemeyer, R., Schlote, S., Schumacher, M., Ivell, R., Fuhrmann, K. and Marten, A., Biol. Reprod., 61 (1999) 512.
4. Duffy, D.M., Stouffer, R.L. and Stewart, D.R., Biol. Reprod., 53 (1995) 447.
5. Steinetz, B.G., Randolph, C. and Mahoney, C.J,. Endocrinology, 130 (1992) 3601
6. Stouffer, R.L., Ottobre, J.S., Molskness, T.A.and Zelinski-Wooten, M.B., Prog. Clin. Biol. Res., 294 (1989) 129.
7. Sherwood, O.D., Downing, S.J., Guico-Lamm, M.L., Hwang, J.J., O'Day-Bowman, M.B. and Fields, P.A., Oxf. Rev. Reprod. Biol., 15 (1993) 143.
8. Quagliarello, J., Goldsmith, L., Steinetz, B., Lustig, D.S. and Weiss, G., J. Clin. Endocrinol. Metab., 51 (1980) 74.
9. Ottobre, A.C., Ramsey, K.R. and Ottobre, J.S., J. Reprod. Fertil., 91 (1991) 313.
10. Seki, K., Uesato, T., Tabei, T. and Kato, K., Endocrinol. Jpn., 32 (1985) 741.
11. Stewart, D.R., Overstreet, J.W., Celniker, A.C., Hess, D.L., Cragun, J.R., Boyers, S.P. and Lasley, B.L., Clin. Endocrinol., 38 (1993) 379.
12. Ghosh, D., Stewart, D.R., Nayak, N.R., Lasley, B.L., Overstreet, J.W., Hendrickx, A.G. and Sengupta, J., Hum. Reprod., 12 (1997) 914.
13. Maas, S., Jarry, H., Teichmann, A., Rath, W., Kuhn, W. and Wuttke, W., J. Clin. Endocrinol. Metab., 74 (1992) 306.
14. Einspanier, A. and Hodges, J.K., J. Endocrinol., 141 (1994) 403.
15. Fehrenbach, A., Einspanier, A., Nicksch, E. and Hodges, J.K., Tissue and Cell, (1995) 27 467.
16. Fehrenbach, A., Hodges, J.K. and Einspanier, A., J. Endocrinol., 161 (1999) 433.
17. Jarry, H., Einspanier, A., Kanngiesser, L., Dietrich, M., Pitzel, L., Holtz, W. and Wuttke, W., Endocrinology, 126 (1990) 2350.
18. Einspanier, A., Jarry, H., Pitzel, L., Holtz, W. and Wuttke, W., Endocrinology, 129 (1991) 3403.
19. Schams, D., J. Physiol. Pharmacol., 43 (1992) 117.
20. Goldsmith, L.T. In MacLennan, Tregear, Bryant-Greenwood (eds) Progress in Relaxin Research,, World Scientific Publishing Co. Ltd., Singapore, 1994, p. 132.

Isolation and bioactivity of relaxin from the tammar wallaby (*Macropus eugenii*).

Ross AD Bathgate, Andrew Siebel, Antonia Claasz, Mary Macris, Geoffrey W Tregear and Laura J Parry
Howard Florey Institute, University of Melbourne, Victoria 3010, Australia

Introduction

Recent cloning of a prepro-relaxin cDNA from the tammar wallaby (Macropus eugenii)[1] indicated that the derived amino acid sequence retains the structural elements common to relaxin peptides from all species, including the three residues that make up the putative receptor binding motif (Fig.1). Although neither marsupial prorelaxin nor relaxin have been extracted and purified from any tissue, early studies demonstrated that crude extracts of corpora lutea (CL) from the pregnant tammar wallaby contain bioactive relaxin [2]. Both the CL and placenta are sites of relaxin mRNA expression during pregnancy in the tammar but transcripts are considerably more abundant in the CL [1]. Within the CL, relaxin gene expression is highest in early and mid-pregnancy, reduced at term and absent postpartum [1]. Antisera to porcine relaxin was used to show immunoreactive relaxin localized to membrane-bound, electron-dense granules in the luteal cell cytoplasm [3]. A large increase in immunoreactive relaxin in the tammar CL is observed from Day 22 of the 26-day gestation but does not coincide with the increase in relaxin gene expression [1]. Moreover, electron-dense granules, the known storage site for relaxin, are not present in the luteal cell cytoplasm in large numbers until Day 10 of gestation [1]. It was suggested that prorelaxin is stored in cytoplasmic granules in the CL during the early stages of gestation and that further processing to form mature relaxin peptide occurs towards the end of gestation when the hormone is released. The objective of this study was to isolate and purify tammar prorelaxin or processed relaxin peptides from the tammar wallaby CL and determine their structure and bioactivity.

Results and Discussion

Relaxin was isolated from pooled CL of pregnant tammar wallabies using a well-defined extraction protocol [4]. Relaxin peptides were purified to homogeneity by reverse phase, followed by size exclusion and ion-exchange, HPLC. No prorelaxin peptide was isolated, however multiple processed relaxin peptides were identified which by mass spectrometer (MS) analysis were shown to differ in the lengths of their B chains (B28, B29, B30, B32 see Fig. 1). One peptide (A24/B28) was further characterized by determination of amino acid content. This data together with the MS results was able to confirm the structure of the native tammar relaxin peptide conformed to the derived amino acid sequence from the cDNA clone [1] (Fig. 1).

G.W. Tregear et al. (eds.), Relaxin 2000, 145–146.
© 2001 *Kluwer Academic Publishers. Printed in the Netherlands.*

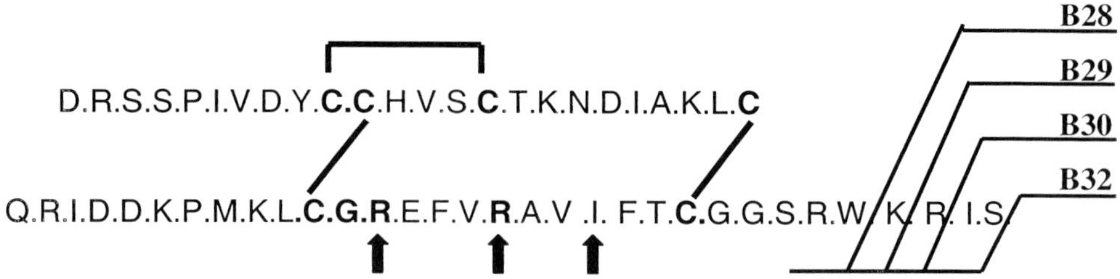

Fig. 1. Primary structure of tammar wallaby relaxin (A24 top/B32 bottom). Residues conserved with relaxin molecules from other species are shown in bold. The three residues implicated in binding of relaxin to its receptor are conserved and are arrowed. The different B chain length relaxin peptides isolated from tammar corpus luteum are marked. Note that the glutamate residue at the end of the B chain is cyclized to pyroglutamic acid in some extracts.

The bioactivity of the tammar wallaby relaxin was tested in two separate assays. Firstly it was tested for its ability to displace ^{33}P-labelled H2 relaxin from rat cortical relaxin receptors. Tammar relaxin (B29) competed for rat cortical relaxin receptors with similar affinity (K_i = 4.94nM) to rat relaxin (K_i = 2.54nM). Secondly it was tested for its ability to stimulate cAMP production from the human monocytic cell line, THP1 [6]. Tammar relaxin (B29) was able to stimulate the production of cAMP in THP1 cells but was much less effective (EC_{50} = 17.5nM) than H2 relaxin (EC_{50} = 0.15nM).

In conclusion, we have confirmed that the structure of tammar wallaby relaxin conforms to that of relaxin of other species. Furthermore, its ability to bind to (rat), and elicit a biological response from (human), relaxin receptors from other mammalian species indicates that the function of relaxin has been conserved in mammalian evolution.

Acknowledgments

Ross Bathgate and Laura Parry are recipients of an NHMRC Howard Florey Centenary Fellowship and an ARC QEII Fellowship respectively. Research was funded by grants from the Australian Research Council (S199980) and a block grant to the Howard Florey Institute (Reg Key 983001) from the NHMRC.

References

1. Parry, L.J., Rust, W. and Ivell R., Biol. Reprod., 57 (1997), 119.
2. Tyndale-Biscoe, C.H., J. Reprod. Fertil., 19 (1969), 191.
3. Parry, L.J., Clark, J.M. and Renfree, M.B., Cell Tiss. Res., 290 (1997), 615.
4. Layden, S.S. and Tregear, G.W., J. Biochem. Biophys. Methods., 31 (1996), 69.
5. Bullesbach, E.E. and Schwabe, C., J. Biol. Chem., 275 (2000), 35276.
6. Parsell, D.A., Mak, J.Y., Amento, E.P., Unemori, E.N., J. Biol. Chem., 271 (1996), 27936.

Possible role of the IGF family for the bovine ovarian function

Dieter Schams[1], Bajram Berisha[1], Werner Amselgruber[2], Michaela Kosmann[1], and Ralf Einspanier[1]

[1]*Institute for Physiology, FML, Technical University Munich, 85350 Freising-Weihenstephan, Germany; and* [2]*Department of Anatomy and Physiology, University of Hohenheim, 70593 Stuttgart, Germany.*

Introduction

There is evidence that the signal transduction pathway of insulin-like growth factors (IGFs) is similar to relaxin [1]. The existence and possible role of bovine relaxin is still not established. Recently the cDNA sequence of preprorelaxin in the one-humped camel was determined [2]. In contrast, there is strong evidence for an important role of IGFs and IGF binding proteins (IGFBPs) in large farm animals [3]. The aim of our studies in the bovine ovary was to evaluate the mRNA expression and protein localisation of IGFs, IGFBPs and growth hormone receptor (GHR) in follicles during final growth and different corpus luteum (CL) stages.

Entire reproductive tracts from German Fleckvieh cows were collected at a local slaughterhouse. The stage of the oestrous cycle was defined by macroscopic observation of the ovaries (follicles and CL). Follicles that were in the final growth stages (4-5 mm in diameter) were classified according to the oestradiol-17β (E_2) content in follicular fluid (FF) and were divided into five classes (E_2 < 0.5; 0.5-5; 5-20; 20-180, and >180 ng/ml FF). The CL were accordingly assigned to the following stages: days 1-2, 3-4, 5-7, 8-12, 13-18 and >18 (after regression). For RNA extraction, CL tissue, granulosa cells (GC), and the theca layer (TI) were separated, and RNA isolated by means of a guanidinium isothiocyanate/phenol extraction method. The expression of mRNA for the ligand and receptor was evaluated by means of reverse transcription-polymerase chain reaction (RT-PCR) and ribonuclease protection assay (RPA). IGF1, IGF2 protein in follicle and CL was localised by means of immunohistochemistry.

Results and Discussion

A relative high expression of IGF1 in TI was observed already before selection (E<0.5 ng/ml). In GC, the mRNA expression was lower before selection, with the tendency to increase afterwards. In contrast, IGF2 was mainly expressed in the TI. The IGF receptor 1 mRNA was present in the TI and GC with increasing levels during the final development of dominant follicles. There is also clear evidence for the local expression of IGFBP1-6 in the TI and GC-compartment during final growth of follicles. A specific regulation for each IGFBP is obvious.

G.W. Tregear et al. (eds.), Relaxin 2000, 147–148.
© 2001 *Kluwer Academic Publishers. Printed in the Netherlands.*

In CL the highest expression of IGF1 was observed during the early luteal phase (d 1-4), followed by a decrease, and again an increase during the mid and late luteal phases of the cyclic CL. IGF2 showed a similar trend. IGFR1 mRNA was highest on d 3-4, followed by a decrease, and again a slight increase. Distinct receptors for IGF1, IGF2 and insulin were present in CL membrane preparations at all stages. IGF1 binding capacity and the affinity of IGFR1 was constant throughout the cycle. Intense immunostaining for IGF1 in CL was observed mainly in large or in small luteal cells. In contrast, IGF2 protein was localized in perivascular fibroblasts and pericytes of the capillaries. Evidence for a mRNA for all six IGFBPs, with distinct differences between the BPs were obtained.

The IGF/GH system in the ovary provides an integrated, cooperating, local network that is regulated by auto/paracrine and endocrine mechanisms to promote growth and differentiation in the ovary. There is now clear evidence for a central role for GH and the IGFs in follicular development and CL function in a number of species. Stimulation of IGFs and IGFBPs and their effects are species specific. Gonadotropins, by enhancing IGF peptide and inhibiting IGFBP production, serve to maximize IGF1 or IGF2 availability to receptors in the process of follicular development and during the lifetime of the CL. It can be speculated that relaxin (if it is expressed in the ovary) has IGF-like effects on follicular growth and CL function.

Acknowledgements

The work was supported by the German Research Foundation (DFG).

References

1. Sherwood, O.D., In Knobil, E. and Neill, J.D. (Eds.), The Physiology of Reproduction. 2nd ed., Raven Press, New York, USA, 1994, p. 861.
2. Hombach-Klonisch, S., Abd-Elnaeim, M., Skidmore, J.A., Leiser, R., Fischer, B. and Klonisch, T., Biol. Reprod., 62 (2000) 839.
3. Schams, D., Berisha, B., Kosmann, M., Einspanier, R., Amselgruber, W.M., Dom. Anim. Endocrinol., 17 (1999) 279.

Identification of consensus sequences in the rat relaxin promoter which bind nuclear proteins

Ping Fu[1], Jennifer Gunnersen[1], Geoffrey Tregear[1] and Peter Roche[2]
[1]*Howard Florey Institute, University of Melbourne, Parkville, Victoria 3010, Australia; and*
[2]*Department of Medical Laboratory Science, RMIT University, Victoria, 3001 Australia*

Introduction

Relaxin is a peptide hormone that plays an important role in reproduction. Regulation of relaxin gene expression occurs *via* both tissue-specific and temporal mechanisms. In the rat, relaxin is mainly produced by the corpus luteum of the ovary during the later stage of pregnancy [1]. The corpus luteum during the menstrual cycles also produces relaxin, but the level is 200-400 fold lower than that in pregnancy [2]. In addition to the ovary, relaxin mRNA has also been detected in a number of rat tissues, including the prostate gland, uterus, pancreas and brain [3].

Although some groups have examined the factors that regulate relaxin hormone secretion, very little is known of the molecular mechanisms regulating the expression of the relaxin gene. There is, however, some indirect evidence suggesting that changes in relaxin peptide levels are regulated at the level of relaxin mRNA. For example, in women the level of relaxin mRNA is directly correlated with the intensity of relaxin immunostaining in luteal cells [4]. In rat, there is a strong correlation between the level of relaxin mRNA and the concentration of immunoreactive relaxin in ovaries during pregnancy and in the estrous cycle [1]. The observed correlations between relaxin mRNA and protein levels imply that relaxin gene expression is regulated, at least in part, at the level of transcription. In order to elucidate the molecular mechanisms regulating the expression of the rat relaxin gene, we have cloned the promoter region of the gene and analysed nuclear protein binding sites in the region.

Results and Discussion

The sequence of rat relaxin promoter has low homology with those of the porcine and human H1 and H2 relaxin promoters. However, six DNA regions conserved (CR) in all four promoters were identified (Table 1). Some of these regions contain binding sites for known transcription factors. To analyse the conserved regions that might be involved in the transcriptional regulation of the rat relaxin gene, *in vitro* protein binding assays (footprinting) were carried out using the ^{32}P-end labelled promoter sequence of rat relaxin. Nuclear protein extracts from the relaxin mRNA-rich tissue (ovaries of 18-20 day pregnant rats), mRNA- intermediated tissue (ovaries of non-pregnant rats), and mRNA-poor tissues (liver and kidney) were examined. Multiple binding sites for nuclear proteins were found within the promoter, including the CR II, III, IV and V regions.

G.W. Tregear et al. (eds.), Relaxin 2000, 149–150.
© 2001 *Kluwer Academic Publishers. Printed in the Netherlands.*

Table 1 Analysis of conserved regions in relaxin (RLX) promoters

Species	Sequences	Name/Location	EMSA results
rRLX pRLX h1RLX h2RLX	TATAAAAG TATAAAAG TACTAAAG TACAAAAG	CR I /-23 bp -27 bp -29 bp -29 bp	The CR I contains a Goldberg-Hogness TATA box element, which has been found to bind transcription factor TFIID.
rRLX pRLX h1RLX h2RLX	GGAAGTGGGTGGAGCT GGAAAGAGGTGGGGCT GGAAAGGGGTGGAGCT GGAAAGGGGTGGAGCT	CR II /-76 bp -82 bp -79 bp -85 bp	A extra slower migrating complex was observed with ovarian extracts from pregnant rats, which was not observed with extracts from normal ovary or liver.
rRLX pRLX h1RLX h2RLX	AGAAGAACGAA AAAAGGAGGAA AAATGAAGGAA AAATGAAGGAA	CR III /-139 bp -154 bp -142 bp -147 bp	A single major complex band was observed with both pregnant and non-pregnant ovary extracts, as well as liver extracts, but not with kidney extracts
rRLX pRLX h1RLX h2RLX	AGCATACAAGGTGGTTTGAGGG AGTATACAAAATTGTCTGAGAC AGTATACAAAGTTGTTTCAGGG AGTATACTAAATTGTTTGAGGG	CR IV /-190 bp -229 bp -218 bp -221 bp	The CR IV contains a sequence similar to the cPRE element. A single major complex band was observed with both ovary and liver nuclear extracts
rRLX pRLX h1RLX h2RLX	TTGTAAGAAGAGAGGTGGTCTCGACTTT TTGTGAGAATATGGATGATTTCAATTTT TTGTAAGAATATTGGCGATCTCAATTTT TTGTAAGAATATGGGTGATCTCAAATTT	CR V /-293 bp -330 bp -297 bp -300 bp	A single complex band was observed with the liver extract, while three bands were seen with ovarian extracts from . pregnant and non-pregnant rats.
rRLX pRLX h1RLX h2RLX	TCAACAGTTTA TCAACAATATA TCAACAATTTA TCAACAATTTA	CR VI /-412 bp -427 bp -439 bp -451 bp	Not yet determined by EMSA.

Binding of nuclear protein to the conserved regions was further demonstrated in electrophoretic mobility shift assay (EMSA) using the ^{32}P-end labelled double-stranded oligonucleotide probes corresponding to the CR sequences. The observed patterns of protein-DNA interactions differed with various sequence elements, as well as with various tissue extracts (Table 1). The results suggest that proteins binding to these sequences may play an important role in regulating the transcription of the rat relaxin gene *in vivo*. Functional evaluation of the conserved elements and investigation of their role in the regulation of relaxin gene expression remains to be performed.

Acknowledgments

This work was carried out at the Howard Florey Institute and was supported by an institute block grant (reg key 983001) from the NHMRC of Australia.

References

1. Sherwood, O.D., In Knobil, E. and Neill, J.D. (Eds.), The Physiology of Reproduction. 2nd ed., Raven Press, New York, USA, 1994, p. 861.
2. Crish, J.F., Soloff, M.S. and Shaw, A.R. Endocrinology, 119 (1986) 1222.
3. Gunnersen, J.M., Roche, P.J., Tregear, G.W. and Crawford, R.J., J. Mol. Endocrinol., 15 (1995) 153.
4. Stoelk, E.M., Chegini, N., Lei, Z.M., Rao, C.V., Bryant-Greenwood, G. and Sanfilippa, J., Biol. Reprod., 44 (1991) 1140.

SECTION 5

Cardiovascular physiology of relaxin

Effects of relaxin on experimentally-induced myocardial infarction. Is this hormone a natural shield against cardiovascular ischemic disease?

Mario Bigazzi[1], Daniele Bani[2] and Tatiana Bani Sacchi[2]

[1]Prosperius Institute, V.le F.lli Rosselli 60-62, I-50144 Florence, Italy and [2]Department of Anatomy, Histology and Forensic Medicine, University of Florence, V.le G.Pieraccini 6, I-50139 Florence, Italy.

Introduction

Cardiovascular disease (CVD) is one of the most common causes of death in women in most developed countries [1,2]. In 1997, CVD accounted for 43% of all deaths in women in the USA [3]. It is well known that women, similarly to men, show parallel increments of CVD with age, but with a clear-cut delay of approximately two decades, roughly corresponding to the span of their fertile life. A very recent longitudinal study of 85,941 nurses in the USA clearly shows an age-dependent difference in the incidence of coronary heart disease, with an incidence of 25 per 100,000 person-years for subjects aging less than 50 years versus 100, 180 and 250 per 100,000 person-years for the groups aging 51-54, 55-59 and 60-64 years respectively [4]. Hence, in women, the time of menopause seems to be the dividing line between two different attitudes to become prone to CVD.

The above data indicate that female sex hormones could be the protective agents against CVD. Hence, postmenopausal hormone replacement therapy seemed to be the logical solution for CVD prevention. Indeed, supplementation with estrogen or with estrogen plus progestins has been widely applied [5] but recent clinical studies have shown no beneficial effects of unopposed estrogen or estrogen/progestin replacement therapy on the risk of myocardial infarction and death by coronary heart disease among women with established coronary atherosclerosis [6] nor did they show any decline in the progression of coronary atherosclerotic lesions [7]. Rather, administration of estrogen and progestins as contraceptive drugs has been shown to increase the risk of thrombosis and CVD [8]. These results have lead researchers to conclude that estrogen and/or progesterone are not the actual protecting agents against CVD in women during fertile life.

Based on these grounds, one is prompted to ask why two apparently opposite phenomena, i.e. the absence of sex hormones due to menopause and the presence of sex hormones due to assumption of the pill, both result in a loss of the natural protection of women against CVD. The explanation of this paradox may be the physiological or drug-induced abrogation of ovulation and corpus luteum (CL) formation. In fact, the CL not only produces ovarian steroids but also relaxin. In this article, we will explore the possibility that relaxin may be the physiological agent responsible for the protection against CVD in women. This possibility raises from the results of our experimental

G.W. Tregear et al. (eds.), Relaxin 2000, 153–157.
© 2001 *Kluwer Academic Publishers. Printed in the Netherlands.*

studies on the effect of relaxin in animal models, which we will summarize as follows.

Since our previous observations made in the '80s, we noticed that relaxin, given alone, without any estrogen priming, had marked vasodilatory properties in some organs, such as the rat mesentery, the mouse mammary gland, and the pigeon crop sac [9-11], and caused a pronounced reduction in clotting time in rats [12]. Furthermore, in the mouse mammary gland [9] we found that the relaxin-induced vasodilation was independent of mast cell degranulation and histamine release, thus suggesting that the relaxin-induced microvessel dilation was exerted through another mechanism, possibly intrinsic in the vessels, which remained unknown at that time.

Based on these grounds we decided to investigate systematically the different components of the cardiovascular system, with the aim to answering the following principal questions:

- Has relaxin a vasodilatory action *per se* and which is the mechanism underlying this action?
- Does relaxin have a direct action on hemostasis?

Results and Discussion

The experiments carried out in our laboratory in the last decade have given the following results:

1. Relaxin increases coronary flow in isolated and perfused rat and guinea-pig hearts in a concentration-dependent fashion [13], an effect clearly due to a reduction of coronary vessel tone. This effect is mediated through a stimulation of the endogenous production of nitric oxide (NO), a well known vasodilatory agent. The results of this study established the action of relaxin on the heart vessels and clarified the mechanism underlying its vasodilatory action.
2. Relaxin increases the intracellular levels of cGMP, which is the mediator of the cell response to NO, in endothelial cells from the umbilical cord in *in vitro* culture. Moreover, relaxin dilates blood capillaries, whose wall is made up solely of endothelium, in mammary gland [14], pigeon crop sac [15], uterus [16], and liver [17]. The results of these studies show that relaxin is effective on the endothelium.
3. Relaxin, applied to bovine artery smooth muscle cells in *in vitro* culture, increases cGMP, attenuates the vasoconstrictor-induced rise in cytosolic calcium, and induces changes in the actin cytoskeleton consistent with cell relaxation. These effects are mediated through a stimulation of NO production due to induction of the synthesis of type II NO synthase [18]. The results of this study strongly suggest an action of relaxin on the cells of the tunica media of blood vessels specifically involved in the control of the vascular tone. Recently, we observed that intracavernous administration of relaxin to men promptly increases blood flow in the penile artery [19], thus confirming the *in vivo* vasodilatory action of relaxin on vessels possessing a smooth muscle cell coat.
4. Relaxin, added to human and rabbit platelets before stimulation with proaggregants, causes a concentration-dependent inhibition of platelet aggregation via a simulation of NO production [20]. Moreover, relaxin administered systemically to rats, causes a

reduction in the number of circulating platelets and induces ultrastructural changes of spleen megacaryocytes consistent with a depression of thrombocytopoiesis [21]. These effects, together with the observation that relaxin induces a decrease in thromboxane A2 production by platelets and a reduction of plasma fibrinogen, tPA and PAI-1 [19], speak in favour of an inhibitory action of relaxin on platelet aggregation and clot formation.

5. Relaxin has beneficial effects on ischemia-reperfusion-induced myocardial injury in two different animal models, i.e. the isolated, perfused guinea pig heart [22] and the rat heart *in vivo* [23], both subjected to transient ligature of the left anterior coronary artery. Relaxin substantially reduces the extension of the injured myocardium as well as the occurrence of severe ventricular arrhythmias, the latter effect probably depending on its ability to inhibit the release of histamine [22], a well-known arrhythmogenic substance [24]. Moreover, relaxin markedly reduces neutrophil immigration into the damaged cardiac tissue and decreases several morphological and biochemical markers of myocardial cell injury (Fig. 1). Because of the ability of relaxin to dilate coronary vessels by a NO-mediated mechanism, it is likely that in the ischemic-reperfused heart in vivo, in which NO production is also stimulated by relaxin, the patency of collateral blood vessels directed to the ischemic tissue will be favored. Consequently, there is a reduction in the degree of ischemia to the myocardial tissue downstream of the coronary artery ligature and, ultimately, prevention of myocardial damage severity upon reperfusion.

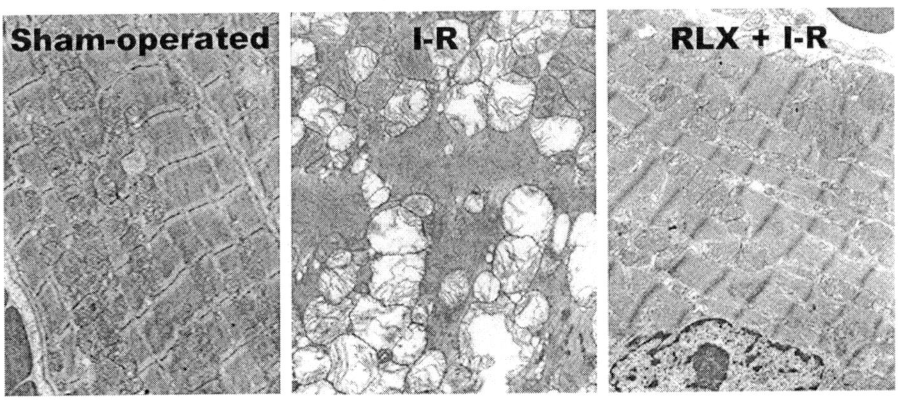

Fig. 1. Representative electron micrographs of myocardial tissue from sham-operated control rats (left), rats subjected to ischemia-reperfusion (centre) and rats given relaxin (RLX) systemically (100 ng i.v.) 30 minutes before being subjected to ischemia-reperfusion (right). Administration of relaxin dramatically decreases the severity of myocardial cell injury (myofibril disarrangement, mitochondrial disruption) caused by ischemia-reperfusion.

Conclusions

Taken together, the above experimental findings demonstrate that relaxin has powerful vasodilatory and anticoagulant properties, supported by its ability to activate the L-arginine-NO pathway. Of note, the coronary system is particularly responsive to the vasoactive action of relaxin, which is responsible for the protection against ischemia-reperfusion-induced myocardial injury. The results obtained by us in experimental myocardial infarction [22,23], together with the observation that the incidence of coronary heart disease is very low in cycling and pregnant women [25] when relaxin production occurs, and becomes markedly elevated after menopause [26], upon cessation of ovarian endocrine function, strongly suggest that relaxin may be the natural shield against cardiovascular ischemic disease in women during their fertile life.

References

1. Wenger, N.K., Speroff, L. and Packard, B., New Engl. J. Med., 329 (1993) 247.
2. Sempos, C.T., Cleeman, J.I., Carroll, M.D., Johnson, C.L., Bachorik, P.S., Gordon, D.J., Burt, V.L., Briefel, R.R., Brown, C.D., Lippel, K., et al., J.A.M.A., 269 (1993) 3009.
3. Hoyert, D.L., Kochanek, K.D. and Murphy, S.L., Natl. Vital Stat. Rep., 47 (1999) 1.
4. Hu, F.B., Stampfer, M.J., Manson, J.E., Grodstein, F., Colditz, G.A., Speizer, F.E. and Willett, W.C., N. Engl. J. Med., 343 (2000) 530.
5. Grodstein, F. and Stampfer, M.J., In Adashi, E.Y., Rock, J.A. and Rosenwaks, Z. (eds.), Reproductive Endocrinology, Surgery and Technology, Vol. 2, Lippincott-Raven, Philadelphia, 1996, p. 1852.
6. Hulley, S., Grady, D. Bush, T., Furberg, C., Herrington, D., Riggs, B. and Vittinghoff, E., J.A.M.A., 280 (1998) 605
7. Herrington, D.M., Reboussin, D.M., Brosnihan, K.B., Sharp, P.C., Shumaker, S.A., Snyder, T.E., Furberg, C.D., Kowalchuk, G.J., Stuckey, T.D., Rogers, W.J., Givens, D.H. and Waters, D., N. Engl. J. Med., 343 (2000) 522.
8. Bonnar, J., Am. J. Obstet. Gynecol., 157 (1987) 1042.
9. Bigazzi,M., Del Mese, A., Petrucci, F., Casali, R. and Novelli, G.P., Acta Endocrinol., 112 (1986) 296.
10. Bani,G., Bani Sacchi, T., Bigazzi,M. and Bianchi, S., Histol. Histopath., 3 (1988) 337.
11. Bigazzi, M., Bani, G., Bani Sacchi,T., Petrucci, F. and Bianchi, S., Acta Endocrinol., 117 (1988) 181.
12. Bigazzi, M. and Bryant-Greenwood G.D., Proc.7th Int. Congr. Endocrinol., Quebec, Canada, July 1-7, 1984.
13. Bani Sacchi, T., Bigazzi, M., Bani, D., Mannaioni, P.F. and Masini, E., Br. J. Pharmacol., 116 (1995)1589.
14. Bani, G., Bani Sacchi, T., Bigazzi, M. and Bianchi, S., Histol. Histopath., 3 (1988) 337.
15. Bianchi, S., Petrucci, F. and Bani, G., Z. Mikrosk.-Anat. Forsch., 102 (1988) 221.
16. Bani, G., Maurizi, M., Bigazzi, M. and Bani Sacchi, T. Biol. Reprod., 53 (1995) 253.
17. Nistri, S. and Bani Sacchi, T., It. J. Anat. Embryol., 104 (1999) 45 (abstr.)
18. Bani, D., Failli, P., Bello, M.G. Thiemermann, C., Bani Sacchi, T., Bigazzi, M. and Masini, E., Hypertension, 31 (1998) 1240.

19. Bigazzi, M., Bani, D., Bani, G. and Bani Sacchi, T., In MacLennan, A.H., Tregear, G. and Bryant-Greenwood, G.D. (eds.) Progress in Relaxin Research, World Scientific Publishing, Singapore, 1995, p. 499.
20. Bani, D., Bigazzi, M., Masini, E., Bani, G. and Bani Sacchi, T., Lab. Invest., 73 (1995) 709.
21. Bani, D., Maurizi, M. and Bigazzi, M., Platelets, 6 (1995) 330.
22. Masini, E., Bani, D., Bello, M.G., Bigazzi, M., Mannaioni, P.F. and Bani Sacchi, T., Endocrinology, 138 (1997) 4713.
23. Bani, D., Masini, E., Bello, M.G., Bigazzi, M. and Bani Sacchi, T., Am. J. Pathol., 152 (1998) 1367.
24. Wolff, A.A, and Levi, R., Circ. Res., 58 (1986) 1.
25. Kannel, W.B. and Abbott, R.D. In Eaker, E.D., Packard, B., Wenger, N.K., Clarkson, T.B. and Tyroler, H.A. (eds.), Coronary heart disease in women, Haymarket-Doyma, New York, USA, 1987, p.208.
26. Manson, J.E., Tosteson, H., Ridker, P.M., Satterfield, S., Hebert, P., O'Connor, G.T., Buring, J.E. and Hennekens, C.H., N. Engl. J. Med., 326 (1992) 1406.

Systemic relaxin administration stimulates angiogenic cytokine expression and vessel formation in a rat myocardial infarct model

Martyn Lewis[1], Usha Deshpande[1], Lorenzo Guzman[3], Beverly H Grove[2], Xinfan Huang[1], Mark E Erikson[2], Lesley B Pickford[3*] and Elaine N Unemori[1]

[1]Departments of Cell Biology, [2]Assay Development, and [3]Pharmacology and Toxicology[†], Connetics Corp., Palo Alto, CA.

[‡] Current address: Coulter Pharmaceuticals Inc., South San Francisco, CA.

Introduction

The myocardium consists of three integrated components: myocytes, extracellular matrix, and a capillary microcirculation. Myocardial injury following vessel occlusion results in the formation of an ischemic zone whose size and position in the myocardium is determined by the anatomical region over which the capillary bed, supplied by the occluded vessel, lies [1]. Ischemia results in the onset of myocyte necrosis and generates an inflammatory response driven by infiltrating neutrophils and macrophages [2,3]. Neutrophils initiate local degradation of matrix at infarct sites by secretion of metalloproteases (MMPs) [4,5]. In an attempt to initiate repair, macrophages and dying myocytes then release TGFβ1 and vascular endothelial growth factor (VEGF) [6]. An early increase in levels of TGFβ1 in the infarct zone results in macrophage and fibroblast chemotaxis and fibroblast proliferation [2,7-9]. These ischemic-induced changes in turn increase VEGF and basic fibroblast growth factor (bFGF) levels [10]. Concomitantly, stress-induced changes to endothelium results in the expression of monocyte chemoattractant protein-1 (MCP-1) and endothelial integrins, which facilitate increased macrophage infiltration [2,11]. The early increase in MMPs, and subsequent increases in VEGF and bFGF levels, are essential requirements for the onset of neoangiogenesis of microvascular endothelial cells in the myocardium.

Traditionally, therapeutic approaches to the management of myocardial ischemia have involved reduction of myocardial oxygen demand or surgical restoration of blood flow by angioplasty or bypass graft. More recently, therapeutic angiogenesis has been designed to promote the development of supplemental collateral vessels as a means of preserving heart function following an ischemic event [12,13]. Collateral vessel formation, increased blood flow, and preservation of aspects of heart function have been demonstrated following local administration of either VEGF or bFGF in animal models of chronic myocardial ischemia [14,15]. Consequently, these molecules are currently being evaluated as therapeutics for inducing vessel development in patients with chronic myocardial ischemia [12,13]. The provision of collateral blood flow to the myocardium

G.W. Tregear et al. (eds.), Relaxin 2000, 159–167.
© 2001 *Kluwer Academic Publishers. Printed in the Netherlands.*

and the subendocardial layers of the heart [16] may mitigate the extent of ischemic myocardial injury caused by coronary vessel occlusion. Animal studies have determined that myocardial necrosis following an acute ischemic event is not inevitable as long as blood flow insufficiency is transient and short-lived [17], with myocardial viability being directly correlated with the availability of collateral circulation [18-20].

Systemic delivery of VEGF has been associated with hypotension, reduced cardiac output and a broad range of vasculopathies including those seen in the brain, lung, and kidney [21-23]. Consequently, current approaches to induce revascularization and reoxygenation have required direct myocardial or pericardial injection of either viral or plasmid DNA vectors expressing the cDNA for VEGF, or of VEGF or FGF protein directly. These avenues of administration have required invasive surgical procedures, and a non-invasive method of promoting angiogenic growth factor expression and a subsequent increase in collateral vessel development would potentially be of enormous importance.

We recently reported rH2 relaxin is able to induce VEGF expression in endometrial stromal cells in vitro and is associated with heavier than usual menstrual bleeding following systemic administration in women [24]. In addition, systemically administered rH2 relaxin induces angiogenesis at ischemic wound sites in rats, potentially mediated by selective upregulation of the potent angiogenic growth factors, VEGF and bFGF in wound macrophages [25]. These observations led to an examination of rH2 relaxin as a potential mediator of neoangeogenesis in the heart following ischemic myocardial injury. As chronic myocardial infarction (MI) induces a severe wound with profound ischemia, we examined the effect of systemic rH2 relaxin administration upon VEGF and bFGF expression, and upon collateral vessel development in a rat model of chronic myocardial ischemia. We have also evaluated relaxin's effect on primary rat and human cardiac cells in vitro to more closely examine the relationship between rH2 relaxin administration and subsequent changes in VEGF and bFGF expression.

Results and Discussion

Chronic MI was induced in female Sprague-Dawley rats by suture-generated occlusion of the left coronary artery, causing profound, acute myocardial damage involving up to 50% of the left ventricular free wall [26] (sutures, in sham-surgery animals, were placed superficially into the muscle adjacent to the coronary artery). As previous studies demonstrated VEGF mRNA expression in the immediate margins surrounding the infarct [27], we sampled this area to assess expression of the $VEGF_{164}$ and $VEGF_{120}$ isoforms, and bFGF. Quantitative reverse-transcriptase-polymerase chain reaction (RT-PCR) analysis demonstrated that mRNA levels of bFGF increase significantly at 7 and 21 days following infarction, as compared to expression in sham hearts. Systemic rH2 relaxin administration to animals with myocardial infarcts caused a significant 2-fold increase in peri-infarct bFGF expression at both 7 and 21 days, compared to peri-infarct expression in vehicle-treated rats ($p<0.05$)(Fig. 1).

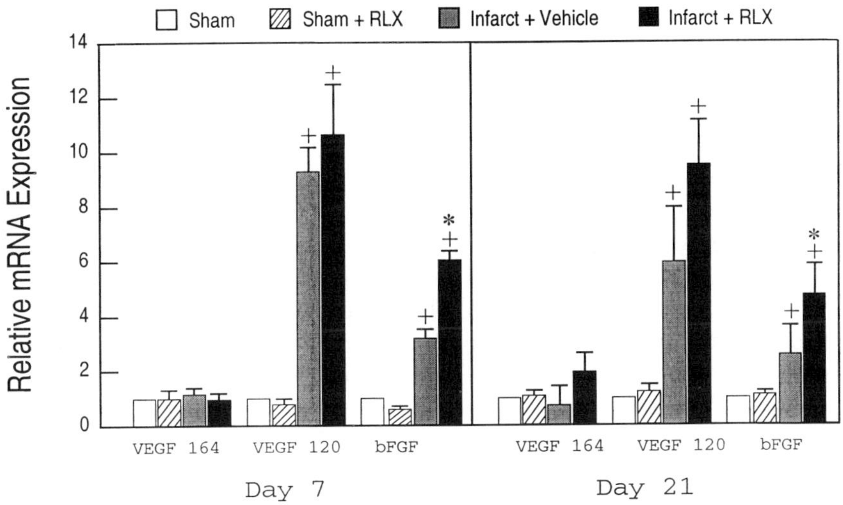

Fig. 1. Quantitative RT-PCR of rat $VEGF_{164}$ and $VEGF_{120}$ isoforms and rat bFGF expression in the heart following infarction and vehicle or rH2 relaxin administration (RLX; 0.1 mg/kg/day) for either 7 or 21 days. Expression levels are relative to glyceraldehyde 3-phosphatase dehydrogenase (GAPDH) transcript levels (not shown). Data are means ± SEM; n = 3; p<0.05 compared to vehicle-treated group; p<0.05 compared to shams.

$VEGF_{120}$ mRNA expression increased dramatically in response to infarction at both day 7 and day 21, in agreement with previously published reports [27]. However, expression of $VEGF_{164}$ mRNA in peri-infarct tissue was not significantly different from the sham group at either time point, and the already high basal level of expression of $VEGF_{164}$ mRNA may suggest a possible role for this isoform in vascular maintenance in the normal heart. While a trend toward a relaxin specific increase, compared to vehicle treated infarcted hearts, was evident by day 21, rH2 relaxin did not significantly alter either $VEGF_{164}$ or $VEGF_{120}$ expression.

The data suggest that the rH2 relaxin-induced expression of angiogenic growth factors is specific for regions of ischemia. Systemic rH2 relaxin administration to sham surgery animals had no effect on VEGF or bFGF mRNA expression in a comparable area of the left ventricle (Fig. 1), nor in the right ventricle of infarct hearts at day 21 (not shown). Furthermore, systemic administration of rH2 relaxin did not induce an alteration in VEGF or bFGF expression in cells harvested from the spleen, bone marrow, peripheral blood (mononuclear cells), or in resident alveolar macrophages (data not shown).

Immunohistochemical analysis of sections from the left ventricle of sham surgery animals indicated a diffuse expression pattern of bFGF in cardiomyocytes and fibroblasts (Fig. 2A). Following infarction, bFGF staining increased specifically in the peri-infarct regions at both 7 and 21 days (Fig. 2B), and was further intensified at both time points following rH2 relaxin treatment (Fig. 2C) with peak staining intensity occurring at day 7. Both cardiomyocytes, identified by colocalization with the protein desmin, and fibroblasts, stained with vimentin, showed enhanced bFGF expression.

VEGF expression was also seen in a homogenous, diffuse staining pattern in cardiomyocytes and fibroblasts in rats with sham surgery (data not shown), similar to staining for bFGF. Following infarction, marked intensification of VEGF staining occurred in the peri-infarct regions at 7 and 21 days in both myocytes and fibroblasts. No differences in intensity of VEGF staining between rH2 relaxin and vehicle-treated hearts were observed. Consistent with previous reports, macrophage-like cells were also present at peri-infarct areas at 7 days, and stained for both VEGF and bFGF, in the presence and absence of rH2 relaxin. We have previously indicated that macrophages, infiltrating into regions of ischemic injury may respond to rH2 relaxin with upregulated expression of VEGF and bFGF [25]. This hypothesis is supported by the presence of macrophage-like cells in the peri-infarct regions of ischemic hearts, and may indicate that inflammatory macrophages within and around the infarct may be one of the cell types responding to rH2 relaxin.

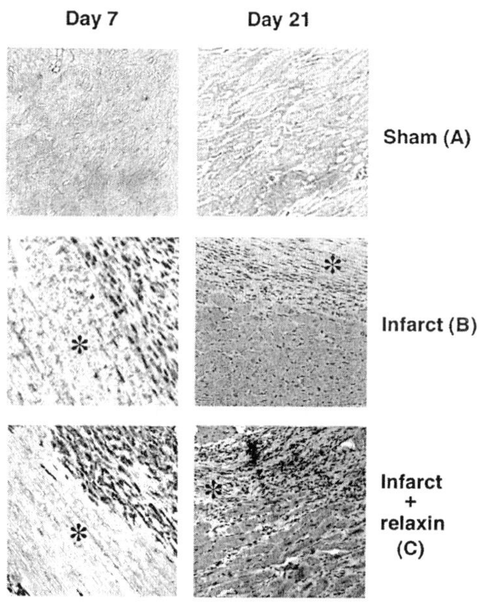

Fig. 2. Immunohistochemical detection of rat bFGF in the left ventricle of sham hearts and in peri-infarct regions. Sections are shown from animals without infarct (sham)(A), with infarct treated with vehicle (B), and infarct treated with rH2 relaxin (C)(* - dead myocardium).

To ascertain whether the rH2 relaxin-induced increase in bFGF expression influenced the number of newly formed collateral vessels, transverse histological sections from 3 levels in each heart, apical, basal, and midway between apical and basal were analyzed. Vessel counts obtained in contiguous fields over the entire infarct and peri-infarct area indicated that systemic administration of rH2 relaxin resulted in a significant increase in the number of thin-walled vessels (no smooth muscle layer) at day 21, with a trend evident at day 7. The number of muscular vessels did not change with ligation or rH2 relaxin administration (data not shown). rH2 relaxin administration did not alter the number of thin-walled vessels in an analogous area of the left ventricle following sham surgery at either day 7 or 21 (Fig. 3). The microcirculation derived from the occluded vessel determines the cellular histology of the infarcted or normal myocardium [28]. The increase in the number of thin-walled vessels would concur with observations indicating that improvement in collateral-dependant flow results from the formation of vessels less than 200 μM in diameter [29]. We then evaluated the effects of rH2 relaxin directly on primary human cardiac cells. Cells were derived from two 19 week-old, and one 20 week-old male fetus and were a mixture of presumptive fibroblasts (85%) and myocytes (15%).

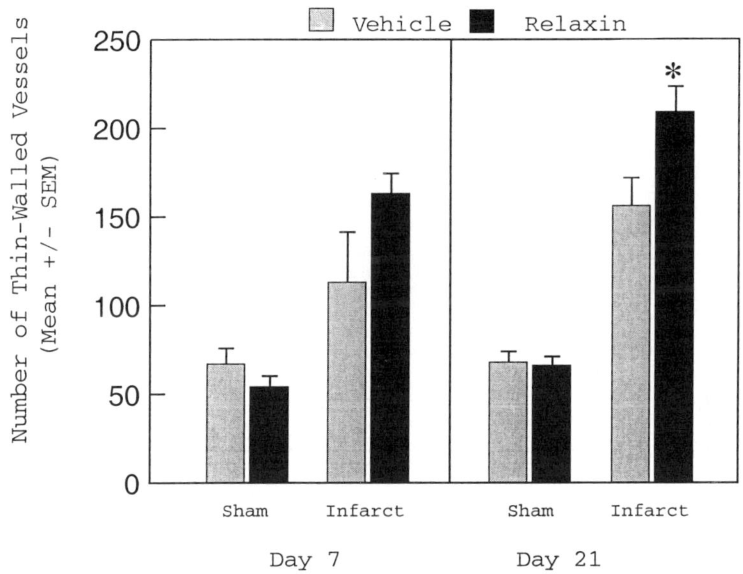

*Fig. 3. Thin-walled vessels in the infarct and infarct-margins, counted in blinded fashion on stained sections of left ventricle, 7 and 21 days post-infarct. Bars = mean ± SEM; *p<0.05 compared to untreated control.*

As observed for rat cardiac fibroblasts (data not shown), the human cardiac cell cultures were also found to respond to rH2 relaxin with significant dose-related increases in levels of secreted VEGF (Fig. 4).

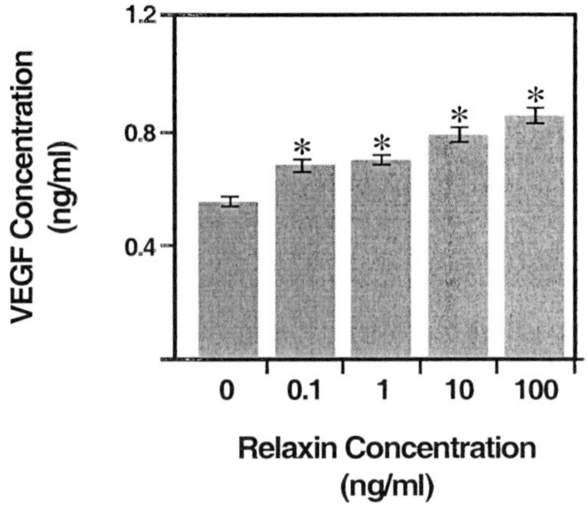

*Fig. 4. Increase of secreted VEGF from human fetal cardiac cell media. Cells were treated with rH2 relaxin (0.1-100 ng/ml) for 24 hr and media was analyzed for human VEGF protein by ELISA. Cultures were treated in triplicate and conditioned media were assayed in quadruplicate. Data are shown as means ± SEM; * p<0.05 compared to untreated control*

As human bFGF expression could not be assayed by ELISA due to the small quantities of this cytokine required in the culture medium, human cardiac cells were analyzed for VEGF and bFGF protein expression by immunocytochemistry (Fig. 5).

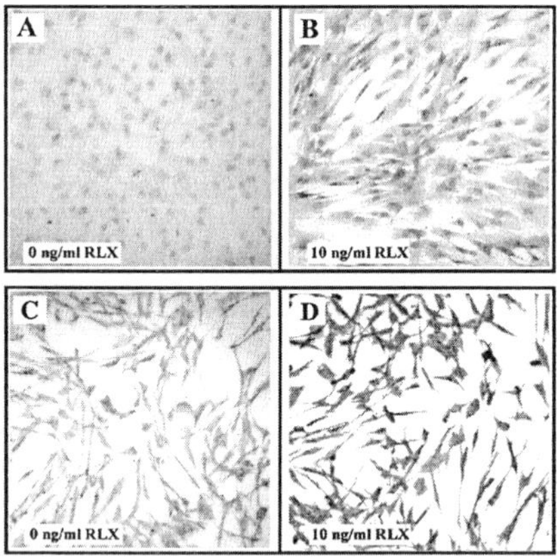

Fig. 5. Immunocytochemical analysis of VEGF and bFGF expression in human cardiac cells. Cells were treated with rH2 relaxin (RLX) at 1-100 ng/ml for 24hr. Untreated (A and C) or rH2 relaxin-treated (10 ng/ml)(B and D) were stained for VEGF (A and B) or bFGF (C and D.)

Cardiac cell monolayers were incubated without or with 1, 10, or 100 ng/ml rH2 relaxin for 24 hr prior to staining with monoclonal antibodies against human VEGF and bFGF. Diffuse and uniform staining was observed in untreated cells stained for both VEGF and bFGF (Fig. 5A and 5C respectively). Following 24 hr treatment with rH2 relaxin at 10 and 100 ng/ml, >90% of cells stained showed more intense cytoplasmic staining for VEGF (Fig. 5B). Nearly 100% of these fetal cells stained more intensely for bFGF expression following treatment with 1, 10, or 100 ng/ml of rH2 relaxin for 24 hr (Fig. 5D).

Human cells were then analyzed to ascertain whether rH2 relaxin could induce an increase in VEGF and bFGF mRNA expression. Total RNA was extracted from human cells following treatment with 1, 10, and 100 ng/ml rH2 relaxin for 24 hr. Quantitative RT-PCR showed rH2 relaxin induced dose-related increases in the expression of transcripts for both VEGF isoforms and bFGF (Fig. 6). $VEGF_{165}$ and $VEGF_{121}$ mRNA showed maximum 2.3-fold and 3-fold increases in expression respectively over untreated controls at the 100 ng/ml dose of rH2 relaxin, respectively ($p<0.05$). Expression of bFGF reached a maximum increase of 2.2-fold over baseline at the 100 ng/ml dose ($p<0.05$). By comparison, when cells were exposed to low oxygen conditions, VEGF expression levels, were increased 3.5-fold over normoxic levels while bFGF expression was unaffected (Fig. 6).

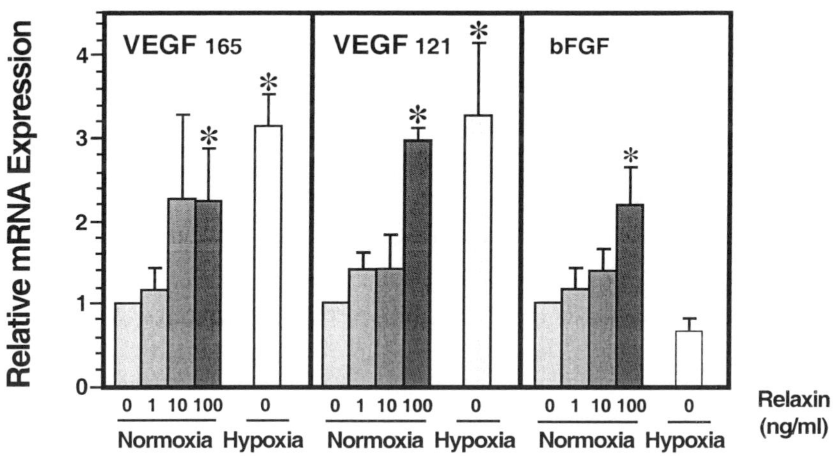

*Fig. 6. Quantitative RT-PCR of the human $VEGF_{165}$ and $VEGF_{121}$ isoform and bFGF expression in human cardiac cells. Cells were cultured for 24 hr either without (under normoxic or hypoxic conditions) or with rH2 relaxin (1-100 ng/ml)(normoxic). VEGF and bFGF expression is graphed relative to GAPDH transcript levels. Data are shown as means ± SEM; n=3; * $p<0.05$, compared to untreated control*

That hypoxia alone does not induce bFGF mRNA expression in human cardiac cells in vitro suggests that an inflammatory component is involved in its upregulation in vivo. Expression of transcripts for human HGF, another potent angiogenic factor [30], was not modulated by rH2 relaxin treatment of these cells (data not shown). rH2 relaxin had no effect on cellular morphology (data not shown).

Our observations suggest that rH2 relaxin may enhance perfusion of ischemic sites by specific VEGF and bFGF induction, thus augmenting collateral vessel formation. The safety profile of systemically administered rH2 relaxin [31], in addition to its ischemic site-specific, angiogenic potential suggest that rH2 relaxin may have a potentially important therapeutic role, not only in ischemic heart disease, but also other peripheral arterial diseases. Clinical trials involving systemic administration of rH2 relaxin [24,32] have not shown accompanying elevations in systemic VEGF levels [24], nor in hypotension [32], as observed following VEGF or bFGF administration [21,33]. A pro-angiogenic, injury site-specific therapy, such as rH2 relaxin, may provide an important alternative to currently available therapies.

Acknowledgements

The authors gratefully acknowledge Edna Malonzano for providing assistance with histochemical analysis of rat heart sections, and Dr. Kenneth Altera for providing histopathology expertise to this project.

References

1. Alexander, R.W., Schlant, R.C. and Fuster, V. (eds.) Hurst's; The heart, arteries and veins. 9th ed., McGraw-Hill, New York, USA, 1998, p.1248.
2. Birdsall, H.H., Green, D.M., Trial J., Youker, K.A., Burns, A.R., MacKay, C.R., LaRosa, G.J., Hawkins, H.K., Smith, C.W., Michael, L.H., Entman, M.L. and Rossen, R.D., Circulation, 95 (1997) 684.
3. Neumann, F-J., Ott, I., Gawaz, M.,Richardt, G., Holzapfel, H., Jochum, M. and Schomig, A., Circulation, 92 (1995) 748.
4. St. John Sutton, M.G., and Sharpe, N., Circulation, 101 (2000) 2981.
5. Romanic A.M., White R.F., Arleth, A.J., Ohlstein, E.H., and Barone, F.C., Stroke, 29 (1998) 1020.
6. Banai, S., Shweiki, D., Chandra, M. and Keshet, E., Circulation, 88(suppl I) (1993) I-542.
7. Battegay, E.J., J. Mol. Med., 73 (1995) 333.
8. Booz, G.W. and Baker, K.M., Cardiovasc. Res., 30 (1995) 537.
9. Chen, M.M., Lam, A., Abraham, J.A., Schreiner, G.F. and Joly, A.H., J. Mol. Cell. Cardiol., 32 (2000) 1805.
10. Stavri, G.T., Zachary, I.C., Baskerville, P.A., Martin, J.F. and Erusalimsky, J.D., Circulation, 92 (1995) 11.
11. Schaper, W. and Ito, W.D., Circ. Res., 79 (1996) 911.
12. Losordo, D.W., Vale P.R., Symes, J.F., Dunnington, C.H., Esakof, D.D., Maysky, M., Ashare, A.B., Lathi, K. and Isner, J.M., Circulation, 98 (1998) 2800.
13. Henry, T.D., B.M.J., 318 (1999) 1536.
14. Unger, E.F., Banai, S., Shou, M., Lazarous, D.F., Jaklitsch, M.T., Scheinowitz, M., Correa, R., Klingbeil, C. and Epstein, S.E., Am. J. Physiol., 266 (1994) H1588.

15.Lopez, J.J., Laham, R.J., Stamler, A., Pearlman, J.D., Bunting, S., Kaplan, A., Carrozza, J.P., Sellke, F.W. and Simons, M., Cardiovasc. Res., 40 (1998) 272.
16.Charney, R.C. and Cohen, M., Am. Heart. J., 126 (1993) 937.
17.Murdock, R.H., Chu, A., Grubb, M. and Cobb, F.R., Am. Phys. Soc., (1985) H783.
18.Rivas, F., Circ. Res., 38 (1976) 439.
19.Cobb, F.R., J.C.I., 57 (1976) 1359.
20.Jennings, R.B. and Reimer, K.A., Art. Cells Blood Subs. and Immob. Biotech., 22 (1994) 253.
21.Tilton, R., Chang, K.C., LeJeune, W.S., Stephen, C.C., Brock, T.A. and Williamson, J.R., Invest. Opthalmol. Vis. Sci., 40 (1999) 689.
22.Yang, R., Bunting, S., Ko, A., Keyt, B.A., Modi, N.B., Zioncheck, T.F., Ferrara, N. and Jin, H., J. Pharmacol. Exp. Ther., 284 (1998) 103.
23.Yang, R., Thomas, G.R., Bunting, S., Ko, A., Ferrara, N., Keyt, B., Ross, J. and Jin H., J. Cardiovasc. Pharmacol., 27 (1996) 838.
24.Unemori, E., Erikson, M.E., Rocco, S.E., Sutherland, K.M., Parsell, D.A., Mak, J. and Grove B.H., Hum. Reprod., 14 (1999) 800.
25. Unemori, E.N., Lewis, M., Constant, J., Arnold, G., Grove, B.H., Normand, J., Deshpande, U., Salles, A., Pickford, L.B., Erikson, M.E., Hunt, T.K. and Huang, X., Wound Rep. Regen., 8 (2000) 361.
26.Selye, H., Bajusz, E., Grasso, S.. and Mendell, P., Angiology, 11 (1960) 398.
27.Li, J., Brown, L.F., Hibberd, M.G., Grossman, J.D., Morgan, J.P. and Simons M., Am. J. Physiol., 270 (1996) H1803.
28.Factor, S.M., Okun, E.M. and Kirk, E.S., Circ. Res., 48 (1981) 640
29.White, F.C., Carroll, S.M., Magnet, A. and Bloor, C.M., Circ. Res., 71 (1992) 1490.
30.Morishita, R., Nakamura, S., Hayashi, S., Moriguchi, A., Nagano, T., Taiji, M., Noguchi, H., Takeshita, S., Matsumoto, K., Nakamura, T., Higaki, J. and Ogihara T., Hypertension, 33 (1999) 1379.
31.Seibold, J.R., Clements, P.J., Furst, D.E., Mayes, M.D., McCloske, D.A., Moreland, L.W., White, B., Wigley, F.M., Rocco, S., Erikson, M., Hannigan, J.F., Sanders, M.E. and Amento E.P., J. Rheumatol., 25 (1998) 302.
32. Seibold, J.R., Korn, J.H., Simms, R., Clements, P.J., Moreland, L.W., Mayes, M.D., Furst, D.E., Rothfield, N., Steen, V., Weisman, M., Collier, D., Wigley, F.M., Merkel, P.A., Csuka, M.E., Hsu, V., Rocco, S., Erikson, M., Hannigan, J., Harkonen, W.S., Sanders, M.E., Ann. Intern. Med., 132 (2000) 871.
32.Unger E.F., Gonvales, L., Epstein, S.E., Chew, E.Y., Trapnell,C.B., Cannon, R.O. and Quyyumi, A.A., Am. J. Cardiol., 85 (2000) 1414.

Is relaxin the "elusive" vasodilator of pregnancy?

Kirk P Conrad[1,2], Lee A Danielson[3], Jacquelien N Novak[1], Laurie J Kerchner[1], Robin E Gandley[1], O David Sherwood[4], Rolando J Ramirez[1], and Pamela A Moalli[1]

[1]Department of Obstetrics, Gynecology and Reproductive Sciences and Magee-Womens Research Institute, and [2]Department of Cell Biology and Physiology, University of Pittsburgh School of Medicine, Pittsburgh, PA 15213 USA; [3]Department of Pathology, University of New Mexico School of Medicine, Albuquerque, NM 87131 USA; [4]Department of Molecular and Integrative Physiology and College of Medicine, University of Illinois at Urbana-Champaign, Urbana, IL 61801 USA

Introduction

Among the most striking changes seen in biology are those associated with pregnancy. In particular, profound vasodilation of *nonreproductive* organs epitomizes the maternal cardiovascular adaptation to *early* gestation in women. Cardiac output, renal blood flow and glomerular filtration rate rise by 40 to 80 %, while vascular resistances plummet and blood pressure declines modestly [reviewed in ref. 1]. Global arterial compliance markedly increases, thus preserving the efficiency of ventricular and arterial coupling, as well as preventing excessive decline of diastolic blood pressure in the face of the marked fall in systemic vascular resistance [SVR; 2]. All of these alterations begin immediately after conception, peak by the end of the first or beginning of the second trimester, and persist throughout most of gestation. It is likely that the circulations of nonreproductive organs such as the kidney serve as "arteriovenous" shunts during early gestation, thereby reducing ventricular afterload which initiates the enormous increase in cardiac output, and subsequently, the expansion of plasma volume - maternal adaptations associated with healthy pregnancies. Furthermore, pressor response to administration of angiotensin II and vascular reactivity to infusion of norepinephrine become attenuated [1]. Insight into the mechanisms responsible for these vasodilatory phenomena may be particularly critical, because in preeclampsia, the attenuation of pressor responsiveness to angiotensin II, the reduced vascular reactivity to norepinephrine, as well as the systemic and renal vasodilation are compromised [1].

The gravid rat model

Current understanding of the vasodilatory mechanisms of normal pregnancy has been facilitated by the use of animal models. In this regard, chronically instrumented, conscious pregnant rats manifest renal, cardiovascular and endocrine changes remarkably similar to those of gravid women. For example, both effective renal plasma flow (ERPF) and glomerular filtration rate (GFR) increase while effective renal vascular resistance (ERVR) declines [3]; cardiac output rises while systemic vascular resistance falls [4]; and pressor responses to exogenous vasoconstrictors diminish [5-8].

G.W. Tregear et al. (eds.), Relaxin 2000, 169–176.
© 2001 Kluwer Academic Publishers. Printed in the Netherlands.

Furthermore, small arteries from gravid rats show diminished vasoconstrictor and myogenic reactivity *in vitro* [e.g., 9-11]. Therefore, the gravid rat has proven to be an excellent animal model in which to explore potential mechanisms for maternal circulatory changes of pregnancy. Other advantages include the relatively short gestational period (~ 22 days), and a hemochorial placentation which resembles that of women.

The molecular basis for renal circulatory adaptations to pregnancy involves the endothelin B receptor subtype and nitric oxide

We tested and rejected the hypothesis that vasodilatory prostaglandins are responsible for the vasodilatory phenomena of pregnancy in rats [6,12]. A more recent study in women reached a similar conclusion [13]. We then proposed that another endothelial factor, endothelium-derived g factor (EDRF), may contribute. The hypothesis was based on our findings that urinary excretion, plasma level and "metabolic production" of cyclic guanosine-3',5'-monophosphate (cGMP), a second messenger of EDRF, are increased in rat gestation [14,15]. (An earlier report [16] showed increased urinary excretion of cGMP in gravid women, although at the time its potential significance was unknown as cGMP was still thought by many to be a mediator of vascular smooth muscle contraction!) Our field of research, as well as many others, received a tremendous boost when EDRF was verified to be nitric oxide [NO;17,18]. Several lines of evidence were then gathered demonstrating increased biosynthesis of NO during pregnancy in rats although the tissue source(s) and NOS isoform(s) were not identified [19].

Because gravid rats manifest renal vasodilation and hyperfiltration [3], as well as increased biosynthesis of NO [19], we next asked whether NO mediates these renal changes of pregnancy. Five lines of evidence supported this hypothesis [5,9]. (1) The NOS inhibitor, L-NAME, equalizes ERPF and ERVR, as well as GFR in gravid and virgin rats. (2) Identical results are observed with a structurally different NOS inhibitor, N^G-monomethyl-L-arginine (NMA). (3) After infusion of L-NAME *in vivo*, aortic and renal cGMP content are reduced to comparable levels in virgin and gravid rats *ex vivo* indicating that comparable inhibition of NO synthase was reached during infusion of L-NAME in the two groups of animals. (4) Infusion of angiotensin II magnifies the differences in ERPF and GFR between virgin and gravid rats demonstrating that the convergence of these parameters in the two groups of animals following NOS blockade is not merely a consequence of non-specific renal vasoconstriction. (5) Reduced myogenic reactivity of small renal arteries isolated from gravid rats is abrogated by NOS blockade.

Next, we began investigating the molecular mechanisms and proximal signals underlying nitric oxide-mediated renal vasodilation and hyperfiltration of pregnancy. By using several experimental approaches, we found that endothelin (ET) via the ET_B receptor subtype on the endothelium mediates the NO-dependent renal vasodilation and hyperfiltration in pregnant rats [20], as well as the NO-dependent reductions in myogenic reactivity of small renal arteries from gravid rats *in vitro* [9] (Fig. 1). Further dissection of the molecular mechanisms of this vascular ET/NO signaling pathway in pregnancy is currently underway in our laboratory.

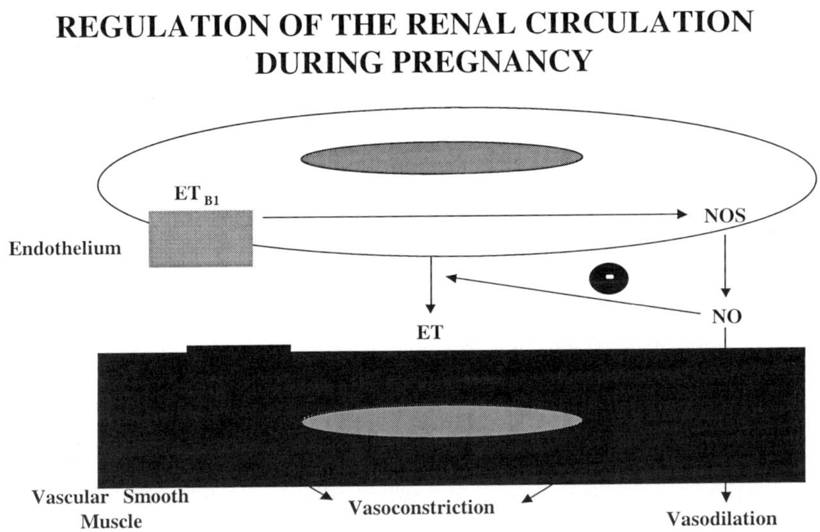

Fig. 1. In the nonpregnant condition, the low vascular resistance of the renal circulation is maintained by endogenous endothelin, activation of the ET_B receptor subtype on the endothelium, and subsequent production of nitric oxide. This molecular pathway is upregulated by pregnancy, thereby producing gestational renal vasodilation and hyperfiltration, as well as reduction in myogenic reactivity of small renal arteries [9,20].

What are the hormonal signal(s) initiating maternal circulatory adaptations to pregnancy: a critical role for the ovarian hormone, relaxin

To come full circle, we asked which "pregnancy hormone(s)" lies upstream of the ET/NO vasodilatory pathway in the renal circulation. Although 17β-estradiol (E) has been traditionally considered as the uterine and systemic vasodilator of pregnancy [21,22], this hormone has little, if any effect on the renal circulation [23-25], which markedly vasodilates so early in pregnancy. Furthermore, neither E nor progesterone (P) reduces blood pressure [26,27], plasma osmolality (reviewed in ref. 28; another early pregnancy adaptation at least in women and rats), or attenuates pressor responses to exogenous vasoconstrictors [26,27] in conscious, ovariectomized rats. Nor does administration of E and/or P to ovariectomized rats increase 24-h urinary excretion of cGMP or NOx (K.P.C., unpublished). Thus, we considered the potential contribution of other hormones besides E with respect to maternal vasodilation during pregnancy. Although progesterone may have some capacity to vasodilate the renal circulation [23,29,30], the pregnancy *protein* hormones also deserve consideration.

In this regard, we hypothesized that relaxin contributes. First, the plasma concentration of relaxin increases rapidly after conception in women (reviewed in ref. 31) corresponding with the marked increase in both GFR and ERPF during the first trimester [1]. A stimulus for secretion of relaxin by the corpus luteum is human chorionic gonadotrophin (hCG) which rises in parallel [31]. Second, circulating relaxin also increases during the luteal phase of the menstrual cycle [31-34] coinciding with a transient, 20% increase in GFR and ERPF [35-40]. Third, the gestational rise in relaxin corresponds with another early physiological adaptation to human pregnancy, namely, osmoregulatory changes [41]. These osmoregulatory changes are mimicked by administering hCG to women in the luteal phase of the menstrual cycle and intact female rats, but not to men or ovariectomized rats suggesting the intermediary role of an ovarian hormone [28,42,43]. Associated with the osmoregulatory changes observed in women after administration of hCG is a 15-20% rise in GFR, which again is not observed in men administered the hormone (J.M. Davison, personal communication). More recently, chronic administration of recombinant human (rH2) relaxin to ovariectomized rats was shown to reduce plasma osmolality and the osmotic threshold for arginine vasopressin release comparable to the gravid condition [44]. Although renal vasodilation and hyperfiltration are observed in gravid rats as early as day 5 of gestation (as well as in pseudopregnancy [45,46]) before detectable increases in ovarian and plasma relaxin occur, there is a marked jump in GFR and ERPF between gestational days 8 and 12 [3], when ovarian and circulating relaxin surge [31]. Conceivably, the early increase in GFR and ERPF in pregnant rats and the comparable renal changes in the pseudopregnant animals may stem from circulating relaxin which is below the sensitivity of assay detection. Further rationale for the proposed role of relaxin in the renal circulatory changes of pregnancy is that chronic administration of the hormone reduces blood pressure and vasoconstrictor responses in the mesenteric circulation of spontaneously hypertensive rats [47,48], while acute treatment increases coronary blood flow and reduces platelet aggregation via nitric oxide and cGMP [49,50].

Based on these pieces of indirect evidence, we investigated the role of relaxin in the renal circulation, and showed that chronic administration of the hormone produces potent renal vasodilation and hyperfiltration, as well as osmoregulatory changes in conscious female rats [51] (Fig. 2).

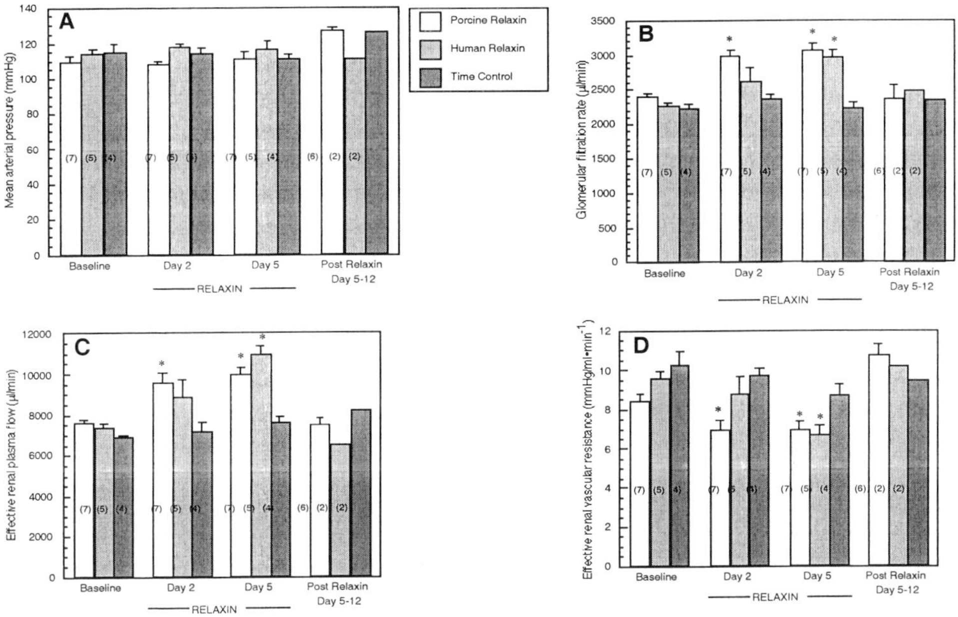

*Fig. 2. Effect of long-term infusion of purified porcine relaxin, rH2 (human) relaxin, or vehicle on mean arterial pressure and renal function. The number in parentheses depict the number of rats tested. *p < 0.05 vs baseline and vehicle. Reprinted with permission from ref. 51.*

Both the renal and osmoregulatory effects of chronic relaxin administration to conscious rats resemble the physiological changes of pregnancy in several respects: (1) marked increases in ERPF and GFR with a mediatory role for nitric oxide, (2) attenuation of the renal circulatory response to angiotensin II, and (3) reduction in plasma sodium concentration and in osmolality [51]. The renal response persists in ovariectomized animals indicating that an intermediary or permissive role for sex steroids is not required. Surprisingly, comparable renal and osmoregulatory responses to relaxin administration are observed in conscious *male* rats [52]. Analogous to midterm pregnancy in conscious rats, relaxin-induced renal vasodilation and hyperfiltration in nonpregnant rats is also mediated by endothelin through the endothelial ET_B receptor subtype and NO [20,52] (Fig.1).

The best (if not only) approach for testing the physiologic role of a hormone is to block production or inhibit the action of the hormone. So far, we have successfully utilized the approach of neutralizing *endogenous* circulating relaxin during rat gestation with monoclonal antibodies and the results support a role for relaxin in the renal and osmoregulatory changes of pregnancy (unpublished). We are currently testing whether ovariectomy and 17β-estradiol/progesterone supplementation during pregnancy will corroborate these findings. Finally, we are also testing the role of relaxin in mediating other major cardiovascular adaptations to pregnancy, i.e., reduced systemic vascular resistance and increased arterial compliance.

Perspectives

Identification of the hormone(s) responsible for the remarkable circulatory changes of gestation is crucial for complete understanding of maternal adaptation to normal pregnancy, and will likely facilitate investigation of preeclampsia in which the vasodilatory response is inappropriate. Because hypertension later in life is less frequent in women who have normal pregnancies(s) [53], and the cardiovascular system during normal pregnancy (low SVR, modestly reduced blood pressure, increased arterial compliance, robust endothelial function, etc.) is a veritable "fountain of youth", we believe that discovery of the pregnancy hormone(s) responsible is a priority as these hormones may provide new treatment(s) for combating increased vascular resistance and arterial stiffness that occur with aging and hypertension. In this regard, we hypothesize that the pregnancy hormone, relaxin, holds tremendous potential. Whether the vasodilatory role of relaxin in rats can be extrapolated to humans is currently being investigated. Nevertheless, it is exciting to also contemplate a potential therapeutic role for relaxin in renal disease not only with respect to its newly described vasodilatory properties [51], but also in light of its well-known matrix-degrading attributes [54].

Acknowledgments

Work in the authors' laboratories was funded by NIH RO1 HD30325 and NIH KO4 HD01098. We thank Vicky McClain for superb secretarial support.

References

1. Lindheimer, M.D., Roberts, J.M., Cunningham, F.G. (eds). Hypertensive Disorders in Pregnancy. Second Edition, Appleton & Lange, 1999.
2. Poppas, A., Shroff, S.G., Korcarz, C.E., Hibbard, J.U., Berger, D.S., Lindheimer, M.D., Lang, R.M, Circulation, 95 (1997) 2407.
3. Conrad, K. P., Kidney Int., 26 (1984) 24.
4. Gilson, G.J., Mosher, M.D. and Conrad, K.P., Am. J. Physiol., 263 (1992) H1911.
5. Danielson, L.A. and Conrad, K.P., J. Clin. Invest., 96 (1995) 482.
6. Conrad, K.P. and Colpoys, M.C., J. Clin. Invest., 77 (1986) 236.
7. Conrad, K.P., Morganelli, P.M., Brinck-Johnsen, T. and Colpoys, M.C. Am. J. Obstet. Gynecol., 161 (1989) 1065.
8. Conrad, K.P. and Russ, R.D., Am. J. Physiol., 262 (1992) R472.
9. Gandley, R.E., Conrad, K.P. and McLaughlin M.K., Am. J., Physiol., 280 (2001) R1.
10. Griggs K.C., Conrad, K.P., Mackey, K. and McLaughlin, M.K., Am. J. Physiol., 265 (1993) F309.
11. Gandley, R.E., Griggs, K.C., Conrad, K.P. and McLaughlin M.K., Am. J. Physiol., 273 (1997) R22.
12. Conrad, K.P. and Dunn M.J., Am. J. Physiol., 253 (1987) F1197.

13. Sorensen, T.K., Easterling, T.R., Carlson, K.L., Brateng, D.A. and Benedetti, T.J., Obstet. Gynecol., 79 (1992) 661.
14. Conrad, K.P. and Vernier, K.A., Am. J. Physiol., 257 (1989) R847.
15. Conrad, K.P., Am. J. Kidney Diseases, 9 (1987) 253.
16. Kopp L., Paradiz, G. and Tucci, J.R., J. Clin. Endocrinol. Metab., 44 (1977) 590.
17. Palmer, R.M.J., Ferrige, A.G. and Moncada, S., Nature (London), 327 (1987) 524.
18. Marletta, M.A., Yoon, P.S., Iyengar, R., Leaf, C.D. and Wishnok, J.S., Biochemistry, 27 (1988) 8706.
19. Conrad, K.P., Joffe, G.M., Kruszyna, H., Kruszyna, R., Rochelle, L.G., Smith, R.P., Chavez, J.E. and Mosher, M.D., FASEB J., 7 (1993) 566.
20. Conrad, K.P., Gandley, R.E., Ogawa, T., Nakanishi, S., and Danielson, L.A., Am. J. Physiol., 276 (1999) F767.
21. Veille, J.C., Morton, M.J., Burry, K., Nemeth, M. and Speroff, L., J. Clin. Endocrinol. Metab., 63 (1986) 721.
22. Magness, R.R., Parker, C.R., and Rosenfeld, C.R., Am. J. Physiol., 265 (1993) E690.
23. Chesley, L.C. and Tepper, I.H., J. Clin Endocrinol. Metab., 27 (1967) 576.
24. Christy, N.P. and Shaver, J.C., Kidney Internat., 6 (1974) 366.
25. Berl, T. and Better, O.S., In Brenner B.M. and Stein J.H, (Eds.), Hormonal Function and the Kidney, Churchill Livingston, New York, Edinburg and London, 1979, p. 194.
26. Novak, K. and Kaufman, S., Am. J. Physiol., 261 (1991) R1164.
27. Conrad, K.P., Mosher, M.D., Brinck-Johnson, T. and Colpoys, M.C., Am. J. Physiol., 266 (1994) R1267.
28. Davison, J.M., Shiells, E.A., Philips, P.R. and Lindheimer, M.D., J. Clin. Invest., 81 (1988) 798.
29. Atallah, A.N., Guimarães, J.A.G., Gebara, M., Sustovich, D.R., Martinez, T.R. and Camano, L., Brazilian J. Med. Biol. Res., 21 (1988) 71.
30. Oparil, S., Ehrlich, E.N. and Lindheimer, M.D., Clin. Sci. Mol. Med., 49 (1975) 139.
31. Sherwood O.D., In Knobil E., Neill J.D., Greenwald G.S., Markert C.L. and Pfaff, D.W. (Eds.), The Physiology of reproduction. 2nd ed., Raven Press, New York, USA, 1994, p. 861.
32. Stewart D.R., Celniker, A.C., Taylor, Jr. C.A., Cragun, J.R., Overstreet, J.W. and Lasley, B.L., Clin. Endocrinol Metab., 70 (1990) 1771.
33. Johnson, M.R, Carter, G., Grint, C. and Lightman, S.L., Acta. Endocrinol., 129 (1993) 121.
34. Wreje, U., Kristiansson, P., Aberg, G., Bystrom, B. and von Schoultz, B., Gynecol. Obstet. Invest., 39 (1995) 197.
35. Davison, J.M. and Noble, M.C.B., Br. J. Obstet. Gynaecol., 88 (1981) 10.
36. Paaby, P., Brøchner-Mortensen, J., Fjeldborg, P., Raffn, K., Larsen, C.E. and Møller-Petersen, J., Acta. Med. Scand., 222 (1987) 281.
37. Paaby, P., Moller-Petersen, J., Larsen, C.E. and Raffin, K., Acta. Med. Scand., 221 (1987) 191.
38. Brøchner-Mortensen, J., Paaby, P., Fjeldborg, P., Raffn, K., Larsen, C.E. and Møller-Petersen, J., Scand J. Clin. Lab. Invest., 47 (1987) 829.
39. Van Beek, E., Houben, A.J.H.M., van Es, P.N., Willekes, C., Korten, E.C.C.M., de Leeuw, P.W. and Peeters, L.L.H., Clin. Sci., 91 (1996) 163.
40. Chapman, A.B., Zamudio, S., Woodmansee, W., Merouani, A., Osorio, F., Johnson, A., Moore, L.G., Dahms, T., Coffin, C., Abraham, W.T. and Schrier, R.W., Am. J. Physiol., 273 (1997) F777.
41. Davison, J.M., Vallotton, M.D. and Lindheimer, M.D., Br. J. Obstet. Gynaecol., 88 (1981) 472.
42. Lindheimer, M.D. and Davison, J.M., Eur. J. Endocrinol., 132 (1995) 133.

43. Lindheimer, M.D., Barron, W.M. and Davison, J.M., Am. J. Physiol., 257 (1989) F159.
44. Weisinger, R.S., Burns, P., Eddie, L.W. and Wintour, E.M., J. Endocrinol., 137 (1993) 505.
45. Atherton, J.C., Bu'lock, D. and Pirie, S.C., J. Physiol., 324 (1982) 11.
46. Baylis C., Am. J. Physiol., 234 (1982) F300.
47. St.-Louis, J. and Massicotte, G., Life Sci., 37 (1985) 1351.
48. Massicotte, G., Parent, A. and St.-Louis, J., Proc. Soc. Exp. Biol. Med., 190 (1989) 254.
49. Bani-Sacchi, T., Bigazzi, M., Bani, D., Mannaioni, P.F. and Masini, E., Br. J Pharmacol., 116 (1995) 1589.
50. Bani D., Bigazzi, M., Masini, E., Bani, G. and Bani-Sacchi, T., Lab Invest., 73 (1995) 709.
51. Danielson, L.A., Sherwood, O.D., and Conrad, K.P., J. Clin. Invest., 103 (1999) 525.
52. Danielson, L.A., Kerchner, L.J., and Conrad, K.P., Am. J. Physiol., 279 (2000) R1298
53. Lindheimer, M.D., Fisher, K.A., Spargo, B.H. and Katz, A.I., Contr. Nephrol., 25 (1981) 71.
54. Unemori, E.N., Pickford, L.B., Salles, A.L., Piercy, C.E., Grove, B.H., Erikson, M.E. and Amento E.P., J. Clin. Invest., 98 (1996) 2739.

Reproductive hormones and blood pressure during pregnancy

Per Kristiansson[1, 2] and J X Wang[2]

[1]Uppsala University, Department of Public Health and Caring Sciences, University Hospital, S-751 85 Uppsala, Sweden; and [2]University of Adelaide, Department of Obstetrics and Gynaecology, South Australia 5011, Australia.

Introduction

The mechanisms involved in the cardiovasular changes during human pregnancy and the complicated aetiology of gestational hypertension are unclear. Reproductive hormones have known effects on the cardiovascular system in non-pregnant state and in animal systems but the effects in human pregnancy are uncertain. Oestrogen is suggested to have both functional and structural haemodynamic effects via a transient vasodilating effect and through increased aortic compliance in postmenopausal women [1]. Progesterone has been reported to reduce blood pressure in hypertensive subjects [2] and to induce vascular relaxation [3]. The polypeptide hormone relaxin predominantly produced by the corpus luteum of the ovary is best known for its connective tissue remodelling actions on the female reproductive system [4]. However, additional effects in other organs and systems are suggested. In several animals and tissues relaxin has been shown to cause powerful vasodilatation [5, 6].

The present study was designed to evaluate changes of systolic and diastolic blood pressure in a healthy population during pregnancy and to study the possible relationship between the changes and the reproductive hormones relaxin, oestrogen and progesterone.

Results and Discussion

The mean systolic blood pressure continued to rise during pregnancy, with the highest increase of 2.1 mmHg between the weeks 8-12 and 13-17 ($p<0.01$). Thereafter the increases between each of the week intervals were smaller and not statistically significant. The overall increase of mean systolic blood pressure from weeks 8-12 to 32-38 was 3.3 mmHg, which was 2.8% above the initial level ($p<0.0001$). The mean systolic blood pressures at the different week intervals were highly inter-correlated ($0.46<r<0.74$, $p<0.0001$) where mean systolic blood pressure at 8-12 weeks was most closely correlated to mean systolic blood pressure at weeks 32-38 ($r=0.56$, $p<0.0001$).

The mean diastolic blood pressure showed a statistically significant ($p<0.001$) decrease of 2.7 mmHg from 69.6 mmHg at weeks 8-12 to the nadir of 66.9 mmHg at weeks 18-23. From this lowest level the diastolic blood pressure increased by 6.5 mmHg to 73.4 mmHg at weeks 32-38 ($p<0.0001$). The overall increase of diastolic blood pressure over the study period was 3.8 mmHg that was 5.4% above the initial level ($p<0.0001$). The mean diastolic blood pressures at the different week intervals were

G.W. Tregear et al. (eds.), Relaxin 2000, 177–182.
© 2001 *Kluwer Academic Publishers. Printed in the Netherlands.*

highly inter-correlated ($0.40<r<0.74$, $p<0.0001$). The mean diastolic blood pressure at weeks 8-12 showed the strongest correlation to mean diastolic blood pressure at weeks 32-38 ($r=0.54$, $p<0.0001$).

Both serum progesterone and relaxin levels at weeks 8-12 showed a weak but statistically significant negative correlation to mean systolic blood pressures at the week intervals 18-23 ($-0.17<r<-0.25$, $p<0.05$), 24-31 ($-0.18<r<-0.22$, $p<0.05$) and 32-38 ($-0.17<r<-0.22$, $p<0.05$), respectively. The same tendency was shown for mean systolic blood pressures at week intervals 8-12 and 13-18, although this was not statistically significant. Progesterone and relaxin levels measured later in pregnancy showed no association with mean systolic blood pressure. Oestradiol concentrations at the different week intervals showed no statistically significant relation to mean systolic blood pressure at any of the week intervals ($-0.06<r<0.13$, n.s.). Increasing parity was weakly and negatively associated with mean systolic blood pressure during pregnancy ($-0.15<r<-0.19$, $p<0.05$) except with those at weeks 32-38. In addition, weight gain during pregnancy showed a positive correlation to mean systolic blood pressure at weeks 32-38 ($r=0.18$, $p=0.01$). There was no statistically significant relation to the total number of blood pressure measurements, the number of previous pregnancies, age, cigarette smoking habits or body mass index in early pregnancy.

While mean diastolic blood pressure at weeks 18-23 showed a weak negative correlation to progesterone levels at weeks 13-17 ($r=-0.16$, $p<0.05$) progesterone at other week intervals as well as relaxin and oestradiol levels did not correlate with diastolic blood pressure at any interval. Body mass index in early pregnancy was significantly correlated to mean diastolic blood pressure at all week intervals ($0.16<r<0.22$, $p<0.01$). No significant relationship was shown between diastolic blood pressure and the total number of blood pressure measurements, age, smoking habits, the number of earlier pregnancies or deliveries.

At week interval 32-38, 4.8% (9/188) of the women had a maximum diastolic blood pressure above 90 mmHg. Within this group the mean relaxin value at week interval 8-12 was statistically significantly lower (650 ng/ml) than among those with a maximum diastolic blood pressure below or equal to 90 mmHg (911 ng/ml, $p<0.05$). With these women with high diastolic blood pressure excluded, the relationship between relaxin and systolic blood pressures as described previously remained significant.

To study the compound effects of mean systolic blood pressure in early pregnancy and the serum markers on systolic blood pressure in late pregnancy, a multivariate regression analysis was performed. The mean systolic blood pressure at weeks 32-38 was used as the dependent variable, and mean systolic blood pressure at weeks 8-12, serum concentrations of progesterone and relaxin at weeks 8-12 as the independent variables. Mean early systolic blood pressure ($p=0.0001$) and serum relaxin ($p=0.01$), but not progesterone, were independently related to the variation of systolic blood pressure in late pregnancy. The R^2 of the model was 0.36. The size of effect of relaxin concentration and mean blood pressure measurement, at weeks 8-12, on systolic blood pressure in late pregnancy is illustrated in Fig 1.

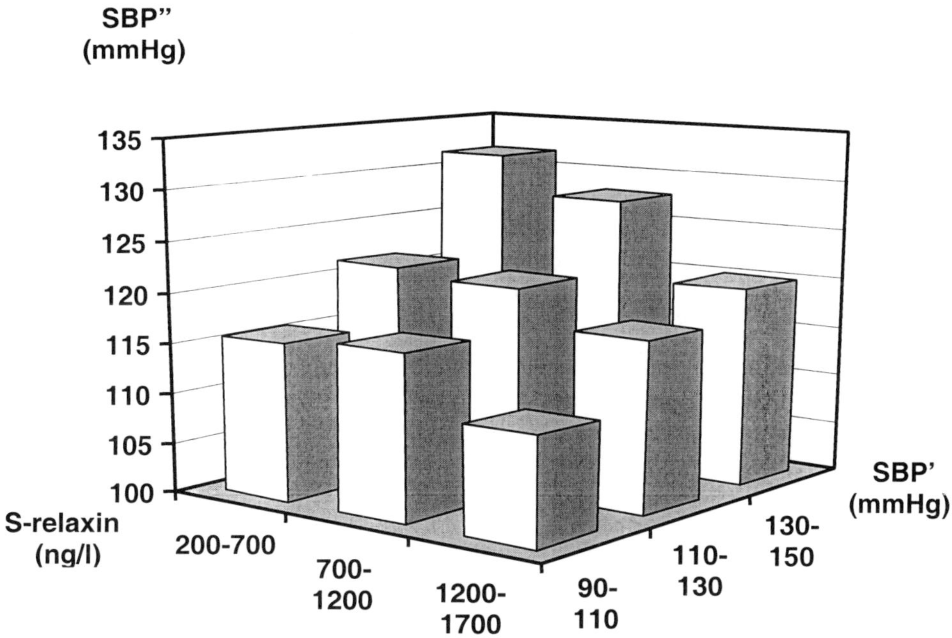

Fig 1. Systolic blood pressure in late pregnancy (SBP'') by systolic blood pressure (SBP') and relaxin concentration in early pregnancy.

Women with highest relaxin values and the lowest systolic blood pressure in early pregnancy displayed the lowest systolic blood pressures in late pregnancy and women with the lowest relaxin values and the highest systolic blood pressures displayed the highest. Also, a multivariate regression analysis, with mean diastolic blood pressure in late pregnancy as the dependent variable with relaxin and progesterone concentrations and body mass index in early pregnancy as independent variables, was performed where BMI ($p=0.003$) was the only factor related to diastolic blood pressure. The R^2 of the model was 0.06.

The present data suggest a hormonal influence on the haemodynamic changes in human pregnancy since the higher concentrations of progesterone and relaxin in early pregnancy were significantly associated with lower mean systolic blood pressure in the second and third trimesters. However, from an observational study causality can not be determined. As expected there was a steady increase in blood pressure from early to late gestation except for a small initial decrease of diastolic blood pressure [7, 8]. However, exclusion criteria in these studies explain the slightly lower blood pressure values compared to the general population of pregnant women. In such a study the systolic blood pressure increased by 3% during pregnancy to 118 ± 11 mmHg and the diastolic blood pressure fell initially with an overall increase by 7.7% to 72.2 ± 8.7 mmHg [9] which are in agreement with the result in the present study.

There have been a few limitations of this study, including the accuracy of blood pressure measurement and small sample size. The blood pressure was measured to the nearest 5 mmHg instead of recommended 2 mmHg [10] and no precaution was taken against digit preference or threshold avoidance [11]. This probably reduced the sensitivity of the study to show any but the strongest effect on blood pressure. This may explain why the known weak association of age with blood pressure during pregnancy was not statistically significant [12]. However, this may also suggest that the chance of false positive in detecting a relationship is small. In addition, the weak relationship between relaxin and systolic blood pressure may be because relaxin is one of several residual or redundant systems that control blood pressure in human pregnancy [13].

No previous study that we know of has related relaxin concentration in pregnant women to blood pressure. In pregnancies achieved from ovum donation an increased incidence of gestational hypertension was showed [14]. This is in accordance with our results, since women pregnant following ovum donation have no or low levels of relaxin detected in the circulation [15]. Furthermore, relaxin is reported to influence the changes in uterine blood flow in early pregnant women [16].

The proposed relationship between early relaxin values and lower systolic blood pressure in later pregnancy may be explained by a vasodilatatory action of relaxin seen in animal studies [5,6]. That relaxin is involved in the regulation of blood pressure in rats is further indicated by the antihypertensive effect shown in female non-pregnant spontaneously hypertensive rats [17]. Relaxin is proposed to exert the vasodilatatory effect with a mediatory role of nitric oxide [6,13,18] which is believed to contribute to the generalised vasodilation in normal pregnancy [19].

Why the relationship is only seen between later systolic blood pressures and early relaxin values is unclear. There are reports about timing of effect of relaxin suggesting that relaxin has a delayed effect. A vasodilatory effect was demonstrated in the renal circulation of conscious rats following prolonged administration of relaxin, but not following short-term infusion [6]. In addition, no effect was detected on systolic or diastolic blood pressure of human relaxin given vaginally prior to labour in a clinical trial of cervical ripening [20]. In the present study, the mean systolic blood pressure showed the greatest increase during the first trimester which may not be determined by the relaxin. The lower increase during later pregnancy may be determined by relaxin in early pregnancy, as shown by the relationship between relaxin and systolic blood pressure in the present study.

Progesterone has also been reported to reduce blood pressure in hypertensive subjects [2] and to induce vascular relaxation [3]. In the present study, the effect of progesterone on the variation of systolic blood pressure could be explained by the variation in relaxin concentration, as shown in the multivariate regression analysis.

About 5% of the present cohort had a diastolic blood pressure above 90 mmHg and may have gestational hypertension. This prevalence is close to the previously reported prevalence of 6.7% for hypertensive disorders of pregnancy [19]. In the present study, women with diastolic blood pressure >90 mmHg had significantly lower relaxin values in early pregnancy as compared to women without hypertension. However, in the whole group of women no relationship was shown between the levels of relaxin and diastolic blood pressure. There might be a relationship between the concentration of relaxin in very early pregnancy, not measured in the present study, and the subsequent diastolic blood pressure.

In conclusion, the hormone relaxin in early pregnancy may be involved in the haemodynamic changes during pregnancy. However the size of the effect may be moderate.

Acknowledgements

We thank Professor A. MacLennan for assistance in the preparation of the manuscript and Genentech Inc., San Francisco, California, USA for generously providing the reagents for the relaxin ELISA. This study was supported by grants from the Swedish Medical Research Council, the Swedish Institute, the Emil Andersson foundation of medical research and Uppsala University.

References

1. Gilligan, D.M., Badar, D.M., Panza, J.A., Quyyumi, A.A. and Cannon, R.O. 3rd, Am. J. Cardiol., 75 (1995) 264.
2. Rylance, P.B., Brincat, M., Lafferty, K., De Trafford, J.C., Brincat, S., Parsons, V. and Studd, J.W., Br. Med. J., 290 (1985) 13.
3. Omar, H.A., Ramirez, R. and Gibson, M., J. Clin. Endocrinol. Metab. 80 (1995) 370.
4. Sherwood,O.D. In Knobil, E. and Neill, J.D. (Eds.), The Physiology of Reproduction. 2nd ed., Raven Press, New York, USA, 1994, p. 861.
5. Bigazzi,M., Del Mese, A., Petrucci, F., Casali, R. and Novelli, G.P., Acta Endocrinol., 112 (1986) 296.
6. Danielson, L.A., Sherwood, O.D. and Conrad, K.P., J. Clin. Invest., 103 (1999) 525.
7. Contard, S., Chanudet, X., Coisne, D. Battistella, P., Marichal, J.F., Pitiot, M., de Gaudemaris, R. and Ribstein, J., Am. J. Hypertens., 6, (1993) 880.
8. Churchill, D., Perry, I.J. and Beevers. D.G., Lancet, 349 (1997) 7.
9. Strevens, H., Wide-Svensson, D. and Ingemarsson, I., Swedish Society of Medicine, Stockholm (1997)
10. Davey, D.G. and MacGillivray, I., Am. J. Obstet. Gynecol., 158 (1988) 892.
11. Steer, P.J., Br. J. Obstet. Gynaecol., 106, (1999) 753.
12. Margulies, M., Voto, L.S., Fescina, R. Lastra, L., Lapidus, A.M. and Schwarz, R., Am. J. Obstet. Gynecol., 156 (1987) 1105.
13. Bani, D., Failli, P. and Grazia Bello, M., Hypertension, 31 (1998) 1240.
14. Salha, O., Sharma, V., Dada, T., Nugent, D., Rutherford, A.J., Tomlinson, A.J., Philips, S., Allgar, V. and Walker, J.J., Hum. Reprod., 14 (1999) 2268.
15. Johnson, M.R., Abdalla, H., Allman, A.C., Wren, M.E., Kirkland, A. and Lightman, S.L., Fertil. Steril., 56 (1991) 59.

16. Jauniaux, E., Johnson, M.R., Jurkovic, D. Ramsay, B., Campbell, S. and Meuris, S., Obstet. Gynecol.. 84 (1994) 338.
17. St-Louis, J. and Massicotte, G., Life Sci., 37 (1985) 1351.
18. Ramsay, B., De Belder, A., Campbell, S. Moncada, S. and Martin, J.F., Eur. J. Clin. Invest., 24 (1994) 76.
19. Churchill, D. and Beevers, D.G., Hypertension in Pregnancy, (1999) BMJ Books.
20. Bell, R.J., Permezel, M. and MacLennan, A., Obstet. Gynecol., 82 (1993) 328.

The effect of relaxin on vascular function in rat aorta

Julianne J Reid[1], Ng-Hi Tran[1], Geoffrey W Tregear[2], and Peter J Roche[1]

[1]*Department of Medical Laboratory Science, RMIT University, Melbourne, Victoria 3001, Australia; and* [2]*Howard Florey Institute, University of Melbourne, Victoria 3010, Australia.*

Introduction

It has been suggested that the reproductive hormone relaxin may be involved in the regulation of cardiovascular tone [1]. Relaxin has been shown to cause increases in coronary blood flow [1] and renal plasma flow [2], which could be reduced by inhibitors of nitric oxide synthase (NOS). However there is little or no information on the vascular actions of relaxin in an isolated blood vessel where experimental conditions can be specifically controlled.

The aim of the present study was to investigate the effect of recombinant human (rH2) relaxin on the functioning of rat isolated thoracic aortic rings, and specifically to determine whether H2 relaxin produces vasodilatation through the production of endothelial nitric oxide (NO) or inducible NO.

Results and Discussion

In endothelium-intact aortic rings precontracted with noradrenaline, a 20-min exposure to rH2 relaxin (0.1 μM) had no effect on tension or on the concentration-dependent response to the endothelium-dependent vasodilator acetylcholine (n=7, $P>0.05$, ANOVA). However, it was demonstrated that rH2 relaxin (0.01 nM - 0.1 μM) caused a concentration-dependent increase in rat atrial rate to a maximal response of 112 $\pm$ 12 beats/min (n=4) as previously described [3], confirming the viability of the peptide.

Longer incubation periods of 6 or 24 hr were also investigated to allow sufficient time for the induction of NO synthase. A 6-hr incubation in 10 μg/ml lipopolysaccharide (LPS), a known stimulator of inducible NO synthase [4], significantly reduced ($P<0.05$, ANOVA) the constrictor response to phenylephrine (PE, 1 nM-10 μM) in endothelium-denuded aortic rings (maximum response at 10 μM PE: control 2.29$\pm$0.13 g; LPS 0.89$\pm$0.16 g; n=6). However exposure to 0.1 μM rH2 relaxin for 6 hr had no significant effect ($P>0.05$, ANOVA) on the response to PE (10 μM PE: control 2.33$\pm$0.09 g; rH2 relaxin 2.38$\pm$0.14 g; n=8).

Similarly, the response curve to PE was reduced by a 24-hr incubation in 10 μg/ml LPS ($P<0.05$, ANOVA; n=3), but was not significantly altered by exposure to 1 μM rH2 relaxin for 24 hr ($P>0.05$, ANOVA; n=19) compared to the corresponding time controls (Fig. 1).

G.W. Tregear et al. (eds.), Relaxin 2000, 183–184.
© 2001 *Kluwer Academic Publishers. Printed in the Netherlands.*

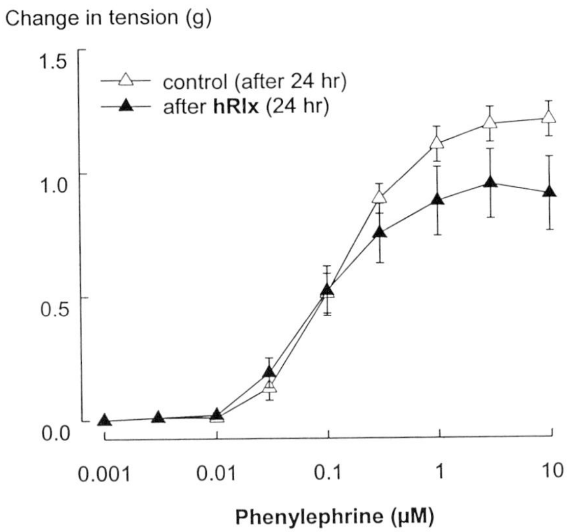

Fig. 1. The constriction to phenylephrine was not significantly altered by a 24-hr incubation in 1 μM rH2 relaxin.

These findings do not provide any evidence that rH2 relaxin can affect the release of endothelial NO or inducible NO in rat thoracic aorta under these experimental conditions. Recent data on renal vasodilation indicates that relaxin may be operating indirectly through endothelin and the endothelin ET(B) receptor subtype to produce NO [5]. Thus further study is required to clarify the effect of relaxin on vascular function, since longer incubation periods or different experimental conditions may be necessary to allow for endothelin mediated NO release.

Acknowledgements

This work was supported by a faculty research grant from the Faculty of Life Sciences, RMIT University, Melbourne.

References

1. Bani, D., Gen. Pharmacol., 28 (1997) 13.
2. Danielson, L.A., Sherwood, O.D., and Conrad, K.P., J. Clin. Invest., 103 (1999) 525.
3. Kakouris, H., Eddie, L.W. and Summers, R.J., Lancet, 339 (1992) 1076.
4. Binko, J., Murphy, T.V. and Majewski H., Clin. Exp. Pharmacol. Physiol., 25 (1998) 120.
5. Danielson, L.A., Kercher, L.J., and Conrad, K.P., Am. J. Physiol. Regul. Integr. Comp. Physiol., 279 (2000) R1298.

Cardiovascular phenotypes of relaxin knockout mice: Importance of gender

Xiao-Jun Du[1], Ling Zhao[2], Xiao-Ming Gao[1], Geoffrey W. Tregear[2]
[1] *Baker Medical Research Institute, Melbourne, Victoria 8008, Australia; and*
[2] *Howard Florey Institute, University of Melbourne, Melbourne, Victoria 3052, Australia*

Introduction

Specific relaxin binding sites present in the heart and blood vessels and chronotropic and inotropic responses to relaxin have been documented [1,2] implying that relaxin exerts cardiovascular properties. The circulation levels of endogenous relaxin, however, are extremely low whilst the doses of relaxin used for demonstrating pharmacological effects are very high. Thus, the biological effects of relaxin on the cardiovascular system remain unclear. We have studied the cardiac phenotype of 15~20 month old male and female mice with disruption of the relaxin gene (-/-) [3], in comparison with wild-type (+/+) and heterozygous (+/-), generated by Zhao *et al.* at the Howard Florey Institute [3]. Under anaesthetized conditions, transthoracic echocardiography was conducted using a HP 5500 echo machine with a 15MHz linear probe, followed by cardiac catheterization via the right carotid artery using a 1.4F Millar micro-tip pressure transducer catheter [4]. Organ weights were obtained and myocardial content of collagen was measured by determining concentration of hydroxyproline in the left ventricle (LV). Each group had 8 to 10 mice.

Results and Discussion

Under anesthetized conditions, heart rate, arterial blood pressure, left ventricular (LV) dimensions at systole and diastole, fractional shortening, ventricular contractility (dP/dt_{max} and dP/dt_{min}) were similar among the 3 groups of both sexes. Intravenous administration of a β-adrenergic agonist isoprotorenol (50~2000 pg per mouse) induced dose-dependent increase in heart rate and dP/dt and the extent of the responses was also not significantly different between relaxin -/- male and female versus other groups. These findings indicate an unaltered cardiac contractile function in aged relaxin -/- mice.

Compared to relaxin +/+ and +/- males, -/- males had 15% increase in LV mass (by both echo and weight), 30% increase in atrial weight, and 10 to 30% increase in wet weights of lungs and liver (both $P<0.05$), but not kidney. Functionally, LV end-diastolic pressure doubled in relaxin -/- males ($P<0.05$, Table 1). All these findings are in keeping with an increased resistance for venous return with modest congestion in atria, lung and liver. Further, hydroxyproline concentration in the myocardium was higher in relaxin -/- males than +/+ (1.82±0.10 vs. 1.56±0.07 μg/mg dry weight, $P=0.065$) and +/- males (1.47±0.08 μg/mg, $P<0.02$), suggesting an increased collagen content in the LV.

G.W. Tregear et al. (eds.), Relaxin 2000, 185–187.
© 2001 *Kluwer Academic Publishers. Printed in the Netherlands.*

Interestingly, none of these parameters was significantly different among the three female groups.

Table 1. Organ wet weights, LV mass from echocardiography and LVEDP from LV catheterization in wild-type (+/+), heterozygous (+/-) and knockout (-/-) mice.

	Heart/TL. (mg/mm)	Atria/TL (mg/mm)	Lung/TL. (mg/mm)	Liver/TL (mg/mm)	Echo LV mass (mg)	LVEDP (mm Hg)
+/+	10.7±0.6*	0.85±0.07*	10.3±0.3*	85±5*	98±11	2.5±0.9*
+/-	10.6±0.4*	0.83±0.05*	9.9±0.3*	92±5*	95±9*	2.6±0.6*
-/-	12.2±0.6	1.27±0.11	11.4±0.4	110±4	124±8	5.0±1.0

TL: tibia length. * $P<0.05$ vs -/- group.

Our findings from the relaxin knockout male mice suggest that a life-time disruption of relaxin results in moderate cardiac hypertrophy, impeded venous return and modest organ congestion. These are most likely due to increased LV collagen content with a resultant increase in LV chamber stiffness. Although endogenous and exogenous relaxin has been known to reduce collagen content in non-cardiac tissues by upregulating matrix metalloproteinase [5,6], this is the first report that a life-time deprivation of relaxin leads to increased collagen content in the heart of male animals. The fact that the phenotype was observed in males but not in females indicates the importance of hormonal state or other gender related-factors on the cardiac actions of relaxin. Interestingly recent studies on other gene-targeted mouse strains have shown that gender greatly influences the development of cardiac phenotypes [7,8].

Acknowledgment

This study was supported by the institute block grants to the Baker Medical Research Institute and Howard Florey Institute from the NHMRC of Australia.

References

1. Osheroff, R.L., Cronin, M.J. and Lofgren, J.A., Proc. Nat. Acad. Sci., 89 (1992) 2384.
2. Tan, Y.Y., Wade, J.D., Tregear, G.W. and Summers, R.J., Br. J. Pharmacol., 127 (1999) 91.
3. Zhao, L, Roche, P.J., Gunnersen, J.M., Gunnersen, J.M., Hammond, V.E., Tregear, G.W., Wintour, E.M. and Beck, F., Endocrinology, 140 (1999) 445.
4. Gao, X.M., Dart, A.M., Dewar, E., Jennings, G.L. and Du, X.D., Cardiovasc. Res., 45 (2000) 330.
5. Zhao, L., Samuel, C.S., Tregear, G.W., Beck, F. and Wintour, E.M., Biol. Reprod., 63 (2000) 697.
6. Unemori, F.N., Pickford, L.B., Salles, A.L., Piercy, C.E., Grove, B.H., Erikson, M.E. and Amento, E.P., J. Clin. Invest., 98 (1996) 2739.
7. Kadokami, T., McTiernan, C.F., Kubota, T., Frye, C.S. and Feldman, A.M., J. Clin. Invest., 106 (2000) 589.

8. Heymans, S., Luttun , A., Nuyens, D., Thelimeier, G, Creemers, E., Moons, L., Syspersin, G.D., Cleutjens, J.P.M., Shipley, M., Angellilo, A., Levi, M., Nübe, O., Baker, A., Keshet, E., Lupu, F., Herbert, J.M., Smits, J.F.M., Shapiro, S.D., Beas, M., Borgers, M., Collen, D., Daemen, M.J.A.P. and Carmeliet, P., Nat. Med. 5(1999):1135.

RELAXIN'
on the Beach...
October, 2000

SECTION 6

Physiology of relaxin I: Brain

Relaxin action on and in the brain

Alastair JS Summerlee and Josh D Silvertown
Department of Biomedical Sciences, Ontario Veterinary College, University of Guelph, Guelph, Ontario N1G 2W1, Canada.

Introduction

The first report describing an action of relaxin on the brain was published in 1984 [1]. Exogenous administration of porcine relaxin in lactating rats suppressed relaxin milk-ejection and it was suggested that relaxin might be inhibiting the release of oxytocin (Fig. 1). Since that time there have been several studies showing that relaxin affects the release of a number of hypothalamo-pituitary hormones [2]. It has been suggested that relaxin may have a physiological role in the timing of birth [3], cardiovascular control, particularly in pregnancy [2] and water balance in pregnancy [4,5]. One of the fundamental question in this area of relaxin research is whether relaxin acts **'on** or **'in'** the brain to have its affects. This paper explores the evidence that suggests that relaxin may act **'on'** the brain by binding to receptors at intermediary sites, notably the pituitary gland or the circumventricular organs. It may also act **'in'** the brain where it may be produced and released locally to have its effects.

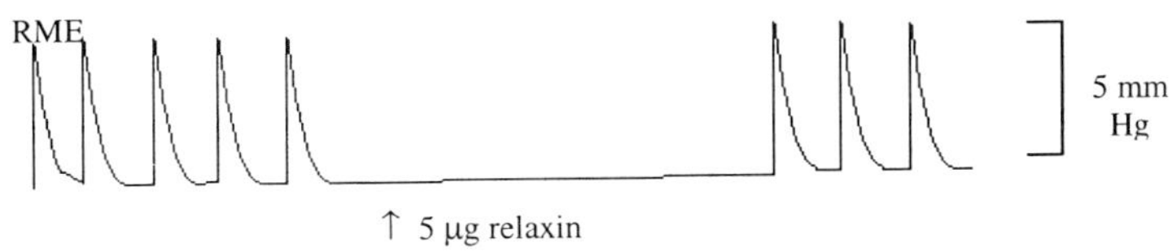

*Fig. 1. The effect of exogenous relaxin (**iv**) on reflex milk-ejection (RME) in an anaesthetised, lactating rat.*

The blood-brain barrier, the pituitary and the circumventricular organs

Three specialized layers, or meninges, surround the brain to form a structure known as the blood-brain barrier [6]. This physical barrier is considered to be an important protective mechanism for the brain as it creates a physical barrier that impedes the entry of all but the smallest molecules. It not only separates the blood from the extracellular fluid of the brain, it also separates the blood and cerebrospinal fluid (*csf*) from the extracellular fluid of the brain. The blood-brain barrier has three components: (1) the capillary endothelial cells fit tightly together and have no pores or fenestrations; (2) there is a thick basement membrane around the endothelial cells; and (3) the capillaries are wrapped in a layer made by the end-feet of astrocytes (one of the types of glia or

G.W. Tregear et al. (eds.), Relaxin 2000, 191–199.
© 2001 *Kluwer Academic Publishers. Printed in the Netherlands.*

support cells in the brain). There are, however, two places where the blood-brain barrier is not complete: the pituitary and the circumventricular organs.

The pituitary gland protrudes from the base of the brain outside the blood-brain barrier. The gland is divided into three parts: a neural or posterior part which is principally associated with the release of oxytocin and vasopressin; an intermediate part linked to opioidergic control; and a glandular or anterior (rostral) part associated with release of a number of gonadotrophic hormones such as follicle-stimulating hormone, luteinizing hormone and prolactin. The pituitary is richly vascularized and there are large pores in capillary walls, which allows relatively free passage of molecules of quite large molecular weight to pass in and out of the blood.

There are several circumventricular organs that are situated in the walls of the cerebral ventricles [7]: the most studied are the subfornical organ (SFO) and the antero-ventral region of the third ventricle (AV3V), which includes the *organosum of the lamina terminalis* (OVLT) – effectively a conglomerate of small centres towards the rostral aspect of the third ventricle. These circumventricular organs sit at the interface between the blood, the *csf* and the brain at the sites where there is a considerable flow of blood and *csf* per minute and a large number of neurone terminals connected to diverse and disparate parts of the brain [8]. The blood-brain barrier in the circumventricular organs is incomplete. The endothelial walls are highly fenestrated, the astrocytic end-feet are absent and the basement membrane is relatively thin compared with other parts of the brain. These organs have a high concentration of angiotensin containing nerve terminals and receptors, and, through the forebrain angiotensin system, have been implicated in mediating cardiovascular control in a number of ways [9].

Evidence that relaxin acts on the brain via the pituitary

There are conflicting views about the action of relaxin at the level of the posterior pituitary. Dayanithi and colleagues [10] showed that relaxin facilitated vasopressin and oxytocin release from potassium-stimulated neural lobe tissue *in vitro* but inhibited the release of both oxytocin and vasopressin from unstimulated tissue. However, using a neurointermediate lobe preparation from lactating rats, O'Byrne et al. [11] failed to demonstrate any affect of electrical stimulation of the lobe on either oxytocin or vasopressin secretion. These differences may be explained by the different preparations used: for example, O'Byrne et al. [11] used a combined neurointermediate lobes from female rats whereas Dayanithi et al. [10] used isolated neural lobes from male rats: the steroid milieu could make influence the findings. The different results might also be explained by the presence of some factor(s) from the intermediate lobe that suppressed the release of oxytocin and vasopressin [11]. Parry et al. [12] showed that the cardiovascular response to relaxin varied with different stages of the reproductive cycle suggesting that even in female rats, the steroid milieu at the time of the experiment might have a mediating role in the action of relaxin either directly or indirectly.

Relaxin may affect the release of hormones from the anterior pituitary but these data are also contradictory. Human recombinant relaxin increases circulating growth hormone and prolactin in rhesus monkeys [13] but this was not confirmed in pigs [14]. In the former case, it is possible that relaxin could be acting directly on the anterior pituitary or indirectly through releasing hormones from the hypothalamus. It has been reported, however, that porcine relaxin can stimulate prolactin release directly from rat adenohypophyseal cells *in vitro* [15].

Finally, there is also controversy about whether or not relaxin could act directly on the pituitary. Early work on relaxin-binding sites [16] provided evidence for the possibility that relaxin might bind to this area directly from the blood stream. However, in more extensive studies [17, 18] where relaxin-binding was identified in different areas of the brain there was no reference to relaxin-binding in the pituitary.

Evidence that relaxin acts on the brain via the circumventricular organs

Relaxin is a powerful dipsogen when injected into the ventricular system of the brain (Fig. 2). There are several lines of evidence to suggest that the actions of relaxin are mediated through the circumventricular organs: lesion of the subfornical organ blocks the action of relaxin on blood pressure [19], reflex milk-ejection [20] and the timing of the onset of birth [21]; there are relaxin-binding sites in the subfornical organ and organum vasculosum of the lamina terminalis [17,18]; ***icv*** treatment with relaxin is followed by an increased expression of *c-fos* mRNA in the subfornical organ [22]; exogenous relaxin activates subfornical organ neurons [5, 23, 24], and finally osmoreceptors for the regulation of thirst and vasopressin secretion are present in the circumventricular organs [8, 25, 26]. It is therefore possible that the dipsogenic action of relaxin is mediated through the circumventricular organs.

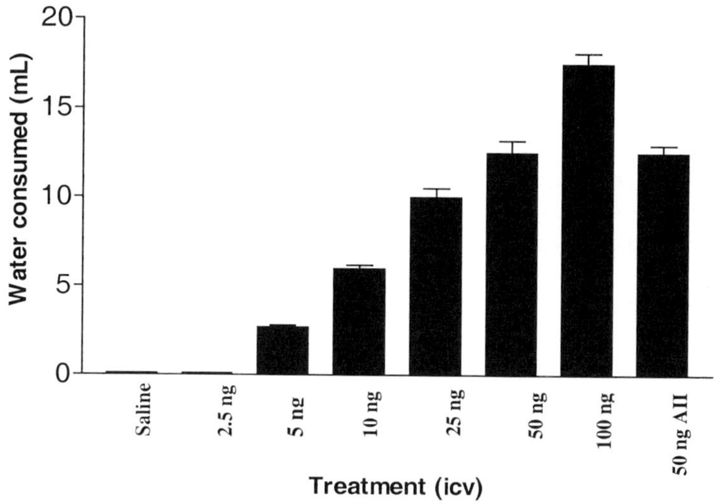

Fig. 2. Dipsogenic effect of different doses of relaxin (ng) injected into the cerebral ventricles of female rats (with permission [30]).

Relaxin-activation of the forebrain angiotensin II system results in the release of vasopressin and oxytocin [27], in addition to the reported effects on drinking behaviour. The release of vasopressin, but not oxytocin, appears to be mediated through the nerve cells of the median preoptic nucleus as cell-body destruction of neurons in this region blocks the vasopressinemia to exogenous relaxin [28]. However, relaxin is not acting directly on cells in the median preoptic nucleus, as these neurons do not respond to topical relaxin *in vitro* (BC Wilson, unpublished observation). Hornsby and Summerlee [29] have shown that an intact median preoptic nucleus is also required for the relaxin-induced dipsogenesis in conscious rats but the role and importance of this nucleus in the control of water intake in pregnancy remains to be established.

Evidence that relaxin acts through the forebrain angiotensin II system is based on the observations that prior treatment with a peptidic angiotensin II receptor antagonist, saralasin, (which blocks AT_1 and AT_2 receptors) inhibits relaxin-stimulated drinking [30]. It also blocks the relaxin-induced release of oxytocin and vasopressin and the pressor response to relaxin [27, 31, 32]. More recently, it has been shown that the specific AT_1 receptor antagonist, losartan, blocks the dipsogenic action of centrally administered relaxin [33]. It is unlikely that the effects of losartan represent a non-specific depressive effect on drinking behaviour because ***icv*** losartan does not inhibit water intake after dehydration, or salt intake after sodium depletion in rats [34].

There are also data implying that relaxin may be physiologically relevant in dipsogenesis during pregnancy and that there may be a different action for circulating relaxin from that which is in the *csf* or brain. Using monoclonal antibodies to neutralize the effects of endogenous relaxin in the periphery, Zhao and colleagues [35] demonstrated that endogenous circulating relaxin had marginal effects on drinking during pregnancy: relaxin-neutralized rats consumed less water during the day compared with intact controls. Recently Omi and colleagues [36] expanded these observations by showing that peripheral injection of exogenous relaxin promotes only moderate increases in water intake during late pregnancy in rats and does not affect drinking during the night. These authors suggested that the more likely explanation was that circulating relaxin might enter the *csf* to have its affect on drinking. However, they did comment that it was possible that relaxin produced within the brain could have a direct, local action on drinking behaviour. There is evidence that relaxin is synthesized within the brain [18, 37]. Furthermore, data from our laboratory showed that passive neutralization of relaxin within the *csf* highlighted a role in the timing of the onset of birth that was different from the role observed when the effects of relaxin were neutralized in the periphery [3].

To examine whether central, as opposed to peripheral relaxin might have a physiological role in drinking behaviour during pregnancy, Summerlee and colleagues [4] investigated the effects of central administration of monoclonal antibodies specific to rat relaxin on drinking behaviour of pregnant rats. Pregnant rats were injected daily with a specific monoclonal antibody raised against rat relaxin from day 12 to day 22 of pregnancy through a chronically implanted ***icv*** cannula. Drinking and eating behaviour was monitored every 12 h during pregnancy and the data were compared with control animals treated with saline alone. There was a significant decrease in water consumed at night in the relaxin-neutralized rats versus the relaxin-intact animals (Fig. 3). In contrast, there was no effect on drinking during the day. The relaxin-neutralized animals also showed a decrease in weight gain during pregnancy.

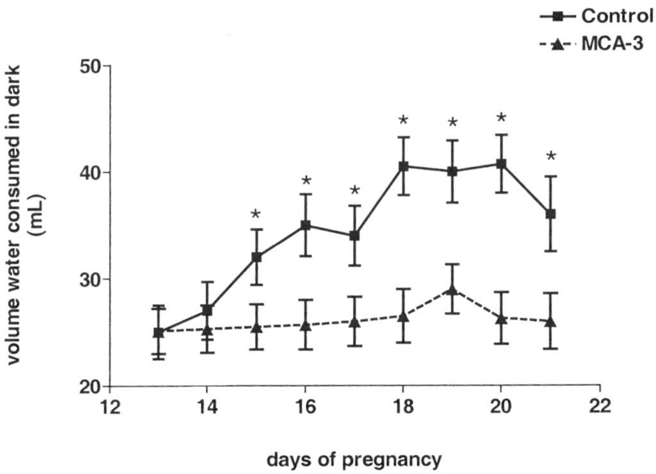

Fig. 3. Effect of neutralizing relaxin in the ventricular system on drinking behaviour in pregnant rats at night (redrawn with permission [3]).

These data indicate that neutralizing the action of relaxin inside the ventricular system of the brain disrupts the normal increase in drinking that occurs in the second half of pregnancy in rats. It implies that relaxin may have a physiological role in water balance in late pregnancy. It also raises questions about whether relaxin acts **'in'** or **'on'** the brain. When compared with data on the effects of systemic relaxin on water balance in pregnancy (see above), they suggest that the actions of relaxin in the brain may be separate and different from the actions in the periphery. There is, however, some evidence that relaxin may also affect blood vessel tone in the periphery in normotensive and hypertensive rats [38]. Original work done on the effects of relaxin on the timing of birth [40, 41] was based on the premise that very high levels of circulating relaxin observed at the end of pregnancy [41 might 'spill' over into the *csf* to have a central effect, i.e., that relaxin of peripheral origin might act on central receptors for relaxin.

Evidence that relaxin acts in the brain

There is some evidence that relaxin acts directly **'in'** the brain. There are relaxin binding sites throughout the brain in sites other than the circumventricular organs [17] and evidence of differential expression of relaxin mRNA at various times in the neonatal period in rats [18]. Preliminary data from my laboratory (Fig. 4.) has shown that direct microinjection of relaxin into the neocortex excites neuronal activity. Neurones show a 5-15-fold increase in activity and remain activated for about 10 minutes after treatment. However, the physiological relevance of relaxin action at this and other sites in the brain is not yet determined.

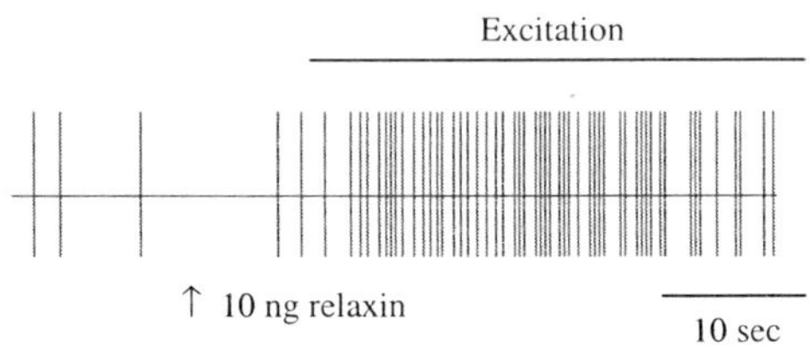

Fig. 4. A short period of excitation following direct injection of relaxin onto a neurone in the neocortex of a rat.

Different effects of relaxin have been reported on the timing of birth in rats depending on whether the central [3] or peripheral effects [41] of relaxin are blocked using specific monoclonal antibodies: blocking the action of circulating relaxin in the blood showed that relaxin did **not** affect the onset of the time of birth in pregnant rats but did affect the course of labour [35] whereas neutralizing the action of relaxin in the ventricular system of the pregnant rats had the opposite effect [3]. The time of the onset of labour was advanced but the course of labour was not affected. These data imply that relaxin in the *csf* not the blood may be important for mediating the actions of relaxin on the brain but there is some evidence that relaxin from the peripheral circulation can reach the brain [42]. It could therefore be argued that experiments neutralizing the effect of relaxin in the *csf* are further proof that peripheral relaxin acts **'on'** the brain by transport into the ventricular system. It does not, however, rule out a direct **'in'** the brain.

Summary

The original observation that peripheral administration of relaxin caused an effect on the brain led to the belief that relaxin produced in the periphery might "spill over into the brain" to have its effects. However, the brain is protected by a blood-brain barrier which would exclude molecules the size of relaxin. Over the past ten years, evidence has accumulated that specialized organs at the interface between the blood, the brain and the *csf* might be critical mediators of relaxin action on the brain and there is evidence that these centres play an important role. However, there are relaxin-binding sites inside the blood-brain barrier and evidence that relaxin is produced in sites remote from the circumventricular organs which may imply a true and independent action of the relaxin in the brain. There is a curious analogy with work on the rennin-angiotensin systems. Originally the dipsogenic action of angiotensin II was reported as a 'spill over' effect into the brain from peripheral injection but it is now accepted that there are completely separate peripheral and central angiotensinergic mechanisms although both have a role in cardiovascular control and dipsogenesis [43]. Perhaps there are similar separate systems for relaxin outside and inside the brain. Finally, it is important to observe that there have been no investigations of the possible effects of relaxin-like factor (RLF) or other members of the insulin family in the brain. With the prospect that there may be overlap in binding between relaxin and RLF [44], this possibility should be investigated.

Acknowledgements

The author acknowledges the support of MRC (UK), NSERC (Canada) and OMAFRA (Ontario) for various aspects of the work reported from the laboratories in Bristol (UK) and Guelph (Canada). This review is dedicated to the memory and friendship of David G Porter and to friendship of the many people who have worked in the laboratories in Bristol and Guelph.

References

1. Summerlee, A.J.S., O'Byrne, K.T., Paisley, A.C., Breeze, M.E. and Porter, D.G., Nature, 309 (1984) 372.
2. Geddes, B.J. and Summerlee, A.J.S., J. Neuroendocrinol., 7 (1995) 411.
3. Summerlee, A.J.S., Ramsey, D.G. and Poterski, R.S., Endocrinology, 139 (1998) 479.
4. Summerlee, A.J.S., Hornsby, D.J. and Ramsey, D.G., Endocrinology, 13 (1998) 2322.

5. McKinley, M.J., Allen, A.M., Burns, P., Colvill, L.M. and Oldfield, B.J., Clin. Exp. Pharmacol. Physiol., 25 (1998) S61.
6. Pardridge, W.M., Endocrine Rev., 28 (1988) 314.
7. Gross, P.M. and Weindl, A., J. Cereb. Blood Flow Metab., 7 (1987) 663.
8. Miselis, R.R., Brain Res., 230 (1981) 1.
9. Saavedra, J.M., Endocrine Rev., 13 (1992) 329.
10. Dayanithi, G, Cazalis, M and Nordmann, J.J., Nature, 325 (1987) 813.
11. O'Byrne, K.T., Eltringham, L., Clarke, G. and Summerlee, A.J.S., J. Endocrinol., 109 (1986) 393.
12. Parry, L.J., Poterski, R.S. and Summerlee, A.J.S., Biol. Reprod., 50 (1994) 622.
13. Bethea, C.L., Cronin, M.J., Haluska, G.J. and Novy, M.J., J. Clin. Endocrinol. Metab., 69 (1989) 956.
14. Samson, W.K. and Anthony R.V., Program of 72nd Annual Meeting of Endocrine Society, Atlanta GA, USA abstract 318.
15. Sortino, M.A., Cronin, M.J. and Wise, P.M., Endocrinology, 124 (1989) 2013.
16. O'Byrne, E.M., Brindle, S., Quintavalla, J., Strawinski, C, Tabachnick, M. and Steinetz, B.G., Ann. N.Y. Acad. Sci. USA, 88 (1982) 6413.
17. Osheroff, P.L. and Phillips, H.S., Proc. Natl. Acad. Sci. USA, 88 (1991) 6413.
18. Osheroff, P.L. and Ho, W.H., J. Biol. Chem., 268 (1993) 15193.
19. Mumford, A.D., Parry, L.J and Summerlee, A.J.S., J. Endocrinol., 122 (1989) 747.
20. O'Byrne, K.T., Eltringham, L. and Summerlee, A.J.S., Brain Res., 405 (1987) 80.
21. Summerlee, A.J.S. and Wilson, B.C., Endocrinology, 134 (1994) 2115.
22. McKinley, M.J., Burns, P., Colvill, L.M., Oldfield, B.J., Wade, J.D., Weisinger, R.S. and Tregear, G.W., J. Neuroendocrinol., 9 (1997) 431.
23. Wilson, B.C, Summerlee, A.J.S. and Ingram, C.I., (*in preparation*)
24. McKinley, M.J. and Oldfield, B.J., Trends Endocrinol. Metab., 9 (1998) 349.
25. McKinley, M.J., Denton, D.A., Leksell, L.G., Mouw, D.R., Scroggins, B.A., Smith, M.H., Weisinger, R.S. and Wright, R.D., Brain Res., 236 (1982) 210.
26. Thrasher, T.N. and Keil, L.C., Am. J. Physiol., 253 (1987) R108.
27. Geddes, B.J., Parry, L.J and Summerlee, A.J.S., Endocrinology, 134 (1994) 1188.
28. Geddes, B.J., Poterski, R.S., Rubenstein, D. and Summerlee, A.J.S., J. Neuroendocrinol., (*submitted*) 2001
29. Hornsby, D.J. and Summerlee, A.J.S., Endocrinology, (submitted) 2000
30. Summerlee, A.J.S. and Robertson, G.F., Endocrine, 3 (1995) 377.
31. Parry, L.J. and Summerlee, A.J.S., Endocrinology, 129 (1991) 47.
32. Parry, L.J., Poterski, R.S. and Summerlee, A.J.S., Biol. Reprod., 50 (1994) 622.
33. Summerlee, A.J.S., Poterski, R.S. and Wilson B.C., (Relaxin Conference 2000).
34. Weisinger, R.S., Blair-West, J.R., Burns, P., Denton, D.A. and Tarjan, E., Peptides, 18 (1997) 977.
35. Zhao, S., Malmgren, C.H., Shanks, R.D. and Sherwood, O.D., Endocrinology, 136 (1995) 1892.
36. Omi, E.C., Zhao, S., Shanks, R.D. and Sherwood, O.D., J. Endocrinol., 153 (1997) 33.
37. Gunnersen, J.M., Crawford, R.J. and Tregear, G.W., Mol. Cell. Endocrinol., 110 (1997) 55.
38. Massicotte, G., St-Louis, J., Parent, A. and Schiffrin, E.L., Can. J. Physiol. Pharmacol., 65 (1987) 2466.
39. Jones, S.A. and Summerlee, A.J.S., J. Endocrinol., 109 (1986) 85.
40. Jones, S.A. and Summerlee, A.J.S., J. Endocrinol., 111 (1986) 99.

41. Sherwood, O.D., Crnekovic, V.E., Gordon, W.I. and Rutherford, J.F., Endocrinology, 107 (1980) 691.
42. Parry, L.J., PhD Thesis, University of Guelph, Guelph, ON, Canada (1991).
43. Fitzsimons, J.T., Physiol. Reviews, 78 (1998) 583.
44. Büllesbach, E.E. and Schwabe, C., J. Biol.Chem., 270 (1995) 16011.

Neurons and neural pathways mediating the actions of circulating relaxin on the brain

Michael J McKinley, Peta Burns, Liana Colvill, Michelle Giles, Puspha Sinnayah, Nana Sunn, Brian J Oldfield and Richard S Weisinger
Howard Florey Institute of Experimental Physiology and Medicine, University of Melbourne, Victoria, 3010, Australia

Introduction

Circulating relaxin, like most (but not all) peptide hormones, probably does not readily pass across the blood-brain barrier. However, relaxin can exert influences on the brain, and *in vitro* autoradiographic binding [1] and physiological studies [2] suggest that it does this by binding to specific receptors in the subfornical organ and organum vasculosum of the lamina terminalis (OVLT), two circumventricular organs (CVOs) lacking a blood-brain barrier [3], that are situated in the anterior wall of the third cerebral ventricle. Included in the brain-mediated responses that circulating relaxin invokes are water drinking, as well as vasopressin and oxytocin secretion from the posterior pituitary [2].

In order to identify the neurons in the CVOs that are responsive to relaxin, we utilised immunohistochemical techniques to detect relaxin-induced increases in expression of Fos, the protein encoded by the immediate early gene *c-fos,* in rats. Increased expression of *c-fos* has been shown to be a reliable indicator of increased neuronal activity in many brain regions [4], including the CVOs [5]. We then investigated whether relaxin-stimulated neurons in the CVOs had efferent axonal connections to the supraoptic and paraventricular nuclei of the hypothalamus. These two nuclei are the sites in the brain where magnocellular neurons synthesise vasopressin and oxytocin for release from their nerve terminals in the posterior pituitary [6]. They also receive afferent neural input originating from neurons in the subfornical organ and OVLT, as well as from neurons in many other brain regions [7]. Our experiments determined whether neurons in the subfornical organ and OVLT that had been labelled with a tracer that had been retrogradely-transported from the supraoptic and paraventricular nuclei, also exhibited Fos-immunoreactivity in response to intravenously infused H2 relaxin. Finally, we studied relaxin-induced water drinking and its possible interactions with angiotensin II, derived from either the circulation or the brain.

Results and Discussion

Adult Sprague-Dawley rats of either sex were anaesthetised with equithesin, then prepared surgically with a polyethylene cannula in the femoral vein. After allowing three days of recovery from this procedure, an infusion of either H2 relaxin or vehicle

G.W. Tregear et al. (eds.), Relaxin 2000, 201–208.
© 2001 *Kluwer Academic Publishers. Printed in the Netherlands.*

was made into the femoral vein for 1 hour. After 90 min, rats were killed, their brains perfused with isotonic saline followed by fixative (4% paraformaldehyde in phosphate buffer) and immersed overnight in 20% sucrose/phosphate buffer. Coronal sections (40 μm) were prepared on a freezing microtome, and these sections were subsequently processed with standard immunohistochemical procedures to detect Fos immunoreactivity, using a primary antibody raised in rabbits (Ab-2; Oncogene Science). A biotinylated secondary antibody directed against rabbit immunoglobulin was used, in conjunction with an avidin-biotin-peroxidase complex and diaminobenzidine as the chromogen, as we have previously described in more detail [8].

Intravenous infusion of synthetic H2 relaxin into Sprague-Dawley rats (n = 5) at 25μg/ml/h for one hour, which would probably produce blood levels of relaxin somewhat higher than those observed in pregnant rats, resulted in a large number of neurons in the subfornical organ, OVLT, supraoptic nucleus and hypothalamic PVN exhibiting Fos-immunoreactivity (Fos-IR) in their cell nuclei. This did not occur in control rats (n = 4) infused intravenously with the vehicle (isotonic 0.15M NaCl solution at 1ml/h). In rats infused with H2 relaxin, there was a characteristic distribution of the Fos-IR in the subfornical organ and OVLT. Many Fos-positive neurons were present in the more peripheral and dorsal parts of the subfornical organ, with an absence of Fos-IR in the central core of the subfornical organ (Fig. 1).

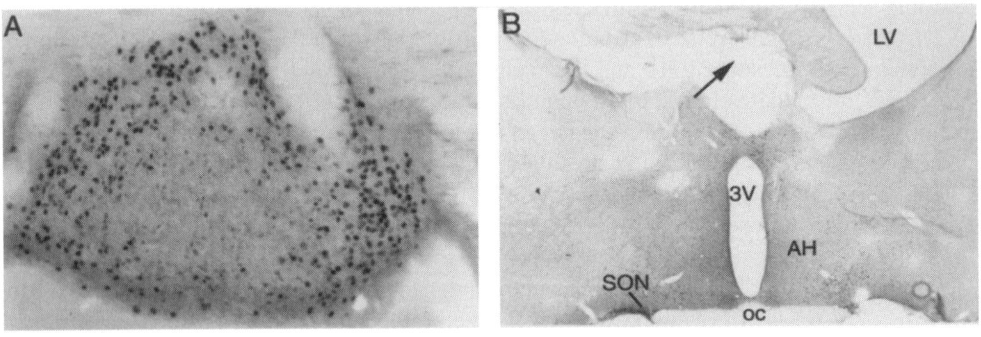

Fig. 1. A. Coronal section of the subfornical organ of a rat that had been infused intravenously with synthetic H2 relaxin (25μg/h), showing Fos-immunoreactivity around the periphery of this structure. Magnification x 150. B. A low power coronal section through the diencephalon of a rat in which the subfornical organ had been ablated (arrow marks the site of the lesion). Abbreviations: AH, anterior hypothalamus; LV, lateral ventricle; OC, optic chiasm; SON, supraoptic nucleus; 3V, third ventricle.

In the OVLT, Fos-IR was present in neurons in its dorsal cap region (Fig. 2). Only a few neurons exhibited Fos-IR in other parts of this CVO. In the hypothalamic paraventricular nucleus (PVN) many neurons within the major magnocellular groups exhibited Fos-positive neurons (Fig. 3), while Fos-IR was observed throughout the entire extent of the supraoptic nucleus (Fig. 3). In some rats, sections were also incubated with primary antibodies raised against either oxytocin or vasopressin, and it was observed that both oxytocin- and vasopressin-containing neurons in the supraoptic nucleus and PVN exhibited Fos-IR after intravenous infusion of H2 relaxin. Other investigators also showed that intravenous injection of porcine relaxin increased Fos-IR in these two nuclei [9].

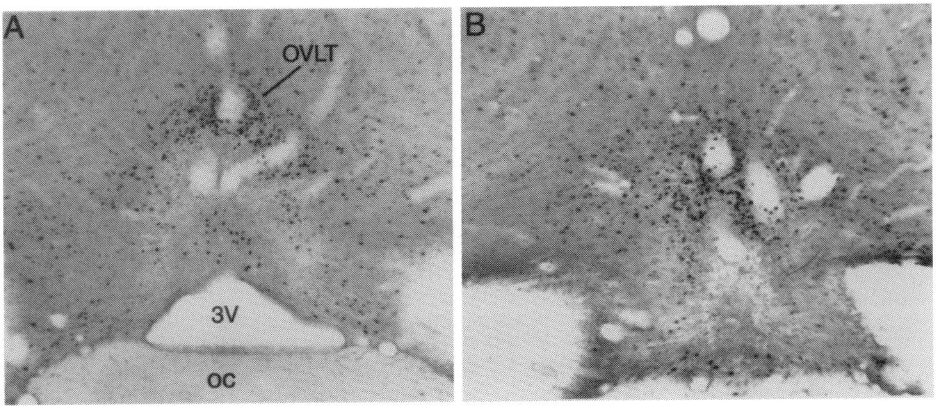

Fig. 2. Coronal sections through the OVLT of rats infused intravenously with synthetic H2 relaxin at 25 μg/h, showing Fos immunoreactivity (seen as small black dots) in the dorsal cap region of the OVLT. A. Section of OVLT from a normal rat. B. OVLT of a rat in which the subfornical organ had been ablated (as shown in Fig. 1.). Abbreviations: OC, optic chiasm; OVLT, organum vasculosum of the lamina terminalis; 3V, third ventricle. Magnification x 100.

In six rats, the subfornical organ had been ablated by electrolysis as previously described [10]. Intravenous H2 relaxin at 25 μg/h for 1 hour still resulted in a considerable number of neurons exhibiting Fos-IR in the dorsal cap of the OVLT (Fig. 2). Fos-IR in response to intravenous H2 relaxin was also observed in the supraoptic and paraventricular nuclei. This result shows that blood-borne relaxin directly activates neurons in the dorsal part of the OVLT, and not secondarily from neural pathways from the subfornical to OVLT, that are known to exist [11]. The relaxin-induced Fos-IR around the periphery of the subfornical organ is also probably due to a direct action of relaxin, because relaxin increases the firing rate of neurons in an *in vitro* preparation of isolated rat subfornical organ (H. Schmid, M. Rauch, M.J. McKinley, unpublished observations).

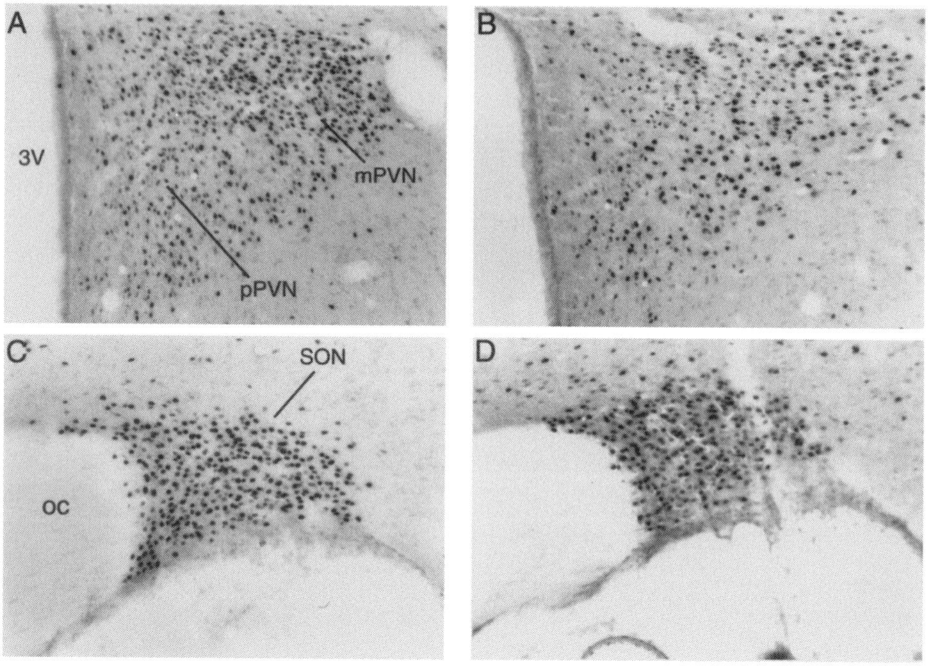

Fig. 3. Coronal sections of the hypothalamic paraventricular nucleus (A and B) and supraoptic nucleus (C and D) of rats that had been infused intravenously with synthetic H2 relaxin at 25μg/h, showing Fos immunoreactivity (seen as black dots). Sections in A and C are from normal animals, while thos in B and D are from a rat in which the subfornical organ had been ablated (shown in Fig. 1). Abbreviations: mPVN, magnocellular region of the paraventricular nucleus; pPVN, parvocellular region of the paraventricular nucleus; OC, optic chiasm; SON, supraoptic nucleus; 3V, third ventricle. Magnification x 150.

In other rats, minute amounts (approximately 50 nl) of the retrogradely transported tracer cholera toxin B sub-unit conjugated to gold particles (CTB-gold) was injected into either the hypothalamic supraoptic or paraventricular nuclei of anaesthetised rats, using stereotaxic surgery. After 1-2 weeks recovery, infusions of H2 relaxin were made into the femoral vein for 1 hour as described above. After 90 min, rats were killed, their brains perfused with fixative and sections prepared on a freezing microtome. The coronal sections were then subjected to histological procedures to visualise the neurons that contained both retrogradely transported CTB-gold and/or Fos using techniques described previously [12]. Injections of CTB-gold into the supraoptic nucleus were successful in 6 rats, and neurons that had been labelled with tracer retrogradely transported from this site were observed throughout the lamina terminalis, but particularly in the periphery of the subfornical organ and dorsal part of the OVLT [13]. A significant proportion of these neurons, as well as containing CTB-gold, also exhibited Fos-IR in their cell nuclei. Similar double-labelling with Fos-IR and CTB-gold also was observed in the subfornical organ and OVLT of rats that had received microinjections of CTB-gold into the hypothalamic PVN [13]. These results indicate that neurons around the periphery of the subfornical organ and in the dorsal cap of the OVLT, which are activated by relaxin infused into the circulation, send axonal projections to the hypothalamic supraoptic nucleus and PVN. It seems most likely that these efferent neural pathways from the two CVOs of the lamina terminalis mediate vasopressin and oxytocin release from axons of neurons in the supraoptic and paraventricular nuclei in response to blood-borne relaxin.

We have previously observed [12] that intravenous infusion of hypertonic saline also stimulates many neurons to express Fos in the dorsal cap of the OVLT, and to a much lesser extent, in the periphery of the subfornical organ. The majority of these osmosensitive neurons in the dorsal cap of the OVLT have efferent projections to the supraoptic nucleus, and we have proposed that they are the main physiological osmoreceptors in the rat brain regulating vasopressin secretion [12]. It is possible that these same osmosensitive neurons in the dorsal cap of the OVLT are also responsive to intravenously infused H2 relaxin, and this could provide a neuroanatomical basis for the proposal [14] that relaxin resets the osmostat during pregnancy.

Angiotensin AT_1 receptors are also abundant on neurons in the subfornical organ and OVLT [15]. Many neurons in these two CVOs express Fos in response to increased blood levels of angiotensin II, and although the majority of stimulated neurons are situated in the core of the subfornical organ and the periphery of the OVLT, some are also observed in the outer parts of the subfornical organ and the dorsal cap of the OVLT [16,17]. As relaxin-sensitive neurons are found in these latter two sites, it is possible that some angiotensin-sensitive neurons also respond to blood-borne relaxin. This may be of relevance to the synergistic action of angiotensin II on relaxin-stimulated drinking in the rat [18], which is discussed in the following paragraphs.

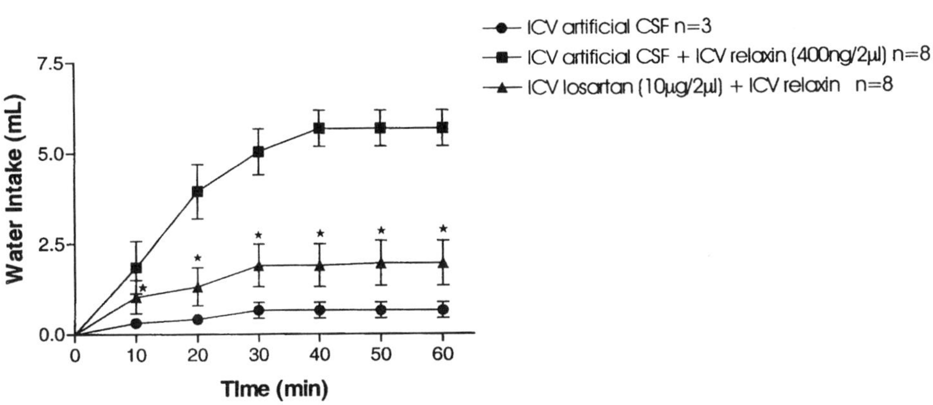

Fig. 4. Cumulative water drinking of male Sprague-Dawley rats in response to injection of synthetic H2 relaxin (400 ng) into the lateral cerebral ventricle, and inhibition of this response by prior intracerebroventricular (ICV) injection of losartan (10µg). Mean ±SEM are shown. Stars indicate significant difference from values obtained with ICV artificial CSF + relaxin.

In recent years there have been reports that water drinking could be stimulated by injections of porcine relaxin into the lateral cerebral ventricle of the rat [19,20]. However, because relaxin is a circulating peptide that is unlikely to cross the blood brain barrier, it was of interest to know whether relaxin administered into the blood-stream could stimulate drinking. Therefore, we infused H2 relaxin at doses of 25-80 μg/kg/h for one hour in water replete Sprague-Dawley rats, and observed a dose-related increase in water drinking within 30-60 min of the commencement of infusions [18]. The water drinking induced by intravenous infusion of H2 relaxin was inhibited by the prior intracerebroventricular (ICV) administration of the angiotensin AT_1-selective receptor antagonist losartan, indicating the involvement of a central angiotensinergic neural pathway [18]. Earlier evidence indicated that ICV administration of saralasin, a non-selective peptide antagonist of angiotensin receptors, blocked vasopressin and oxytocin secretion as well as water drinking stimulated by centrally-administered relaxin [19,21], and we have observed that centrally administered losartan also blocks drinking induced by ICV H2 relaxin (Fig. 4). Thus, it seems likely that neural pathways that are stimulated by either peripherally- or centrally-administered relaxin in the rat, utilise angiotensin as a neurotransmitter or neuromodulator. The site of this role of angiotensin is yet to be identified.

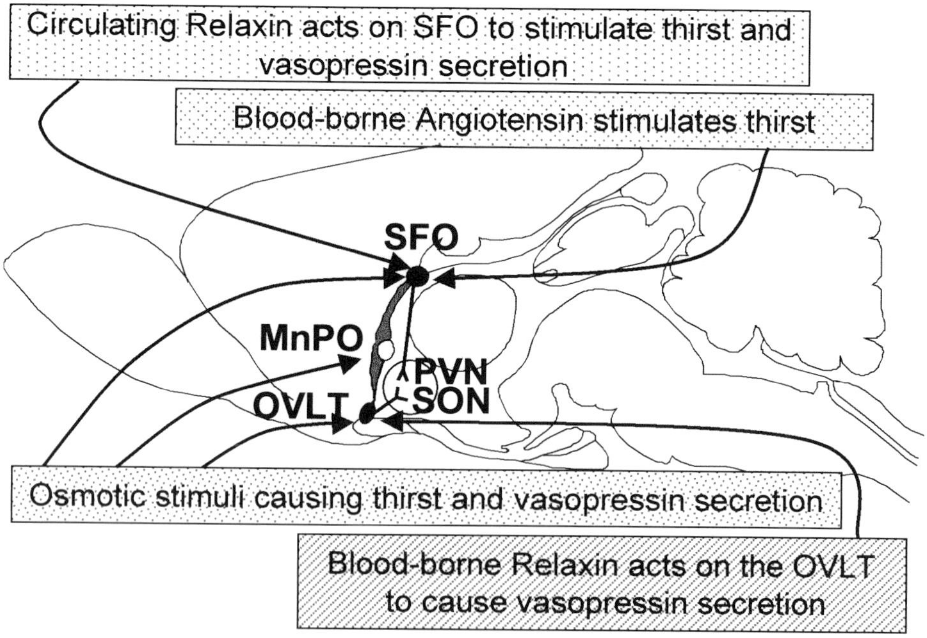

Fig. 5. Diagram of actions of blood-borne relaxin, angiotensin II or systemic hypertonicity on the circumventricular organs of the lamina terminalis to influence body fluid homeostasis.

Quite distinct from the role that brain angiotensin may play in the neural circuitry subserving relaxin-induced responses, is a possible synergy between circulating angiotensin II and relaxin in stimulating thirst. We have observed that water drinking induced by intravenous infusion of H2 relaxin is potentiated by peripheral administration of angiotensin II. The dose of angiotensin II required for this potentiation was below the threshold-dose required to induce drinking, and should have resulted in blood levels of angiotensin II similar to those that may arise during a state of hypovolemia [22]. As both relaxin and angiotensin II levels in blood are elevated during pregnancy [23], it is possible that they have a synergistic role in maintaining water intake during pregnacy. Treatment of pregnant rats with antisera to neutralise circulating relaxin has shown a modest reduction of fluid intake, indicating that relaxin may be a dipsogenic hormone in the pregnant animal [24,25]. Another consequence of this synergistic action of these two circulating hormones on thirst is the possibilty that stronger drinking responses will occur in the pregnant animal in response to hypovolemic or hypotensive states where blood angiotensin levels increase [22].

In conclusion, these observations in rats show that circulating relaxin directly stimulates neurons in two regions of the brain which lack a blood-brain barrier, the subfornical organ and the OVLT. Moreover, specific subgroups of neurons in these circumventricular organs are stimulated, leading to the activation of neural pathways that subserve water drinking and oxytocin and vasopressin secretion. These neural pathways that emanate from the periphery of the subfornical organ and the dorsal cap of the OVLT probably utilise central angiotensin II as either a neurotransmitter or modulator at some point. In addition to the involvement of central angiotensin in relaxin-induced influences on body fluid balance, blood-borne angiotensin II which also acts on neurons in the subfornical organ and OVLT, may potentiate the effects of relaxin on neural pathways subserving thirst. The activity of the renin-angiotensin system increases during the course of pregnancy, and it is possible that relaxin and angiotensin II act in concert to maintain fluid intake during pregnancy despite the hypo-osmolality that results.

Acknowledgments
This work was supported by an Institute Block Grant 983001 from the National Health and Medical Research Council of Australia, the G Harold and Leila Y Mathers Charitable Foundation, and the Robert J Jr and Helen C Kleberg Foundation.

References

1. Osheroff, P.L. and Phillips, H.S., Proc. Nat. Acad. Sci. USA., 88 (1991) 6413.
2. Geddes, B.J. and Summerlee, A.J.S., J. Neuroendocrinol., 7 (1995) 411.
3. McKinley, M.J., McAllen, R.M., Mendelsohn, F.A.O., Allen, A.M., Chai, S.Y. and Oldfield B.J., Front. Neuroendocrinol., 11 (1990) 91.
4. Sagar, S.M., Sharp, F.R., and Curran, T., Science, 240 (1988) 1328.
5. McKinley, M.J., Badoer, E., Vivas L. and Oldfield, B.J., Brain Res. Bull., 37 (1995) 131.
6. Silverman, A.J. and Zimmerman, E.A., Annu. Rev. Neurosci., 6 (1983) 357-380.

7. Armstrong, W.E., In Paxinos, G. (Ed) The Rat Nervous System, Second Edition, Academic Press, San Diego, USA, 1985, p. 377.
8. McKinley, M.J., Burns, P., Colvill, L.M., Oldfield, B.J., Wade, J.D., Weisinger, R.S. and Tregear, G.W., J. Neuroendocrinol., 9 (1997) 431.
9. Heine, P.A., Di, S., Ross, L.R., Anderson, L.L. and Jacobson, C.D., Neuroendocrinol., 66 (1997) 38.
10. Weisinger, R.S., Denton D.A., Di Nicolantonio, R., Hards, D., McKinley, M.J., Oldfield, B.J. and Osborne P.G., Brain Res., 526 (1990) 23.
11. Miselis, R.R., Shapiro, R.E. and Hand, P.J., Science, 205 (1979) 1022.
12. Oldfield, B.J., Hards, D.K. and McKinley, M.J., Neurosci., 60 (1994) 255.
13. Sunn, N., McKinley, M.J. and Oldfield, B.J., J. Neuroendocrinol. (In Press 2001).
14. Weisinger, R.S., Burns, P., Eddie, L.W. and Wintour, E.M., J. Endocrinol., 137 (1992) 505.
15. Allen, A.M., Allen, A.M., Oldfield B.J., Giles, M.E., Paxinos G., McKinley M.J. and Mendelsohn, F.A.O. In Quirion, R., Bjorklund, A. and Hokfelt, T. (Eds.), Handbook of Chemical Neuroanatomy, Vol 16: Peptide Receptors, Part I, Elsevier Science, Amsterdam, Netherlands, 2000, p. 79.
16. McKinley, M.J., Badoer, E. and Oldfield, B.J., Brain Res., 594 (1992) 295.
17. McKinley M.J., Allen A.M., Burns, P., Colvill, L.M. and Oldfield, B.J., Clin. Exp. Pharmacol. Physiol., 25 (1998) S61.
18. Sinnayah,P., Burns, P., Wade, J.D., Weisinger, R.S. and McKinley, M.J., Endocrinology, 140 (1999) 5082.
19. Summerlee, J.S. and Robertson, G.F., Endocrine, 3 (1995) 377.
20. Thornton, S.N. and Fitzsimons, J.T., J. Neuroendocrinol., 7 (1995) 165.
21. Geddes, B.J., Parry, L.J. and Summerlee, A.J.S., Endocrinology, 134 (1994) 1188.
22. Mann, J.F.E., Johnson, A.K. and Ganten, D., Am. J. Physiol., 238 (1980) R372.
23. Sherwood, O.D., In Knobil. E., Neill, J.D. (Eds), The Physiology of Reproduction, Second edition, Raven Press, New York, U.S.A., 1994, p. 861.
24. Zhao, S., Malmgren, C.H., Shanks, R.D. and Sherwood, O.D., Endocrinology, 136 (1995) 1892.
25. Summerlee, A.J.S., Hornsby, D.J., and Ramsey, D.G., Endocrinology, 139 (1998) 2322.

Identification of relaxin and relaxin responsive cells in the rat brain

Tanya C D Burazin, Pamela J Davern, Michael J McKinley and Geoffrey W Tregear
Howard Florey Institute, University of Melbourne, Parkville, Victoria 3052, Australia.

Introduction

Relaxin is best known as an ovarian hormone, however, in addition to its sites of action in the female reproductive tract, it also exerts actions in the brain to influence the cardiovascular system and body fluid homeostasis [1, 2]. It has been suggested that relaxin may be produced in the brain and may have a local central mode of action. Furthermore, the actions of relaxin in the brain and central nervous system (CNS) may be distinct from that of relaxin in the systemic circulation [3]. Interestingly, the distribution of structures showing relaxin-like immunoreactivity within the CNS has not previously been described. At this time, there does not appear to be any reports in the literature on the presence and distribution of relaxin in nuclei of the brain, other than a preliminary report from our laboratory [4].

The most convincing evidence for different roles of relaxin in the CNS compared to the peripheral nervous system come from studies using monoclonal antibodies raised against rat relaxin [3, 5]. Intravenous administration (iv) of relaxin monoclonal antibody disrupts the birthing process but not the timing of birth. In contrast, when administered intracerebroventricularly (icv), the timing but not the process of birth is disrupted. In addition, the drinking pattern normally observed in the second half of pregnancy in rats, is also altered. Injection of relaxin directly into the brain completely negates reflex milk ejection [6], increases blood pressure [7], and directly influences the secretion of neuropeptides from the hypothalamus and neural lobe of the pituitary [8-10].

Several lines of evidence suggest that circumventricular organs (CVOs) may mediate the central actions of circulating relaxin (i) the effect of relaxin on blood pressure and on the timing of birth are inhibited by ablating the subfornical organ (SFO) [11, 12] (ii) relaxin binding sites are located in the SFO and organum vasculosum of the lamina terminalis (OVLT) [13-15] which contain osmoreceptors that regulate thirst and vasopressin secretion [16-18] and (iii) intracerebroventricular infusion of relaxin is accompanied by expression of the immediate early gene, c-*fos*, within the SFO [19]. Expression of c-*fos* is a marker of increased neuronal activity [20]. Relaxin can induce dose-dependent drinking behaviour when injected into the cerebral ventricles of rats [21, 22] and when infused acutely into the blood stream [23].

CVOs are a group of highly vascularised structures, which occupy midline locations at the interfaces between the cerebrospinal fluid spaces and the brain parenchyma [24]. They lack a blood brain barrier and are intimately associated with

G.W. Tregear et al. (eds.), Relaxin 2000, 209–214.
© 2001 *Kluwer Academic Publishers. Printed in the Netherlands.*

surrounding brain areas, allowing them to communicate information between the blood and the brain [25, 26]. Information concerning the neuronal afferent and efferent pathways to and from the CVOs and possible transmitters involved in these systems, have revealed that CVOs are highly specialised organs, differing with respect to structure, neuronal connections and functional roles [27, 28].

Nitric oxide synthase (NOS), has been found to be abundant within neurons in both the subfornical organ and OVLT [29, 30]. In addition, nitric oxide (NO) may act as a neurotransmitter in the hypothalamus [31-33]. Thus, the distribution of the enzyme which catalyzes nitric oxide production, NOS [34], was determined by staining for NADPH-diaphorase [35] in combination with immunohistochemistry for Fos (Fos-IR), following infusion into the femoral vein of synthetic human (H2) relaxin at 25 μg/kg/h. In separate experiments the distribution of relaxin mRNA and immunoreactivity within specific nuclei of untreated male rat brain were determined using *in situ* hybridisation histochemistry with oligonucleotide probes and previously characterised antibodies coupled to a more sensitive tyramide signal amplification method.

Results and Discussion

Following intravenous administration of H2 relaxin, activated cells were detected in two CVOs in the brain. A characteristic and specific distribution of Fos-IR was detected within the dorsal cap of the OVLT and the outer annulus of the SFO, as has been reported previously [36]. In contrast, Fos-IR was absent from the inner core of the SFO, suggesting that cells in this region of the SFO are not activated by H2 relaxin.

In addition, the existence of NOS within CVOs has been described in the OVLT and SFO. Within the OVLT a mixed population of cells were detected; cells activated by relaxin that do not contain NOS; cells that do contain NOS but are not activated by relaxin; and cells that are activated by relaxin *and* contain NOS. Within the SFO, cells activated by H2 relaxin were almost all NADPH-diaphorase positive (data not shown). Thus, it seems that NO may play a role in mediating the effects of relaxin within CVOs. Given that NO plays a key role in the control of vascular tone, it is possible that NO may function in the OVLT and SFO to alter blood flow and/or vessel permeability [37, 38]. This would increase the probability of interaction of humoral signals in the plasma with their receptors. In addition, double-labelling experiments have made it possible to determine the co-localisation of NOS with other peptides of interest. NOS co-exists with vasopressin (AVP) and oxytocin (OT), as well as other peptides in cell bodies of magno- and parvo-cellular neurosecretory neurons [37]. Thus, NO may influence the release of hormones/peptides that are found to be co-localised with NOS, via a direct action, or indirectly via the other co-existing peptides.

Initial studies of relaxin-like immunoreactivity (relaxin-IR) utilised polyclonal antibodies raised in rabbit against rat relaxin and conventional horseradish peroxidase-DAB detection methods. A preliminary report from this laboratory described relaxin-IR within the arcuate nucleus only [4]. The results presented here utilised a tyramide signal amplification kit (TSA™), which allowed a more sensitive detection of relaxin-IR fibres and cells in rat brain. The TSA reaction involves the activation of biotinyl tyramide by horseradish peroxidase which results in the deposition of numerous biotin molecules close to the antibody-HRP complexes, which was further labelled with streptavidin-HRP to produce significant signal amplification. Relaxin-IR was detected in cells or fibres of the anterior olfactory bulb, piriform cortex, septohippocampal nucleus (Fig. 1), limbs of the horizontal and vertical diagonal band of Broca, throughout the median preoptic region, lining the third ventricle, hypothalamic and thalamic paraventricular nuclei, and scattered within the arcuate nucleus.

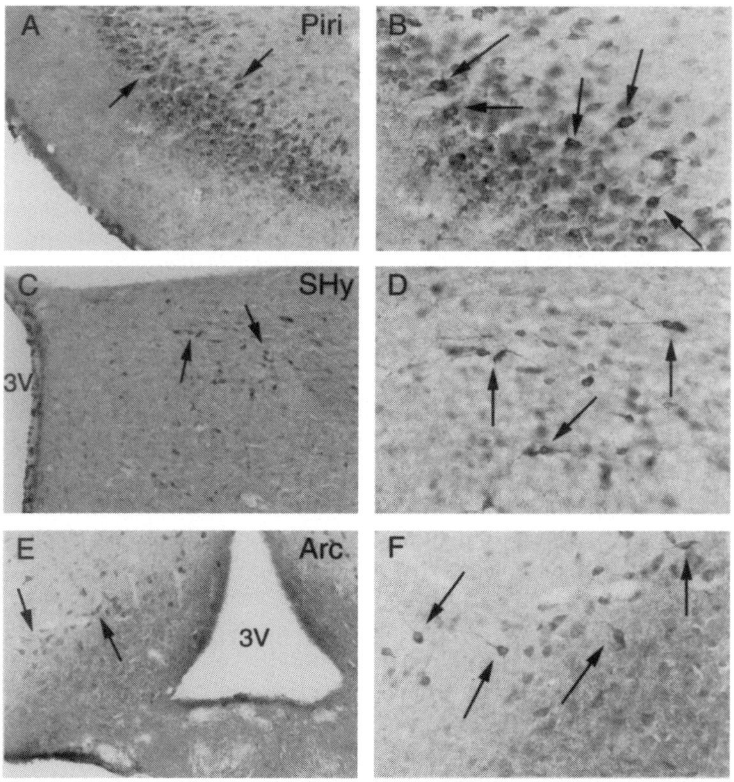

Fig. 1. Relaxin-like immunoreactivity (relaxin-IR) in the rat brain detected using a TSA™ detection method. Brightfield micrographs reveal the localisation of relaxin-IR in the piriform cortex (A, B), septohypothalamic nucleus (C, D), and in the arcuate nucleus (E, F).

Previously, Northern Blot analysis using ^{32}P-labelled DNA probes determined that ovary form rats in late pregnancy contained far greater levels of relaxin mRNA than brain [13]. Using ^{35}S-labelled riboprobes and *in situ* hybridisation histochemistry, relaxin mRNA was detected in the rat brain [13]. Using ^{35}S-labelled *oligonucleotide* probes, we were able to confirm the distribution of relaxin mRNA in the anterior olfactory bulb, tenia tecta, piriform cortex, orbital cortex and neocortex (Fig. 1). However, we were not able to detect relaxin mRNA in the hippocampus or dentate gyrus, although oligonucleotide probes were utilised which are less sensitive than riboprobes. None-the-less, these results further strengthen the argument that relaxin is synthesized in normal rat brain.

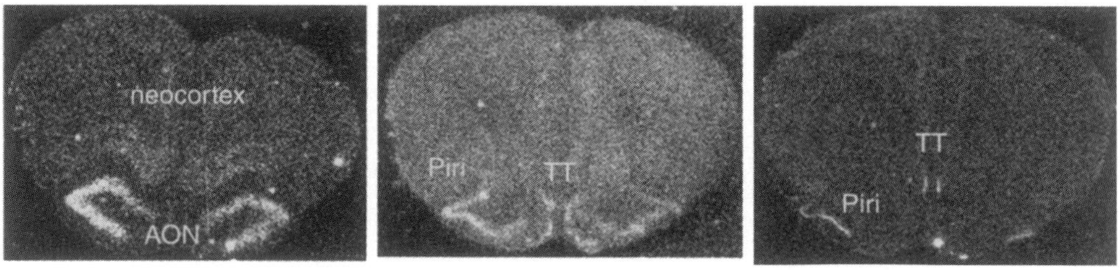

Fig. 2. Distribution of relaxin mRNA in coronal sections of rat brain detected using ^{35}S-labelled oligonucleotides. Images are digitized photographic prints of X-ray film autoradiograms showing high levels of relaxin expression in the anterior olfactory nucleus (AON), moderate levels in the tenia tecta (TT) and piriform cortex (Piri), with lower levels in the rostro and medial neocortex.

All the components of a functional relaxin system exist in the brain. Relaxin mRNA is less widely distributed than relaxin receptors. Only three regions of the brain have been identified to contain both relaxin mRNA and receptors; the anterior olfactory nucleus, dentate gyrus and the neocortex. Relaxin is thought to be produced in the cell body and transported through processes to nerve terminals. Thus, relaxin binding sites may represent sites of release, which may be some distance away from the sites of relaxin mRNA synthesis. Alternatively, the apparent lack of relaxin mRNA throughout the brain could be due to very low levels of relaxin being present, beyond the limit of detection.

The olfactory system is involved in a number of important functions such as reproductive and maternal functions, neuroendocrine regulation, memory and olfaction just to name a few [39]. Anatomical projections of the main olfactory bulb have been described to the anterior olfactory nucleus, tenia tecta and the piriform cortex, and the output pathways of these nuclei relay back to the olfactory bulb [40-42]. The most direct and dense projections to the hypothalamus arise from neurons in the anterior olfactory nucleus and the piriform cortex, and these terminate in the lateral hypothalamic area [39]. The olfactory bulb sends projections to the neocortex and it is postulated that these sites may integrate olfactory and gustatory signals. The arcuate nucleus has a well established role in control of anterior pituitary function [43], and relaxin has been postulated to be released into the capillary plexus of the external lamina of the median eminence which links the hypothalamus and the anterior pituitary, to modulate the production and release of adenohypophysial hormones [4].

Thus, it seems reasonable that separate role(s) of relaxin exist in the CNS since; relaxin mRNA is localised in several discrete regions of the rat brain [13]; relaxin binding sites have been demonstrated in a number of brain regions [15]; it is unlikely that relaxin in the systemic circulation would reach these areas of the brain; administration of a relaxin antibody into the ventricular system of the brain disrupts the timing of birth, but not the process [3, 5]; and injection of monoclonal relaxin antibodies into the ventricular system of the brain affects the drinking response in pregnant rats by blocking the increase in night time drinking.

Our future directions include investigating the possible co-localisation of relaxin with other neurotransmitters and neuropeptides in the olfactory system and hypothalamus. Direct application of relaxin into regions known to contain relaxin binding sites may provide further information about other cells that are activated in the brain, and in order to better determine the function(s) of relaxin in the brain, it is of great importance to determine the nature of the cells containing relaxin and relaxin mRNA. Are they neuronal or glial in nature?

Acknowledgements

This work was supported by an Institute Block Grant 983001, to the Howard Florey Institute from the National Health and Medical Research Council of Australia.

References

1. Sherwood, O.D. In Knobil, E. and Neill, J.D. (Eds.), The Physiology of Reproduction, Raven Press, New York, 1994, p. 861.
2. Geddes, B.J. and Summerlee, A.J.S., J. Neuroendocrinol., 7 (1995) 411.
3. Summerlee, A.J.S., Ramsey, D.G. and Poterski, R.S., Endocrinology, 139 (1998) 479.
4. Gunnersen, J.M., Crawford, R.J. and Tregear, G.W., Mol. Cell Endocrinol., 110 (1995) 55.
5. Zhao, S., Malmgren, C.H., Shanks, R.D. and Sherwood, O.D., Endocrinology, 136 (1995) 1892.
6. Summerlee, A.J.S., O'Bryne, K.T., Jones, S.A. and Eltringham, L., J. Endocrinol., 115 (1987) 347.
7. Parry, L.J. and Summerlee, A.J.S., Endocrinology, 129 (1991) 47.
8. Dayanithi, G., Cazalis, M. and Nordmann, J.J., Nature, 325 (1987) 813.
9. Way, S.A. and Leng, G., J. Endocrinol., 132 (1992) 149.
10. Jones, S.A. and Summerlee, A.J.S., J. Endocrinol., 111 (1986) 99.
11. Mumford, A.D., Parry, L.J. and Summerlee, A.J.S., J. Endocrinol., 122 (1989) 747.
12. Summerlee, A.J.S. and Wilson, B.C., Endocrinology, 134 (1994) 2115.
13. Osheroff, P.L. and Ho, W.-H., J. Biol. Chem., 268 (1993) 15193.
14. Osheroff, P.L., Ling, V.T., Vandlen, R.L., Cronin, M.J. and Lofgren, J.A., J. Biol. Chem., 265 (1990) 9396.
15. Osheroff, P.L. and Phillips, H.S., Proc. Natl. Acad. Sci. USA, 88 (1991) 6413.
16. Miselis, R.R., Brain Res., 230 (1981) 1.
17. McKinley, M.J., Denton, D.A., Leksell, L.G., Mouw, D.R., Scoggins, B.A., Smith, M.H., Weisinger, R.S. and Wright, R.D., Brain Res., 236 (1982) 210.

18. Thrasher, T. N. and Keil, L. C. Am. J. Physiol., 253 (1987) R108.
19. McKinley, M. J., Burns, P., Colvill, L. M., Oldfield, B. J., Wade, J. D., Weisinger, R. S and Tregear, G. W. J. Neuroendocrinol., 9 (1997) 431.
20. Sagar, S. M., Sharp, F. R. and Curran, T. C. Science, 240 (1988) 1328.
21. Summerlee, A. J. S. and Robertson, G. F. Endocrine, 3 (1995) 377.
22. Thornton, S. N. and Fitzsimons, J. T. J. Endocrinol., 7 (1995) 165.
23. Sinnayah, P., Burns, P., Wade, J. D., Weisinger, R. S. and McKinley, M. J. Endocrinology, 140 (1999) 5082.
24. Hofer, H. Verh. Deutsch Zool. Ges., 8 (1958) 2.
25. Wislocki, G. B. and Leduc, E. H. J. Comp. Neurol., 96 (1952) 371.
26. McKinley, M. J. and Oldfield, B. J. In Paxinos, G. and Watson, G. (Eds.), The Human Nervous System, Academic Press, New York, 1990, p. 415.
27. Lind, R. W., Swanson, L. W. and Ganten, D. Brain Res., 321 (1984) 209.
28. Korf, H. W. and Møller, M. In Mess, B., Ruzsas, C., Tima, L. and Pèvet, P. (Eds.), The pineal gland: Current state of pineal research, Elsevier, Amsterdam, 1985, p. 47.
29. Alm, P., Skagerberg, G., Nylén, A., Larsson, B. and Andersson, K.-E. Exp. Brain Res., 117 (1997) 59.
30. Jurzak, M., Schmid, H. and Gerstberger, R. In Pleschka, K. and Gerstberger, R. (Eds.), Integrative and Cellular Aspects of Autonomic Functions: Temperature and Osmoregulation., John Libbey Eurotext, Paris, 1994, p. 451.
31. Garthwaite, J., Charles, S. L. and Chess-Williams, R. Nature, 336 (1988) 385.
32. Garthwaite, J., Garthwaite, G., Palmer, R. M. J. and Moncada, S. Eur. J. Pharmacol., 172 (1989) 413.
33. Dawson, T. M., Dawson, V. L. and Snyder, S. H. Ann. Neurol., 32 (1992) 297.
34. Palmer, R. J. M., Ashton, D. S. and Moncada, S. Nature, 333 (1988) 664.
35. Dawson, T. M., Bredt, D. S., Fotuhi, M., Hwang, P. M. and Snyder, S. H. Ann. Neurol., 32 (1991) 297.
36. McKinley, M. J., Allen, A. M., Burns, P., Colvill, L. M. and Oldfield, B. J. Clin. Exp. Pharmacol. Physiol., 25 (Suppl.) (1998) S61.
37. Yamada, K., Emson, P. and Hökfelt, T. J. Chem. Neuroanat., 10 (1996) 295.
38. Rauch, M., Schmid, H. A., DeVente, J. and Simon, E. J. Neurosci., 17 (1997) 363.
39. Shipley, M. T., McLean, J. H. and Ennis, M. In Paxinos, G. (Eds.), The Rat Nervous System, Academic Press, Sydney, 1995, p. 899.
40. Shipley, M. T. and Adamek, G. D. Brain Res. Bull., 12 (1984) 669.
41. De Olmos, J., Hardy, H. and Heimer, L. J. Comp. Neurol., 181 (1978) 213.
42. Luskin, M. B. and Price, J. L. J. Comp. Neurol., 216 (1983) 264.
43. Swanson, L. W. and Sawchenko, P. E. Ann. Rev. Neurosci., 6 (1983) 269.

Effect of Relaxin on the Expression of Angiotensin II Receptors in the Subfornical Organ of Rats during Pregnancy

Alastair JS Summerlee, Roman S Poterski and Brian C Wilson
Department of Biomedical Sciences, Ontario Veterinary College, University of Guelph, Guelph, Ontario N1G 2W1, Canada.

Introduction

Relaxin is a powerful dipsogen [1] and may be important in water balance in pregnancy. Neutralizing the action of endogenous relaxin within the ventricular system of the brain blocks the normal nighttime drinking response observed in pregnant rats [2] and there appears to be a differential response to central relaxin during pregnancy [3]. Evidence suggests that relaxin acts on the brain through the circumventricular organs, in particular the subfornical organ [4]. The central actions of relaxin are mediated through the angiotensin II receptors [5], which are also present in the subfornical organ. Therefore, experiments were done to investigate the possibility that the changing response to relaxin during pregnancy may be due to differential expression of the AT_1 receptors in the subfornical organ.

Results and Discussion

The effect of relaxin on expression of the AT_1 receptors was studied in two groups of pregnant rats: Group 1 (saline-treated ***icv***) and Group 2 (relaxin-neutralized ***icv*** using a rat specific monoclonal antibody developed in the laboratory [6]). Three animals from each group were euthanized on each of days 10-22 of pregnancy and on days 2 and 10 of lactation. Tissue was amalgamated from punch biopsies of the subfornical organs from each group and subjected to RT-PCR analysis of AT_1 receptors using primers developed by J. Woods, University of Guelph. AT_1 receptor expression was seen on days 12-18 of pregnancy with the highest level seen on day 16 with no evidence of expression before Day 12 and only slight expression after day 18 (Fig. 1). In contrast, treatment with the monoclonal antibody blocked expression of AT_1 receptors (Fig. 2).

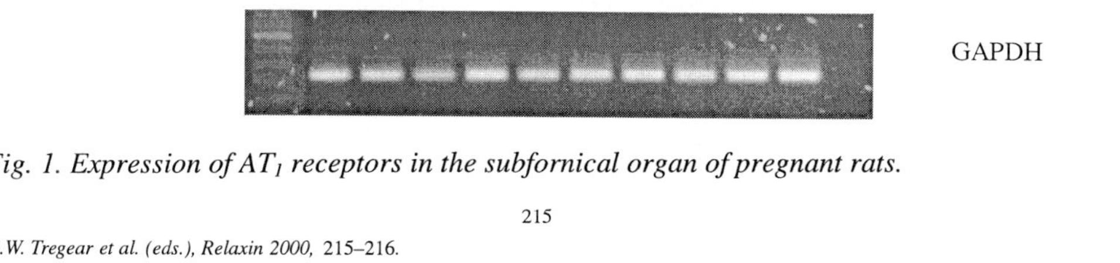

Fig. 1. Expression of AT_1 receptors in the subfornical organ of pregnant rats.

G.W. Tregear et al. (eds.), Relaxin 2000, 215–216.
© 2001 *Kluwer Academic Publishers. Printed in the Netherlands.*

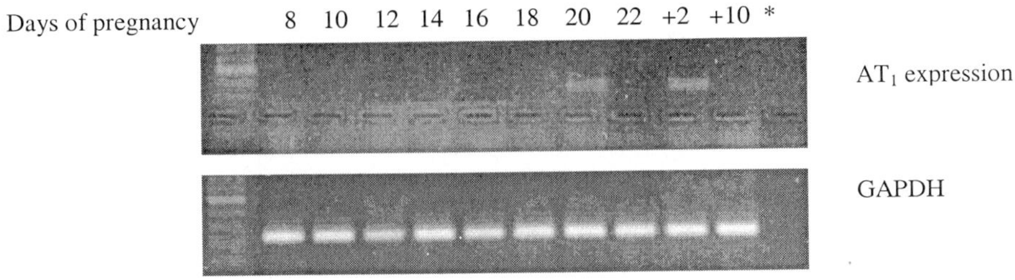

Fig. 2. AT_1 receptor expression in the subfornical organ of relaxin-neutralized pregnant rats.

In situ hybridization studies were also done to reveal presence of AT_1 receptors using a probe developed by J. Cirello, University of Western Ontario, Canada. The number of grains per cell indicating the density of the identified AT_1 receptors was determined. Data indicate that there was a progressive increase in AT_1 receptors during pregnancy with peak values on day 16 (Fig. 3). These data are in accord with the AT_1 receptor expression data above.

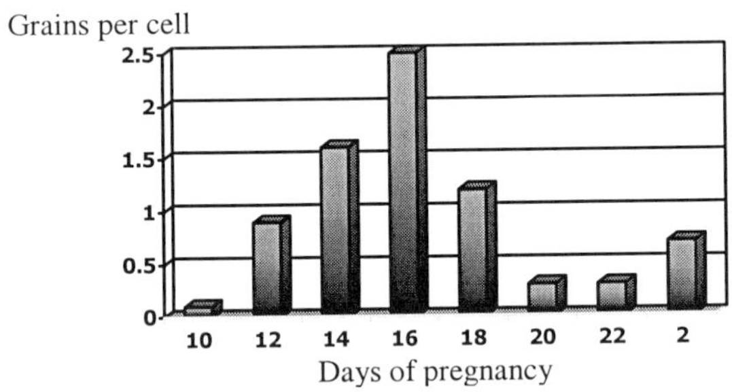

Fig. 3. AT_1 receptor gene expression in pregnant rats

The data show that AT_1 receptor expression increases in pregnancy and there may be a causal relationship between plasma relaxin and AT_1 receptor expression in pregnant rats.

Acknowledgements

This project was sponsored by NSERC (Canada). The drinking experiments were, of course, supported by Fr. E. Ixenet.

References

1. Summerlee, A.J.S. and Robertson, G.F., Endocrine, 3 (1995) 377.
2. Summerlee, A.J.S., Hornsby, D.J. and Ramsey, D.G., Endocrinology, 139 (1998) 2322.
3. Parry, L.J., Poterski, R.S. and Summerlee, A.J.S., Biol. Reprod., 50 (1994) 622.
4. Mumford, A.D., Parry, L.J. and Summerlee, A.J.S., J.Endocrinol., 122 (1989) 747.
5. Geddes, B.J., Parry, L.J. and Summerlee, A.J.S., Endocrinology, 134 (1994) 1188.
6. Summerlee, A.J.S., Ramsey, D.G. and Hornsby, D.J., Endocrinology, 139 (1998) 479.

Relaxin and relaxin receptors in brain

Yean Yeow Tan[1,2], Geoffrey W Tregear[1] and Roger J Summers[2]
[1]*Howard Florey Institute, University of Melbourne, Parkville, Victoria 3010 and*
[2]*Department of Pharmacology, Monash University, Clayton, Victoria 3800, Australia.*

Introduction

Recent studies have shown that in addition to its role as a hormone of pregnancy, relaxin also has physiological functions in the central nervous system [1, 2, 3]. In the rat brain, relaxin binding sites have been located in the organum vasculosum of the lamina terminalis (OVLT) and the subfornical organ (SFO) [4] and centrally administered relaxin increases neuronal Fos expression in these areas [5]. Relaxin binding was also found in a deep layer of rat cerebral cortex [4, 6], a site which is inaccessible to circulating relaxin. However, relaxin mRNA has been detected in rat cortex [6], leading to the hypothesis that relaxin may be produced, released and act locally in this area of the brain.

Quantitative receptor autoradiography with [^{33}P]-labelled human gene 2 (H2) relaxin (B33) [7] was used in this study to investigate the binding characteristics of relaxin receptors in rat brain and to localize relaxin binding sites in human brain. We have also extracted peptides from rat brain using protocols previously described [8] and performed protein dot blot and Western blot analysis using polyclonal anti-rat relaxin antibody to determine whether relaxin is produced and stored in this tissue.

Results and Discussion

[^{33}P]-H2 Relaxin (B33) bound to the rat OVLT, SFO and the 5th layer of the cerebral cortex. Characterization of the binding sites in the cerebral cortex showed that the binding was saturable to a single population of non-interacting sites (Fig. 1). Competition studies demonstrated that H2 relaxin (B29) and porcine relaxin compete for the binding sites in cerebral cortex with similar affinities while rat relaxin competes with a significantly lower affinity ($p<0.02$) (Fig. 1).

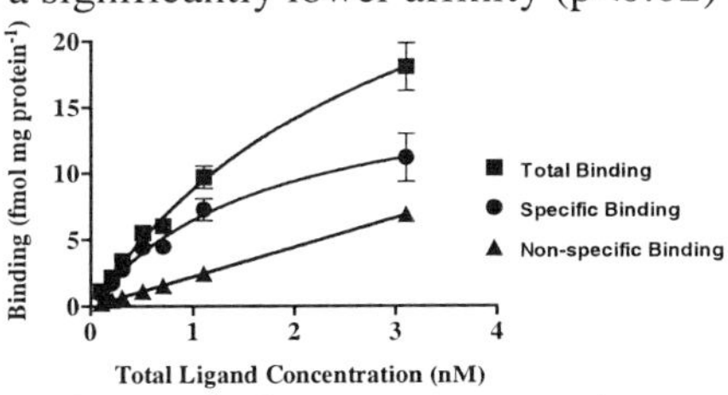

Relaxin	Binding Affinity (pK_i)
H2 (B33)	8.79±0.04 (pK_D)
H2 (B29)	9.17±0.06
porcine	8.90±0.03
rat	8.22±0.07*

*Fig. 1. The graph shows a representative saturation isotherm demonstrating saturable relaxin binding to a single population of non-interacting sites in rat cerebral cortex. The table compares binding affinities of H2, porcine and rat relaxins in rat cerebral cortex (*p<0.02) measured using quantitative receptor autoradiography.*

G.W. Tregear et al. (eds.), Relaxin 2000, 217–218.
© 2001 *Kluwer Academic Publishers. Printed in the Netherlands.*

[^{33}P]-H2 relaxin (B33) binding sites were also detected in the deep layer of human parietal (Fig. 2a), entorhinal and frontal cortex but not in cerebellum, red nucleus, substantia nigra, cerebral peduncle or hippocampus.

The results from protein dot blot analysis showed that immunoreactive material was detected in the rat cerebral cortical, and to a lesser extent the rat ventricular (negative control) extracts eluted from a C_{18} cartridge with 40% acetonitrile in 0.1% aqueous trifluoroacetic acid. When these fractions were separated with SDS-PAGE, followed by a Western blot, two immunoreactive bands (between 20.1 and 30 kDa and <14.4 kDa) were detected in the cerebral cortical extract while only one (<14.4 kDa) was present in the ventricular extract (Fig. 2b). These preliminary results suggest that prorelaxin (approximately 18 kDa) may be stored in the rat cerebral cortex. Further characterization of these immunoreactive bands is required to elucidate their identity.

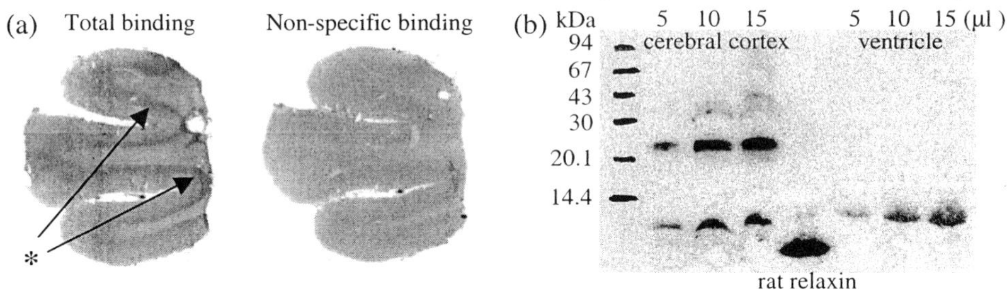

Fig.2. (a) An autoradiogram showing binding of [^{33}P]-h2Rlx (B33) to a deep layer of human parietal cortex (*) and (b) a Western blot showing large and small molecular weight immunoreactive proteins detected in rat cerebral cortical extract (µl extract) but only small molecular weight immunoreactive protein detected in cardiac ventricle extract.

Acknowledgments

We gratefully thank Ms. Courtney Bond (Pharmacology, Monash University) for the competition data on H2 (B29) relaxin. Supported by an Institute Block Grant Reg Key 983001 from the NHMRC to the Howard Florey Institute and by the National Heart Foundation of Australia.

References

1. Weisinger, R.S., Burns, P., Eddie, L.W. and Wintour, E.M., J. Endocrinol., 137 (1993) 505.
2. Dayanithi, G., Cazalis, M. and Nordmann, J.J., Nature, 325 (1987) 813.
3. Way, S.A. and Leng, G., J. Endocrinol., 132 (1992) 149.
4. Osheroff, P.L. and Phillips, H.S., Proc. Natl. Acad. Sci. USA, 88 (1991) 6413.
5. McKinley, M.J., Burns, P., Colvill, L.M., Oldfield, B.J., Wade, J.D., Weisinger, R.S. and Tregear, G.W., J. Neuroendocrinol., 9 (1997) 431.
6. Osheroff, P.L., and Ho, W.-H., J. Biol. Chem., 268 (1993) 15193.
7. Tan, Y.Y., Wade, J.D., Tregear, G.W. and Summers, R.J., Br. J. Pharmacol., 127 (1999) 91.
8. Layden, S.S. and Tregear, G.W., J. Biochem. Biophys. Methods, 31 (1996) 69.

SECTION 7

Physiology of relaxin II: Male

Relaxin knockout mice demonstrate delayed growth and have underdeveloped male reproductive tracts

Chrishan S Samuel[1,2] and Edward P Amento[1,2,3]
[1]Molecular Medicine Research Institute, Mountain View, CA 94043 and Departments of [2]Dermatology and [3]Medicine, Stanford University School of Medicine, Stanford, CA 94305, USA.

Introduction

Relaxin is a 6kDa polypeptide hormone, which belongs to the insulin family of peptides [1] and potentially has a number of functions in mammals. While relaxin is produced by both males and females in a range of mammalian species, the hormone is best known for its actions on the female reproductive tract and pelvic girdle, specifically during pregnancy, in preparation for parturition [2]. Increasing efforts to understand the role of relaxin in the male have led to the identification of the peptide or binding sites for its receptor in the prostate [3], seminal fluid [4] testis cells [5] and spermatozoa [6]. These findings in turn, have led to a number of studies demonstrating the ability of relaxin to increase sperm motility [7] as well as aid the penetration of sperm into the cervical mucus [8].

The relaxin knockout mouse model was established recently and an initial phenotypic study of pregnant relaxin null mice demonstrated deficiencies in mammary gland and nipple development [9]. In the present study, we have further characterized the phenotype of the relaxin gene-knockout mouse by examining the effect of relaxin deficiency on neonatal and early adult rapid growth and on the development of the male reproductive tract.

Results and Discussion

All relaxin wildtype (+/+), relaxin heterozygous (+/-) and relaxin null mutant (-/-) mice used in these studies were generated from relaxin +/- parents. The animals were genotyped by using mouse DNA, lysed from tail tissue as described by Zhao and colleagues [9].

Body weight measurements: The body weights of relaxin +/+, relaxin +/- and relaxin -/- mice were measured from birth to adulthood, at 1week, 1 month, 2 months and 3 months of age ($n \geq 20$ for each group). No significant difference in the body weight of male and female animals was noted between the groups at 1 week of age. By 1 month of age, however, both relaxin -/- males and females were significantly smaller ($p<0.05$) than their relaxin wildtype counterparts, weighing approximately 1.5-2g less. Relaxin -/- mice continued to be significantly smaller ($p<0.05$) than relaxin +/+ mice at 2 months of age, although the difference between the groups had diminished. By 3months of age, there was no difference between the adult mice of each group. No

G.W. Tregear et al. (eds.), Relaxin 2000, 221–223.
© 2001 *Kluwer Academic Publishers. Printed in the Netherlands.*

difference in body weights was noted between relaxin +/+ and relaxin +/- animals or between relaxin +/- and relaxin -/- mice at any of the ages studied.

Male reproductive tract: Male reproductive tract tissues (which included the testis, epididymis, prostate and seminal vesicle) were collected from relaxin +/+ and relaxin -/- male mice at 1week, 1month and 3 months of age ($n \geq 10$ for each group). No significant difference in the overall weight of the male reproductive tract was noted at 1 week to 1 month of age. At 3 months of age, however, the reproductive tract of relaxin -/- mice weighed significant less ($p<0.05$) than reproductive organs derived from relaxin +/+ mice. At this age, each component of the male reproductive tract (testis, epididymis, prostate and seminal vesicle) derived from relaxin -/- mice was significantly smaller and weighed less than its relaxin wildtype counterpart.

Histology: Histological examination of the testis, epididymis, prostate and seminal vesicle of relaxin +/+ and relaxin -/- mice was performed on tissues of animals from 1 week to 3 months of age. Male reproductive tissues were weighed, measured for size and placed into 10% formalin. Samples were then processed and embedded into paraffin blocks before being cut (4μm sections) and placed onto glass slides for staining. Tissues from 1 week, 1 month and 3 month mice were then stained with H&E (hematoxylin and eosin), and Masson trichrome stain for collagen.

We found decreased sperm accumulation and increased cell death in the testis of 1-3 month old relaxin -/- mice, increased collagen in the testis, epididymis and prostate and differences in the size and compactness of testis and epididymis (1 week to 3 months), compared to similar tissues obtained from relaxin +/+ mice. In relaxin -/- mice, the level of sperm in the testis and epididymis of adult (3 month) animals resembled that of immature (1 month) relaxin +/+ mice. Furthermore, the absence of relaxin correlated with a dense extracellular matrix in all tissues that was evident as early as 1 month of age.

Based on the observation of increased cell death in the testis of relaxin -/- mice, further histological evaluation by immunohistochemistry, using specific antibodies to detect cell apoptosis and cell proliferation was performed. Cell apoptosis was investigated using Bax (which accelerates apoptotic death by inducing cytokine deprivation) and Caspase-9 (a central death protease, which induces apoptosis via a separate pathway to Bax). At both 1 and 3 months of age, we noted an increased ($p<0.05$) number of apoptotic cells in the testis and epididymis of relaxin KO mice, compared to that seen in relaxin wildtype mice. Using an antibody to the proliferating cell nuclear antigen (PCNA) we did not detect any difference in proliferating cells in tissues derived from relaxin +/+ and relaxin -/- mouse tissues.

These findings implicate a role for relaxin in normal mammalian growth and development, which may be consistent with a role for relaxin in other processes involving an accelerated remodeling of connective tissues in physiological and pathological conditions. Relaxin deficient mice had delayed body growth, underdeveloped male reproductive tracts and increased apoptosis in all tissues examined. Separate reports have identified relaxin-like factor (RLF) expression in the testis [10], while the establishment of RLF knockout mouse has further demonstrated a potential role for RLF in male sexual differentiation as well as germ cell development

[11]. It should be noted though, that the changes observed in this study were induced by mice carrying the active RLF gene, but without a functional relaxin gene, indicating that the changes we noted were associated with the lack of relaxin. While relaxin knockout mice reproduced successfully, a study of the success rate of insemination by male relaxin knockout has not been done. Based on our observations, we postulate that relaxin plays a previously unexpected role in normal mammalian growth, may function as an anti-apoptotic factor as well as influence male fertility.

References

1. Schwabe, C. and McDonald, J.K., Science, 197 (1977) 914.
2. Sherwood, O.D. In Knobil, E. and Niell, J. (Eds.), The Physiology of Reproduction. 2nd ed., Raven Press, New York, USA, p861.
3. Ivell, R., Hunt, N., Khan-Dawood, F. and Dawood, M.Y., Mol. Cell. Endocrinol., 66 (1989) 251.
4. Winslow, J.W., Shih, A., Bourell, J.H., Weiss, G., Reed, B., Stults, J.T. and Goldsmith, L.T., Endocrinology, 130 (1992) 2660.
5. Min, G. and Sherwood, O.D., Biol. Reprod., 59 (1998) 401.
6. Carrell, D.T., Peterson, C.M. and Urry, R.L., Endocrine Res., 21 (1995) 697.
7. Weiss, G., Biol. Reprod., 40 (1989) 197.
8. Brenner, J.H., Lessing, J.B., Schoenfeld, C., Amelar, R.D., Dubin, L. and Weiss, G., Fertil. Steril., 42 (1984) 92.
9. Zhao, L., Roche, P.J., Gunnersen, J.M., Hammond, V.E., Tregear, G.W., Wintour, E.M. and Beck, F., Endocrinology, 140 (1999) 445.
10. Ivell, R., Rev. Reprod., 2 (1997) 133.
11. Adham, I., Emmen, .M., Engel, W., Molec. Cell. Endocrinol., 160 (2000) 11.

Relaxin expression and binding in the rat prostate

David J Hornsby, Roman S Poterski and Alastair JS Summerlee
Department of Biomedical Sciences, Ontario Veterinary College, University of Guelph, Guelph, Ontario N1G 2W1, Canada.

Introduction

Many of the key actions of relaxin involve connective tissue remodelling [1] including effects on the cervix, skin, nipples and vagina. Much of the work to date has focussed on relaxin in the female but it is also present in the prostate [2]. The prostate gland undergoes a number of structural changes during life. Prostatic hyperplasia and tumours are relatively common in men over 45 years of age and in older males in other species [3]. The etiology of both hyperplasia and carcinoma is not known. Hyperplasia precedes, and is a prerequisite for, prostate cancer in men and male dogs [4]. It begins with a marked remodelling of the connective framework of the gland and significant angiogenesis [5]. Therefore, experiments were done to investigate the possibility that relaxin of prostatic origin might be associated with changes in prostate gland structure in male rats. The expression of relaxin in the prostate, the presence of hormone in the tissue and the occurrence of relaxin-binding sites was studied in aging rats.

Results and Discussion

In situ hybridization for relaxin was carried out in prostatic tissue collected from three groups of rats: Group 1 - prepubertal rats (100-120g), Group 2 - mature animals (220-250g) and Group 3 - senile rats (>450g). Tissue was pre-screened for evidence of hyperplasia and results from these animals discarded. A ^{35}S-labelled probe was used in the studies. Prostatic tissue from adult post-pubertal (Group 2) and senile rats (Group 3) only showed positive hybridization. Control (sense) probes did not show any hybridization.

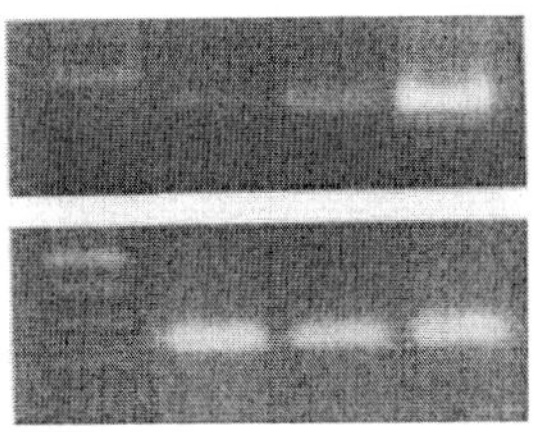

Fig. 1. Expression of relaxin gene transcripts in prostate of prepubertal rats (Lane 1), adult rats (Lane 2) and senile rats (Lane 3).

G.W. Tregear et al. (eds.), Relaxin 2000, 225–227.
© 2001 Kluwer Academic Publishers. Printed in the Netherlands.

Relaxin gene transcripts identified by RT-PCR were seen in prostatic tissue in all three groups of rats (Fig. 1). Relaxin mRNA (1.1 kb) was detected in glandular tissue, vascular and smooth muscle elements by Northern blotting (RS Poterski, unpublished observations). No attempt was made to determine the relative levels of mRNA in the current study but there were qualitative differences observed: the lowest amount of transcript was observed in tissue from the prepubertal rats (Group 1) and the highest amount in the senile animals (Group 3). These data imply that expression of relaxin increases with age in prostatic tissue.

A

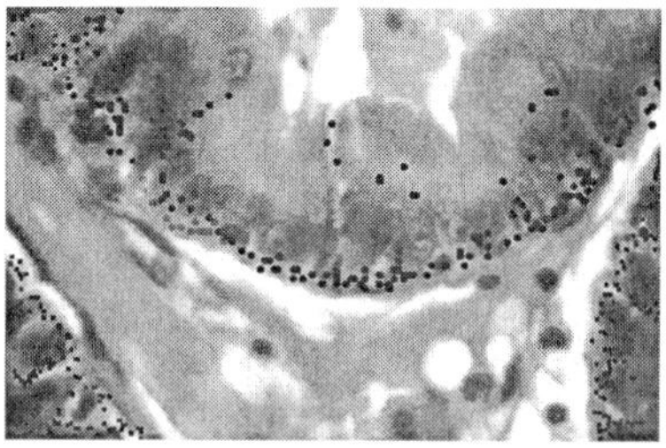

B

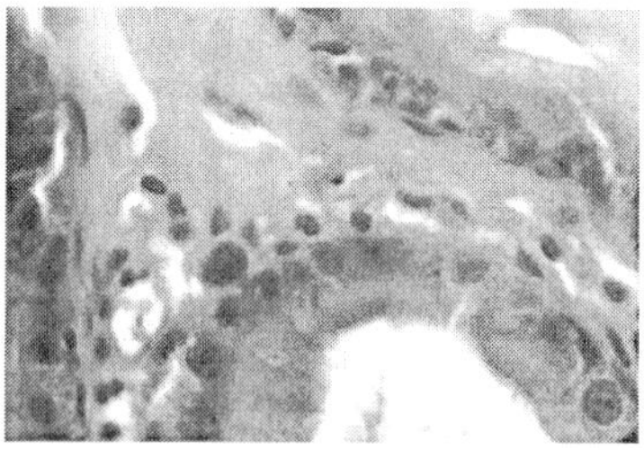

Fig. 2. Binding of biotinylated relaxin to prostatic tissue in senile rats (A). Binding was displaced in the presence of excess relaxin (B)

The presence of putative relaxin-binding sites was determined using biotinylated relaxin. Biological activity of the biotinylated relaxin was confirmed in the mouse pelvic ligament assay and the mean number of biotin molecules per molecule of biotinylated relaxin was determined to be 2.3. Putative relaxin-binding sites were observed in prostatic tissue from adult (Group 2) and senile (Group 3) rats but no evidence of binding was identified in prepubertal (Group 1) animals (Fig. 2). No signal was obtained in sections of prostate treated with unmodified relaxin and the binding was observed to be specific: it was displaced by excess unmodified relaxin but not displaced by excess insulin.

The data imply that expression, presence and binding-sites for relaxin increase with age in the rat prostate. Whether or not this is a causal or associative relationship with prostatic hyperplasia remains to be determined but it raises the possibility that relaxin might be a useful marker in detecting early senescent change in the prostate of aging men.

Acknowledgements

We acknowledge support from NSERC (Canada) and The Mikado in making this possible.

References

1. Sherwood, O.D., In Knobil, E. and Neill, J.D. (Eds.), The Physiology of Reproduction. 2nd ed., Raven Press, New York, USA, 1994, p.681.

2. Ivell, R., Hunt, N., and Khan-Dawood, M.Y., Mol. Cell. Endocrinol., 66 (1989) 251.
3. Barret-Connor, E., Garland, C., McPhillips, J.B., Khaw, K.T. and Wingard, D.L., Cancer Res., 50 (1990) 169.
4. Normua, A., Heilbrun, L.K., Stemmerman, G.N. and Judd, H.L., Cancer Res., 48 (1988) 3515.
5. Lissbrandt, I.F., Stattin, P., Damber, J.E. and Bergh, A., Prostate, 33 (1997) 38.

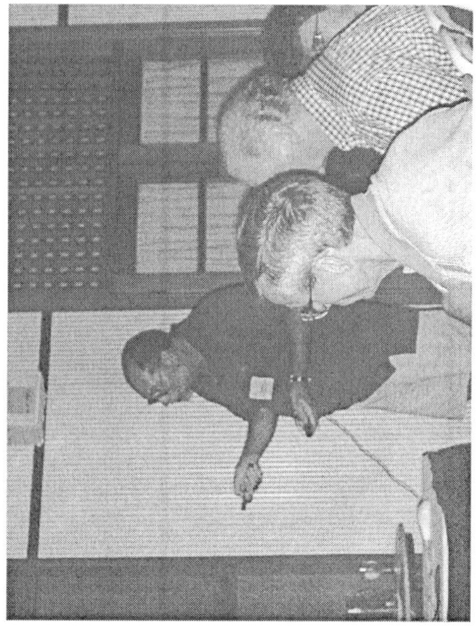

BIACORE

SECTION 8

Preparation of relaxin & related peptides

Chemical synthesis of relaxin analogues: Current status and future developments

Geoffrey W Tregear, Marc N Mathieu, Nicola F Dawson, Kathryn J Smith, Antonia A Claasz, Andrew B Clippingdale, Ping Fu and John D Wade
Howard Florey Institute, University of Melbourne, Victoria 3010, Australia

Introduction

Knowledge of the chemistry of relaxin has increased steadily since the first determination of its primary structure in the early 1970s. The two peptide chain/three disulfide bonds structure mirrors that of insulin and established the concept of the so-called insulin superfamily of peptides and indicated that insulin and relaxin both evolved from a common ancestral gene [1]. Advances in molecular biology and protein microisolation techniques have since led to the identification of more than six other protein members of this superfamily (Table 1). All possess the same insulin/relaxin cystine pairing pattern of one A-chain intramolecular bond and two intermolecular bonds. It is probable that with the recent completion of sequencing of the human genome several more insulin-like proteins will be identified. Indeed, a recent genomic analysis of the nematode, *Caenorhabditis elegans*, suggested the presence of at least ten insulin-like proteins, three of which are predicted to contain an additional disulfide bond or an A-chain intramolecular bond that is substituted by a hydrophobic interaction [2]. Like insulin, relaxin and the other members of this family are synthesized on the ribosome as preprohormones in which the A- and B-chains are joined together via a connecting C-peptide that is proteolytically removed following peptide folding [3]. The exceptions, insulin-like growth factors I and II, do not have excised C-peptides. It is of considerable interest that at a length of approximately 100 residues, the relaxin C-peptide is far longer than thought necessary for efficient folding and suggests that it might be a precursor to smaller, biologically active peptides. Indeed, recent findings have shown that insulin C-peptide plays a role in a broad range of biological activities [4].

Much of the substantial body of acquired knowledge of relaxin's structure-function relationships, biological effects and physicochemical behaviour was acquired using native porcine relaxin which is available in abundant quantities in the ovaries of pregnant sows. However, the growing realization that there is considerable species variation in biological activity of relaxin led to the need to obtain protein material by other means. By far the most common source of relaxin has been from solid phase chemical peptide synthesis although the complex structure of the protein has long prevented its ready production in good quantities. More recently, use of recombinant DNA expression methods for production of relaxin has substantially facilitated further

G.W. Tregear et al. (eds.), Relaxin 2000, 231–236.
© 2001 *Kluwer Academic Publishers. Printed in the Netherlands.*

study. This review undertakes to describe the current methods of preparation of relaxin as well as of recent biological results stemming from the products.

Table 1. Current known members of the insulin superfamily

Member	Abbreviation(s)
insulin	INSL
insulin-like growth factor I	IGF I
insulin-like growth factor II	IGF II
H1 relaxin	H1 rlx
H2 relaxin	H2 rlx
relaxin-like factor	RLF, INSL3
placentin	INSL4
insulin 5	INSL5
insulin 6	INSL6

Chemical synthesis of relaxin and related peptides

The assembly and purification of relaxin peptides and subsequent chain combination is difficult because of the chemical and physical properties of the individual A- and B-chains and overall yield of target product are often low. The isolated B-chain of most relaxin sequences is particularly difficult to handle because of poor solubility and a tendency to aggregate. Selection of the appropriate chemical strategy and choice of resin derivative for the assembly of relaxin chains is often crucial to the success of the synthesis and is very sequence dependent. The so-called Fmoc approach which utilizes base-sensitive N^{α}-protecting groups is generally more convenient than the alternative Boc method that relies on acid sensitive amino terminal protection and the use of anhydrous hydrogen fluoride to cleave the peptide from the resin. Both methods have chemical limitations that require considerations prior to their use for the synthesis of the relaxin peptides. For example, subtle structural rearrangements of C-terminal cysteine residues can be a problem with the Fmoc procedure that highlights the need for sensitive characterization protocols. We have found that the synthesis and chain combination of relaxin-related peptides (e.g. rat and human INSL3) proceeds more efficiently than the corresponding relaxin sequences which, in turn, probably reflects the influence of sequence-dependent structural conformations. The choice of synthetic strategy and tactics is dependent on analysis of sequence and is often arrived at through trial and error. In our experience, there is no preferred recipe for relaxin chemical synthesis. Currently there are three main approaches for the linking of the synthetic A- and B-chains through formation of the inter- and intra-molecular disulfide bonds.

(i) Two-chain approach:

The first successful assembly of relaxin (porcine) adopted an approach similar to that used for the preparation of insulin, *viz.*, separate syntheses of the A- and B-chains followed by their combination in solution at high pH [5]. However, it was observed at this early stage that the physicochemical properties of the B-chain were substantially different to that of the corresponding insulin chain. Particularly noteworthy was the poor solubility of the peptide which, in turn, necessitated the use of a shortened, more soluble version for subsequent easier purification and handling. This characteristic has been observed for nearly all B-chains examined to date and has long caused overall yields of relaxin to be lower than expected [6]. An additional difficulty is observed using shortened or point-mutated chains for combination; such peptides usually have subtly altered secondary structures that adversely impact on the success of combination. Therefore relaxin analogues are generally best prepared by other means (see below). Nevertheless by use of the two-chain approach, many species of relaxin have been successfully prepared and have allowed a significant contribution to the study of the structure and function of the hormone [7-9].

(ii) Regioselective disulfide bond formation approach:

With the advent of a series of novel cysteine side chain protecting groups for use in solid phase peptide synthesis, it has now become possible for the synthesis of multicystine peptides to be prepared by way of individual and selective stepwise formation of each disulfide bond. There are several variations of this approach but one developed by Schwabe and colleagues permitted the total synthesis of H2 relaxin in good overall yield [10]. The experimental protocol employed the use of the Fmoc-solid phase synthesis of the A-chain that could then be manipulated to generate an intramolecular disulfide bond. This peptide was then combined with an appropriately S-protected Boc-solid phase synthesized B-chain via separate formation of the two interchain cystines. This approach has been used to prepare a variety of relaxin and relaxin-related peptides [11-14]. Later, an excellent variation of the regioselective methodology was developed by Kiso *et al* in which a single solid phase chemistry (Fmoc) was used to prepare insulin in high yield (Fig. 1A) [15]. The two chains were separately synthesized such that the thiols that subsequently form the three cystines were protected by the Npys, Acm and Bu^t groups respectively. The first cystine between the two chains was formed by thiolysis, the second by iodine oxidation and the third and final cystine, the A-chain intramolecular bond, was formed by reaction with silyly chloride. This approach was recently used for the preparation of marmoset monkey relaxin but overall yields are still low due to severe losses of material during purification in the intermediate disulfide formation steps [Clippingdale, unpublished]. Variations of this approach have been successfully employed for the preparation of both bombyxin, an insulin-like neuropeptide from the brain of the silk worm (*Bombyx mori*) [16], and relaxin (Echner *et al.*, this volume). Further refinement of this general procedure is likely to allow more ready and successful preparation of insulin superfamily members as well as substantially modified analogues.

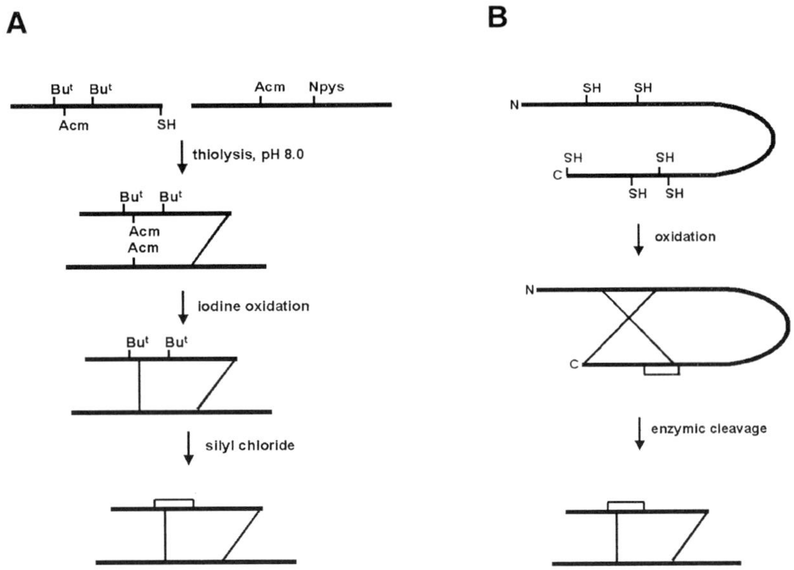

Fig. 1. Alternative approaches to the chemical synthesis of relaxin and relaxin-related peptides. A. Regioselective disulfide bond approach via method of Akaji et al [15]. B. Mini-C proprotein approach [19].

(iii) Mini-proprotein approach:

A curious aspect of the ribosomal synthesis of members of the insulin superfamily is the diversity of lengths of connecting (C) peptides that range from approximately 30 residues in the case of insulin to more than 100 in relaxin. It is questionable that such lengths are needed for efficient folding and disulfide bond formation of the prohormones. Indeed, studies by Chang *et al* [17] showed that replacement of the insulin C-peptide with a short turn-forming 5 amino acid sequence resulted in an enhanced refolding yield. Furthermore, B-chain joined head-to-tail to the A-chain will also fold correctly [18]! On the basis of this information, researchers at Genentech found that recombinant DNA-derived relaxin could be produced in high yield (Fig. 1B) [19]. Although this concept has not yet been used for the chemical synthesis of relaxin, there is no compelling reason why this cannot be exploited in the near future.

Other relaxin analogues

One of the principal goals of relaxin research is to prepare analogues possessing higher and longer-lasting activity. Although there is a substantial body of related experience with insulin, very little progress has been made with relaxin. In an attempt to delineate the active site of the peptide, a two chain/two disulfide bond peptide corresponding to the so-called tryptic fragment T7 of H2 relaxin was chemically synthesized [20]. It was shown to be devoid of relaxin-like activity in the rat isolated heart assay. More recently, a template-assisted synthetic peptide (TASP) was prepared in which four copies of the helical segment of the relaxin B-chain was "bundled" onto a cyclic peptide template. Such TASP molecules are reported to contain substantial α-helix. However, the relaxin TASP had little such secondary structure and it also lacked biological activity (Mathieu *et al.*, this volume) [21].

Recent relaxin structure and function relationships

Most of what we currently know about structure and biological function of the relaxin molecule has been derived from the study of chemically synthesized analogues and an examination of the amino acid sequences of the relaxins from various species. The central B-chain region encompassing the two invariant arginine residues is recognized as the key structural domain required for bioactivity [22] but other elements of the B-chain and linkage with the A-chain are also essential. Recently, Büllesbach and Schwabe [23] used chemical synthesis to prepare analogues of H2 relaxin in a study of relaxin binding site geometry. Their results reveal a triangular contact region of the relaxin suface involving three crucial binding residues in the B-chain, namely, Arg (B13), Arg (B17) and Ile (B20).

Future perspectives

Although the chemical synthesis of relaxin continues to be an arduous and challenging endeavour, there is no question that the technique has provided a wealth of information on the individual amino acids and structural domains in the relaxin molecule responsible for biological activity. Yet, modern solid phase peptide synthesis has allowed significant advances in the study of the peptide to occur. It is probable that constant evolution of chemical methods, *viz*, new amino acid side chain protecting groups, solid supports, synthetic strategy, will allow more ready preparation of relaxin and relaxin-related peptides in higher yields. Such approaches will undoubtedly be supplemented by improved protein expression methods. The resulting acquisition of additional structure-function knowledge, will facilitate the design of analogues, most probably non-peptide mimetics, having more potent and longer lasting action for therapeutic application.

Acknowledgments

This work was carried out at the Howard Florey Institute and was supported by an institute block grant (reg key number 983001) from the NHMRC of Australia.

References

1. Sherwood, O.D., In Knobil, E. and Neill, J.D. (Eds.), The Physiology of Reproduction. 2nd ed., Raven Press, New York, USA, 1994, p. 861.
2. Duret, L., Guex, N., Peitsch, M.C. and Bairoch, A., Genome Res., 8 (1998) 348.
3. Hudson, P., Haley, J., Cronk, M., Shine, J. and Niall, H.D., Nature, 291 (1981) 127.
4. Ido, Y., Vindigni, A., Chang, K., Stramm, L., Chance, R., Heath, W.F., DiMarchi, R.D., DiCera, E. and Williamson, J.R., Science, 277 (1997) 563.
5. Tregear, G.W., Du, Y-C., Fagan, C., Reynolds, H., Scanlon, D., Jones, P., Kemp, B. and Niall, H.D., In Rich, D.H. and Gross, E. (Eds.), Peptides: Synthesis, Structure and Function. Pierce Publ., Illinois, USA, 1981, p.249.
6. Wade, J.D. and Tregear, G.W., Meths. Enzymol., 289 (1997) 637.
7. Hudson, P., Haley, J., John, M., Cook, M., Crawford, R., Haralambidis, J., Tregear, G., Shine, J. and Niall, H., Nature, 301 (1983) 628.
8. Wade, J.D., Layden, S.S., Lambert, P.F., Kakouris, H. and Tregear, G.W., J. Protein Chem., 13 (1994) 315.
9. Wade, J.D., Lin, F., Talbo, G., Otvos, L., Tan, Y-Y. and Tregear, G.W., Biomed. Peptides Prot. Nucl. Acids, 2 (1997) 89.
10. Büllesbach, E.E. and Schwabe, C., J. Biol. Chem., 266 (1991) 10754.
11. Büllesbach, E.E. and Schwabe, C., Biochem Biophys. Res. Commun., 196 (1993) 311.
12. Steinetz, B.G., Büllesbach, E.E., Goldsmith, L.T., Schwabe, C. and Lust, G., Biol. Reprod., 54 (1996) 1252.
13. Büllesbach, E.E., Steinetz, B.G. and Schwabe, C., Biochemistry, 35 (196) 9754.
14. Büllesbach, E.E. and Schwabe, C., Eur. J. Biochem., 241 (1996) 533.
15. Akaji, K., Fujino, K., Tatsumi, T. and Kiso, Y., J. Am. Chem. Soc., 115 (1993) 11384.
16. Maruyama, K., Nagata, K., Tanaka, M., Nagasawa, H., Isogai, A., Ishizaki, H and Suzuki, A., J. Prot. Chem., 11 (1992) 1.
17. Chang, S-G., Kim, D-Y., Choi, K-D., Shin, J-M.. and Shin, H-C., Biochem. J., 329 (1998) 631.
18. Hua, Q-X., Hu, S-Q., Jia, W., Chu, Y-C., Thompson-Burke, G., Wang, S-H., Wang, R-Y., Katsoyannis, P.G. and Weiss, M.A., J. Mol. Biol., 277 (1998) 103.
19. Vandlen, R., Winslow, J., Moffat, B. and Rinderknecht, E. In MacLennan, A.H., Tregear, G.W. and Bryant-Greenwood, G.B. (Eds.), Progress in Relaxin Research. World Scientific Publishing Co. Ltd., Singapore, 1995, p.59.
20. Wade, J.D., Tan, Y-Y., Otvos, L., Summers, R.J. and Trgear, G.W. In Fields G.B., Tam, J.P. and Barany, G. (Eds.), Peptides for the New Millenium, Kluwer Academic Publishers, Dordrecht, 2000, p. 664.
21. Mathieu, M.M., Wade, J.D., Tan, Y-Y., Summers, R.J. and Tregear, G.W., J. Peptide Sci., 6 (2000) 235.
22. Büllesbach, E.E., Yang, S. and Schwabe, C., J. Biol. Chem., 267 (1992) 22957.
23. Büllesbach, E.E. and Schwabe, C., J. Biol. Chem., 275 (2000) 35276.

A new synthesis of human 2 relaxin using selective bridging

Hartmut Echner[1], Franz P Armbruster[2], Wolfgang Voelter[1]

[1]Dept. of Physik. Biochemie, Physiologisch-chemisches Institut, University of Tübingen, Hoppe-Seyler-Str. 04, D-72076 Tübingen, Germany; and [2]Immundiagnostik AG, Wiesenstr. 04, D-64625 Bensheim, Germany

Introduction

Relaxin is a polypeptide hormone produced by the corpora lutea of the ovary during pregnancy. It consists of two peptide chains (A and B) with one intra-chain and two inter-chain disulfide bridges in analogy to insulin.

*p*Q-L-Y-S-A-L-A-N-K-C-C-H-V-G-C-T-K-R-S-L-A-R-F-C

D-S-W-M-E-E-V-I-K-L-C-G-R-E-L-V-R-A-Q-I-A-I-C-G-M-S-T-W-S

Fig. 1: Primary structure of human relaxin 2.

The problems related to the unusual solubility and structural characteristics of the isolated B-chain and the resulting chromatographical behavior and the poor yields recovered from combination of the unprotected A- and B-chain prompted us to develop a new synthetic route using a new cysteine protecting group which allows us to introduce the intra-chain cystine bridge to the fully protected A-chain. Both A- and B-chain can be combined afterwards in their protected forms.

Results and Discussion

The combined use of base labile Fmoc amino protecting group and arylalkylthioether type side-chain protection for cysteine has found wide application in SPPS [1-4]. Differences in the chemical stability of the protected thiols are prerequisite for the selective introduction of more than one disulfide bond in a peptide. We prepared the 9-methylanthracenyl (Anm) residue as a new arylalkylthioether type masking group selectively removable by treatment with methylmercaptane sodium salt in presence of the well-established Mmt and Trt cysteine protecting groups.

A- and B-chain were synthesized successfully according to Fmoc/tBu-strategy using the quasiorthogonal Mmt, Trt and Anm protecting groups for the cysteine residues. After syntheses, the N-terminal Fmoc group of the B-chain was exchanged by the Boc residue and both chains were cleaved from the resin in their fully protected forms to preliminarily introduce the intra-chain cystine bridge to the protected A-chain, followed by the selective combination of both chains. The two peptide chains in their linear form were obtained from batchwise automated SPPS in 0.11 mMol (A-chain) and 0.16 mMol (B-chain) scale using Fmoc-Cys(Trt)-Trt- and Fmoc-Ser(tBu)-Trt- Tentagel-

G.W. Tregear et al. (eds.), Relaxin 2000, 237–238.
© 2001 *Kluwer Academic Publishers. Printed in the Netherlands.*

R-resin (Rapp Polymere, Tübingen, Germany) respectively. Couplings (10 eq AA) of 30 min duration were performed using TBTU/DIEA in DCM/NMP/DMF (1:1:1, v/v/v) at RT. All cysteine residues were incorporated via symmetrical anhydrides. Gly[12,24] (B-chain) were incorporated as backbone protected Hmb derivatives. Anm protected Cys[10,15] (A-chain) residues were unmasked using methylmercaptane sodium salt in DMF/HMPT overnight, at RT and cyclized applying DMSO/DMF/H_2O (15:25:10, v/v/v, pH 8.5). Fully protected A- and B-chain were separated from the resins within 2h by a mixture of HOAc/TFA/DCM (50:10:40, v/v/v). Mmt residues masking Cys[11] (A-chain and B-chain) were removed from both peptides (TFA/DCM/TIPS (1.5/97.5/1; v/v/v; 1h) and the A-chain was activated by di-*tert*-butylazodicarboxylate in DMF for inter-chain bridging. Finally, oxidation with I_2 (5 min in TCM/TFE) was employed to achieve the second inter-chain disulfide bond. Employing TFA/phenol/thioanisole/anisole/H_2O/TIPS (12 : 0.5 : 0.6 : 0.3 : 0.3 : 0.1) for 3.5 h at RT yielded the free unprotected H2 relaxin (Fig. 2).

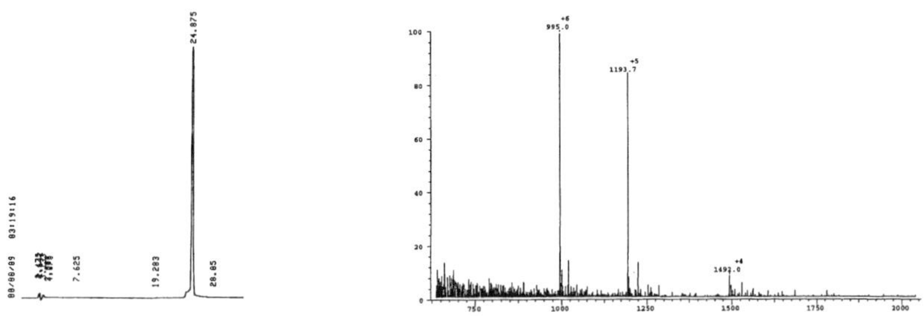

Fig. 2: HPLC and ESI-MS (Finnigan MAT TSQ 700) of synthetic H2 relaxin. Stationary phase: Nucleosil 300 C_{18} 5µm, eluent: A: 0.70% TFA/H_2O; B: 0.58% TFA/60% CH_3CN, gradient: 20 $\rightarrow$ 50% B in 31 min; $\lambda = 214$ nm, 1 ml/min.

Acknowledgements

The authors thank Dr. M. Deeg for the preparation of the ESI-mass spectra and Dr. H.-J. Kohlbau for the preparation of the manuscript.

References

1. Atherton, E., Pinori, M. and Sheppard, R.C., J. Chem. Soc. Perkin Trans., I (1985) 2057.
2. Atherton, E. and Sheppard, R.C., J. Chem. Soc. Perkin Trans. I, (1985) 2025.
3. Echner, H. and Voelter, W., In Innovations and Perspectives in Solid Phase Peptide Synthesis & Combinatorial Chemical Libraries, Canterbury, England, 1992, p. 371.
4. Echner, H. and Voelter, W., In Maia, H.L.S., (ed.) Peptides 1994, ESCOM, Leiden, 1995, p. 157.

N^{α}-*mono*-biotinylated rat relaxin: synthesis and activity

Marc N Mathieu[1], Courtney P Bond[2], John D Wade[1], Roger J Summers[2] and Geoffrey W Tregear[1]
[1]*Howard Florey Institute, University of Melbourne, Victoria 3010, Australia; and*
[2]*Departmentof Pharmacology, Monash University, Clayton, Victoria 3168, Australia.*

Introduction

In relaxin, the disulfide crosslink pattern between the two chains is identical to that of insulin. Within species, the only invariant residues are the six cysteines and the glycines adjacent to these, and the two arginines in the B-chain. Despite considerable sequence variations, all relaxins are predicted to retain a similar tertiary structure consisting of a central B-chain α-helix and two α-helices in the A-chain separated by a pronounced turn (for literature ref. refer to the present book).

The biotin-avidin system is a powerful tool in the identification of hormone receptors. A crucial step is the preparation of a specifically biotinylated hormone that is able to bind to its receptor while leaving the biotinyl group free for interaction with avidin [1]. Introducing a biotin label at the N-terminus of the B-chain should not modify the tertiary structure of relaxin or its activity, as the interaction with the receptor involves the two arginine residues located at the opposite side of the molecule (Fig.1) [2]. However, the synthesis and purification of relaxin presents a major challenge because of the differing chemical and physical properties of the individual A- and B-chains [3]. We describe here an improved method to synthesise this monolabelled peptide and measure its biological activity using the isolated rat atria bioassay. It is compared with native and synthetic rat relaxins obtained from previous studies [4].

Z^{1}SGALLSEQ**C**^{10}CHIG**C**TRRSI^{20}AKL**C**^{24}

Biotin **-** R^{1}VSEEWMDQV^{10}IQV**C**GRGYAR^{20}AWIEV**C**GASV^{30}GRLAL^{35}

Fig.1. The primary structure of rat relaxin containing the biotin label on the N-terminus of the B-chain.

Results and Discussion

The biotin label was introduced at the N-terminus of the B-chain using standard coupling protocols (Fig. 1), while the peptide was still on the resin. Combination with the A-chain was carried out at pH 10.5 using the same conditions as previously described [5]. After 5 h, in addition to several new products corresponding to various forms of oxidised A- and B-chains, some relaxin had already formed. The reaction was

G.W. Tregear et al. (eds.), Relaxin 2000, 239–240.
© 2001 *Kluwer Academic Publishers. Printed in the Netherlands.*

left for a further 18 h in order to complete both the conformational alignment and disulfide bond formation, prior to final work-up and purification (Fig. 2). Analytical RP-HPLC and MALDI-TOF MS (data not shown) of the synthetic peptide showed both a single peak. Mono-biotin labelled rat relaxin exhibited very similar circular dichroism (CD) spectra in water (results not shown) compared with native and synthetic relaxin from the same species.

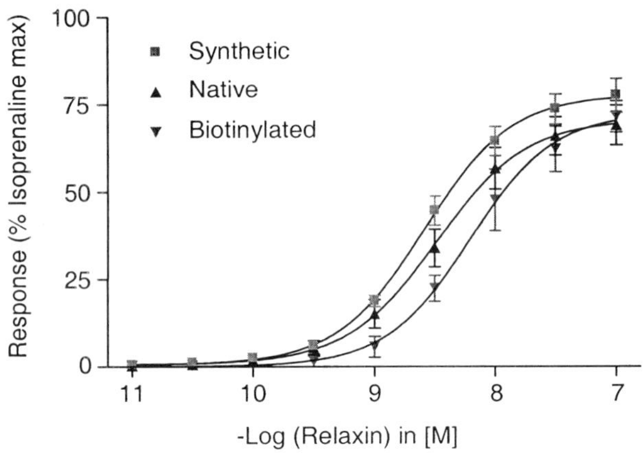

Fig.2. Effects of native, synthetic and biotinylated rat relaxin in producing positive chronotropic responses in the isolated rat atrial bioassay [4].

In the rat isolated atrial bioassay, a comparison of the pEC_{50}s of the synthetic mono-labelled biotinylated relaxin (8.2 ± 0.03) with those reported for native (8.52) and synthetic (8.60) rat relaxins [4] showed that biotinylation of rat relaxin had little effect on bioactivity and further confirmed that the N-terminal region of the B-chain is not involved in the interaction of relaxin with its receptor(s).

Acknowledgments

This work carried out at the Howard Florey Institute was supported by an Institute block grant (Reg. key number 983001) from the NHMRC of Australia.

References

1. Büllesbach, E.E. and Schwabe, C., Int. J. Peptide Protein Res., 35 (1990) 416.
2. Büllesbach, E.E., Yang, S. and Schwabe, C., J. Mol. Biol., 267 (1992) 22957.
3. Wade, J.D. and Tregear, G.W., Methods in Enzymology, 289 (1997) 637.
4. Tan, Y.Y., Wade, J.D., Tregear, G.W. and Summers, R.J., Brit J. Pharm., 123 (1998) 762.
5. Wade, J.D., Layden, S.S., Lambert, P.F., Kakouris, H. and Tregear, G.W., J. Protein Chem., 13 (1994) 315.

Design, synthesis and biological activity of template-assembled peptidic mimetics of rat relaxin

Marc N Mathieu[1], John D Wade[1], Yean-Yeow Tan[2] Roger J Summers[2] and Geoffrey W Tregear[1]
[1]*Howard Florey Institute, University of Melbourne, Victoria 3010,Australia; and* [2]*Department of Pharmacology, Monash University, Clayton, Victoria 3168, Australia.*

Introduction

A recent exciting development in the study of structure-function relationship is to design and construct small peptide domains of the native protein important for biological activity. They provide the basis for the rational design of shortened, conformationally restrained mimetics which possess either antagonist or agonist effects. These single-chain domain peptides should provide a useful starting point for the design of smaller non-peptide organic mimetics. The use of the Template Assembled Synthetic Protein (TASP) concept in which a peptidic template is used as a folding device for covalently attached peptides will be described [1].

Our aim was to prepare relaxin mimetics based on this approach using the most conserved and important region for the binding of the hormone to the receptor. Depending on the chemistry applied to attach the peptides to the template, the reacting moiety was placed either at the C-terminus or N-terminus respectively. This allowed the choice of orientation relative to the carrier molecule. The template consisted of a linear decapeptide with two proline-glycine turns and four evenly spaced lysine residues that were functionalized with an appropriate chemical moiety. The two template-assembled mimetics of relaxin were then analysed by circular dichroism spectroscopy and tested for characteristic biological activity using the rat isolated atria bioassay [2].

Results and Discussion

The assembly of peptides on template molecules according to the TASP concept has been shown to induce or stabilise specific conformations of peptides and therefore to modify their biological and pharmacokinetic properties [3]. The peptides we synthesised represent truncated rat relaxin B-chain α-helix peptide fragments (region B15-B25) in "long" and "short" versions (Fig. 1). Both sequences include a tryptophan residue, which was replaced by a leucine (B22) to prevent any side reactions during synthesis and steric hindrance during the assembly. In this case, the original sequence was unable to assemble as a tetramer in reasonable yield.

The conformational studies revealed fully unordered structure under physiological conditions for both peptides alone (results not shown). Upon attachment to the template a β-sheet type structure was measured in the case of the long fragment and no ordered structure for the short fragment. This might explain the results of the

G.W. Tregear et al. (eds.), Relaxin 2000, 241–242.
© 2001 *Kluwer Academic Publishers. Printed in the Netherlands.*

bioassays indicating partial antagonistic effects on the chronotropic responses of H2 (B29) relaxin for both template-assembled mimetics of relaxin (results not shown). Additionally, both compounds failed to shift the chronotropic response curves for relaxin on the chronotropic response in right atria. It is feasible therefore that the antagonist effects seen in the left atria may result from interference with relaxin signalling rather than an effect exerted directly on relaxin receptors [4].

In conclusion, the TASP concept provides a valuable tool for producing antagonist mimetics of relaxin. Future work will attempt to design a helix-forming peptide following attachment to the template.

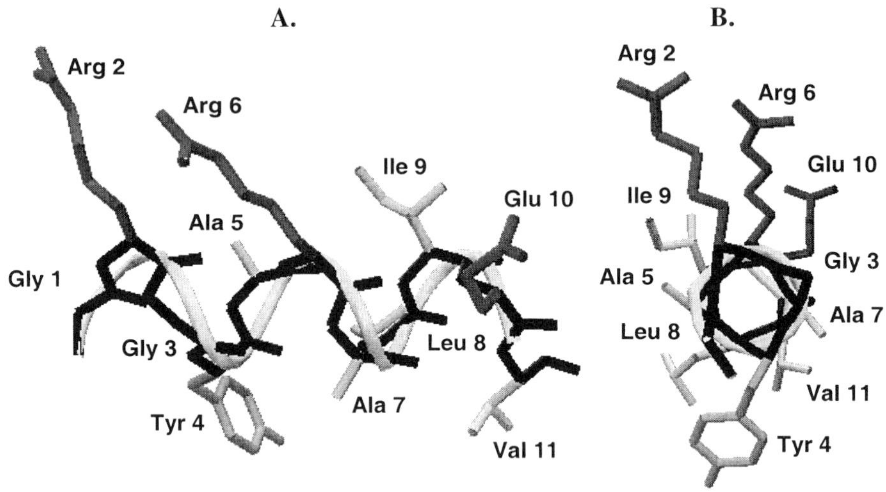

Fig. 1. Homology model of the rat relaxin B-chain sequence attached at four sites to the peptidic template ("long" version). A. View from the side. B. View from the N-terminus. Model based on the crystal structure of human relaxin.

Acknowledgments

This work carried out at the Howard Florey Institute was supported by an Institute block grant (Reg. key number 983001) from the NHMRC of Australia.

References

1. Vuilleumier, S. and Mutter, M., Angew. Chem. Int. Ed. Engl., 28 (1989) 535.
2. Tan, Y.Y., Wade, J.D., Tregear, G.W. and Summers, R.J., Brit. J. Pharmacol., 123 (1998) 762.
3. Grouzmann, E., Buclin, T., Martire, M., Cannizzaro, C., Doerner, B., Razaname, A. and Mutter, M., J. Mol. Biol., 272 (1997) 7699.
4. Mathieu, M.N., Wade, J.D, Tan, Y.-Y., Summers, R. J. and Tregear, G. W., J. Peptide Sci., 6 (2000) 235.

Chemical synthesis and relaxin activity of analogues of ovine Insulin 3 containing specific B-chain residue replacements

Antonia A Claasz[1], Ross A Bathgate[1], Nicola F Dawson[1], Roger J Summers[2], Laszlo Otvos, Jr.[3], Geoffrey W Tregear[1] and John D Wade[1]

[1]Howard Florey Institute, University of Melbourne, Victoria 3010, Australia; [2]Department of Pharmacology, Monash University, Clayton, Victoria 3168, Australia; and [3]The Wistar Institute, 3601 Spruce Street, Philadelphia, PA 19104, USA

Introduction

Insulin 3 (INSL3, RLF) is a new member of the insulin/relaxin family of peptide hormones. It possesses all the structural features of this peptide family including a two chain structure (A and B chains) linked by two interchain and one intrachain disulfide bond in the A chain. Close examination of the INSL3 sequence indicates that it is more relaxin-like than insulin-like and circular dichroism (CD) spectroscopy measurements of the synthetic sheep peptide indicate a secondary structure similar to relaxin [1]. Furthermore, the INSL3 B-chain contains an Arg-XXX-Arg, a cassette that is also present in the relaxin B-chain (Fig. 1) where it is known to be essential for relaxin bioactivity [2]. However, in INSL3 this cassette is displaced towards the C-terminus by four residues.

Native sheep:	**AAQEAPEKLCG**	***HHF***	**VRALV**	***R***	**LCGGPRWSSEDG**
A:		***R***			
D:		***REL***			
E:		***R***		***A***	
G:				***A***	

Fig. 1. Sequence of native sheep INSL3 B-chain and modifications of B-chains in analogues. The INSL3 Arg cassette is underlined. Substituted residues are shown in italics.

While INSL3 itself does not have relaxin-like activity, replacement of the His residue in position B12 with Arg (Fig. 1) to give an INSL3 with an Arg-XXX-Arg cassette in the same position as in relaxin (when sequences are aligned by cysteine residues), results in an analogue having weak relaxin-like activity in isolated rat atria [1]. This peptide was labelled analogue A. Another analogue was also produced, that contains the entire relaxin Arg cassette present in the relaxin molecule (analogue D). This peptide was no more active than analogue A (Tan *et al.*, this volume). It was noted that the extra Arg residue present in INSL3 (from the original Arg cassette) may cause steric hindrance, thus reducing the activity of this peptide. To further investigate this, we undertook synthesis of two new INSL3 analogues where this additional (B20) Arg

G.W. Tregear et al. (eds.), Relaxin 2000, 243–246.
© 2001 *Kluwer Academic Publishers. Printed in the Netherlands.*

residue was substituted for a less bulky Ala residue. The first analogue (labeled analogue E) had the same His^{12} replacement as in analogue A plus the additional Arg residue replaced by an Ala. Analogue G simply had the latter modification only. More recently, it has been shown that Ile^{20} in the human relaxin B-chain, or Val^{20} in porcine relaxin B-chain, is vital for receptor binding [3]. Interestingly, this corresponds to a Val residue (B19) in INSL3. Therefore, as the Arg (B20) residue is adjacent to this Val residue, it may have the additional effect of hindering the interaction of this Val residue with the relaxin receptor. Hence removal of this Arg may better expose the Val residue allowing it to interact with the receptor.

Results and Discussion

Synthesis of the two novel INSL3 analogues was undertaken by continuous flow Fmoc solid-phase peptide synthesis of the separate A- and B-chains. A-chains were assembled manually on a CRB Pepsynthesiser, using Fmoc-His(Trt)-Novasyn PA 500 (Novabiochem, Switzerland) as support, and B-chain synthesis was carried out using an automated MilliGen 9050 synthesiser on PAC-PEG-PS support. Peptides were then cleaved and sidechain deprotected by 3h treatment with TFA in the presence of scavengers. Crude peptides were purified using a by RP-HPLC on a Vydac C_{18} column. Disulfide bonds were then formed by chain combination in pH 10.5 buffer at 4°C. The reaction was monitored by RP-HPLC, and terminated by addition of TFA when no starting B-chain was remaining (after approximately 24 h). Preparative RP-HPLC was then used to isolate and purify the product, and the products were then characterised by mass spectrometry and amino acid analysis.

CD spectra were performed in order to compare the overall secondary structure of the analogues to relaxin (data not shown). The results showed that while there were no major differences in the structures of INSL3 analogues compared to relaxin, the analogues were slightly less helical in nature.

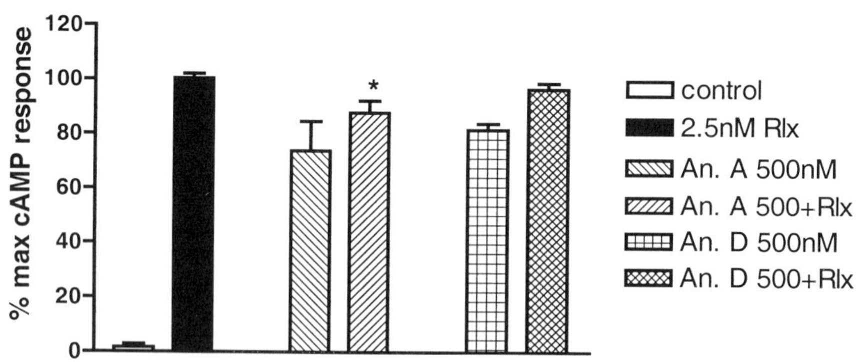

Fig. 2. cAMP response in THP-1 cells to Analogues A and D (p<0.05 as compared to 2.5nM Relaxin (Rlx)).*

To assay for relaxin-like activity analogues were tested for their ability to stimulate cAMP production in THP-1 cells, which express the human relaxin receptor. Assays were performed as previously published [4], using a Biotrak cAMP ELISA kit. Concentration-response curves to H2 relaxin (B29) were performed as a control in each ELISA. A maximum response was produced at 2.5nM, so this concentration was used when assaying for antagonist activity. Data was presented as mean ± SEM as a percentage of this maximum relaxin response.

As represented in Fig. 2, analogues A and D do produce an increase in cAMP in THP-1 cells, both producing a response at ~80% of the maximum relaxin response (at 2.5nM) at concentrations of 500nM, confirming that these peptides possess only weak relaxin-like activity. This data also indicated that the presence of the relaxin Arg cassette in its entirety is not necessary, since analogue A can produce an equal response with only one amino acid substitution. Analogue A also appears to possess some antagonist activity. Analogue D appears to depress the relaxin response slightly, though not significantly. Clearly further investigation is required to draw any firm conclusions regarding this effect.

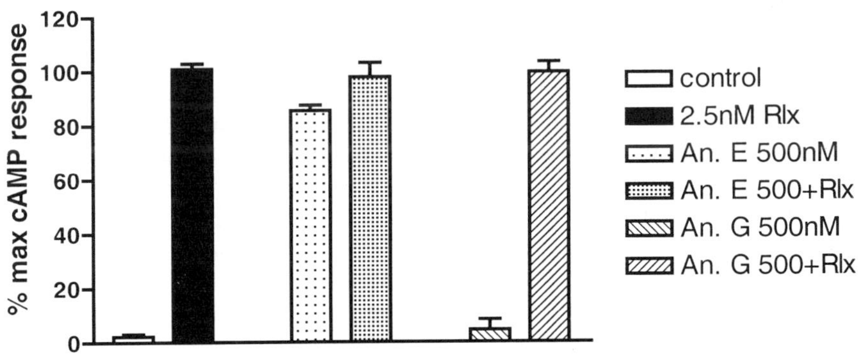

Fig. 3 cAMP response in THP-1 cells to Analogues E and G.

Analogue E caused around 80% of the maximum response to 2.5nM relaxin at 500nM (Fig. 3.), similar to the result observed for analogues A and D (Fig.2.). It therefore appears that removal of the additional Arg residue from the original Arg cassette in INSL3, does not improve relaxin-like bioactivity. However, at this concentration, the removal of the Arg^{20} residue abolishes the antagonist effect seen with analogue A. Analogue G did not have any relaxin-like activity in the relaxin bioassay, nor did it exhibit any antagonist activity at this concentration. This data provides further proof that the Arg-X-X-X-Arg cassette is required in the B13-17 position, though this on its own affords these analogues with only weak relaxin-like activity. It also suggests that removal of the additional Arg^{20} alone does not give these INSL3 analogues full relaxin-like activity, with or without the Arg cassette in the relaxin conformation.

Conclusions

While the B-chain Arg cassette and the Ile/Val B^{20} residue have been identified as vital components of the relaxin binding motif, forming an extended Arg-X-X-X-Arg-X-X-Ile/Val binding cassette, these residues may not be the sole determinants of relaxin bioactivity. Insl3 analogues A, D and E all contain this motif and possess only weak relaxin activity. Another key factor for relaxin-like activity may be structural components of the entire relaxin molecule. The relaxin B-chain is mainly helical in structure, and the vital Arg and Val/Ile residues are located in the α-helix. It would be reasonable to assume that the overall structure of the B-chain, and thus the conformation of these residues, plays a vital role in the binding and bioactivity of relaxin. Indeed, our earlier observation that the INSL3 analogues were less helical in nature compared to relaxin could partially explain their depressed bioactivity. Furthermore, since the relaxin B-chain on its own does not have bioactivity, the A-chain also clearly plays a role, if only to provide a scaffold to hold the B-chain in its correct conformation. Perhaps the INSL3 and relaxin A-chains differ sufficiently in structure, thus influencing the conformation of the B-chains, which in turn may contribute to the reduced activity of these analogues.

An interesting element of this study is the discovery of the antagonist activity of analogue A. At a concentration of 500nM analogue D also showed some antagonist activity although this was not significant. It would seem that the presence of an extra Arg cassette has a role in the antagonism of relaxin activity. If the Arg^{20} residue is substituted by Ala, antagonist activity is lost, as in analogues E and G. It should be noted that native INSL3, which also contains this cassette, does exhibit some antagonist activity in rat atrial bioassays at a concentration of 1μM (Tan *et al.*, this volume), further supporting this hypothesis. Native INSL3 was not tested in this study, but clearly further studies need to be performed, perhaps targeting other B-chain residues, in order to identify all the residues required both for relaxin-like bioactivity and antagonism.

Acknowledgements

The work was carried out at the Howard Florey, and was supported by an Institute Block Grant Reg Key Number 983001 from the NHMRC. We would like to thank Kathryn Smith for amino acid analysis and Marc Mathieu for molecular modelling.

References

[1] Dawson, N.F., Macris, M., Tregear, G.W., Wade, J.D., Tan, Y.Y., Summers, R.J. and Otvos, L., J. Peptide Res., 53 (1999) 542.
[2] Büllesbach, E.E., and Schwabe, C., Int. J. Peptide Protein Res., 32 (1988) 361.
[3] Büllesbach, E.E., and Schwabe, C., J. Biol. Chem., 275 (2000) 35276.
[4] Parsell, D.A., Mak, J.Y., Amento, E.P. and Unemori, E.N., J. Biol. Chem., 271 (1996) 27936.

Chemical synthesis and biological activity of ovine Insulin 3, a relaxin structural homologue

Nicola F Dawson[1], Mary Macris[1], Roger J Summers[2], Yean-Yeow Tan[1,2], Geoffrey W Tregear[1] and John D Wade[1]

[1]Howard Florey Institute, University of Melbourne, Victoria 3010, Australia and [2]Department of Pharmacology, Monash University, Clayton, Victoria 3800, Australia

Introduction

Relaxin, a two-chain, three disulfide-bonded peptide, is produced primarily by the ovary during pregnancy. Its principle roles in many mammalian species include softening the cervix and remodelling the interpubic ligaments in preparation for birth. No relaxin peptides have been identified in ruminants, despite target tissues of these animals being responsive to exogenous relaxin from other species [Bathgate *et al.*, this volume]. A peptide isolated from the Leydig cells in pig testis was shown to possess striking structural similarity to relaxin and which is also expressed in the ovary [1]. The primary structure of sheep Leydig cell insulin-like peptide, or INSL3, was determined from isolated mRNA from sheep ovary and shown to consist of a 26-residue A-chain, a 32-residue B-chain and three disulfide bonds in the same disposition as for relaxin and insulin [2]. We undertook the chemical synthesis of this native peptide and two INSL3 analogues, each analogue containing modifications to their B-chains to more closely mimic the human relaxin Gene 2 (H2) B-chain (Fig 1). After comprehensive chemical characterization, we tested all three peptides for any relaxin-like biological activity.

1 A.T.A.V.N.P.A.R.H.C.C.L.S.G.C.T.R.Q.D.L.L.T.L.C.P.H

2 A.A.Q.E.A.P.E.K.L.C.G.H.H.F.V.R.A.L.V.R.L.C.G.G.P.R.W.S.S.E.D.G
3 A.A.Q.E.A.P.E.K.L.C.G.**R**.H.F.V.R.A.L.V.R.L.C.G.G.P.R.W.S.S.E.D.G
4 A.A.Q.E.A.P.E.K.L.C.G.**R.E**.F.V.R.A.L.V.R.L.C.G.G.P.R.W.S.S.E.D.G
5 D.S.W.M.E.E.V. I .K.L.C.G.R.E.L.V.R.A.Q. I.A.I. C.G.M.S.T.W.S

*Fig. 1: Sequences of native ovine INSL3 and INSL3 analogues: 1) native A-chain, 2) native B-chain, 3) analogue **3** B-chain, 4) analogue **4** B-chain, and 5) native H2 relaxin B-chain (residue changes in bold).*

Results and Discussion

The A- and B-chains of the native INSL3 peptide and each of the two analogues, **3** and **4**, were synthesized by conventional continuous flow Fmoc solid phase methodology. Chain combination for all three INSL3 peptides were conducted in a 4 A-chain:1 B-chain ratio in 10% MeOH/ 0.15 M CAPS/ 1 M GdHCl/ 2.5 mM DTT/ pH 8.0 buffer at

G.W. Tregear et al. (eds.), Relaxin 2000, 247–248.
© 2001 *Kluwer Academic Publishers. Printed in the Netherlands.*

4°C. Progress was monitored by analytical RP-HPLC and in all three peptides, chain combination was completed within 24 hours. Final yields of purified products were as follows: 28% native INSL3, 27% INSL3 analogue **3** and 32% INSL3 analogue **4**. Mass spectrometry identification of fragments generated by trypsin digestion confirmed correct disulfide bond pairing of synthetic INSL3 ie. A10/A15, A11/B10 and A24/B22. In a rat isolated atrial assay [3], at high concentrations, the native INSL3 peptide did not produce positive chronotropic or inotropic responses, but may have weak relaxin antagonist activity. INSL3 analogue **3**, did produce positive chronotropic and inotropic responses, as did INSL3 analogue **4** to a lesser degree (Fig 2). Neither of the modified INSL3 peptides, however, had any antagonist effects on H2 relaxin activity.

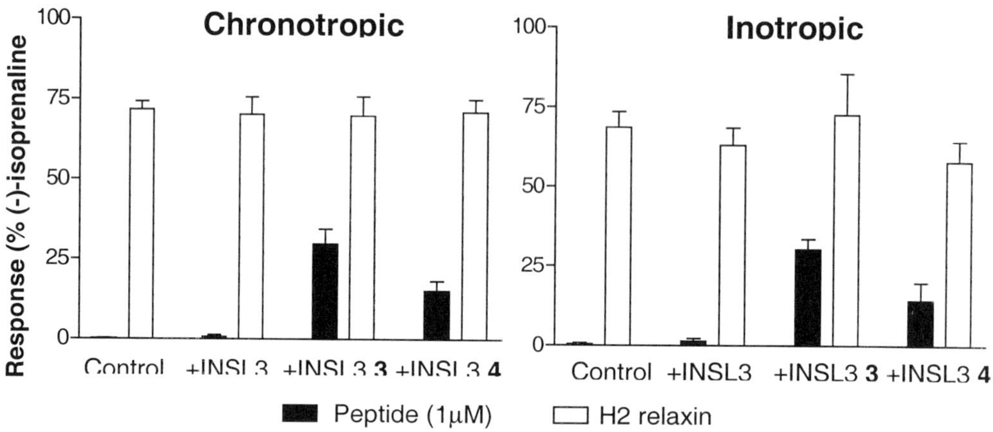

Fig. 2. Rat isolated atrial assay for relaxin-like activity of native INSL3, INSL3 analogue **3** *and INSL3 analogue* **4** *in the absence and presence of 10nM H2 relaxin.*

These results show that, at physiological doses, ovine INSL3 itself is not a relaxin-like hormone, and that modifying the INSL3 B-chain to more closely resemble H2 relaxin produces some relaxin-like activity. This activity of the modified peptide suggests that ovine INSL3 has similar 3-D structure to H2 relaxin.

Acknowledgements

Work carried out at the Howard Florey Institute was supported by an Institute block grant (reg key 983001) from the NHMRC of Australia. Work carried out at the Department of Pharmacology was supported by the NHF of Australia.

References

1. Adham, I.M., Burkhardt, E., Benahmed, M. and Engel, W., J. Biol. Chem., 268 (1993) 26668.
2. Roche, P.J., Butkus, A., Wintour, E.M. and Tregear, G.W., Mol. Cell Endo., 121 (1996) 171.
3. Tan, Y.Y., Wade, J.D., Tregear, G.W. and Summers, R., Brit. J. Pharmacol., 123 (1998) 762.

Solid phase syntheses of rat prorelaxin C-peptides (1-73) and (76-101)

Nicola F Dawson[1], Yean-Yeow Tan[1,2], Roger J Summers[2], Mary Macris[1], Laszlo Otvos, Jr.[3], Geoffrey W Tregear[1] and John D Wade[1]

[1]Howard Florey Institute, University of Melbourne, Victoria 3010, Australia; [2]Department of Pharmacology, Monash University, Clayton, Victoria 3168, Australia; and [3]The Wistar Institute, 3601 Spruce Street, Philadelphia, PA 19104, USA.

Introduction

Like insulin, relaxin is made on the ribosome as a preprohormone that undergoes subsequent proteolytic processing to yield the native two-chain peptide [1]. Curiously, the connecting C-peptide is approximately 100 residues in length, compared to the 25 residue corresponding peptide in proinsulin and 45 residues in INSL 3. Studies with insulin have shown that C-peptide mimics as short as one residue are sufficient to allow efficient chain folding and correct disulfide bond pairing [2]. Furthermore, recombinant DNA-derived human Gene 2 prorelaxin may be readily and efficiently folded and oxidized using an "mini" C-peptide of length 13 residues [3]. Thus the large length of the native relaxin C-peptide is unnecessary for the correct folding of the A- and B-chains and suggests that it may perhaps have an additional function or serve as a precursor to biologically active peptides. Support for this concept has recently arisen following the surprising finding that, in pharmacological doses, human proinsulin C-peptide has an insulin-like action [4].

The primary structure of rat prorelaxin shows that the 101-residue C-peptide contains a pair of basic residues within its sequence. One possibility is that during the processing of the prohormone by specific peptidases, cleavage at this pair also occurs to generate two discrete peptides of length 73 and 26 residues respectively. To determine if either of these peptides may produce a biological response similar to the situation that occurs with insulin, these two fragments of rat prorelaxin C-peptide were synthesised and assayed for both relaxin-like or relaxin-potentiating activity.

Results and Discussion

The solid phase synthesis of the two fragments of rat prorelaxin C-peptide was successfully achieved using the continuous flow Fmoc methodology. Both peptides were purified by preparative RP-HPLC in overall yield of approximately 7 and 30% respectively. The purity of the resulting products were subjected to comprehensive chemical characterization including by MALDI-TOF mass spectroscopic analysis that gave values close to the calculated [C-peptide (1-73): found MH^+ 7,904.0; calc. MH^+ 7907.8; C-peptide (76-101): found MH^+ 2,761.8; calc. MH^+ 2,761.9].

G.W. Tregear et al. (eds.), Relaxin 2000, 249–250.
© 2001 *Kluwer Academic Publishers. Printed in the Netherlands.*

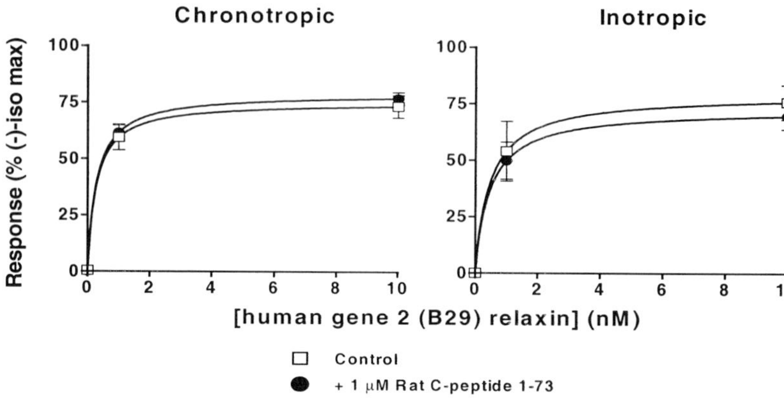

Fig. 1. The effect of synthetic rat relaxin C-peptide (1-73) (1 μM) on chronotropic and inotropic responses to H2 relaxin (1 and 10 nM) in rat isolated right and left atria. n = 12 for controls and 4 for other experiments.

The peptides were assessed for characteristic relaxin-like activity in the rat isolated heart assay. Neither peptide elicited a significant response nor did they either enhance or inhibit relaxin activity (for example, Fig. 1). The function of the C-peptide (if any) thus far remains unknown. However, the extraordinary variation in primary structure (53% between rat and human) and length (up to ten residues) between known relaxin C-peptides suggests that caution is required before postulating a general role other than prohormonal packaging. The detection by radioimmunoassay of high levels of prorelaxin C-peptide in seminal plasma [5] suggests an alternative site of examination for a potential biological role for the peptide.

Acknowledgments

An Institute Block Grant Reg Key 983001 from the NHMRC supported the work carried out at the Howard Florey Institute.

References

1. Sherwood, O.D., In Knobil, E. and Neill, J.D. (Eds.), The Physiology of Reproduction. 2nd ed., Raven Press, New York, USA, 1994, p. 861.
2. Busse, W-D. and Carpenter, F.H., Biochem., 15 (1976) 1649.
3. Vandlen, R., Winslow, J., Moffat, B. and Rinderknecht, E. In MacLennan, A.H., Tregear, G.W. and Bryant-Greenwood, G.D. (Eds.) Progress in Relaxin Research, Global Publication Services, Singapore, 1995, p.59.
4. Ido, Y., Vindigni, A., Chang, K., Stramm, L., Chance, R., Heath, W.F., DiMarchi, R.D., Di Cera, E. and Williamson, J.R., Science, 277 (1997) 563.
5. Borthwick, G.M., Borthwick, A.C., Grant, P. and MacLennan, A.H. In MacLennan, A.H., Tregear, G.W. and Bryant-Greenwood, G.D. (Eds.) Progress in Relaxin Research, Global Publication Services, Singapore, 1995, p.251.

Chemical synthesis and relaxin activity of rat Insulin 3

Kathryn J Smith[1], Ross A Bathgate[1], Antonia A Claasz[1], Nicola F Dawson[1], Laszlo Otvos Jr[2], Geoffrey W Tregear[1] and John D Wade[1]

[1]*Howard Florey Institute, University of Melbourne, Victoria 3010, Australia; and*
[2]*The Wistar Institute, 3601 Spruce Street, Philadelphia, PA 19104, U.S.A.*

Introduction

Insulin 3 (INSL3; also called the relaxin-like factor, RLF, or Leydig cell insulin-like peptide, Ley I-L) is a recently identified member of the insulin-relaxin superfamily of peptide hormones. It is primarily expressed in the Leydig cells of the testis and thecal cells of the ovary [1]. Recently, two independent groups developed knockout mice for INSL3 [2,3]. The major phenotype in these animals was the lack of testis descent in the male and an alteration in oestrous cycle length in females, both resulting in a dramatic decrease in fertility.

Although the INSL3 cDNA sequence has now been cloned from a number of species the primary structure of the circulating INSL3 peptide is not known. The rat INSL3 cDNA has been recently cloned [4] and the predicted structure of the INSL3 prohormone shows high homology to that of predicted INSL3 structures from other species. Based on what is known of relaxin and insulin prohormone processing in various mammalian species, we have deduced the mature rat INSL3 peptide sequence. The predicted processing results in the production of a two-chain peptide (A- & B-chains) containing two inter- and one intra-chain disulfide bonds (Fig.1).

A-Chain SV**A**T**N**AVHR**CCL**T**GC**T**QQDL**LG**LC**PH

B-Chain EPP**E**ARA**KLCGHHLVR**ALVRV**C** **GG**PRW **S**PEA

Fig.1. The predicted structure of the rat INSL3 peptide. Residues that are conserved between species are in ***bold*** *and those important for INSL3 binding are boxed [5]. The relaxin-like receptor binding motif is underlined.*

The primary purpose of this study was to, for the first time, synthesize rat INSL3. Furthermore, as INSL3 shows high structural similarity to relaxin and it has a similar although displaced relaxin-like binding motif (Fig.1), its relaxin activity will be tested. The two chains were each assembled by Fmoc solid-phase peptide synthesis, purified by conventional RP-HPLC and the combined in solution at high pH [6]. The purity and identity of the chain-combined peptide was confirmed by comprehensive chemical characterisation including mass spectrometry.

G.W. Tregear et al. (eds.), Relaxin 2000, 251–252.
© 2001 *Kluwer Academic Publishers. Printed in the Netherlands.*

Results and Discussion

A major problem in INSL3/relaxin syntheses is poor recovery. The chemical synthesis of the two chains and subsequent formation of the three disulphide bonds (Fig.1) presents a difficult challenge. Both A- and B-chains were synthesised by standard SPPS methods. After cleavage and deprotection the resulting peptides were purified by RP-HPLC in overall yields of 8.4 and 8.9% respectively which are significantly lower than reported for ovine INSL3 [6]. The purified A- and B-chains were then combined in CAPS buffer, pH 10.5 at 4°C to form the three disulphide bonds. Following purification, MALDI-TOF MS showed a single species with MH^+ value of 6114.06 (calc. MW 6113) (Fig.2). Secondary structural analysis by CD spectroscopy showed the peptide to have a remarkably high α-helix content (35%) for a synthetic peptide, when dissolved in TFE (data not shown). The lack of relaxin activity of the peptide was confirmed by its inability to stimulate cAMP production in a relaxin receptor expressing cell line, THP1, and its poor ability to compete for relaxin receptor binding sites in the rat cerebral cortex (Ki = 3.33μM). Future studies will involve using the synthesis rat INSL3 to determine its physiological role in the rodent.

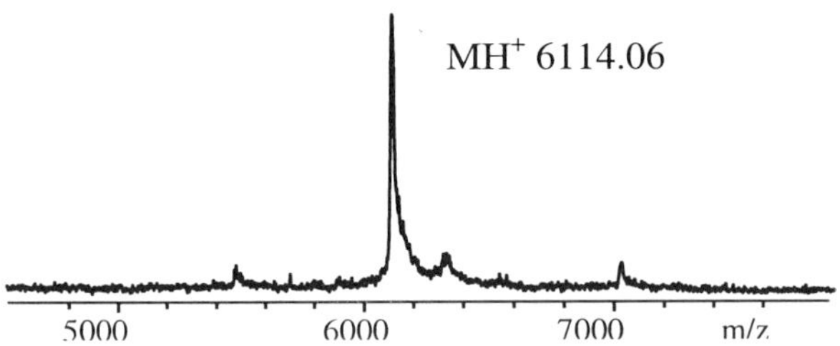

Fig.2. MALDI-TOF MS analysis of synthetic rat INSL3

Acknowledgments

This work was carried out at the Howard Florey Institute and was supported by an Institute block grant (registration key number 983001) from the NHMRC of Australia.

References

1. Ivell, R., Reviews in Reproduction, 2 (1997) 133-138.
2. Nef, S. and Parada, L.F., Nat. Genet., 22 (1999) 295.
3. Zimmerman, S., Steding, G., Emmen, J.M.A., Brinkmann, A.O., Nayernia, K., Holstein, A.F., Engel, W., and Adham, I.M., Mol. Endicrinol., 13(5) (1999) 681.
4. Speiss, A-N., Balvers, M., Tena-Sempere, M., Huhtaniemi, I., Parry, L. and Ivell, R., Mol. Reprod. Dev., 54 (1999) 319.
5. Büllesbach, E.E. and Schwabe, C., Biochemistry, 38 (1999) 3073.
6. Dawson, N.F., Tan, Y-Y., Macris, M., Otvos, Jr. L., Summers, R.J., Tregear, G.W. and Wade, J.D., J. Peptide Res., 53 (1999) 542.

SECTION 9

Relaxin bioassays

Structure-activity relationships of relaxin and relaxin-like factor/insulin 3

Yean Yeow Tan[1,2], Nicola F Dawson[1], Andrew R Kompa[2], Courtney Bond[2], Antonia Claasz[1], John D Wade[1], Geoffrey W Tregear[1] and Roger J Summers[2]

[1]*Howard Florey Institute, University of Melbourne, Parkville, Victoria 3010 and*
[2]*Department of Pharmacology, Monash University, Clayton, Victoria 3800, Australia.*

Introduction

Relaxin is a structural homologue of insulin. It does not cross react with insulin but acts on its own receptors to produce distinctive functional responses [1,2]. For relaxin, the primary active site consists of two invariant arginine residues at positions 13 and 17 within the central B-chain α-helix (numbered according to human gene 2 (H2) (B29) relaxin B-chain) (Fig. 1). The side chains of these residues project away from the core molecule of relaxin into the surrounding water to act as a dual prong to interact with relaxin receptors and produce the biological responses [3]. Substitutions of these residues have been shown to reduce or abolish binding and functional activity of this peptide [3]. Recently, it was also demonstrated that the isoleucine residue in B20 of H2 relaxin is involved in the binding of relaxin to its receptor, suggesting that relaxin receptor binding sites form a triangular contact with the surface of the relaxin molecule [4].

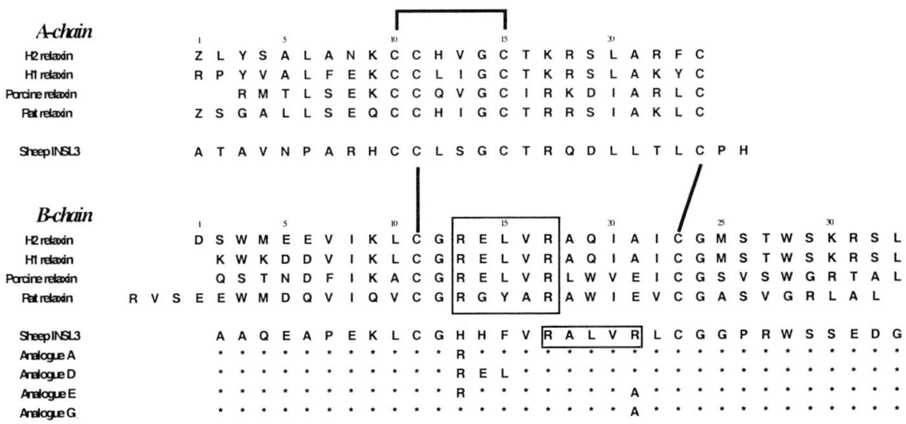

Fig. 1. The amino acid sequence of H2, H1, porcine and rat relaxins, sheep insulin-like factor 3 (INSL3) and analogues investigated in this study. The cysteine residues for disulphide bond formation are aligned and the arginine cassette important for relaxin binding and bioactivity are boxed.

G.W. Tregear et al. (eds.), Relaxin 2000, 255–260.
© 2001 *Kluwer Academic Publishers. Printed in the Netherlands.*

The sheep genome contains a relaxin-like gene sequence that is incapable of producing RNA that can be translated to a functional peptide [5]. Nevertheless, in this species there is a peptide which is structurally similar to relaxin and insulin in the testicular Leydig cells [6]. This peptide has been described as Leydig cell insulin-like peptide (Ley I-L), relaxin-like factor (RLF) or insulin-like factor 3 (INSL3). Sheep INSL3 has two arginine residues in the B-chain α-helix with n, n+4 configuration, but they are offset towards the carboxyl terminus by precisely four residues [6] (Fig. 1). Therefore, the side chain of these arginine residues would project out from the 2nd and the 3rd loop of the B-chain α-helix instead of the 1st and the 2nd loop. Studies have shown that human INSL3 lacks relaxin-like bioactivity but enhances H2 relaxin activity in mouse assays at high concentrations [7].

In this study, we used the rat atrial bioassay [2] and quantitative receptor autoradiography [8], and a high throughput human THP-1 cell binding assay [9] to investigate and compare the activity of relaxin and analogues which have had residues in the B-chain arginine binding region modified to determine the structure-activity relationships. We also investigated the sheep INSL3, synthesized and tested analogues with amino acid substitutions in the B-chain α-helix to determine regions of the sheep INSL3 molecule that are important for interaction with the relaxin receptor and hence the production of relaxin-like activity.

Results and Discussion

In the rat atrial functional assay, H2 relaxin (B29 and B33) and porcine relaxin have similar potency in producing positive chronotropic and inotropic responses. However, rat relaxin is significantly less active than H2 relaxin and porcine relaxin ($p<0.05$, $n=6$) (Table 1). These results confirm earlier observations that rat relaxin is also less active than human and porcine relaxins in the mouse interpubic ligament assay [10]. Similar results were also observed using quantitative receptor autoradiography with rat atrial sections, and rat relaxin had significantly lower binding affinity compared to human and porcine relaxins ($p<0.02$, $n=6$) (Table 1). In the human THP-1 cell binding assay, preliminary studies indicated that in contrast to the rat atrial assays, there was a higher binding affinity for H2 (B33) relaxin than H2 (B29) relaxin (Table 1). This was not observed in a previous binding study in THP-1 cells [9]. However this study [9] used a single high concentration (100 nM) of H2 (B29) relaxin or H2 (B33) relaxin that produced near maximal displacement of binding in contrast to the study reported here that uses a range of concentrations to determine complete concentration/binding relationships. The results from THP-1 cell binding assay also showed that the binding affinity of rat relaxin was less than those of H2 (B29 and B33) relaxin and porcine relaxin (Table 1).

Table 1. Comparison of the potency and binding affinity of H2, porcine and rat relaxins in rat isolated atrial bioassay, receptor autoradiography and human THP-1 cell binding. Rat relaxin is significantly less active than H2 and porcine relaxins in producing positive chronotropic ($p<0.005$) and inotropic (†$p<0.05$) responses and in competing for the relaxin receptors in rat atrial sections (‡$p<0.02$, all $n=6$).*

Relaxin	Potency (pEC_{50}) Chronotropic	Potency (pEC_{50}) Inotropic	Binding affinity Rat atria (pK_D/pK_i)	Binding affinity THP-1 cell (pK_D/pK_i)
H2 (B33)	9.35±0.07	9.05±0.08	8.92±0.09	9.22±0.34
H2 (B29)	9.06±0.10	9.03±0.07	9.17±0.08	8.38±0.15
Porcine	9.12±0.07	8.91±0.03	8.90±0.02	8.27±0.22
Rat	8.52±0.08*	8.60±0.09†	8.35±0.09‡	8.06±0.31

The anomalies in potency between H2/porcine and rat relaxins may be due to the 3 amino acid residues which sit in between the n, n+4 arginine residues in the B-chain. H2 relaxin and porcine relaxin have an identical sequence (-Glu-Leu-Val-), in contrast to that in rat relaxin (-Gly-Tyr-Ala-) (Fig. 1). A recent published study showed that the amino acid residues in B14 and B16, that is -Glu- and -Val- for H2 relaxin and porcine relaxin, and -Gly- and -Ala- for rat relaxin, are not important for activity since amino acid substitution at these positions did not affect the bioactivity of H2 relaxin [4]. Nevertheless, studies have yet to be conducted to determine whether the amino acid residue in B15 is important in predicting the potency of a relaxin. It has also recently been identified that B20 isoleucine in H2 relaxin is important in determining bioactivity [4]. Whereas there is relatively little effect of substituting B20 valine for isoleucine, there is a loss of 3 orders of magnitude of bioactivity with B20 alanine substitution. Interestingly, sheep INSL3 and all of the modified B-chain analogues used in this study have B20 valine similar to that in porcine relaxin, suggesting that this is not a factor in the reduced activity observed with INSL3 analogues.

We also investigated the bioactivity of synthetic H1 relaxin and the effects of B-chain shortening in the bioactivity of this peptide. H1 (B2-29 and B2-33) relaxins are active in eliciting positive chronotropic and inotropic responses in rat atria and the potency values obtained were similar to those of H2 relaxin. Shortening the B-chain of H1 relaxin caused a progressive loss of bioactivity (Fig. 2). Thus, H1 (B2-26) relaxin has 3 amino acid residues removed from the carboxyl terminal and was clearly less potent than H1 (B2-29) relaxin. Further removal of another 2 amino acid residues from the carboxyl terminal (H1 (B2-24) relaxin) totally abolished the bioactivity of the peptide (Fig. 2). In human THP-1 cell binding assay, preliminary studies also showed this pattern where the B2-29 and B2-33 forms of H1 relaxin compete for relaxin binding sites with highest affinity, followed by B2-26 and B2-24 forms. The binding result for B2-24 is rather interesting because at high concentrations, it was capable of competing for [^{33}P]-H2 (B33) relaxin binding in the THP-1 cells but the same analogue fails to produce any functional response in the rat atria (Fig. 2). Whether this implies that the

B2-24 form of H1 relaxin is an antagonist or that it behaves differently on human *versus* rat relaxin receptors has yet to be confirmed. Previous studies with porcine relaxin have indicated that up to seven residues of the carboxyl terminal of B-chain can be removed before the bioactivity of the peptide is affected [11].

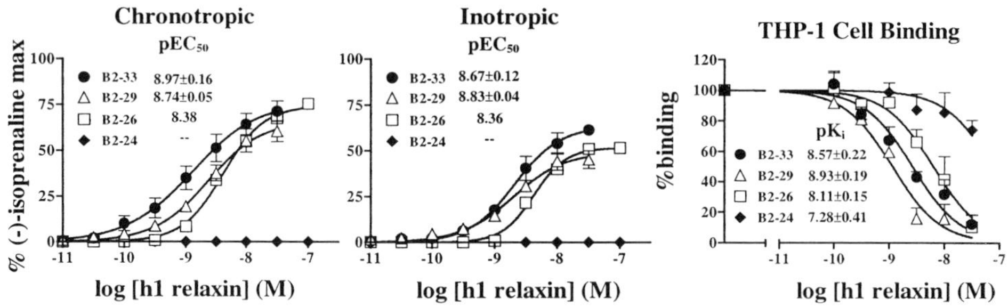

Fig. 2. The effects of shortening H1 relaxin in producing positive chronotropic and inotropic responses in rat isolated atria. H1 (B2-33), (B2-29) and (B2-26) relaxins are active in this bioassay but there is a decrease in potency (n=6 except for (B2-26) where n=2), while (B2-24) form is inactive. Preliminary results from human THP-1 cell binding show similar trend but (B2-24) is capable of competing for binding sites at concentrations which do not produce corresponding biological responses in rat atria.

Sheep INSL3 alone at 1 μM did not have positive chronotropic and inotropic effects on the rat isolated atria. However, its presence appeared to consistently reduce the activity of H2 (B29) relaxin (Fig. 3). Experiments with higher concentrations of INSL3 will be required to confirm this finding. In the receptor autoradiographic studies, sheep INSL3 only began to compete with [^{33}P]-H2 (B33) relaxin at 1 μM (not shown), suggesting that sheep INSL3 may interact with the rat atrial relaxin receptors at high concentrations.

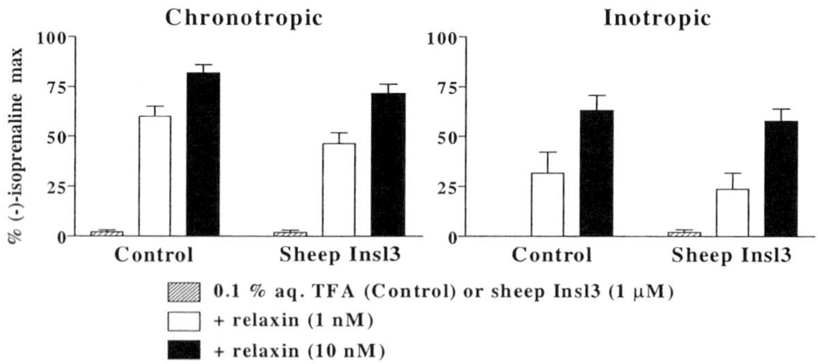

Fig. 3. The effects of sheep INSL3 and H2 (B29) relaxin on positive chronotropic and inotropic responses in rat isolated atria. Sheep INSL3 at 1 μM did not have a significant stimulant effect but it reduced both the chronotropic and inotropic effects of 1 nM and 10 nM of H2 (B29) relaxin added cumulatively.

Sheep INSL3 has n, n+4 arginine residues in the 2nd and the 3rd loop of the B-chain α-helix instead of the 1st and the 2nd. We therefore synthesized an analogue, Analogue A, which has the histidine residue in the 1st loop replaced with an arginine (Fig. 1). Analogue A at 1 μM showed moderate relaxin-like activity in producing positive chronotropic and inotropic responses in the rat isolated atria and it competes for the relaxin receptors in rat atria with higher affinity compared to native INSL3 (Fig. 4). This result confirmed that the presence of n, n+4 arginine residues is important for relaxin-like activity but they must be in correct position and configuration. Further modifications were made to produce Analogue D — which now has the residues in the arginine binding cassette identical to that of H2 relaxin and porcine relaxin (Fig. 1) — to determine whether we could regain the full relaxin-like activity. However, Analogue D was less active than Analogue A in functional assay but had binding affinity similar to Analogue A (Fig. 4). Analogue E, which is a derivative of Analogue A but with the arginine residue in the 3rd loop substituted by an alanine (Fig. 1), was synthesized to investigate whether the lack of activity shown by Analogues A and D is due to the presence of an additional arginine residue in the 3rd loop which causes steric hindrance, thus preventing proper interaction of these peptides with relaxin receptors. Analogue E had positive chronotropic activity weaker than Analogue A and Analogue D but had no positive inotropic activity in rat isolated atria (not shown). Analogue G, which has arginine at the 3rd loop of the B-chain α-helix replaced by alanine (Fig. 1) and was synthesized as a control, had no relaxin-like activity.

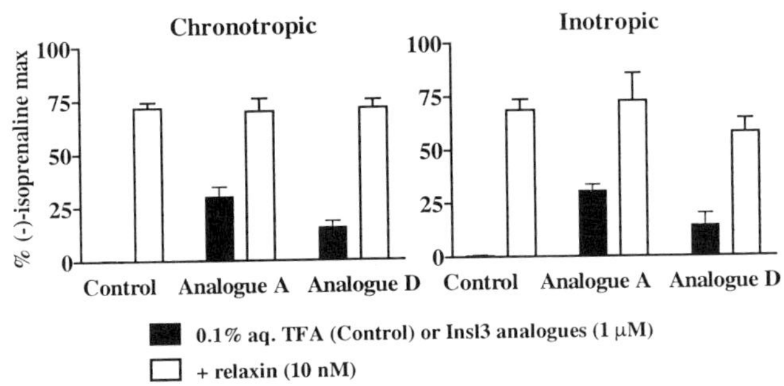

Peptide	Binding Affinity Rat Atria (pK_i)
Sheep Insl3	4.69±0.11
Analogue A	5.15±0.03*
Analogue D	5.29±0.09*
Analogue E	4.70±0.30
Analogue G	4.50±0.25

Fig. 4. Effects of Analogue A and Analogue D and H2 (B29) relaxin on chronotropic and inotropic responses in rat isolated atria. Analogue A and Analogue D both had moderate positive chronotropic and inotropic effects at 1 μM. The addition of 10 nM H2 (B29) relaxin increased the responses further and matching those of control, indicating that Analogue A or Analogue D were not acting synergistically with relaxin. The table compares binding affinities of sheep INSL3 and analogues in rat atria ($p<0.003$ compared to sheep INSL3, $n=6$) measured using quantitative receptor autoradiography.

Overall, these results indicate that the position of arginine binding cassette is important for relaxin-like activity but other as yet unknown regions beyond this core of the peptide molecule are important in affecting the potency. We believe that Ile/Val in B20 may not be the answer since a valine residue is present in the corresponding position of sheep INSL3 and all the analogues. Even with Analogue E which has all the amino acid residues required for binding, which are two arginines in correct position and the valine in B20, there is only weak relaxin-like activity. Clearly, these observations suggest that we need to investigate other regions of the relaxin molecules for bioactivity.

In summary, we showed that H2 relaxin and porcine relaxin are more potent than rat relaxin in rat atria bioassay and also have higher binding affinity values for the relaxin receptors in this tissue. Progressive shortening of the H1 relaxin B-chain causes reduction and obliteration of peptide bioactivity but not the binding activity as shown in the human THP-1 cell binding assay. Further investigation into this phenomenon is in progress. We also showed that sheep INSL3 is not a sheep relaxin and amino acid substitution studies indicated that the arginine binding cassette in relaxin B-chain is important for relaxin bioactivity but only if it is in the correct position and configuration.

Acknowledgments

Supported by an Institute Block Grant Reg Key 983001 from the NHMRC to the Howard Florey Institute and by the National Heart Foundation of Australia.

References

1. Osheroff, P.L., Ling, V.T., Vandlen, R.L., Cronin, M.J. and Lofgren, J.A., J. Biol. Chem., 265 (1990) 9396.
2. Tan, Y.Y., Wade, J.D., Tregear, G.W. and Summers, R.J., Br. J. Pharmacol., 123 (1998) 762.
3. Büllesbach, E.E., Yang, S. and Schwabe, C., J. Biol. Chem., 267 (1992) 22957.
4. Büllesbach, E.E. and Schwabe, C., J. Biol. Chem., 275 (2000) 35276.
5. Roche, P.J., Crawford, R.J. and Tregear, G.W., Mol. Cell. Endocrinol., 91 (1993) 21.
6. Roche, P.J., Butkus, A., Wintour, E.M. and Tregear, G.W., Mol. Cell. Endocrinol., 121 (1996) 171.
7. Büllesbach, E.E. and Schwabe, C., J. Biol. Chem., 270 (1995) 16011.
8. Tan, Y.Y., Wade, J.D., Tregear, G.W. and Summers, R.J., Br. J. Pharmacol., 127 (1999) 91.
9. Parsell, D.A., Mak, J.Y., Amento, E.P. and Unemori, E.N., J. Biol. Chem., 271 (1996) 27936.
10. Sherwood, O.D., Endocrinology, 104 (1979) 886.
11. Tregear, G.W., Fagan, C., Reynolds, H., Scanlon, D., Jones, P., Kemp, B., Niall, H.D. and Du, Y.-C., In: Peptides, Synthesis, Structure, Function: Proceedings of the Seventh American Peptide Symposium, ed. Rich, D.H. and Gross, E., Rockford, IL: Pierce Chemical Co. (1982) 249.

Relaxin bioassays

Mark E Erikson and Shu-Hui Liu
Connetics Corporation, Palo Alto, CA, 94303, USA

Introduction

A well-characterized biotechnology product is a protein whose identity, purity, impurities, potency, and quantity can be determined and controlled. Implicit in this definition is the ability to define the biological activity of these molecules by a potency or bioassay. The bioassay plays, in many cases, the paramount role in determining whether the biologic displays correct tertiary and quaternary structure. Additionally, while a panel of physicochemical tests provides information on the structure of protein products and impurities, a biological assay is the only one that provides a direct measure of function. Proper folding of denatured proteins from inclusion bodies within E. coli is necessary for recognition by cognate receptors. This recognition is accompanied by the realization that a critical measure of growth factors and cytokines is the ability of the product to transduce a signal within the cell.

In the development of a bioassay that reproducibly measures relaxin's bioactivity, efforts have been directed at developing assays for the quantitative detection of the following analytes or activities:

- inhibition of collagen production
- production of matrix metalloproteins (MMP's)
- production of tissue inhibitor of MMP's (TIMP's)
- production of angiogenic factors
- regulation of VEGF promoter activity
- production of prolactin
- production of insulin-like growth factor binding-protein 1
- production of or inhibition of the production of cytokines
- cGMP production
- nitric oxide generation
- adenylate cyclase activation
- protein kinase A activation
- protein phosphorylation
- measures of changes in overall metabolic activity utilizing microphysiometry, intracellular metabolic stains, or intracellular pH dyes
- specific activity of intracellular enzymes – i.e., esterase, aromatase
- calcium flux/mobilization.

In general, while useful in elucidating the biological activities of rH2 relaxin, most of the methods failed as bioassay candidates due to a lack of reproducible,

G.W. Tregear et al. (eds.), Relaxin 2000, 261–270.
© 2001 *Kluwer Academic Publishers. Printed in the Netherlands.*

quantifiable response, lack of high throughput ability, or difficulty in procuring large quantities of relevant cells. In addition, the development of an optimal assay has been hampered by the failure to clone the relaxin receptor and by the lack of sufficient information on the signal transduction cascade following relaxin's binding to a cell surface receptor.

In this chapter, two candidate bioassays will be described. In this discussion, the specific performance of the assays in the context of their potential role in a drug development program will be evaluated.

Microphysiometry

The Cytosensor microphysiometer (Molecular Devices, Sunnyvale, Calif.) has been used to study both receptor pharmacology and signaling [1,2]. Microphysiometry is not dependent upon specific transduction pathways and thus is an extremely versatile tool for the study of ligand-receptor interaction when little is known about the pathway subsequently utilized. In addition, the response is measured in real-time. These two factors offer advantages over traditional assays, which usually measures a specific cellular product with an assay time ranging from several hours to days.

The microphysiometer quantifies cellular metabolic activity by measuring changes in the extracellular acidification rate. Most, if not all, physiological changes, such as receptor activation, alter the rate of energy metabolism. This results in the formation of acid metabolites (e.g. lactate and H^+), which are then secreted from the cell in order to maintain a constant intracellular pH. The microphysiometer uses silicon semiconductor sensors to detect these changes in extracellular pH, reflecting changes in metabolic rate. By use of a capsule, insert, and spacer, cells are sandwiched between two polycarbonate membranes forming a small chamber in close proximity to the sensor. A peristaltic pump controls perfusion of the cells with media with low buffering capacity. In a cyclical fashion, the pump is turned off (typically for 30 s, with the pump on for 90 s), thus allowing the accumulation of acid metabolites at the sensor surface, and the acidification rate is determined. By means of a valve, drug samples can be perfused over the cells for controlled periods, allowing pharmacological studies. An important aspect of the microphysiometer is its ability to measure, in a non-invasive manner, decreases in metabolic activity, as well as the easier-to-measure increases in metabolic activity which occur as a result of receptor activation. The microphysiometer has, therefore, been used successfully to monitor activation of a wide variety of cell surface receptors that couple to different effectors, e.g., muscarinic M1, M2, M3, M4, and M5 receptors [3] or D2, D3, or D4 receptors [4], intracellular targets such as nuclear hormone receptors [5], and signal transduction pathways. The microphysiometer is particularly suited to the study of signal transduction pathways, as activation or inhibition of a wide variety of effectors and transduction processes will affect metabolic rate. With the use of activators or inhibitors of a specific kinase or phosphatase, for example, the microphysiometer can also be used to study the interaction and cross-talk between signal pathways. As almost all signal transduction modifying agents have nonspecific effects, appropriate controls must be included in the experimental design to allow for reliable interpretation of the data. The kinetics of the extracellular acidification response (ECAR) can provide some clues to the type of signal transduction mechanisms involved. However, a degree of caution must be exercised, as the ECAR is an integrated response often consisting of several components [6]. For example, the interleukin-1a (IL-1a)-induced ECAR in transfected Chinese Hamster Ovary (CHO) cells expressing the type1 interleukin-1 receptor (IL-1 RI) is due to activation and interplay of protein kinase A (PKA), protein kinase C (PKC), and phosphatidyl-3 (PI-3) kinase components [7].

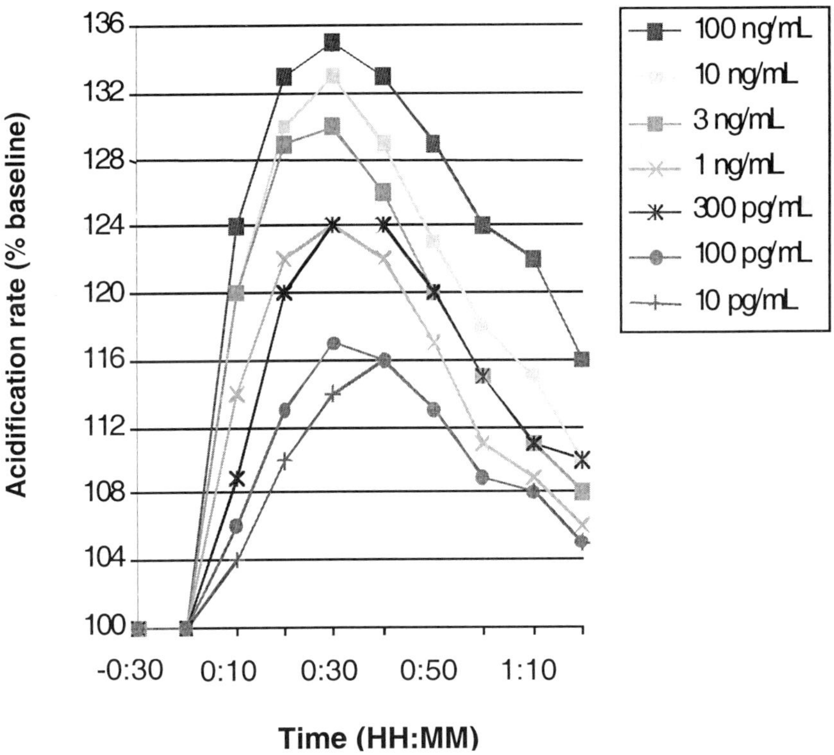

Fig. 1. Response of human endometrial stromal Cells to rH2 relaxin as measured by extracellular acidification rate. NHE cells were cultured for 24 hours on semi-porous polycarbonate membranes in serum-free media. Cells (10^5 /chamber) were equilibrated for 60 min, treated for 30 min with a phosphodiesterase inhibitor (100 mM RO-20-1724), and then exposed, still in the presence of the phosphodiesterase inhibitor, to increasing concentrations of rH2 relaxin for 30 minutes.

The microphysiometer was initially used to measure the effect of rH2 relaxin on normal human endometrial (NHE) stromal cells. These are a primary cell line from a well documented target tissue for relaxin, the endometrium. Fig. 1 presents typical response curves for NHE cells in the presence of rH2 relaxin. As can be seen, the response is rapid, being measurable in the first few minutes of exposure, and a fairly broad dose response is evident. Removal of rH2 relaxin from the medium allowed for a slow decrease in the response, and in general, there was not a return to the original baseline level of activity. As they are primary cells, NHE cells tend to lose their responsiveness to rH2 relaxin over time in culture and are, therefore, of limited utility in the development of a long-term bioassay necessary for serial monitoring of rH2 relaxin during the course of manufacturing and drug stability studies.

Another cell type with high affinity binding sites for rH2 relaxin is the THP-1 cell line (a human monocytic cell line), a line readily available commercially [8]. Fig. 2 shows representative binding data for ^{32}P-rH2 relaxin on THP-1 cells. While Scatchard analysis revealed a receptor with a dissociation constant of 102 pM, receptor density, at approximately 275 per cell, is extremely low. The Scatchard plot suggests that the rH2 relaxin binds to a single class of receptor on THP-1 cells. Both the mature B29 form of rH2 relaxin and the extended B33 form used for labeling reproducibly displaced 60–80% of the total radioactive rH2 relaxin that was bound. In contrast, human insulin, IGF-I, IGF-II, and RLF each typically displaced less than 20% of the bound ^{32} P-relaxin. In addition, neither estrogen nor progesterone appeared to regulate the receptor expression on THP-1 cells.

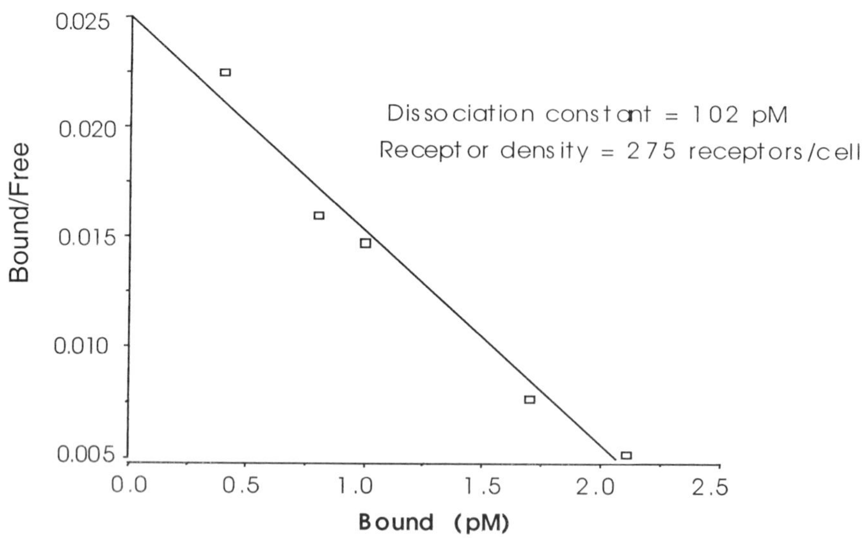

Fig. 2. Scatchard analysis of ^{32}P-rH2 relaxin binding to THP-1 cells. Triplicate samples (1 X 10^6 cells/tube) were incubated for 60 min at room temperature with increasing concentrations (20–400 pM) of labeled relaxin in the presence or absence of excess (1.4 mM) unlabeled relaxin. The number of cells present at the end of the experiment was determined by counting cells in parallel samples. Scatchard analysis yielded a dissociation constant (Kd) of 102 pM and a receptor density of 275 receptors/cell. [8]

THP-1 cells presented several advantages over the NHE cells with regard to the development of a reliable, reproducible bioassay. Inter-laboratory and inter-assay variation tends to be much lower when cell lines are used instead of freshly isolated or primary, passage-dependent cells, and stable cell lines are the most commonly used systems for bioassays. THP-1 cells do not appear to lose responsiveness to rH2 relaxin over a number of passages in culture. Additionally, the cells are non-adherent and grow rapidly (doubling time of < 18 hours), allowing shorter and less labor-intensive set-up time for the assays. For these reasons, these cells replaced the NHE cells in routine use in the microphysiometer. (Fig. 3)

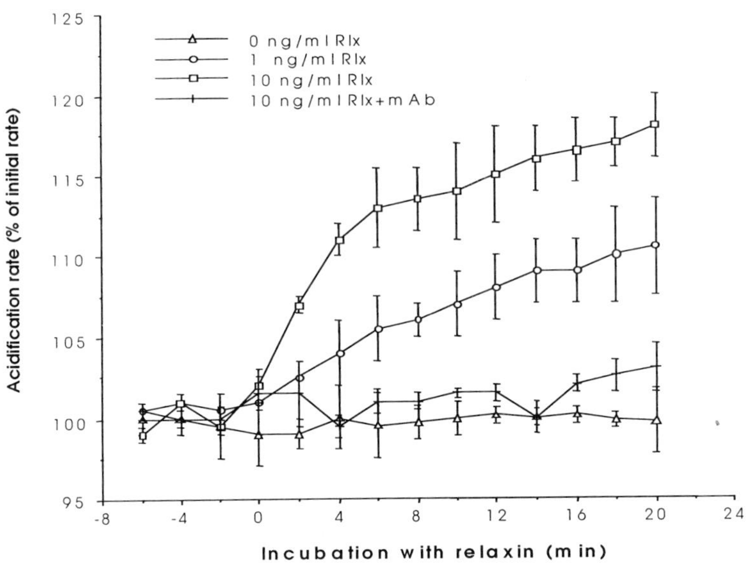

Fig.3. rH2 relaxin induces increased extracellular acidification in THP-1 cells. Cells (10^5 /chamber) were equilibrated for 60 min, treated for 30 min with a phosphodiesterase inhibitor (100 mM RO-20-1724), and then exposed, still in the presence of the phosphodiesterase inhibitor, to increasing concentrations of relaxin (RLX) for the times indicated. To demonstrate specificity, cells were treated with 10 ng/ml relaxin in the presence of a 10-fold molar excess of a neutralizing anti-relaxin antibody (mAb8). The values shown are mean ± SEM values calculated from 2–4 independent experiments.

An example of how this bioassay is utilized in drug development is its utility in evaluating responses to rH2 relaxin administration in a clinical setting. During the course of developing rH2 relaxin for the treatment of systemic sclerosis, it was discovered that approximately 30% of patients receiving rH2 relaxin at a 25 μg/kg/day dose developed anti-rH2 relaxin antibodies, while 60% of those receiving 100 μg/kg/day developed antibodies. While no antibody-related (i.e. allergic) reactions were observed, further characterization of these antibodies was necessary. The ability of these antibodies to neutralize rH2 relaxin bioactivity was tested utilizing the microphysiometer. Protein G-purified immunoglobin was isolated from serum samples known to contain rH2 relaxin-binding antibodies; these serum samples were collected from subjects two weeks following stopping drug treatment. Consistent with its serum half-life of one to two hours, no rH2 relaxin was detectable in these samples, either by rH2 relaxin ELISA or bioassay. THP-1 cells were then incubated for 30 minutes in the presence of immunoglobulin alone, rH2 relaxin alone, or rH2 relaxin pre-incubated with immunoglobulin. Cells were pre-incubated with immunoglobulin to control for possible effects of immunoglobulin binding directly to the surface of the THP-1 cells. Fig. 4 shows that rH2 relaxin alone at 1, 5, and 10 ng/ml produced a dose-related increase in the acidification rate, with a 10% increase above baseline apparent at the 10 ng/ml dose. Also shown are the results following incubation of rH2 relaxin (10 ng/ml) with immunoglobulin preparations from two antibody-positive patients and one rH2 relaxin antibody-negative patient. To date, preincubation of rH2 relaxin with immunoglobulin from any antibody-positive patient failed to inhibit the response of THP-1 cells to rH2 relaxin. In contrast, the response of THP-1 cells to rH2 relaxin can be completely inhibited by preincubation with either a monoclonal antibody specific for human rH2 relaxin (Fig. 3), or an affinity-purified polyclonal antibody used in the rH2 relaxin ELISA (data not shown).

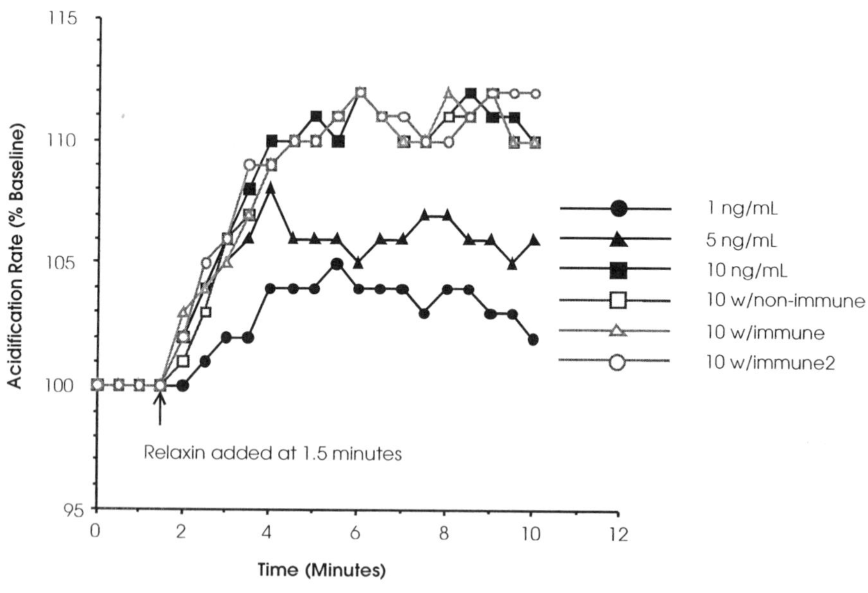

Fig. 4. Response of THP-1 cells to rH2 relaxin is not neutralized by human anti-rH2 relaxin antibodies. Cells (10^5 /chamber) were equilibrated for 60 min, treated for 30 min with protein G

purified immunoglobulin from the serum of rH2 relaxin antibody positive patients, non-immune patients, or media in the presence of 100 mM RO-20-1724, and then exposed, still in the presence of the phosphodiesterase inhibitor, to increasing concentrations of rH2 relaxin in media alone or which had been preincubated with immunoglobulin from the same patient.

The rapid, dose-dependent response of THP-1 cells to rH2 relaxin proved fairly reproducible when measured by microphysiometry. As with the NHE cell response, the removal of rH2 relaxin from THP-1 cells resulted in a gradual lowering of acidification rates, but rarely to the original baseline levels. In addition, a second round of rH2 relaxin stimulation often showed a greater response to a given concentration of rH2 relaxin, suggesting that each capsule of THP-1 cells could be used to reliably test rH2 relaxin only once. Therefore, in the context of a high throughput assay needed to support expanding development and manufacturing efforts, the microphysiometry has serious disadvantages. In addition, the cost of the cell capsules and the required set-up time made this a prohibitively expensive and labor intensive method for what needed to be routine potency measurements.

cAMP assay

A large number of hormones, neurotransmitters, and other signal substances activate adenyl cyclase upon binding to their cognate cell surface receptors, and utilize adenosine 3',5' cyclic monophosphate (cAMP) as an intracellular second messenger. Cyclic AMP regulates a number of different cellular processes such as cell growth and differentiation, ion channel conductivity, synaptic release of neurotransmitters, and gene transcription. Relaxin-dependent changes in cAMP production were reported as early as 1978 in mouse pubic symphysis [9] and in 1980 in the rat uterus and cervix [10, 11, and 12]

The cAMP bioassay for rH2 relaxin currently used by Connetics is a fairly standard cell-based assay, which is based on the method of Fei [13], with modifications. Two major modifications were: (1) the fortuitous isolation of THP-1 cells showing a greater than 4 fold increase in responsiveness to rH2 relaxin; and (2) the inclusion of a parallel-line analysis in the generation of results. In addition, over the course of several years' optimization of the assay, the precision and robustness of the method have been improved. Briefly, THP-1 cells, in the presence of a phosphodiesterase inhibitor added to stabilize cAMP, are incubated for 30 minutes with dilutions of a rH2 relaxin reference standard or test samples. rH2 relaxin samples being analyzed for bioactivity are typically tested at concentrations of 0.5, 1.0, 2.0, and 4.0 ng/ml. Following centrifugation and a wash step, cells are lysed using a proprietary lysing reagent, and intracellular cAMP levels are measured using the cAMP Biotrak EIA system (Amersham Pharmacia Biotech Inc., Piscataway, NJ), according to the manufacturer's instructions. Data analysis for newly manufactured lots of rH2 relaxin are carried out by comparison with a reference (gold) standard using a parallel-line analysis. The result is reported out as a percent relative potency compared to the reference material, with a single number representing the mean of three separate assays. The concentrations of cAMP generated by either reference standard or test samples are generally significant and fall within the linear portion of the dose response. The specificity of the assay can be shown by abrogation of the response by the inclusion of a monoclonal antibody specific for rH2 relaxin, which blocks binding and biological activity of the molecule.

While the inclusion of a continuous cell line has decreased overall assay variability, the challenge to validation of this method for quality control of manufactured lots of rH2 relaxin, as with most cell based assays, lies in inter-assay reproducibility. If this assay is to be used to verify that a specific lot of manufactured rH2 relaxin is potent and comparable to previous lots, the assay must be rigorous and reproducible. Typically, highly purified, well-defined products are dosed based upon mass rather than potency. This is a scientifically valid approach since biological assays are better at determining that two lots (a test lot versus a reference) are comparable than they are at assigning a specific potency unitage [14]. However, if the product is a well-characterized biologic with a battery of rigorous physical/chemical tests applied to lot release, the bioassay becomes a verification assay.

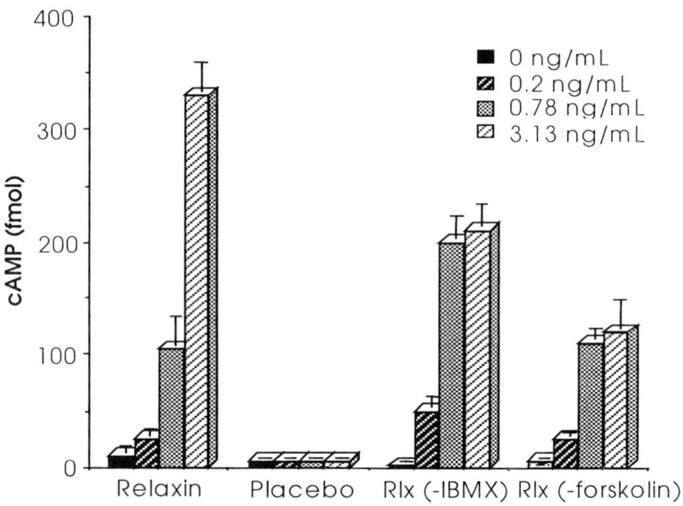

Fig. 5. rH2 relaxin induces an increase in intracellular cAMP in THP-1 cells. THP-1 cells (8 X 10^4 / well) were incubated at 37°C for 30 minutes with placebo or varying concentrations of rH2 relaxin in the presence or absence of 1µM forskolin and 50µM IBMX. CAMP levels in cell lysates were determined using the cAMP Biotrak EIA system from Amersham Pharmacia Biotech. The data are mean + SD values from 3 separate wells (Parsell et al., 1996).

The cAMP response of cells (measured as optical density) to the reference material and to the test sample are each plotted versus the rH2 relaxin concentration on a log scale and fitted with a linear regression as shown in Figure 6. The data are then subjected to a parallel line analysis and the difference between the lines on the horizontal axis is reported as % relative potency, relative to the reference. The design of the bioassay had to take into account factors that introduce variability into the assay, and both the validation procedure itself and analysis of the assay had to test for variability. The validation protocol carried out on the cAMP bioassay included analyses of intra-assay, inter-assay and inter-laboratory precision. Further, a titration of the test material has to be compared with a titration of a reference material, with particular attention paid to the linear portion of the dose-response curve. Therefore, the validation protocol was set up to determine the accuracy, linearity, and range of the method. Validity testing also took into account the parallelism of dose-response lines. The statistical techniques applied in determining parallelism were variance ratio or F tests. Single point assays (ED_{50}s alone) do not allow a comparison of the shapes of the dose-response curves and are not valid.

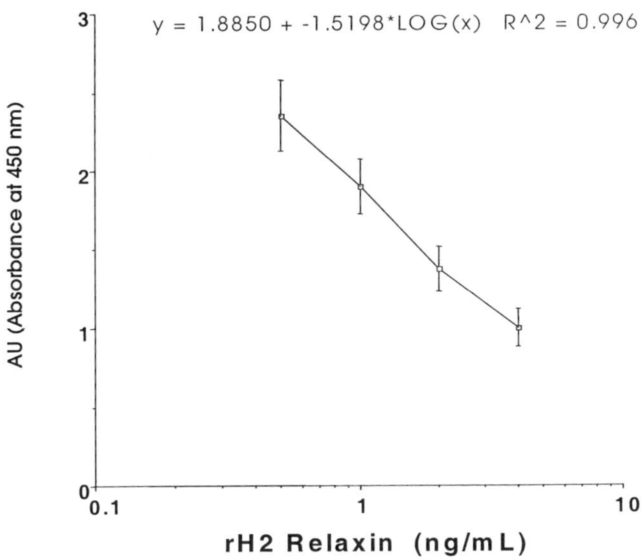

Fig. 6. Typical rH2 relaxin bioassay standard curve. THP-1 cells (4 X 10^4 / well) were incubated at 37°C for 30 minutes with varying concentrations of rH2 relaxin in the presence of forskolin and IBMX. cAMP levels in cell lysates were determined using the cAMP Biotrak EIA system from Amersham Pharmacia Biotech. The data are ± SD values from 10 separate assays.

The validated assay met acceptance criteria for accuracy with a mean recovery of 100% (n=60) with a relative standard deviation of 20%. The slope of the accuracy curve (observed versus expected) was 0.97 over a range of 50% to 150% relative potencies. The intra-assay and inter-assay precision (%CV) were 11% and 7%, respectively. In addition, no bias or significant differences were observed in precision between results produced on different days or by different analysts in different labs.

The assay is now in routine use for lot release, stability testing, comparability testing, and formulation development, and has been run on over 300 samples (representing over 900 separate assays).

References

1. Owicki, J.C., Parce, J.W., Kercso, K.M., Sigal, G.B., Muir, V.C., Venter, J.C., Fraser, C.M. and McConnell, H.M., Proc. Natl. Acad. Sci., 187 (1990) 4007.
2. McConnell, H.M., Owicki, J.C., Parce, J.W., Miller, D.L., Baxter, G.T., Wada, H.G. and Pitchford, S., Science, 257 (1992) 1906.
3. Wood, M.D., Murkitt, K.L., Ho, M., Watson, J.M., Brown, F., Hunter, A.J. and Middlemiss, D.N., Br. J. Pharmacol., 126 (1999) 1620.
4. Coldwell, M.C., Boyfield, I., Brown, A.M., Stemp, G. and Middlemiss, D.N., Br J. Pharmacol., 127 (1999) 1135.
5. Redish, D.M., Raley-Susman, K.M. and Sapolsky, R.M., Horm. Metab. Res., 25 (1993) 264.
6. Smart, D. and Wood, M.D., Biochem. Cell Biol., 78 (2000) 281.
7. Smart, D., Hammond, E.A. and McKnight, A.T., Br. J. Anaesth., 81 (1998) 97.
8. Parsell, D.A., Mak, J.Y., Amento, E.P. and Unemori, E.N., J. Biol. Chem., 271 (1996) 27936.
9. Braddon, S., Endocrinology, 102 (1978) 1292
10. Cheah, S.H. and Sherwood, O. D., Endocrinology, 106 (1980) 1209.
11. Sanborn, B.M., Kuo, H.S., Weisbrodt, N.W. and Sherwood, O.D., Endocrinology, 106 (1980) 1210.
12. Judson, D.G., Pay, S. and Bhoola, K.D., J. Endocrinol., 87 (1980) 153.
13. Fei, D.T.W, Gross, M.C., Lofgren, J.L., Mora-Worms, M. and Chen, A.B., Biochem. Biophys. Res. Commun., 170 (1990) 214.
14. Burn, J.H., Phys. Rev., 1 (1930) 146

Bioassays for relaxin and relaxin-related peptides

Yean Yeow Tan[1,2], Courtney P Bond[2], John D Wade[1], Geoffrey W Tregear[1] and Roger J Summers[2]
[1]Howard Florey Institute, University of Melbourne, Parkville 3010 and [2]Department of Pharmacology, Monash University, Clayton 3800, Victoria, Australia.

Introduction

The bioactivity of relaxin, isolated as native peptide from the ovaries of various species and synthesised as the peptide predicted from gene sequence, has been characterised in several functional bioassays. The *in vivo* guinea pig pubic symphysis palpation [1] and the mouse interpubic ligament [2] bioassays are well documented to measure relaxin-like activity. However, these bioassays are expensive, cumbersome and require a large number of animals to obtain a dose-response relationship. The isolated atrial bioassay is a more recently developed functional bioassay [3], where relaxin produces dose-dependent positive chronotropic and inotropic responses in rat right and left atria respectively. The isolated atrial assay is relatively easy to set up and since a complete dose response curve can be completed in a single preparation, it allows the dose-response relationship of relaxin to be determined with small numbers of animals. In this study, we have used the isolated rat atrial bioassay and a quantitative autoradiographic binding assay using rat atrial sections [4] to characterise the bioactivity of human, porcine and rat relaxin analogues. To assess the interaction of relaxin analogues with a human relaxin receptor, we used a receptor binding assay with human THP-1 cells [5].

Results and Discussion

Human relaxin produced potent chronotropic and inotropic responses in rat isolated atria. Both analogues of human relaxin, H2 relaxin (B29) and H2 relaxin (B33), were equally potent in chronotropic and inotropic responses. Porcine relaxin and prorelaxin were also as potent as the human relaxin analogues, whereas native and synthetic rat relaxin were significant less potent as shown by the pEC_{50} values in Table 1.

Table 1. pEC_{50} values of chronotropic and inotropic responses to human, porcine and rat relaxins ($p<0.05$).*

Relaxin	**Potency (pEC_{50})**	
	Chronotropic	Inotropic
H2 (B29)	9.06 ± 0.10	9.03 ± 0.07
H2 (B33)	9.35 ± 0.07	9.05 ± 0.08
porcine	9.12 ± 0.07	8.91 ± 0.03
prorelaxin	9.01 ± 0.06	8.95 ± 0.06
rat (synthetic)	8.60 ± 0.05*	8.69 ± 0.09*
rat (native)	8.52 ± 0.08*	8.60 ± 0.09*

G.W. Tregear et al. (eds.), Relaxin 2000, 271–272.
© 2001 *Kluwer Academic Publishers. Printed in the Netherlands.*

The results obtained using quantitative autoradiography supported the results using the atrial bioassay and showed that rat relaxin competed for binding with significantly less affinity than either H2 (B33), H2 (B29) or porcine relaxin (Table 2a). Preliminary data obtained using the THP-1 binding assay also confirmed that rat relaxin has a lower affinity in competing for binding at the human relaxin receptor than the other relaxin analogues (Table 2b). However, H2 (B33) relaxin appeared to have a significantly higher affinity for the human receptor compared to H2 (B29) and porcine relaxin. Further research is in progress to determine if there is a difference between human and rat receptor in their ability to recognise relaxin analogues.

Table 2. Binding affinities for H2 (B33), H2 (B29), porcine and rat relaxin. (a) shows the results from the quantitative autoradiography in rat atria while (b) shows preliminary data from the THP-1 cell binding assay.

(a)

Relaxin	Binding affinity (pK_i)
H2 (B33)	8.92 ± 0.09
H2 (B29)	9.26 ± 0.01
porcine	8.90 ± 0.02
rat	8.35 ± 0.09*

(b)

Relaxin	Binding affinity (pK_i)
H2 (B33)	9.07 ± 0.17
H2 (B29)	8.36 ± 0.14
porcine	8.29 ± 0.20
rat	7.93 ± 0.14

The results obtained from these assays correlate well with earlier published results using the guinea pig public symphysis palpation and the mouse pubic ligament assays [2,6], where rat relaxin was less bioactive than porcine relaxin in both bioassays. This indicates the assays used in this study are effective while also being sensitive, reproducible, uncomplicated, graded assays suitable for investigating relaxin-like activity of various relaxin analogues.

Acknowledgments

The work, in collaboration with the Howard Florey Institute, was supported by an Institute block grant (reg. key no. 983001) from the NHMRC of Australia. CB is a Monash University Faculty of Medicine Scholar.

References

1. Kroc, R.L., Steinetz, B.L. and Beach, V.L., Ann. NY Acad. Sci., 75 (1959) 942.
2. Steinetz, B.G., Beach, V.L. and Kroc, R.L., In Dorfman, R.I. (ed.) Methods in Hormone Research, Acadmeic Press, New York, 1969, p. 481.
3. Kakouris, H., Eddie, L.W. and Summers, R.J., Lancet 339 (1992) 1076.
4. Tan, Y.Y., Wade, J.D., Tregear, G.W. and Summers, R.J., Brit. J. Pharm., 127 (1999) 91.
5. Parsell, D.A., Mak, J.Y., Amento, E.P., and Unemori, E.N., J. Biol. Chem., 271 (1996) 27936.
6. Sherwood, O.D., In Knobil, E. and Neill, J.D. (eds.), The Physiology of Reproduction. 2nd ed., Raven Press, New York, 1994, p.861.

A highly sensitive homologous human relaxin ELISA

Franz P Armbruster[1], Ina Maier[1], H-J Grön[1], Susanne M Bailer[1], Thomas Dschietzig[2], Karl Stangl[2], Hartmut Echner[3], Theodor H Lippert[4], H Seeger[4] and Wolfgang Voelter[3]

[1]*Immundiagnostik GmbH, Bensheim, Germany;* [2]*Charité, Department of Cardiology, Berlin, Germany;* [3]*University of Tuebingen, Department of Physical Biochemistry, Tuebingen, Germany; and* [4]*University of Tuebingen, Department of Obstetrics & Gynaecology, Tuebingen, Germany*

Introduction

Relaxin is a peptide hormone of the insulin family with pleiotropic effects [1]. While the physiological role of relaxin during human reproduction has been studied for many years, very recent findings indicate that due to its vasodilatory, cardioprotective, chronotropic, and central hemodynamic properties, relaxin may also have important implications on the cardiovascular system [2, 3].

A highly sensitive homologous ELISA for human (H) relaxin is presented which for the first time allows quantitative detection of relaxin in men and non-pregnant women. In a pilot study, H relaxin was detected in plasma samples of men. In patients with chronic heart failure (CHF) we found significantly elevated H relaxin concentrations. Baseline H relaxin in patients with severe CHF was 2.2 – 2.4 times higher than in patients with moderate CHF, and 8 – 15 times higher than in healthy individuals. This sandwich-type-immunoassay is based on two specific polyclonal antibodies raised in rabbits against recombinant human gene 2 relaxin (rH2 relaxin). For antibody characterization a kit from CBS, Cambridge, UK with overlapping hexapeptides representing the whole peptide sequences of the A- and B-chains of H2 relaxin was used [4]. The experimentally determined epitopes overlap with the theoretical epitopes predicted based on the algorithms of Jameson and Wolf [ref. in 4].

Results and Discussion

Two different populations of antibodies against H2 relaxin were used to develop the highly sensitive ELISA with a detection limit of 1.2 pg/ml (OD (zero) +2SD). The linear range is 1.2 – 250 pg/ml. As standard material, rH2 relaxin was used (kindly provided by Genentech, USA). The recovery of H2 relaxin in serial dilutions of spiked samples is 97±4.5%.

The assay is performed as follows: Serum, plasma, urine or seminal plasma should be stored at -20°C. To avoid matrix effects in plasma these samples should be diluted 1:3, urine samples should be diluted 1:5. Microtiter plates are coated overnight at 4°C with 1 μg rabbit anti H2 relaxin (Protein G purified) in 0.005 M carbonate buffer, pH 9.6 and subsequently blocked with 200 μl PBS pH 7.4, 2% BSA. After a washing

G.W. Tregear et al. (eds.), Relaxin 2000, 273–274.
© 2001 *Kluwer Academic Publishers. Printed in the Netherlands.*

step, 100 μl of standard (0, 4, 8, 16, 3, 62, 125, 250 pg/ml) or sample material are added to the cavities and incubated for 24 h at 4°C. After washing, 100 μl of biotinylated rabbit anti H2 relaxin antibody (Protein G purified) in PBS are added and incubation proceeds for 2 h at 4°C on a shaker. Washing the cavities is followed by incubation with 100 μl of streptavidin-POX for 1 h at 4°C on a shaker. Following a washing step, 100 μl of substrate solution (TMB) is added and incubated for 10-30 min at room temperature until colour difference is sufficient. 50 μl of stop solution (2N H_2SO_4) are added and extinction is measured immediately at 450 nm (reference wavelength 690 nm).

The characteristics of this assay are as follows: the standard range is 1.2 – 250 pg/ml and the detection limit is 1.2 pg/ml ($OD_{(zero)}$ + 2 SD, n = 6) (Fig. 1). Recovery is 97 ± 4.5% (plasma). The intra- and inter-assay values (CV) are 9.6 % (n = 18) 15.3 pg/ml and 10.2 % (n = 12) 15.6 pg/ml respectively.

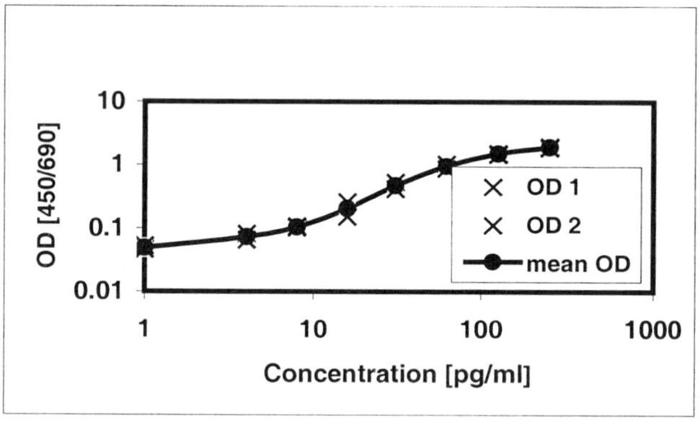

Fig 1. Typical calibration curve.

Conclusion

A highly sensitive homologous ELISA for H2 relaxin for the first time allows quantitative detection of relaxin in men and non-pregnant women. Based on this assay, H2 relaxin concentration was followed during CHF. While healthy controls show a mean concentration of H2 relaxin of 2.1 ± 1.6 pg/ml, in patients with CHF, the H2 relaxin levels are 13.3 ± 4.0 pg/ml. Thus, this assay provides a tool suitable for physiological and pathophysiological research aimed at monitoring H2 relaxin.

References

1. Bani, D., Gen. Pharmac., 28 (1997) 13.
2. Danielson, L.A., Sherwood, O.D. and Conrad, K.P., J. Clin. Invest., 103 (1999) 525.
3. Bani, D., Massini, E., Bello, M.G., Bigazzi, M. and Sacchi, T.B., Am. J. Pathol., 152 (1998) 1367.
4. Meisenbach, M., Armbruster, F.P., Becker, S., Grön, H.J., Grübler, G., Lippert, T.H., Paulus, G. and Voelter, W. In P.T.P. Kaumaya and R.S. Hodges (eds.) Peptides: Chemistry, Structure and Biology, Mayflower Worldwide, United Kingdom, 1996, p. 818.

SECTION 10

Relaxin-induced signal transduction

Relaxin and scaffolding proteins in signalling crosstalk

Barbara M Sanborn, Kimberly L Dodge, Chun-Ying Ku and Caiping Yue

Department of Biochemistry and Molecular Biology, University of Texas Houston Medical School, Houston TX 77030 USA

Introduction

One of the earliest reported biochemical effects of relaxin was its ability to increase cAMP in pubic symphysis, myometrium and cervix [1-3]. In myometrium, relaxin has been shown to increase calcium efflux, decrease myosin light chain phosphorylation, activate calcium-sensitive maxi-K channels (K_{Ca}), decrease oxytocin-stimulated rat myometrial contractions, decrease oxytocin-stimulated phosphatidylinositide turnover and decrease oxytocin-stimulated intracellular calcium elevation and calcium entry from the extracellular [4]. Protein kinase A inhibitors attenuate effects of relaxin on myometrial contraction, stimulation of K_{Ca} channels, and inhibition of oxytocin-mediated phosphatidylinositide turnover, intracellular calcium elevation and calcium entry (Fig. 1) [4-7]. We have explored in more depth the mechanisms by which relaxin elevates cAMP and inhibits oxytocin-mediated phosphatidylinositide turnover in the myometrium.

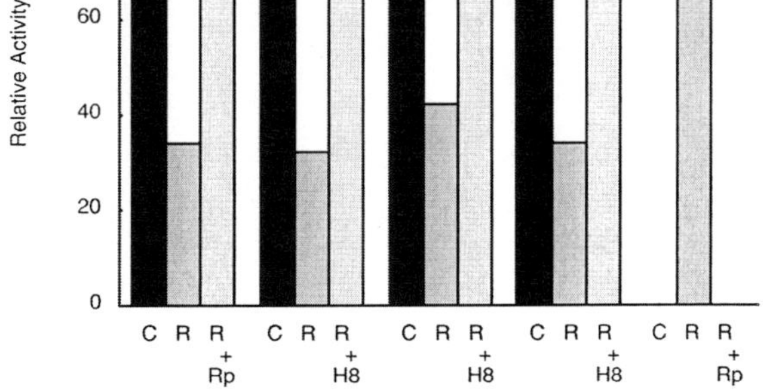

Fig. 1. Composite data from a number of studies in which protein kinase A inhibitors attenuated the effects of relaxin (R) on contraction (expressed as integrated area, Int. Area), inhibition of oxytocin-mediated intracellular calcium (Ca_i) elevation in the presence (Ca_e+) and absence (Ca_e-) of extracellular calcium, phosphatidylinositide (PI) turnover, and stimulation of K_{Ca} channel open probability (Po,,K_{Ca}). Data are expressed as % of control or maximal increase over control (stimulation of K_{Ca} channel). Adapted from [4-7].

G.W. Tregear et al. (eds.), Relaxin 2000, 279–283.
© 2001 *Kluwer Academic Publishers. Printed in the Netherlands.*

Results and Discussion

Although most actions of relaxin can be reversed by cAMP antagonists or protein kinase A inhibitors, it is not entirely clear how relaxin activates cAMP. The elevation of cAMP in myometrium exhibits a significant time lag and is less robust than that of isoproterenol [2,10]. We also demonstrated a weak activation of myometrial plasma membrane adenylate cyclase activity by relaxin [11]. The properties of these responses suggest they could have been indirect. Consistent with this idea, Kuznetsova *et al.* [12] report that the ability of human relaxin to increase adenylate cylase in human myometrium is inhibited by the tryosine kinase inhibitor tryphostin 47 and potentiated by Gpp[NH]p. We have recently found that the tyrosine kinase inhibitor genistein can reverse the effect of relaxin on oxytocin-stimulated phosphatidylinositide turnover in myometrial cells (Fig. 2). These data are consistent with a number of other reports involving tyrosine phosphorylation in the mechanism of action of relaxin [12-15] If the relaxin receptor is shown to stimulate tyrosine kinase activity, activation of adenylate cyclase may be indirect, as suggested for a number of other systems [16-18].

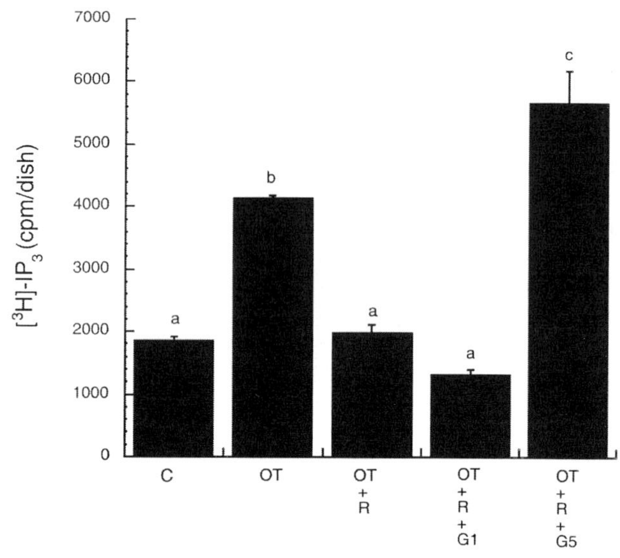

Fig. 2. Genistein reverses the effect of relaxin on oxytocin-stimulated phosphatidylinositide turnover in PHM1-41 myometrial cells. Cells labeled with ^{3}H-inositol were preincubated for 30 min with 1 (G1) – 5 (G5) μg/ml genestein and then 1 μg/ml porcine relaxin (R) for 15 min before stimulation with 200 nM oxytocin (OT) for 5 min and ^{3}H-phosphoinositides isolated [19].

The inhibitory effect of relaxin and cAMP on contractant-induced phosphatidylinositide turnover appears to be at the level of phospholipase C (PLC). The inhibitory effect of cAMP is seen with several receptors coupled to the heterotrimeric protein complex $G\alpha_q$ and is seen in cells other than myometrium, indicating that it is not cell or receptor-specific [19]. This does not rule out additional effects of cAMP on receptor function, however. We have not found phosphorylation of $G\alpha_q$, but have determined that phosphorylation of PLCβ_3 on Ser^{1105} is responsible for the inhibitory action of protein kinase A on $G\alpha_q$-stimulated $PLC\beta_3$ activity [20]. Mutation of this residue eliminates the inhibitory effect of protein kinase A.

Relaxin also inhibits oxytocin-stimulated calcium entry into myometrial cells [5]. In the PHM1-41 myometrial cell line, contractant-stimulated calcium entry is not inhibited by nifedipine and may be the result of capacitative calcium entry [21]. Several putative capacitative calcium Trp proteins are present in myometrium (Gupta et al, unpublished observations). In addition, diacylglycerol, a product of the PLC pathway, and direct interaction with the endoplasmic reticulum inositol 1,4,5-trisphosphate receptor have been implicated in capacitative calcium entry [22,23]. Therefore, relaxin may inhibit this entry as a result of its inhibitory effects on phosphatidylinositide turnover mediated through the action of protein kinase A.

Protein kinase A is localized to organelles in cells by association with A Kinase Anchoring Proteins or AKAPs [8]. Some AKAPs can bind more than one protein and act as scaffolding proteins [24,25]. We have identified plasma membrane-associated 86 kD and 150 kDa AKAPs in human and rat myometrial cells, respectively [26]. Importantly, the ability of relaxin to inhibit oxytocin-stimulated phosphatidylinositide turnover in myometrial cells was attenuated by S-Ht31, a cell-permeable amphipathic helix peptide that mimics the protein kinase A/AKAP interaction site, but not by P-S-Ht31 in which the helix is disrupted by a proline residue (Fig. 3A). The effect was maximal by 5 min (Fig. 3B) and was dose-dependent. Similar effects were observed using forskolin or exogenous CPT-cAMP [26]. These data suggest that the protein kinase A activated by cAMP generated by relaxin is bound to an AKAP and that this interaction is critical for the action of relaxin. S-Ht31 releases both protein kinase A regulatory and catalytic subunits and enzymatic activity from myometrial plasma membranes [26]. These data provide an explanation for our earlier finding of endogenous protein kinase A in rat myometrial plasma membrane [9].

In pregnant rat myometrial membranes, both protein kinase A and protein phosphatase 2B co-precipitate with AKAP150 [27]. Protein phosphatase 2B can remove phosphate from $PLC\beta_3$, thus potentially removing protein kinase A-mediated inhibition. Interestingly, the relative distribution of protein kinase and protein phosphatase 2B changes near the end of pregnancy in the rat, concomitant with a loss in the ability of cAMP to inhibit oxytocin-stimulated phosphatidylinositide turnover [27]. These data provide strong support for a functional role of the plasma membrane AKAP in an important cAMP signaling pathway in myometrium that attenuates pathways that stimulate contraction.

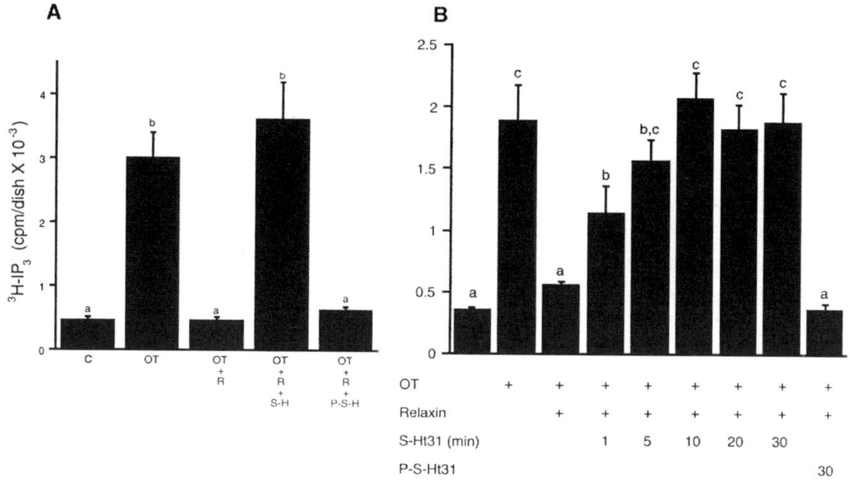

Fig. 3. A. *The AKAP interaction antagonist S-Ht31 (S-H), but not the control peptide P-S-Ht31 (P-S-H), disrupted the ability of relaxin (R) to inhibit oxytocin-stimulated (OT) phosphatidylinositide turnover in PHM1-41 myometrial cell [26]. B. The AKAP interaction antagonist S-Ht31 exhibits a time-dependent ability to attenuate the inhibitory effect of relaxin on oxytocin-stimulated phosphatidylinositide turnover [26]. Different letters indicate significant differences between groups at p<0.05. Reprinted with permission.*

These data are consistent with a mechanism of action of relaxin that involves either a direct or indirect stimulation of adenylate cyclase, an increase in cAMP, activation of protein kinase A localized in the myometrial plasma membrane by the AKAP scaffolding protein, and phosphorylation and inhibition of the ability of PLCβ_3 to couple to Gα_q. This action would become less effective near term in the rat as less protein kinase A is associated with the myometrial plasma membrane AKAP. It is likely that other actions of relaxin that utilize PKA may also involve AKAP scaffolding proteins as well.

Acknowledgements

The authors acknowledge the contributions of Dr. D. Carr and Dr. G. Weber to this work and the generous gifts of reagents from Drs. S. McKnight, J. Scott, M. Simon, J. Hepler, and P. Sternweis. This work was supported in part by NIH HD-09618 and T32-HD07325.

References

1. Braddon, S.A., Endocrinology, 102 (1978) 1292.
2. Sanborn, B.M., Weisbrodt, N.W., Kuo, H.S. and Sherwood, O.D., Endocrinology, 106 (1980) 1210.

3. Cheah, S.H. and Sherwood, O.D., Endocrinology, 106 (1980) 1203.
4. Sanborn, B.M., Anwer, K., Monga, M., Oberti, C., Meera, P., Stefani, E., Toro, L. and Wen, Y. In MacLennan, A.H., Traegear, G.W. and Bryant-Greenwood, G.D. (Eds.), Progress in Relaxin Research, Global Publications Services, Singapore, 1995, p 289.
5. Anwer, K., Hovington, J.A. and Sanborn, B.M., Endocrinology, 124 (1989) 2995.
6. Anwer, K., Hovington, J.A. and Sanborn, B.M., Biol. Reprod., 43 (1990) 851.
7. Meera, P., Anwer, K., Monga, M., Oberti, C., Stefani, E., Toro, L. and Sanborn, B.M., Am. J. Physiol., 269 (1995) C312.
8 Dodge, K. and Scott, J.D., FEBS Lett., 476 (2000) 58.
9. Wen, Y., Anwer, K., Singh, S.P. and Sanborn, B.M., Endocrinology, 131 (1992) 1377.
10. Hsu, C.J., McCormack, S.M. and Sanborn, B.M., Endocrinology, 116 (1985) 2029.
11. Sanborn, B.M., Kuo, H.S., Weisbrodt, N.W. and Sherwood, O.D., In Anderson, R.R. (Ed.), Relaxin, Plenum Press, New York, 1982, p 273.
12. Kuznetsova, L., Plesneva, S., Derjabina, N., Omeljaniuk, E. and Pertseva, M., Regul. Pept., 80 (1999) 33.
13. Ohleth, K.M. and Bagnell, C.A., Biol. Reprod., 53 (1995) 1286.
14. Palejwala, S., Stein, D., Wojtczuk, A., Weiss, G. and Goldsmith, L.T., Endocrinology, 139 (1998) 1208.
15. Bani, D., Failli, P., Bello, M.G., Thiemermann, C., Bani, S.T., Bigazzi, M. and Masini, E., Hypertension, 31 (1998) 1240.
16. Nair, B.G., Parikh, B., Milligan, G. and Patel, T.B., J. Biol. Chem., 265 (1990)
17. Berg, K.A., Maayani, S., McKay, R. and Clarke, W.P., J. Neurochem., 64 (1995) 220.
18. Poppleton, H., Sun, H., Fulgham, D., Bertics, P. and Patel, T.B., J. Biol. Chem., 271 (1996) 6947.
19. Dodge, K. and Sanborn, B.M., Endocrinology, 139 (1998) 2265.
20. Yue, C., Dodge, K.L., Weber, G. and Sanborn, B.M., J. Biol. Chem., 273 (1998) 18023.
21. Monga, M., Campbell, D.F. and Sanborn, B.M., Am. J. Ob. Gyn., 181 (1999) 424.
22. Kiselyov, K., Xu, X., Mozhayeva, G., Kuo, T., Pessah, I., Mignery, G., Zhu, X., Birnbaumer, L. and Muallem, S., Nature, 296 (1998) 478.
23. Hofmann, T., Obukhov, A.G., Schaefer, M., Harteneck, C., Gudermann, T. and Schultz, G., Nature, 397 (1999) 259.
24. Coghlan, V.M., Perrino, B.A., Howard, M., Langeberg, L.K., Hicks, J.B., Gallatin, W.M. and Scott, J.D., Science, 267 (1995) 108.
25. Faux, M.C. and Scott, J.D., J. Biol. Chem., 272 (1997) 17038.
26. Dodge, K.L., Carr, D.W. and Sanborn, B.M., Endocrinology, 140 (1999) 5165.
27. Dodge, K.L., Carr, D.W., Yue, C. and Sanborn, B.M., Mol. Endocrinol., 13 (1999) 1977.

Mechanism of action of relaxin in human cervix

Laura T Goldsmith, Gerson Weiss and Smita Palejwala
Department of Obstetrics, Gynecology & Women's Health, New Jersey Medical School, Newark, New Jersey, 07103, USA

Introduction

In many mammalian species the cervix must undergo marked connective tissue remodeling for successful parturition to occur. In spite of the significance of this process, the physiological and biochemical mechanisms responsible remain poorly defined. In various mammalian species, relaxin is a powerful agent in the reorganization of reproductive tract connective tissue. In several species, relaxin is obligatory for normal delivery due to this marked ability to rearrange connective tissue in the pubic symphysis and cervix. In women, elevated maternal circulating relaxin levels during human pregnancy are significantly correlated with an increased risk of premature delivery [1,2]. However, in spite of the importance of relaxin to normal reproductive function and its potential to be an etiologic agent in preterm delivery in women, knowledge regarding the mechanisms by which relaxin alters human cervical connective tissue is limited. The maintenance of connective tissue architecture requires a precise balance between the production of matrix metalloproteinases, which degrade the extracellular matrix, and the endogenous tissue inhibitors of metalloproteinases, which regulate the activity of the metalloproteinases. The present studies were designed to determine the effects of relaxin on regulation of human cervical matrix metalloproteinase expression and to elucidate the mechanism of action of relaxin in these effects using an established *in vitro* model of human pregnancy cervix, lower uterine segment fibroblasts (hLUSF) [3].

Results and Discussion

Since the human cervix is comprised primarily of Type I collagen, we evaluated the effect of relaxin upon expression of interstitial collagenase (MMP-1), its activator stromelysin (MMP-3), and its endogenous inhibitor TIMP-1 in LUSF. The isolation, primary culture and characterization of the cells has been described in detail previously [3]. Cells were plated and maintained in DMEM supplemented with fetal bovine serum and antibiotics (complete medium) overnight in an atmosphere of 5% CO_2 /95 % air. Medium was removed from the cells and replaced with complete medium containing 1 μM 17-β estradiol. This was repeated 24 and 48 hr later for a total of 72 hr. of estradiol-priming. Medium was then removed and replaced with serum-free DMEM supplemented with 0.2% lactalbumin hydrolysate without or with various amounts of relaxin. Cells in replicate wells were incubated with phorbol-12-myristate-13-acetate (PMA, 10 nM) as a positive control. After 65 hours,

G.W. Tregear et al. (eds.), Relaxin 2000, 285–289.
© 2001 *Kluwer Academic Publishers. Printed in the Netherlands.*

cells were removed from the plate and counted. Conditioned medium was maintained frozen at -20 C until Western Blotting was performed. Cells from replicate wells were used to prepare RNA for Northern Blotting. Levels of proMMP-1, proMMP-3 and TIMP-1 were determined by Western blotting using specific, well characterized antibodies and methods previously described [3].

Relaxin significantly increased expression of procollagenase protein from hLUSF. Both 1 ng/ml and 10 ng/ml relaxin significantly increased procollagenase protein expression to 204 ± 26 % ($p = 0.02$) and 211 ± 21 %, ($p = 0.006$) of control respectively. Relaxin (10 ng/ml) also caused an increase in the expression of the steady state levels of the 2.2 kb procollagenase mRNA transcript to 221 % above that of control. Relaxin also significantly increased expression of prostromelysin protein from hLUSF. Both 1 ng/ml and 10 ng/ml relaxin significantly increased prostromelysin expression to 241 ± 46 % ($p = 0.04$) and 198 ± 31 % ($p = 0.04$) of control, respectively. Relaxin at 10 ng/ml also caused an increase in the expression of the steady state levels of the 1.9kb prostromelysin mRNA transcript to 192 % above that of control.

Relaxin significantly reduced expression of TIMP-1 protein from hLUSF. Relaxin at 0.5 ng/ml, 1 ng/ml or 10 ng/ml significantly decreased TIMP-1 expression to 67 ± 9 % ($p = 0.02$), 50 ± 7 % ($p = 0.001$) and 45 ± 5 % ($p < 0.001$) of control, respectively. In contrast to its effect at the level of the protein, relaxin at 10 ng/ml did not affect steady state levels of the 0.9 kb TIMP-1 mRNA transcript.

These data demonstrate that relaxin is a positive regulator of matrix metalloproteinase expression in hLUSF cells; it causes an increase in metalloproteinase activity and decreases activity of the endogenous inhibitor. These data suggest a role for relaxin in remodelling of human cervical connective tissue and provide a mechanism by which elevated circulating relaxin concentrations may increase prematurity risk.

Relaxin action upon certain target organs, for example rat uterus, is synergistic with the actions of progesterone. However, other data suggest that this may not be the case in other target organs. To determine whether relaxin and progesterone have similar or different actions upon metalloproteinase expression in human cervix, we determined the effects of progesterone upon hLUSF expression of proMMP-1, proMMP-3 and TIMP-1. Treatment of hLUSF with 0.1, 1, 10, 100 or 1000 nM progesterone significantly decreased procollagenase protein expression to 63 ± 4, 58 ± 4, 54 ± 10, 52 ± 5 and 57 ± 3 (M ± SEM, n=4 experiments, $p<0.05$ at all amounts) % of control, respectively. Expression of prostromelysin protein levels was similarly inhibited by these amounts of progesterone to 59 ± 13, 53 ± 9, 58 ± 8, 60 ± 4 and 64 ± 8 (M ± SEM, n=4 experiments, $p<0.05$ at all amounts) % of control, respectively. In contrast, expression of TIMP-1 protein levels was not affected by treatment with progesterone at these doses. Levels remained at 100 ± 4, 93 ± 9, 96 ± 11, 106 ± 9 and 117 ± 11 (M ± SEM, n=4 experiments, $p>0.05$ at all amounts) % of control, respectively. These data demonstrate that progesterone has effects that are opposite to those of relaxin in hLUSF. These data allow for the hypothesis that progesterone may to some extent block or prevent the full effect of relaxin upon the cervix, especially at high concentrations. A greater effect of relaxin may occur when progesterone concentrations are decreased or when the action of progesterone is decreased. Thus, in women, relaxin action upon the cervix may be more robust at term pregnancy when it is thought that progesterone biological activity is decreased. At very high concentrations relaxin may overcome this blockage by progesterone causing an increase in prematurity risk.

Since a hallmark of human cervical ripening at term pregnancy is leukocyte infiltration, we determined the effect of relaxin upon proMMP-2, which degrades basement membrane collagens, elastin, laminin and fibronectin. Treatment of the cells with 1 ng/ml or 10 ng/ml Relaxin caused a significant increase in MMP-2 expression to 210 ± 31 (M ± SEM, n=4 experiments, p=0.039) and 193 ± 18 % (p=0.014) of control, respectively. The increase in MMP-2 expression caused by treatment of the cells with 0.5 ng/ml Relaxin was less pronounced (154 ± 25 %, p=0.11). In contrast, MMP-2 protein levels were not affected by incubation with progesterone at either 0.1, 1, 10, 100, or 1000 nM. Levels remained at 109 ± 20, 117 ± 15, 107 ± 13, 94 ± 7 and 97 ± 10 (M ± SEM, n=4 experiments, p>0.05 at all amounts) % of control, respectively. These data suggest a mechanism by which relaxin may foster leukocyte infiltration in the human cervix, an effect which could serve to enhance the direct positive regulatory effect of relaxin upon MMP expression since leukocytes are a source of various cytokines, many of which themselves have MMP stimulatory activity.

Certain cytokines and other biochemical stimulators of MMP expression in various cell types utilize protein kinase C in mediating their effects. To determine if protein kinase C mediates relaxin action upon hLUSF MMP expression, cells were incubated without or with a selective inhibitor of protein kinase C or a structurally similar molecule which has no inhibitory activity, used as a negative control [6]. Inhibition of protein kinase C activity does not influence relaxin stimulation of proMMP-1 and proMMP-3 protein levels. The highly selective PKC inhibitor used in these experiments markedly inhibited PMA stimulated proMMP-1 and proMMP-3 expression, yet had no effect on relaxin stimulated levels. The structurally similar inactive compound demonstrated no effect upon relaxin regulated expression of proMMP-1 or proMMP-3 or upon PMA stimulated proMMP-1 and proMMP-3 expression. These data suggest that relaxin stimulation of MMPs does not involve protein kinase C. We have previously demonstrated that relaxin does not affect intracellular concentrations of cAMP [3]. Thus, it is unlikely that either of these mediators are involved in relaxin regulation of cervical MMPs.

We have previously demonstrated marked displacement of ^{125}I-relaxin binding to hLUSF with increasing concentrations of unlabelled relaxin [3]. To provide additional information regarding the biochemical nature of the relaxin binding site, we determined the effect of the non-hydrolyzable GTP analog, GTP-γ–S upon relaxin binding. No alteration of relaxin binding was observed in the presence of the GTP analog suggesting that the relaxin receptor is not a heterotrimeric G protein associated receptor. Although various data demonstrate that relaxin action is associated with the cAMP/adenylate cyclase/PKA pathway in several cell types, no previous data have demonstrated coupling of a relaxin-receptor complex to a heterotrimeric G protein.

We have previously demonstrated that the response of hLUSF to relaxin is associated with an increase in phosphorylation of a ~220kDa molecular weight protein, likely the relaxin receptor [3]. To advance our understanding of the mechanisms of relaxin signalling in the cervix and test the hypothesis that the relaxin receptor is a tyrosine kinase receptor, the effects of inhibition of receptor tyrosine kinase activity on relaxin stimulated tyrosine phosphorylation of the ~220kDa protein were studied. hLUSF cells treated with relaxin in the presence of 5μM HNMPA, a potent, specific receptor tyrosine kinase inhibitor [4,5], demonstrated a significant inhibition of Relaxin stimulated tyrosine phosphorylation of the ~220 kDa protein to 33.3 ± 8% (±SEM, p=0.004, n=4 experiments) and 39.2 ± 7% (p=0.003, n=4 experiments) of uninhibited, relaxin-stimulated levels at 15 and 30 min of treatment, respectively, similar to non-relaxin treated levels at the same treatment times. In contrast, two other receptor tyrosine kinase inhibitors, selective for PDGF (AG18) and EGF (AG1478) had no effect upon Relaxin stimulated tyrosine phosphorylation. These data support the concept that the relaxin receptor is a tyrosine kinase receptor, distinct from the PDGF and EGF receptors.

The most well characterized signalling cascade utilized by tyrosine kinase receptors bound to ligand involves the successive activation of Ras, Raf, mitogen-activated protein (MAP) kinase kinase 1 and 2 (MEK) and MAP kinases (ERK1 and ERK2). To elucidate the components of the pathway which is utilized in relaxin signalling, we studied the effect of specific inhibition of synthesis of the serine/threonine kinase c-Raf.

Transfection of hLUSF with a highly specific c-Raf mRNA targeted antisense oligonucleotide [7] resulted in effective inhibition of c-Raf protein levels to 48.6 ± 10% (±SEM, p=0.036) (n=3 experiments, each performed in triplicate) and 61.2 ± 1.5% (p=0.0016) of mismatched oligonucleotide treated levels in untreated and relaxin treated cells, respectively. Inhibition of c-Raf protein resulted in significant down regulation of relaxin stimulated proMMP-1 to 60.5 ± 3.3% (p=0.007), and relaxin stimulated proMMP-3 protein levels to 56 ± 8.8% (p=0.037), of those of relaxin treated cells transfected with the mismatched oligonucleotide control. Reduction of c-Raf protein levels also resulted in a reversal of relaxin inhibition of TIMP-1 protein levels to 146 ± 3.3% (p=0.005) of those of relaxin treated cells transfected with the mismatched oligonucleotide control. Expression of proMMP-1, proMMP-3 and TIMP-1 protein levels in non-relaxin treated cells was not altered by inhibition of c-Raf protein levels. These data point to the involvement of c-Raf in relaxin signalling and suggest the utilization of a Raf-MEK-ERK kinase cascade. However, very recent data suggest that c-Raf can also activate a MEK independent pathway.

In summary, the evidence that relaxin is a positive regulator of the MMPs which affect Type I collagen, in contrast to the negative effect of progesterone, and that relaxin stimulates proMMP-2, provide potential biochemical mechanisms to explain the relationship between increased circulating maternal levels of relaxin and the increased risk of prematurity. Also, these findings allow for the formulation of a new hypothesis regarding the role of the relative actions of relaxin and progesterone at term delivery. In addition, these data, in concert with our previous results, provide additional evidence that the human relaxin receptor is unlikely to be a heterotrimeric G protein associated receptor. The collective data do suggest that the relaxin receptor is likely to be a tyrosine kinase receptor which, when bound to relaxin, activates a signalling pathway mediated by the c-Raf protein. Determination of whether downstream signalling events are MEK dependent or independent requires further study.

Acknowledgements

This work was supported in part by NIH Grant HD 22338

References

1. Petersen, L.K., Skajaa, K. and Uldbjerg, N., Br. J. Obstet. Gynecol., 99 (1992) 292.
2. Weiss, G., Goldsmith, L.T., Sachdev, R., Von Hagen, S. and Lederer, K., Obstet. Gynecol., 82 (1993) 821.
3. Palejwala, S., Stein, D.E., Wotjczuk, A., Weiss, G. and Goldsmith, L.T., Endocrinology, 139 (1998) 1208.
4. Saperstein, R., Vicario, P., Strout, H.V., Brady, E., Slater, E., Greenlee, W., Ondeyka, D., Patchett, A. and Hangauer, D., Biochem., 28 (1989) 5694.
5. Baltensperger, K., Lewis, R., Woon, C.W., Vissavajjhala, P., Ross, A. and Czech, M., Proc. Natl. Acad. Sci. USA, 89 (1992) 7885.
6. Meyer T., Regenass U. and Fabbro D., Int. J. Cancer, 43 (1989) 851.
7. Monia, B.P., Johnston, J.F., Geiger, T., Muller, M. and Fabbro, D., Nature Medicine, 6 (1996) 668.

Myometrial relaxant action of relaxin

Michael Hollingsworth[1], Sarah Rudkin[2] and Sandra Downing[3]
[1]School of Biological Sciences, G38 Stopford building, Manchester University, Oxford Road, Manchester M13 9PT, UK; [2]Parexcel, The Straddle, Victoria Quays, Sheffield S2 5SY; and [3] Biology Department, York University, P.O. Box 373, York YO10 5YW, UK .

Characteristics of the uterine/myometrial relaxant action

The uterine[1] relaxant action of relaxin-containing extracts of porcine corpora lutea was discovered 50 years ago [1]. However, the primary signal transduction event underlying this action is unresolved.

Relaxin is a uterine relaxant *in vivo* and will relax uteri/myometria *in vitro* in several species [2, 3], although porcine and H1 and H2 relaxin do not relax human myometrium [4-7]. Relaxin is one of the most potent relaxants discovered with a concentration to produce 50% inhibition of longitudinally-cut myometrium from the rat of the order of 1-10 nM [8, 9]. The action is rapid in onset, occurring within seconds or minutes, and reversible [9]. These characteristics suggest an action mediated by a target site on the plasma membrane rather than a genomically-mediated action. This site of action is supported by the observation that relaxin is more potent as a relaxant of spasms thought to involve extracellular calcium (Ca^{2+}) than those due to release of Ca^{2+} from internal stores [8]. The uterine relaxant action of relaxin exhibited desensitisation *in vivo* and *in vitro* in a number of studies [10-14], although not all [15]. This phenomenon could offer clues to the signal transduction pathway but to date the mechanism of desensitisation has only been studied to a limited extent [12, 13].

This review will discuss the evidence for four potential early events in the signal transduction pathway:

- cyclic 3'5' adenosine monophosphate (cAMP) – protein kinase A (PKA) pathway
- L-arginine – nitric oxide pathway
- potassium channel opening
- tyrosine kinase linked receptor

Occasional reference will be made to non-uterine tissues as it is likely that signal transduction pathways will be common across many of the effects of relaxin. It is uncertain whether the actions of relaxin with slower onset and offset (e.g. uterine, cervical and mammary tissue growth and cervical softening [3] may be mediated by a common pathway with the actions that have a faster time course.

[1] In this review "uterus" will be used for the tissues containing endometrium while "myometrium" indicates that the endometrium had been removed and/or myometrial cells used. The source of relaxin is the pig unless stated otherwise.

G.W. Tregear et al. (eds.), Relaxin 2000, 291–299.
© 2001 *Kluwer Academic Publishers. Printed in the Netherlands.*

Cyclic 3'5' adenosine monophosphate (cAMP) – protein kinase A (PKA) pathway

There is evidence to support the involvement of the cAMP-PKA pathway in the relaxant action of relaxin but also much contradictory data. The evidence has been predominantly biochemical by measurement of the changes in the content of the second messenger, cAMP, or of activity of enzymes in the pathway. As the relaxin receptor has not yet been isolated, it has not been possible to demonstrate whether the relaxin receptor is coupled to G-proteins. However, the blockade of some of the biochemical events by pertussis toxin [16] is suggestive of a G-protein involvement.

Several studies have shown that relaxin can increase the cAMP content of isolated uteri [17 - 20], myometrial strips [21, 22] or myometrial cells in culture [23, 24] from the rat. However, often the changes in cAMP content followed, rather than preceded, the relaxant action and were only seen in the presence of a phosphodiesterase inhibitor, whereas relaxation occurs in the absence of such inhibitors.

In vivo there was an even clearer dissociation of mechanical and biochemical events. Administration of an i.v. bolus dose of 50 μg/kg relaxin in the rat, which produced a marked and prolonged inhibition of uterine contractions, induced only a small increase (three-fold) in uterine cAMP concentrations that was of short duration [25, Fig. 1]. A smaller dose of relaxin, 5 μg/kg, still induced a prolonged relaxation but without any changes in uterine cAMP content. By contrast, i.v. doses of salbutamol, an agonist at ß-adrenoceptors, which induced similar inhibitions of uterine contractions, produced marked (19-fold) increase in cAMP concentrations.

H2 relaxin increased the adenylyl cyclase activity of a membrane fraction of human myometrium [26]. The activation was enhanced by Gpp[NH]p, a non-hydrolyzable analogue of GTP, suggesting the involvement of G-proteins. However, human relaxin is not a relaxant of human myometrium [6,7].

cAMP activates cAMP-dependent protein kinase A (PKA). Early studies showed that relaxin enhanced the activity of PKA [27] but the effect was maximal at 5 min, that is slower than for the mechanical effect. Relaxin was able to produce a small increase in the PKA activity ratio, a measure of the active form as a fraction of the total PKA, in isolated uterus of the rat [28, Fig. 2]. However, the increase was smaller than seen with salbutamol. In addition, relaxin was a relaxant of the separated longitudinal myometrial layer but did not induce any change in the PKA activity ratio in that layer. An alternative approach has been the use of PKA inhibitors. The PKA inhibitor H-8 blocked the ability of relaxin to inhibit the oxytocin-induced rise in intracellular calcium ions in myometrial cells from the rat [29]. H-8 also blocked the relaxant effect of relaxin on isolated myometrial strips from the rat [30], although in a later study H-8 was not a selective inhibitor of relaxin in rat uterus but also antagonised levcromakalim, a potassium channel opener [28]. S-Ht31, a compound that inhibits PKA binding to an anchoring protein, reduced the relaxin inhibition of the effect of oxytocin on phosphatidylinositol turnover [31].

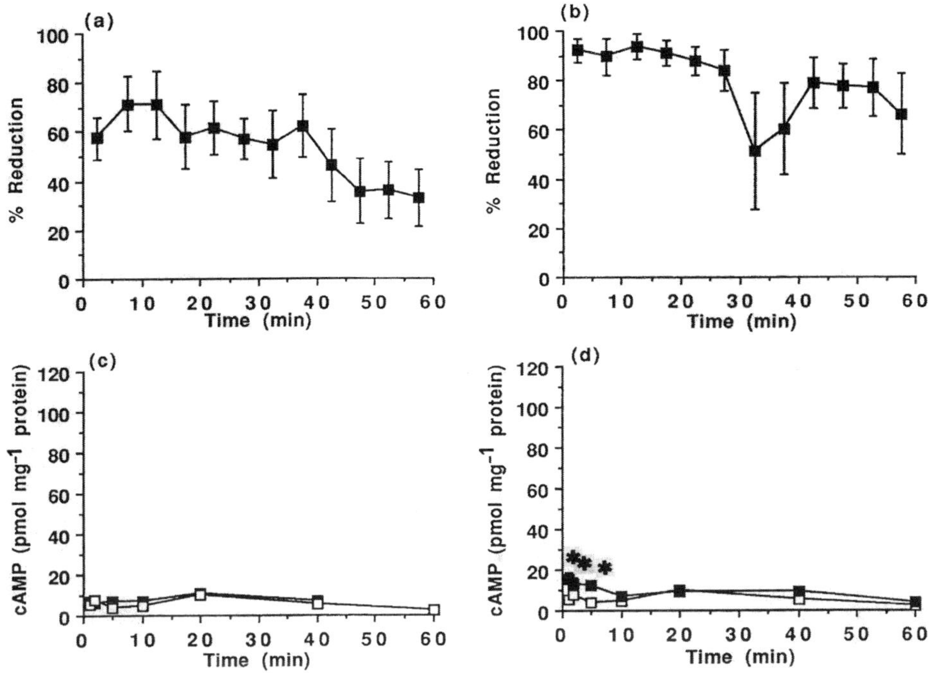

*Fig. 1. Effect of (a) 5 and (b) 50 ug porcine relaxin kg^{-1}, i.v. (■) on uterine contractions and, in a separate of rats, on uterine cAMP concentrations, (c) and (d), respectively. Saline-injected controls for cAMP measurements are shown (□). Ordinates are % reduction in integral of uterine contractions (a,b) and uterine cAMP concentrations (c,d). Values are means ± SEM, n = 4-7. *P < 0.05 versus control for cAMP measurements (two-way ANOVA). Reproduced from ref. 25 with permission.*

There is substantial support for the idea that the relaxant action of agonists at β_2-adrenoceptors, such as salbutamol, is mediated via the cAMP-PKA pathway [32, 33]. Tolerance occurs to the relaxant action of salbutamol [34], which is probably initially mediated by phosphorylation of the receptor and uncoupling of the receptor from the adenylyl cyclase complex [35]. However, there was no cross tolerance between salbutamol and relaxin [12]. Therefore, it can be concluded that either the mechanism of action of the two compounds differs or that tolerance occurs to each compound prior to any common biochemical pathway.

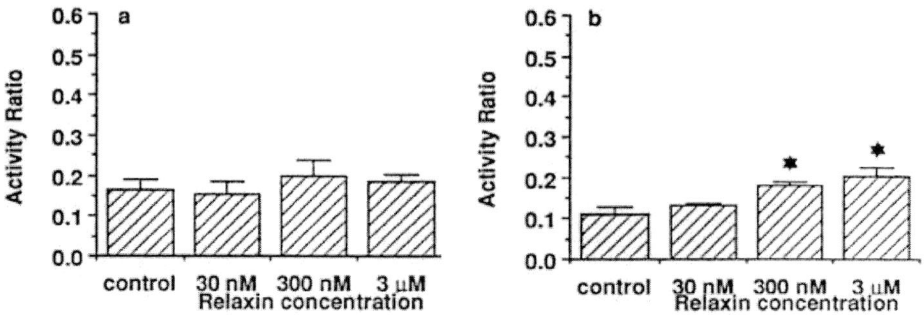

*Fig. 2. The effect of porcine relaxin on the cAMP-dependent protein kinase A activity ratio in tissue extracts from (a) longitudinal myometrium and (b) intact uterus from non-pregnant rats treated with oestradiol. Bars represent the mean values and the vertical lines the SEM of five tissues. *$P < 0.05$ indicates a value that is significant different from the control value.*

It is difficult to reconcile the contradictory results described above. As porcine and human relaxin can induce an increase in the cAMP content of human endometrial cells [36, 37], it is possible that the biochemical changes recorded in whole uteri reflect events in the endometrium rather than the myometrium in several studies. One speculation is that relaxin might act as a uterine relaxant via a cAMP-independent manner in longitudinal myometrium in the rat but relaxin possibly interacts with the circular myometrium or endometrium via a cAMP-dependent mechanism to release a relaxant factor [28]. However, some studies have demonstrated biochemical changes in myometrial cells. It should be recognised that relaxin has several effects in uteri, in addition to relaxation, such as promoting glycogen deposition and growth plus connective tissue remodelling [3]. Possibly some of the changes in the cAMP-PKA pathway relate to those non-relaxant mechanisms.

L-arginine – nitric oxide pathway

The L-arginine – nitric oxide pathway is thought to play a key physiological role in the regulation of vascular smooth muscle tone and to mediate the effects of several vasodilators. Relaxin is a dilator of isolated coronary arteries from the guinea and rat [38]. It has been proposed that this action is mediated by this pathway as it is associated with a rise in concentration of the second messenger cyclic 3'5' guanosine mono-phosphate (cGMP) in vascular smooth muscle cells in culture [39, 40]. Also, relaxin increased the release of nitrites from vascular smooth muscle cells and nitric oxide synthase inhibitors blocked its ability to increase coronary flow [38].

There is a limited data supporting such a mechanism in uteri. Injection of relaxin into mice increased the activity of endothelial-type nitric oxide synthase in myometrium and endothelium [41]. In addition, relaxin, as a relaxant of isolated uteri from the mouse, was antagonised by the nitric oxide synthase inhibitor, N^G-nitro-L-arginine. However, there is a significant amount of data that does not support the idea that this path plays a key role. Several groups have failed to find an increase in the content of cGMP of rat uteri following incubation with relaxin [18, 19, 24]. Also, relaxin, as a relaxant of isolated uterus from the rat, was not antagonised by N^{ω} nitro l-arginine methyl ester, an inhibitor of nitric oxide synthase [28], methylene blue, an inhibitor of cGMP, nor potentiated by M&B22948, a cGMP phosphodiesterase inhibitor (Hollingsworth, unpublished). There is actually significant data that this pathway is not operative in uteri of the non-pregnant rat as drugs such as sodium nitroprusside, whose actions appear to be mediated by the generation of nitric oxide, are not relaxants [42, Hollingsworth unpublished]. However, compounds acting via the nitric oxide pathway are relaxants of uteri from the late pregnant rat [43, 44] so maybe relaxin could act via this pathway in uteri of pregnant animals. Clearly there is a need to reconcile these contradictory data.

Opening of potassium channels

Movement of potassium ions (K^+) through potassium channels plays a key role in terminating action potentials and regulating the resting membrane potential of tissues such as myometrium [45, 46]. There are many types of K^+ channel [47, 48], several of which are susceptible to modulation by hormones and drugs [49].

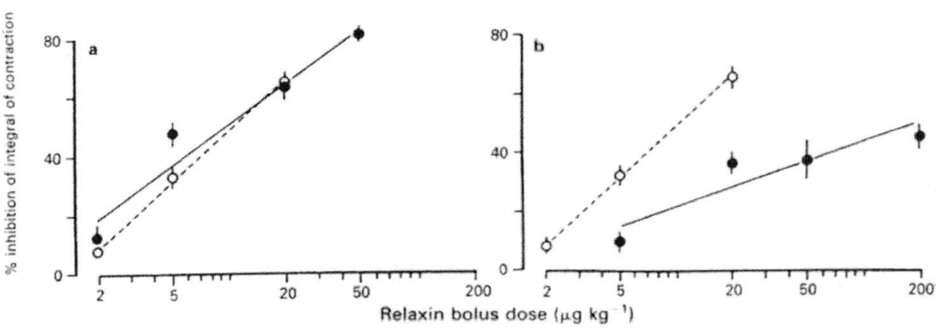

Fig. 3. Effect of porcine relaxin (a) on uterine contractions before (O) and after (●) infusion of vehicle (0.02N sodium hydroxide in 4% glucose, 5 ml kg^{-1} i.v.), n = 40 rats pre-infusion, 10 rats per point post-infusion. Effect of relaxin (b) on uterine contractions before (O) and after (●) infusion of glibenclamide (20 mg kg^{-1}, i.v.), n = 40 rats pre-infusion, 10 rats post-infusion. Ordinate scale is inhibition of integral of uterine contraction over 60 min as % integral for 60min preceding i.v. relaxant dose. Abscissa scale is dose of relaxin on a log scale. Reproduced from ref. 53 with permission.

One of the characteristic features of compounds that open K^+ channels is that they inhibit tension development of smooth muscle induced by low but not high concentrations of K^+ [50]. Relaxin was reported to inhibit uterine spasm induced by 12.5% K^+ (~20 mM) but not that induced by 25% K^+ (~40 mM) [51]. This observation, coupled with the interest in Manchester in drugs that are openers of K_{ATP} channels are uterine relaxants [52], led to the hypothesis that relaxin might be a K^+ channel opener. Specifically it was considered that relaxin might be an opener of ATP-sensitive K^+ channels with a similar mechanism of action to drugs such as cromakalim (and its active isomer levcromakalim). Evidence supporting this mechanism came from the finding that pre-treatment with glibenclamide, a blocker of ATP-sensitive K^+ channels, selectively antagonised relaxin and cromakalim, but not salbutamol, as a uterine relaxant in the conscious rat [53, Fig. 3]. Glibenclamide also reversed established relaxin and cromakalim induced inhibition of uterine contractions but not inhibition by salbutamol. Subsequent studies on isolated uteri and myometrium from the rat were conducted to investigate further this mechanism. However, relaxin inhibited spasms of isolated uterus of the rat to both low and high concentrations of K^+, in contrast to earlier findings, produced only a very small increase in efflux of $^{42}K^+$ from myometrial strips and did not induce hyperpolarization [22, 54]. These observations do not support an action via the ATP-sensitive K^+ channel or via any K^+ channel. In addition, glibenclamide was not an antagonist of relaxin as a uterine relaxant in the isolated uterus of the rat [54, Fig. 4]. Clearly there was a discrepancy between the observations of glibenclamide antagonism of relaxin between *in vivo* and *in vitro*. Further investigation *in vitro* to attempt to mimic the *in vivo* conditions, such as using circularly-cut uterus, uteri from ovariectomized rats and spontaneous tension development, still failed to produce conditions where glibenclamide antagonised relaxin [9]. Therefore, *in vitro* there was no support for the idea that relaxin is a uterine relaxant via ATP-sensitive K^+ channels. Of course, it is conceivable that the mechanism of action of relaxin or glibenclamide differs between *in vivo* and *in vitro* preparations.

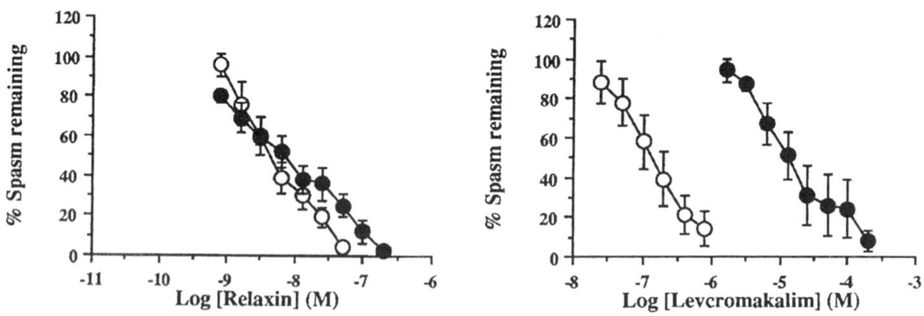

Fig. 4. The effect of glibenclamide (10 μM) on the inhibition of oxytocin (0.2 nM)-induced spasm evoked by porcine relaxin and levcromakalim in isolated uteri from the non-pregnant, 17β-oestradiol benzoate pre-treated rat prepared for the recording of tension in the longitudinal muscle layer. Effects are shown in the absence (O) and presence (●) of glibenclamide (10 μM).

There is one report that H2 relaxin can stimulate the opening of the large Ca^{2+}-activated K^+ channel in a myometrial cell line from pregnant women as measured by the patch clamp technique [56]. However, H2 relaxin is not a relaxant of human myometrium [6, 7] so it is difficult to see the relevance of this finding to its relaxant action. Relaxin was antagonised by Rp-cAMPS, a PKA antagonist [56] suggesting that K^+ channel opening was stimulated via a PKA-mediated mechanism. Many years ago it was shown that a number of agonists at β-adrenoceptors produced a modest hyperpolarization [57, 58], an action probably due to an increase in K^+ conductance [59]. It is likely that the ability of agonists at β-adrenoceptors to increase K^+ channel opening is not fundamental to their relaxant action in smooth muscle [60] and that probably applies to relaxin's myometrial relaxant action as well.

Tyrosine kinase linked receptor

Relaxin has structural similarities to insulin and insulin-like growth factors [3] so, assuming nature is conservative, it is possible that the receptors and the initial signal transduction processes are similar also. The insulin receptor is a large transmembrane protein consisting of α and β subunits with the β subunits possessing intrinsic tyrosine kinase activity [61]. Interaction of insulin with its receptor activates the tyrosine kinase and consequentially a cascade of enzymes that are responsible for the immediate metabolic actions. The longer-term actions of insulin start from the same point but involve activation of transcription factors and subsequent alteration of the rate of RNA transcription. There has been only limited investigation of this pathway as a potential mediator of relaxin's actions. Low concentrations of H2 relaxin has been shown to stimulate tyrosine phosphorylation in fibroblasts from human lower uterine segments [62]. Relaxin did not increase cAMP content in these cells, even in the presence of a phosphodiesterase inhibitor. In contrast, tyrphostin 47, a purported inhibitor of tyrosine kinase, selectively reduced the increase in cAMP content of homogenates of human myometrium produced by H2 relaxin [26].

Concluding remarks

There is not unequivocal support for any of the four processes described above, although the evidence for the cAMP-PKA pathway is strongest. It is possible that relaxin is releasing a substance from endometrium or circular smooth muscle, via the cAMP-PKA or the L-arginine-nitric oxide pathways, that is a myometrial relaxant. However, there are relaxin receptors on myometrial cells so a direct action is perhaps more likely. An alternative hypothesis is that there is an early primary event after relaxin's interaction with its target site that leads to the subsequent second messenger and enzyme activity changes described above. The nature of these secondary changes could vary between tissues. It is possible that activation of tyrosine kinase could be the key primary event [26] and could explain both the rapid and the slower onset actions, e.g. myometrial relaxation and connective changes in the cervix respectively. Currently there is insufficient data to favour any specific mechanism.

Acknowledgements

I am very pleased to acknowledge the support of Action Research, the Medical Research Council and the Wellcome Trust for our research plus a Wellcome Trust travel grant to MH to attend the conference.

References

1. Krantz, J.C., Bryant, H.H. and Carr, C.J., Surg. Gynecol. Obstet., 90 (1950) 372.
2. Downing, S.J. and Hollingsworth, M., J. Reprod. Fert., 99 (1993) 275.
3. Sherwood, O.D., In Knobil, E. and Neill, J.D. (Eds), The Physiology of Reproduction. 2nd ed., Raven Press, New York, 1994, p.861.
4. MacLennan, A.H., In MacLennan, A.H., Tregear, G.W. and Bryant-Greenwood, G.D. (Eds.), Progress in Relaxin Research. 2nd International Congress on the hormone relaxin. Global Publishing Services, Singapore, 1994, 263.
5. MacLennan, A.H., Grant, P., Ness, D. and Down, A., J. Reprod. Med., 31 (1986) 43.
6. MacLennan, A.H. and Grant, P., J. Reprod. Med., 36 (1991) 630.
7. Peterson, L.K., Svane, D., Uldbjerg, N. and Forman, A., Obstet. Gynec., 78 (1991) 757.
8. Hughes, S.J. and Hollingsworth, M., Br. J. Pharmac., 116 (1995) 3028.
9. Hughes, S.J. and Hollingsworth, M., Gen. Pharmac., 29 (1997) 829.
10. Chamley, W.A. and Parkington, H.C., J. Physiol., 353 (1984) 51.
11. Cheah, S.H. and Sherwood, O.D., Endocrinology, 109 (1981) 2076.
12. Downing, S.J. and Hollingsworth, M., J. Endocrinology, 135 (1992) 17.
13. Downing, S.J. and Hollingsworth, M., J. Reprod. Fert., 99 (1993) 121.
14. Wiqvist, N., Acta Endocrinol Suppl., 46 (1959) 3.
15. Downing, S.J. and Sherwood, O.D., Endocrinology, 116 (1985) 1206.
16. Anwer, K., Hovington, J.A. and Sanborn, B.M., Endocrinology, 124 (1989) 2995.
17. Cheah, S.H. and Sherwood, O.D., Endocrinology, 106 (1980) 1203.
18. Sanborn, B.M., Kuo, H.S., Weisbrodt, N.W. and Sherwood, O.D., Endocrinology, 106 (1980) 1210.
19. Judson, D.G., Pay, S. and Bhoola, K.D., J. Endocrinol., 87 (1980) 153.
20. Grazi, R.V., Goldsmith, L.T., Schmidt, C.L., Von Hagen, S. and Weiss, G., Am. J. Obstet. Gynec., 159 (1988) 1402.
21. Sanborn, B.M. and Sherwood, O.D. Endocrin. Res. Commun., 8 (1981) 179.
22. Osa, T., Inoue, H. and Okabe, K., Br. J. Pharmac., 104 (1991) 950.
23. Hsu, C.J., McCormack, S.M. and Sanborn, B.M., Endocrinology, 116 (1985) 2029.
24. Kramer., S.M., Gibson, U.E.M., Fendley, B.M., Mohler, M.A., Drolet, D.W. and Johnston. P.D., In Vitro Cell. Dev. Biol., 26 (1990) 647.
25. Downing, S.J., McIlwrath, A. and Hollingsworth. M., J. Reprod. Fert., 96 (1992) 857.
26. Kuznetsova, L., Plesneva, S., Derjabina, N., Omeljaniuk, E. and Pertseva, M., Regul. Peptid., 80 (1999) 33.
27. Kemp, B. and Niall, H.D., In Bryant-Greenwood, G.D., Niall, H.D.S. and Greenwood, F.C. (Eds.), Relaxin. Elsevier, New York, USA, 1981, 273.
28. Hughes, S.J., Hollingsworth, M. and Elliott, K.R.F., J. Reprod. Fert., 109 (1997) 289.
29. Anwer, K., Hovington, J.A. and Sanborn, B.M., Biol. Reprod., 43 (1990) 851.
30. Sanborn, B.M., Anwer, K., Monga, M., Wen, Y., Singh, S.P., Meera, P., Oberti, C., Toro. and Stefani, E., In MacLennan, A.H., Tregear, G.W. and Bryant-Greenwood, G.D. (Eds.),

Progress in Relaxin Research. 2nd International Congress on the hormone relaxin. Global Publishing Services, Singapore, 1994, 289.
31. Dodge, K.L., Carr, D.W. and Sanborn, B.M., Endocrinology, 140 (1999) 5165.
32. Diamond, J., In Carstein, M.E. and Miller, J.D. (Eds.), Uterine Function. Molecular and Cellular Aspects. Plenum Press, New York, USA, 1990, 249.
33. Hollingsworth, M., In Garfield, R.E. and Tabb, T.N. (Eds.), Control of Uterine Contractility. CRC Press, Boca Raton, Florida, 1994, 401.
34. Abel, M.H. and Hollingsworth, M., Br. J. Pharmac., 88 (1986) 577.
35. Clark, R.B., Knoll., B.J. and Barber, R., Trends in Pharmac. Sci., 20 (1999) 279.
36. Chen, G., Rong Huang. J. and Tseng, L., Biol. Reprod., 39 (1988) 519.
37. Fei, D.T.W., Gross, M.C., Lofgren, J.L., Mora-Worms, M. and Chen A.B., Biochem. Biophys. Res. Commun., 170 (1990) 214.
38. Bani-Sacchi, T., Bigazzi, M., Bani, D., Mannaioni, P.F. and Masini, E., Br. J. Pharmac., 116 (1995) 1589.
39. Bani, D., Gen Pharmac., 28 (1997) 13.
40. Bani, D., Failli, P., Bello. M.G., Thiemermann, C., Sacchi. T.B., Bigazzi, M. and Masini, E., Hypertension, 31 (1998) 1240.
41. Bani, D., Baccari, M.C., Nistri, S., Calamai, F., Bigazzi, M. and Sacchi, T.B., Endocrinology, 140 (1999) 4434.
42. Nakatsu, K. and Diamond, J., Can. J. Physiol. Pharmac., 67 (1989) 251.
43. Yallampalli, C., Izumi, H., Byam-Smith, M. and Garfield, R.E., Am. J. Obstet. Gynec., 170 (1993) 175.
44. Norman, J., Pharmac. Ther., 70 (1996) 91.
45. Parkington, H.C. and Coleman, H.A. In Carstein, M.E. and Miller, J.D. (Eds.), Uterine Function. Molecular and Cellular Aspects. Plenum Press, New York, USA, 1990, 195.
46. Sanborn, B.M. Sem. Perinat., 19 (1995) 31.
47. Yost, C.S., Anaesthiology, 90 (1999) 1186.
48. Alexander, S.P.H. and Peters, J.A., Trends in Pharmac. Sci. Suppl 2000 11th ed., 106.
49. Edwards, G. and Weston, A.H., Prog. Drug. Res., 49 (1997) 93.
50. Edwards, G. and Weston, A.H., Pharmac. Ther., 48 (1990) 237.
51. St-Louis, J., Can. J. Physiol. Pharmac., 59 (1981) 507.
52. Piper, I., Minshall, E., Downing, S.J., Holingsworth, M. and Sadraei, H., Br. J. Pharmac., 101 (1990) 901.
53. Downing, S.J. and Hollingsworth, M., Br.J. Pharmac., 104 (1991) 71.
54. Hughes, S.J. and Hollingsworth, M., Br. J. Pharmac., 117 (1996) 1435.
55. Hughes, S.J. and Hollingsworth, M., Gen Pharmac., 29 (1997) 829.
56. Meera, P.K., Anwer, K., Monga, M., Oberti, C., Stefani, E., Toro, L. and Sanborn, B.M., Am. J. Physiol., 269 (1995) C312.
57. Diamond, J. and Marshall, J.M., J. Pharmac. Exp. Ther., 168 (1969) 13.
58. Diamond, J. and Marshall, J.M., J. Pharmac. Exp. Ther., 168 (1969) 21.
59. Marshall, J.M., Fed. Proc., 36 (1977) 2450.
60. Small, R.C., Chiu, P., Cook., S.J., Foster, R.W. and Isaac, L., Clin. Exp. Allergy, 23 (1993) 802.
61. Rang, H.P., Dale, M.M. and Ritter, J.M. Pharmacology. 4th ed., Churchill Livingstone, Edinburgh, 1999, 389.
62. Palwejwala, S., Stein, D., Wojtczuk, A., Weiss, G. and Goldsmith, L., Endocrinology, 139 (1998) 1208.

The use of biosensor technology to search for orphan biomolecules: towards the relaxin receptor

Edouard Nice[1,2], Bruno Catimel[1], Julie Rothacker[1,2], Ross A Bathgate[3], Marc Mathieu[3], Antonia A Claasz[3], Geoffrey W Tregear[3] and John D Wade[3]

[1]The Ludwig Institute for Cancer Research, Melbourne Tumour Biology Branch; [2]The CRC for Cellular Growth Factors; and [3]Howard Florey Institute, University of Melbourne, Parkville, Victoria 3052, Australia

Introduction

Although it is 75 years since relaxin was first described [1], and almost 20 years since the cDNA sequences encoding for rat, porcine and human relaxin were reported [2,3,4], the isolation and characterisation of the cognate receptor/s, despite extensive investigation [5,6], has to date defied elucidation. Since the receptor would appear to be the key to a fuller understanding of the molecular mechanisms of action of relaxin [6], which is fundamental for structure-based design of agonists and antagonists with therapeutic potential, new approaches for its isolation warrant investigation.

Autoradiography using radiolabelled porcine relaxin has demonstrated the presence of relaxin binding sites in the symphysis pubis, uterus, and ovary [7] as well as specific areas of the brain [8] and heart [9]. Chemical crosslinking of ^{32}P labelled human relaxin to primary human uterine cells [6], primary rat atrial cardiomyocytes [6] or the human monocytic cell line THP-1 [10] have all indicated specific binding to a protein of approximately 220kDa. Additionally, incubation of human uterine fibroblasts with relaxin showed a significant increase in tyrosine phosphorylation of a protein with the same apparent Mr [11], suggesting that the receptor could be a member of the protein tyrosine kinase family. Scatchard analysis has, indicated that the interaction between relaxin and the receptor is of relatively high affinity (K_D = 100pM – 400pM [10]), whilst the number of receptors is low, ranging from approximately 200 receptors per cell in rat atrial cardiomyocytes and THP-1 cells to around 1000 receptors per cell in human endometrial cells [10]. Such low levels will, of course, complicate the purification and characterisation of the receptor.

One approach which has been successfully used for the identification of a number of low abundance "orphan proteins" is the combination of micropreparative HPLC with affinity-based biosensors which measure protein-protein interactions in real time [12 – 19]. In the original approach, one of the binding partners (e.g. recombinantly produced receptor, monoclonal antibody, synthetic or recombinant ligand) was immobilised onto the sensor surface which was used to screen a wide range of biological samples for the presence of a positive specific signal indicating the presence of an interactant. The biosensor was then used as a specific affinity detector to identify positive fractions produced during multidimensional micropurification (Fig. 1). In an

G.W. Tregear et al. (eds.), Relaxin 2000, 301–308.
© 2001 *Kluwer Academic Publishers. Printed in the Netherlands.*

extension of this approach we have recently shown [20, 21] how cuvette-based biosensors (eg IAsys Autoplus), which have both increased surface area and facile recovery of desorbed sample compared with flow based instruments (eg BIAcore), may be used as micropreparative affinity systems to perform automated ligand fishing directly on the sensor surface with recovery of sufficient material for subsequent sensitive and specific down stream analysis (eg microLC, SDS-PAGE, Western blotting, microsequence analysis, mass spectrometry) and may thus form part of a proteomic-type strategy for the characterisation of proteins and protein complexes.

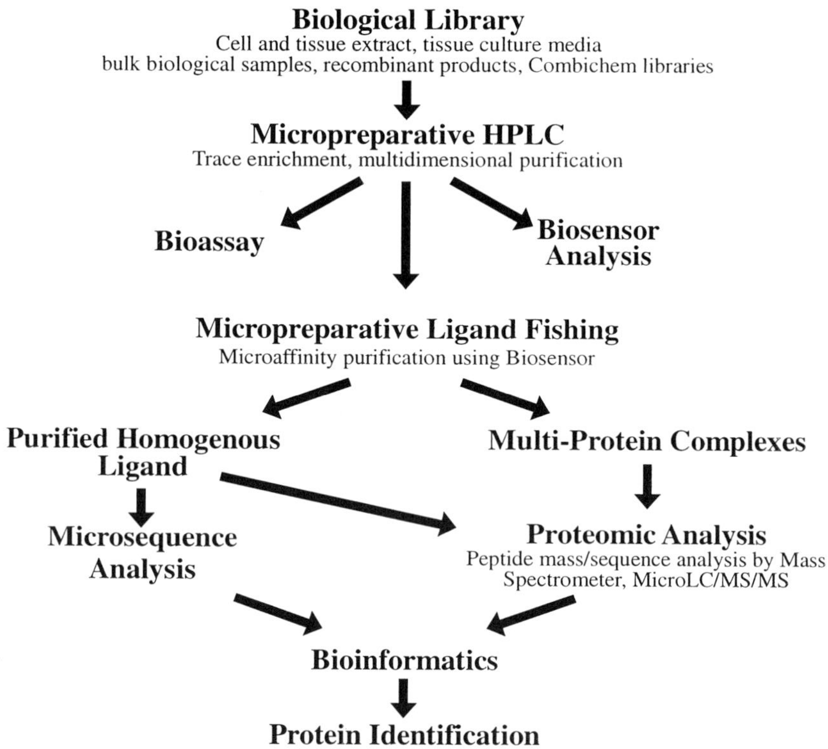

Fig. 1. Strategies for the identification and characterisation of binding partners to orphan proteins.

Importantly such surfaces appear to show significantly reduced non-specific binding when working with trace components compared with microaffinity columns based on the same chemistry, presumably due to the absence of the large surface area chromatographic support and the relatively hydrophilic nature of a number of the sensor surfaces (eg carboxymethyl dextran).

In this report we will describe the optimisation of such biosensor-based protocols and discuss their potential application to the identification and characterisation of the relaxin receptor.

Results and Discussion

Ligand Immobilisation:

Fundamental to successful biosensor studies is the ability to immobilise the target protein in a biologically active orientation to yield a homogeneous, stable sensor surface. Chemistries commonly used in affinity chromatography and cross-linking studies are generally applicable [reviewed in 22, 23]. Suitable strategies including amine and thiol coupling or the use of affinity capture techniques with high affinity (eg biotin-avidin, $K_D=10^{-15}M$). The bioactive surface can be used for multiple analyses if suitable biocompatible operating conditions can be defined (running buffer, pH, regeneration conditions etc).

The widely used amine coupling chemistries (eg NHS/EDC) have the potential disadvantage of causing heterogenous surfaces if multiple lysine residues are present and additionally may result in loss of biological recognition if these lysine residues are located at, or near to, the binding site. However, for screening and ligand searching these chemistries have been shown to frequently result in a surface which shows some activity since it is likely that at least some of the molecules will be coupled in an active orientation [24]. It is, however, useful to have some way of interrogating the surface to demonstrate reactivity. In cases where the known binding partner has not been identified the use of conformationally dependent or neutralising antibodies (ideally of defined epitope) may offer some guidance [16].

Immobilisation strategies for relaxin:

Relaxin is a 6-kDa polypeptide consisting of an A and B chain, and having both inter- and intra-chain disulphide bonds. In the case or rat relaxin (Fig. 2), the A chain has a pyroglutamic acid residue at the amino terminus which renders it unavailable for amine coupling. The receptor-binding site has been shown to involve two arginine residues (R16, R20) and an isoleucine residue (I 23) in the B chain (Fig. 2) [25].

We have recently [26] (Mathieu *et al.*, this volume) described the synthesis of N^{α}-mono-biotinylated rat relaxin (Fig. 2). This protein, which was designed to enable immobilisation in a unique orientation onto a streptavidin sensor surface to complement an affinity-based receptor searching strategy, had essentially full biological activity in the rat isolated atrial bioassay. The protein readily immobilised onto the sensor surface, giving an immobilisation level of 1490RU (equivalent to1.5ng /mm^2 [27], giving a total of 2.4ng immobilised on the 1.6mm^2 surface of the BIAcore chip (Figure 3A)). This surface gave a strong positive signal when challenged with a neutralising antibody (MCA-1) (Figure 3B) or an anti-rat relaxin polyclonal antibody (data not shown). Interestingly a unique orientation can also be obtained with this compound by immobilisation using amine coupling chemistry via the lysine residue in the A chain (K 22), since this is the only reactive amine group present in the molecule.

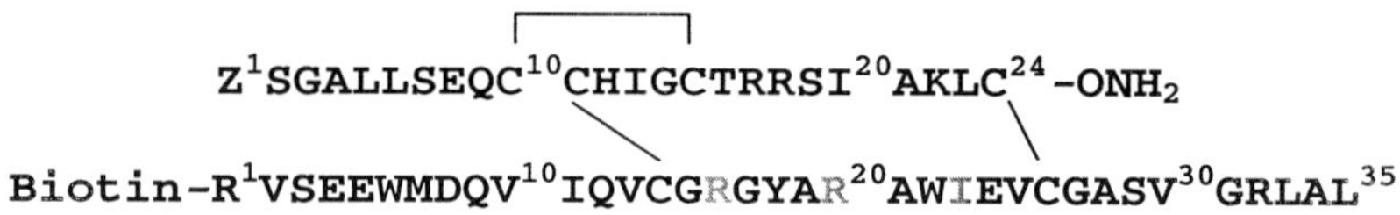

Fig. 2. Schematic of biotinylated rat relaxin. Residues in the B chain implicated in receptor binding [25] are shown in grey.

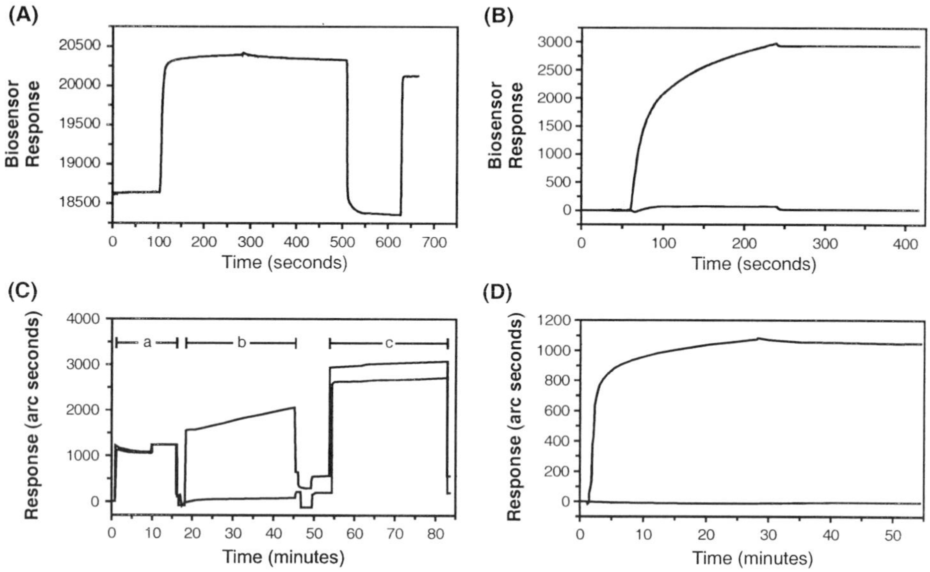

Fig. 3. The generation of relaxin surfaces for biosensor analysis (A) Immobilisation of biotinylated rat relaxin onto a streptavidin (BIAcore SA) surface. The net increase in signal after immobilisation was 1490RU (B) The interaction of a neutralising antibody with the rat relaxin surface described in Figure 3A. (C) Immobilisation of human relaxin onto a IAsys CM5 carboxymethyl dextran cuvette using amine coupling chemistry (a) Activation of the surface with NHS/EDC (b) Binding of human relaxin (c) Deactivation of residual active esters using triethylamine. The net increase in signal after immobilisation was 500 arc seconds. The lower trace shows the generation of a control surface (D) Challenge of the human relaxin surface (Figure 3C) with BL32 polyclonal anti-human relaxin antibody. The lower traces in panels B and D show the corresponding signal obtained with a control surface.

We are currently attempting a similar synthesis to generate biotinylated H2 relaxin (A. Claasz, personal communication). However, in the absence of this analogue, we have also immobilised human relaxin using NHS/EDC amine coupling chemistry to determine whether a reactive surface is generated. In this case the amino terminus of the A chain is also blocked by the presence of a pyroglutamic acid, but immobilisation could occur at sites involving the amino terminus of the B chain, and/or lysines 9 and 17 in the A chain and/or lysine 9 in the B chain. In particular involvement of lysine 9, which is in close proximity to the receptor binding site [25] could interfere with binding. The IAsys Autoplus sensor was used, and signal of 500 arc seconds was obtained corresponding to 0.83ng/mm^2 [28] giving a total of 13.28ng of human relaxin immobilised on the 16mm^2 preparative cuvette. In this case surface reactivity was probed with anti-human relaxin antibody (BL32 IgG) (Figure 3D). Whilst the surface was reactive showing that relaxin had been efficiently immobilised this does not guarantee reactivity towards the receptor since BL32 is a polyclonal antibody of undefined epitope.

Cell based biosensor studies:

We have recently shown that the cuvette-based design of the IAsys Autoplus is compatible with analysis of the interaction between cells and immobilised target proteins. We therefore decided to see whether we could use the biosensor to show specific binding of human relaxin receptor containing cells (THP-1) to a human relaxin surface. In this case a planar carboxylate surface was used to allow the cells to enter the optimum detection region of the evanescent field. Human relaxin was immobilised using similar amine coupling chemistry to that shown in Fig. 3C. Again the surface reactivity was confirmed using the BL32 antibody (Figure 4A).

It can be seen (Figure 4B) that a positive signal was obtained compared with that observed in the blank cuvette when THP-1 cells (1.25 x 10^6) were injected over the human relaxin surface. Importantly, this signal could be significantly ablated by competition with relaxin confirming the specificity of the interaction (Figure 4C). It is encouraging to see that such a strong specific signal could be so readily detected with cells that have such a low receptor number [10]. Interestingly, in a previous study [10], Parsell et al successfully used the Cytosensor [29], a cell based biosensor which measures extracellular acidification following receptor activation, to show the interaction of relaxin with the receptor on THP-1 cells, demonstrating the sensitivity of these techniques.

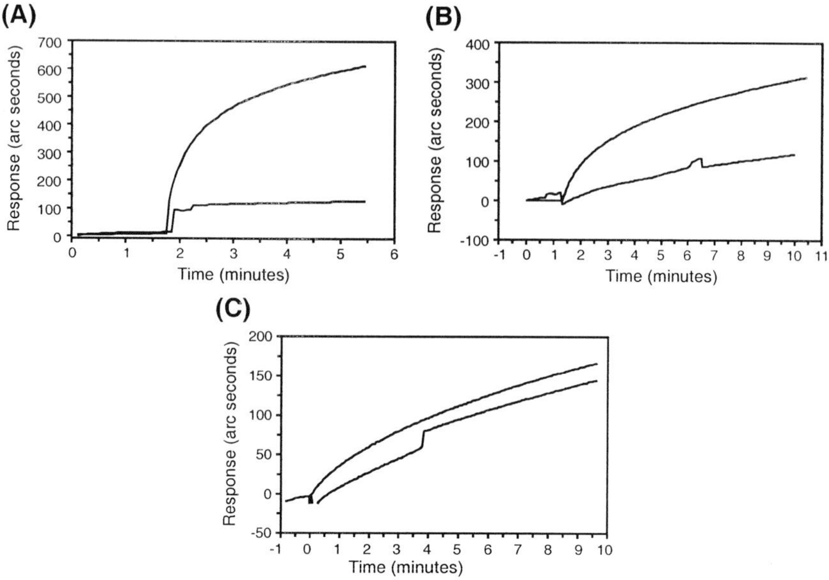

Fig. 4. Analysis of the binding of THP-1 cells to immobilised human relaxin using a cuvette-based biosensor (IAsys Autoplus). (A) Reactivity of the surface towards the BL32 antibody. (B) The response to THP-1 cells (C) Ablation of the cellular response by competition with soluble human relaxin. The lower trace in each panel is the corresponding signal obtained with a control cuvette.

The ability to detect specific binding of relaxin receptor on intact cells to the amine coupled sensor surface shows that, as we had anticipated, in spite of potential heterogeneity due to a reactive N-terminus and the presence of multiple lysine residues, at least part of the surface is in an active orientation. This surface can now be used for rapidly screening cells for the presence of suitable levels of receptor for use in subsequent isolation protocols. Perhaps, more importantly, it also offers the potential to use the sensor to optimise the generation of suitable membrane preparations and investigate alternative receptor solubilisation protocols. The ability to define suitable solubilisation conditions is often rate limiting in the purification of membrane-bound receptors [30]. Previous studies [5], made using porcine relaxin tracer and mouse brain membrane preparations, used Triton X100 (0.1%) for solubilisation. However, comparison of chromatographic size exclusion fractions in the presence or absence of excess cold relaxin suggested a specific binding component with an apparent Mr 65,000, considerably lower that the Mr suggested by cross-linking studies as detailed above [6,10].

Conclusions

The experiments described herein illustrate the potential of using biosensor-based technology for the detection, and hopefully eventually the isolation and characterisation, of the relaxin receptor. Such strategies have been successfully used previously in the isolation and characterisation of a number of other orphan biomolecules present at low levels in bulk biological tissue [12 – 19]. The major hurdle at present would appear to be the low number of receptors present on potential target cells coupled with the generation of a suitable soluble receptor preparation for biosensor-based affinity purification that binds with reasonable affinity to the ligand. If these problems can be overcome the combination of micropreparative HPLC, preparative receptor fishing and proteomics-based approaches might prove the key to this longstanding problem.

Acknowledgements

The authors would like to thank Mr Peter Robert (Advanced Labs, Melbourne, Australia) and Mr John Aherne (Affinity Sensors, Cambridge, UK) who kindly made available the IAsys Autoplus used in some of these studies.

References

1. Hisaw, F., Proc. Soc. Exp. Biol. Med., 23 (1926) 661.
2. Hudson, P., Haley, J., Cronk, M., Shine, J. and Niall H., Nature, 291(1981) 127
3. Haley, J., Hudson, P., Scanlon, D., John, M., Cronk, M., Shine, J., Tregear, G. and Niall, H., DNA., 1 (1982) 155
4. Hudson, P., Haley, J., John, M., Cronk, M., Crawford, R., Haralambidis, J., Tregear, G., Shine, J. and Niall, H., Nature, 301(1983) 628.
5. Büllesbach, E.E., Steinetz, B.G. and Schwabe, C., In MacLennan, A.H., Tregear, G.W. and Bryant-Greenwood, G.D. (Eds.). Progress in relaxin research, Global Publications Services, Singapore, 1994, p145.
6. Osheroff, P.L. in MacLennan, A.H., Tregear, G.W. and Bryant-Greenwood, G.D. (Eds.). Progress in relaxin research, Global Publications Services, Singapore, 1994, p159.
7. Yang, S., Rembiesa, B., Bullesbach, E.E. and Schwabe, C., Endocrinology, 130 (1992) 179.
8. Osheroff, P.L. and Phillips, H.S., Proc. Natl. Acad. Sci. USA, 88 (1991) 6413.
9. Osheroff, P.L., Cronin, M.J. and Lofgren, J.A., Proc. Natl. Acad. Sci. USA, 89(1992) 2384
10. Parsell, D.A., Mak, J.Y., Amento, E.P. and Unemori E.N., J. Biol. Chem., 271 (1996) 27936.
11. Palejwala, S., Stein, D., Wojtczuk, A., Weiss, G., and Goldsmith, L.T., Endocrinology, 139 (1998) 1208.
12. Nice, E.C., Catimel, B., Lackmann, M., Stacker, S., Runting, A., Wilks, A., Nicola, N. and Burgess, A.W., Lett. Peptide Sci., 4 (1997) 107.

13. Catimel, B., Ritter, G., Welt, S., Old, L.J., Cohen, L., Nerrie, M.A., White, S.J., Heath, J.K., Demediuk, B., Domagala, T., Lee, F.T., Scott, A.M., Tu, G.F., Ji, H., Moritz, R.L., Simpson, R.J., Burgess, A.W. and Nice, E.C., J. Biol. Chem., 271 (1996) 25664.
14. Bartley, T.D., Hunt, R.W., Welcher, A.A., Boyle, W.J., Parker, V.P., Lindberg, R.A., Lu, H.S., Colombero, A.M., Elliott, R.L., Guthrie, B.A., Holst, P.L., Skrine, J.D., Toso, R.J., Zhang, M., Fernandez, E., Trail, G., Varnum, B., Yarden, Y., Hunter T. and Fox, G.M., Nature, 368 (1994) 558.
15. Stitt, T.N., Conn, G., Gore, M., Lai, C., Bruno, J., Radziejewski, C., Mattsson, K., Fisher, J., Gies, D.R., Jones, P.F., Masiakowski, P., Rian, T.E., Tobkes, N.J., Chen, D.H., DiStefano, P.S., Long, G.L., Basilico, C., Goldfarb, M.P., Lemke, G., Glass D.J. and Yancopoulos, G.D., Cell, 80 (1995) 661.
16. Lackmann, M., Bucci, T., Mann, R.J., Kravets, L.A., Viney, E., Smith, F., Moritz, R.L., Carter, W., Simpson, R.J., Nicola, N.A., Mackwell, K., Nice, E.C., Wilks, A.F. and Boyd, A.W., Proc. Natl. Acad. Sci. USA, 93 (1996) 2523.
17. Davis, S., Aldrich, T.H., Jones, P.F., Acheson, A., Compton, D.L., Jain, V., Ryan, T.E., Bruno, J., Radziejewski, C., Maisonpierre P.C. and Yancopoulos, G.D., Cell, 87 (1996) 1161.
18. Sakano, S., Serizawa, R., Inada, T., Iwama, A., Itoh, A., Kato, C., Shimizu, Y., Shinkai, F., Shimizu, R., Kondo, S., Ohno, M. and Suda T., Oncogene, 13 (1996) 813.
19. Fitz, L.J., Morris, J.C., Towler, P., Long, A. Burgess, P., Greco, R., Wang, J., Gassaway, R., Nickbarg, E., Kovacic, S., Ciarletta, A., Giannotti, J.A., Finnerty, H., Zollner, R., Beier, D.R., Leak, L.V., Turner K.J. and Wood, C., Oncogene, 15 (1997) 613.
20. Catimel, B., Domagala, T., Nerrie, M., Weinstock, J., White, S., Abud, H., Heath, J., and Nice, E., Prot. Peptide Lett., 6, (1999) 319.
21. Catimel, B., Weinstock, J., Nerrie, M., Domagala, T., and Nice, E.C., J.Chromatogr., 869, (2000) 261.
22. Nice, E.C. and Catimel, B., Bioessays, 21 (1999) 339.
23. Catimel, B., Rothacker, J., and Nice, E., J. Biochem. Biophys. Meth., (2000) in press.
24. MacBeath, G. and Stuart, L., Schreiber, S.L., Science, 289 (2000) 1760.
25. Büllesbach, E.E. and Schwabe, C., J. Biol. Chem., 275 (2000) 35276.
26. Mathieu, M.N., Wade, J.D., Catimel, B., Bond, C. P., Nice, E.C., Summers, R.J., Otvos, L. Jr. and Tregear, G.W., J. Peptide Res., 57 (2001) 374.
27. Stenberg, E., Persson, B., Roos, H. and Urbaniczki, J., J. Colloid Interface, 43 (1990) 513.
28. Edwards, P.R., Lowe, P.A. and Leatherbarrow, R.J., J. Mol. Recognit., 10 (1997) 128.
29. McConnell, H.M. Owicki, J.C., Parce, J.W., Miller, D.L., Baxter, G.T., Wada, H.G. and Pitchford, S., Science, 257 (1992) 1906.
30. Demoliou-Mason, C.C. and Barnard, E.A. In Hulme E.C. (ed.) Receptor biochemistry: A practical approach, Oxford University Press, Oxford, UK., 1990, p105.

Relaxin signal transduction couples tyrosine phosphorylation to cAMP upregulation.

Olaf Bartsch, Bettina Bartlick and Richard Ivell
Institute for Hormone and Fertility Research, University of Hamburg, Grandweg 64, 22529 Hamburg, Germany.

Introduction

Although the hormone relaxin has been known for many decades, it is a surprising fact that we know neither any details of the structure or even basic form of the receptor for this hormone, nor do we have any substantial knowledge about how the informational content of a relaxin stimulus is conveyed within the cell to exert the many phenotypic effects observed *in vivo* and *in vitro*. What we do know is that at some point, this intracellular information is transduced into an elevation in cAMP, and that many of the changes observed in cellular phenotypes are due in part at least to this elevated cAMP. So, for example, Sanborn and colleagues have shown that the suppression of contractility in myometrial smooth muscle cells caused by relaxin is due to cAMP upregulation, in turn inhibiting phospholipase C [1, and this volume]. Similarly, the major phenotypic changes associated with decidualization in human endometrial stromal cells and which are induced by relaxin, can be mimicked by the membrane-permeable dibutyryl-cAMP [2-4]. Paradoxically, however, we would assume that relaxin is like the closely related members of this hormone family, insulin and IGF1, and function via a membrane receptor with tyrosine kinase activity. Yet time course studies of relaxin stimulation of cells in culture show that the hormone causes an increase in cAMP within seconds or minutes only, findings which until now have been uniquely associated with membrane receptors of the heptahelical kind directly coupled to Gs proteins and hence adenylyl cyclase. In the vast literature on signal transduction to date there are no mechanisms for any hormone system described which would allow such a rapid elevation in cAMP starting from a classical tyrosine kinase receptor.

We have made use of two cell-based bioassay systems for relaxin, the monocyte cell line THP-1 [5], and primary cultures of human endometrial stromal cells (ESC) from the cycle [6], in order to analyse the intermediate steps between relaxin stimulation at the cell membrane and subsequent downstream events in intracellular signal transduction. Both cell systems have been described by us in detail elsewhere [7]. Since these two cell types have quite discrete ontogenetic origins, any information confirmed on relaxin signal transduction in both cell types is likely to reflect basic mechanistic aspects of the relaxin receptor itself, rather than a phenotypic specialization of the cells. While most of the experiments described below refer to the THP-1 cells, similar experiments have been carried out using ESC cells with essentially the same results (not shown).

G.W. Tregear et al. (eds.), Relaxin 2000, 309–315.
© 2001 *Kluwer Academic Publishers. Printed in the Netherlands.*

Relaxin stimulates cAMP production in THP1 and ESC cells.

In order to confirm that relaxin is acting in these two cell types via a high affinity membrane receptor, dose-response curves were generated for pure porcine relaxin (generously provided by Dr. O.D. Sherwood) measuring cAMP production as end-point (Fig.1). For the THP-1 cells growing as suspension cultures total cAMP was measured as being the sum of intracellular cAMP and that secreted into the culture medium. For the primary cultures of ESC cells growing as confluent monolayers, intracellular cAMP was measured after briefly washing cells free of culture medium. Previous time-course experiments had shown that maximum cAMP accumulation was measured at 20-30 minutes, although initial increases in cAMP production are detectable already after 1 minute. For both cell types identical dose-response curves are generated (Fig.1) showing an EC_{50} for porcine relaxin of 1-2nM close to the Kd for the relaxin receptor estimated by Scatchard analysis [8,9], and confirming that these cells are responding to relaxin via a typical high affinity membrane receptor.

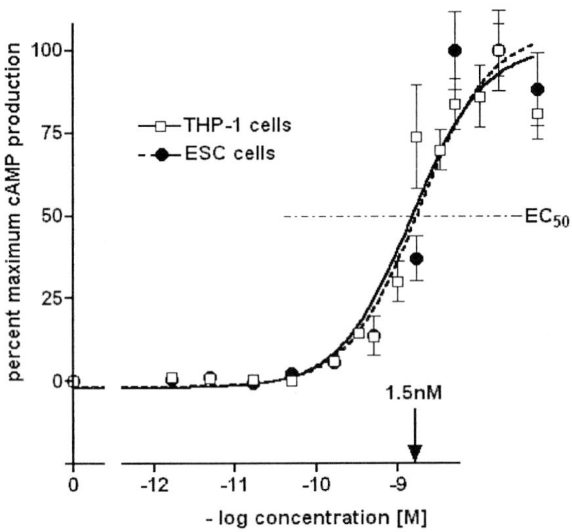

Fig.1. Dose response curves for the effect of relaxin to elevate cAMP production in THP-1 cells (total cAMP; open squares) and ESC cells (intracellular cAMP; filled circles).

As one might imagine, however, the cAMP response of these cells to relaxin is not typical. This is well illustrated by results additionally using the activator of adenylyl cyclase, forskolin (Fig.2). Normally, when cells are treated with forskolin, by-passing the involvement of membrane receptors and G-proteins, the concentration of the resultant cAMP generated is greater than that normally induced by classic G-protein coupled receptors. For THP-1 cells, not only is forskolin alone less effective than relaxin, but together there is a dramatic synergism (Fig. 2). This shows that relaxin cannot be acting upstream of adenylyl cyclase in a G-protein coupled fashion since then such effects would be at best additive. A further curiosity is that THP-1 cells appear to be completely insensitive to moderate amounts of VIP (Fig.2), even though we and others have shown that these cells indeed possess VIP receptors, which couple via G-proteins to adenylyl cyclase and in membrane preparations induce the production of cAMP.

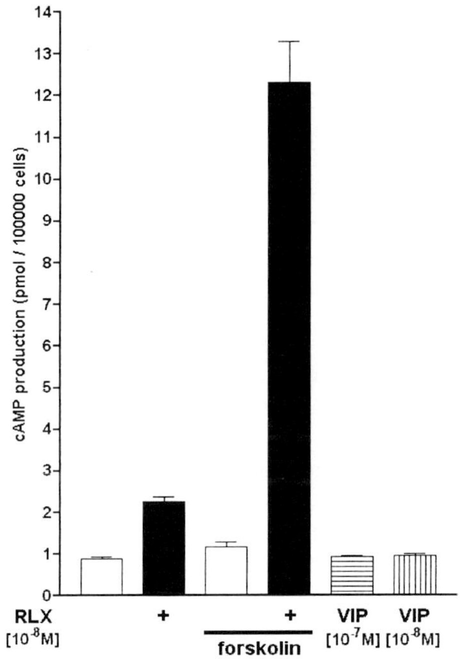

Fig.2. The effect of relaxin (RLX) and VIP on cAMP production by THP-1 cells in the presence or absence of additional forskolin (2μM preincubation for 30 min).

Relaxin acts via inhibition of a phosphodiesterase downstream of adenylyl cyclase.

Both these results strongly support the notion that relaxin is acting downstream of adenylyl cyclase to elevate cAMP. A candidate target enzyme here would be a phosphodiesterase. These enzymes are responsible for degrading cAMP within the cell. Indirect support for this idea is provided by the results shown in Fig.3. When THP-1 cells are treated with an optimal concentration (1mM) of the general phosphodiesterase inhibitor isobutyl methyl xanthine (IBMX) there is a very similar increase in cAMP to that seen upon treating the cells with relaxin. This shows firstly that THP-1 cells have high endogenous levels of phosphodiesterase. Secondly, inhibiting phosphodiesterase with IBMX causes a similar increase in cAMP to that seen with relaxin alone. A second piece of evidence to support this idea, comes from studies with the isoflavone genistein

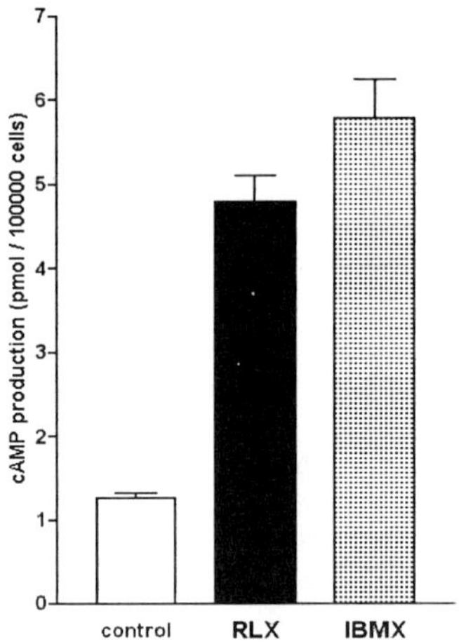

Fig.3. Comparison of the effect of the phosphodiesterase inhibitor IBMX (1mM) with that of relaxin (RLX, 10nM) on cAMP production by THP-1 cells.

(Fig. 4), which is usually considered a general inhibitor of tyrosine kinases. In the absence of IBMX, not only does genistein have no effect on relaxin-induced cAMP; in fact it causes a marked synergistic increase in cAMP levels (Fig. 4A). However, addition of IBMX, while potentiating the relaxin-induced as well as basal levels of cAMP, completely negates the stimulatory effect of genistein (Fig.4B). Indeed, genistein now appears to be inhibitory, in keeping with its role as a possible inhibitor of tyrosine kinase. This result is easy to interpret once it is understood that genistein also can have an inhibitory effect on certain types of phosphodiesterase [10-12], where it can

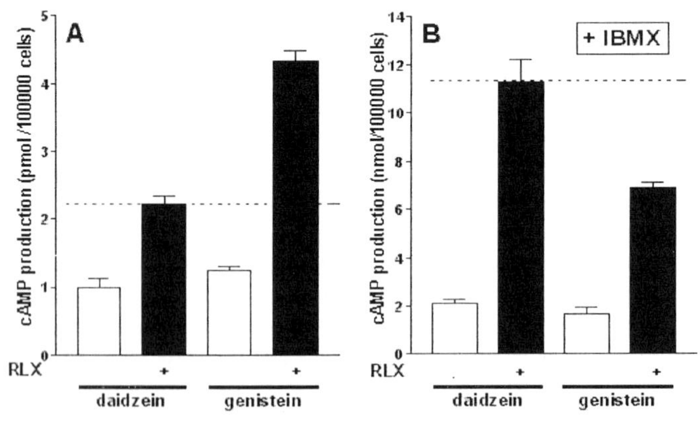

Fig.4. Effect of the isflavone genistein or its inactive analogue daidzein (both 100µM) on the stimulatory effect of relaxin (RLX, 10nM) on cAMP production by THP-1 cells in the absence (A) or presence (B) of 1mM IBMX.

modulate the phosphorylation status of the enzyme. Taking all these results together, therefore, strongly points to an intracellular phosphodiesterase as being a major target for relaxin signal transduction, and by inhibiting this enzyme relaxin is able to cause a rapid and sustained elevation of cAMP.

The relaxin receptor is linked to tyrosine kinase activity.

The next question is how does relaxin inhibit the phosphodiesterase? Earlier studies had suggested that, like receptors for insulin and IGF1, also that for relaxin might be a membrane-type tyrosine kinase. Stimulation of cells by relaxin leads to the rapid tyrosine phosphorylation of a >200kD membrane protein [13]. We have attempted to address this issue using the THP-1 bioassay system, by treating cells with a variety of specific tyrosine kinase and other enzyme inhibitors, measuring the cAMP generated as

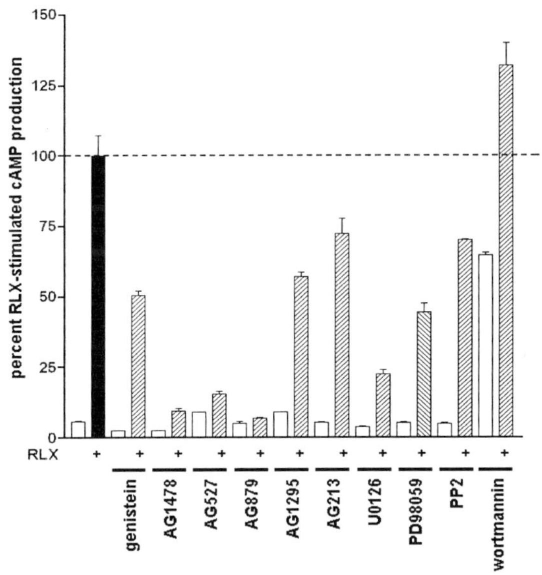

Fig.5. The effect of various inhibitors of tyrosine kinases (all at 100μM concentration) on the relaxin-stimulated (RLX) production of cAMP by THP-1 cells in the presence of 1mM IBMX.

end-point. Fig.5 summarizes these findings, where an optimal concentration of inhibitor (determined elsewhere) is applied to the cells in the presence of the phosphodiesterase inhibitor IBMX to prevent the cAMP generated from being degraded. As already described above, genistein has a mild inhibitory effect in the presence of IBMX. Surprisingly the potent tyrphostins AG 213 and AG 1295 are without effect although these are considered to inhibit a broad range of tyrosine kinases, including those of the insulin and IGF1 receptors. Marked inhibition is achieved, however, with the tyrphostins AG 527, AG 879 and AG 1478, which are described as more specific for

EGF- or NGF- receptor type tyrosine kinases. That other enzymes are involved in the signal transduction pathway stimulated by relaxin has been suggested by Goldsmith and colleagues [13, and this volume], who have postulated a role for raf, ras and MAP kinase. We can confirm this, in that we see a significant inhibition in cAMP production in THP-1 cells using the MAP kinase inhibitors PD98059 and U0126. In contrast, the src tyrosine kinase inhibitor PP2, the PI 3-kinase inhibitor wortmannin, and the protein phosphatase inhibitor okadaic acid (not shown) are all without effect.

Summary and Conclusions

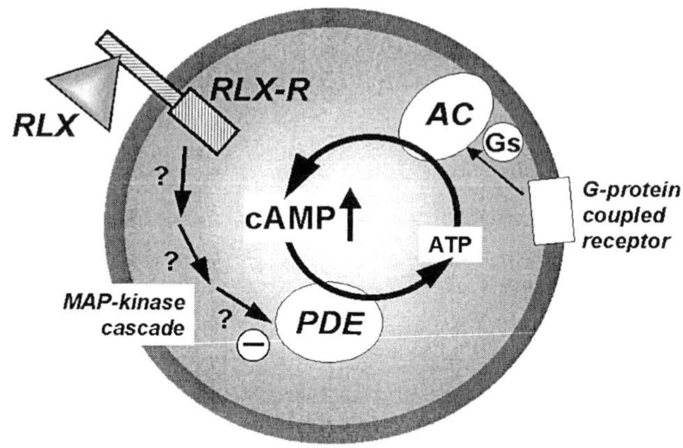

Fig. 6. Schematic illustration of the hypothetical route of signal transduction from the relaxin receptor (RLX-R) to cause an increase in cAMP. AC, adenylate cyclase, PDE, phosphodiesterase.

Combining all the results that we and others have obtained concerning the action of relaxin inside cells leads to a relatively homogeneous pattern of informational transfer. Although the relaxin receptor has still not been cloned, it seems likely that we are dealing with a novel subtype of tyrosine kinase receptor. Unusually this receptor is able to cause a massive and dose-dependent increase in intracellular cAMP, not by activating adenylyl cyclase, but by targetting a phosphodiesterase downstream of cAMP generation (Fig. 6). The fact that relaxin-stimulated cAMP production can still be modulated after inhibiting phosphodieseterase activity using IBMX, suggests that more than one kind of phosphodiesterase enzyme is involved. Indeed, it has recently been shown that THP-1 cells have a new type of phosphodiesterase which is resistant to IBMX action [14]. Between the relaxin receptor and its phosphodiesterase target are probably one or more intermediate enzymes, representing a kinase cascade probably involving MAP-kinase. This pathway appears to explain all the results so far obtained for THP-1, ESC and other cell types, and provides a mechanism by which relaxin is able to cause the rapid, dose-dependent and sustained elevation of cAMP which is so characteristic of relaxin-stimulated cells. This cAMP in turn appears to be responsible for most of the physiological responses generally associated with the peptide hormone *in vivo* and *in vitro*.

Acknowledgements

We gratefully acknowledge many discussions and helpful suggestions from our Hamburg colleagues, in particular Dr Ralph Telgmann, Dr Birgit Gellersen and Dr Rasoul Zarreh-Hoshyari-Khah. We should especially like to thank Dr Dave Sherwood for generously providing us with highly pure porcine relaxin for our experiments, and Professor Freimut Leidenberger for his continued support of this difficult project. This research was in part financed by the Deutsche Forschungsgemeinschaft (grant Iv7/9-1).

References.

1. Sanborn, B.M., Anwer, K., Monga, M., Wen, Y., Singh, S.P., Meera, P., Oberti, C., Toroc, L. and Stefani, E. In MacLennan, A.H., Tregear, G.W. and Bryant-Greenwood, G.D. (Eds.) Progress in Relaxin Research, Global Publications, Singapore, (1995), p. 289.
2. Tang, B., Guller, S. and Gurpide, E., Endocrinology, 133 (1993) 2197.
3. Tang, B., Guller, S. and Gurpide, E., Ann. N. Y. Acad. Sci., 734 (1994) 19.
4. Telgmann, R. and Gellersen, B., Hum. Reprod. Update, 4 (1998) 472.
5. Parsell, D.A., Mak, J.Y., Amento, E.P. and Unemori, E., J. Biol. Chem., 271 (1996) 27936.
6. Fei, D.T.W., Gross, M.C., Lofgren, J.L., Mora-Worms, M. and Chen, A.B., Biochem. Biophys. Res. Com., 170 (1990) 214.
7. Zarreh-Hoshyari-Khah, M.R., Bartsch, O., Einspanier, A., Pohnke, Y. and Ivell, R., Regul. Peptides, 97 (2001) 139.
8. Osheroff, P.L. In MacLennan, A.H., Tregear, G.W. and Bryant-Greenwood, G.D. (Eds.) Progress in Relaxin Research, Global Publications, Singapore, (1995), p. 159.
9. Büllesbach, E.E., Steinetz, B.G. and Schwabe, C. In MacLennan, A.H., Tregear, G.W. and Bryant-Greenwood, G.D. (Eds.) Progress in Relaxin Research, Global Publications, Singapore, (1995), p. 145.
10. Ogiwara, T., Chik, C.L. and Ho, A.K., J. Endocrinol., 53 (1997) 1271.
11. Ho, A.K., Wiest, R., Ogiwara, T., Murdoch, G. and Chik, C.L., J. Neurochem., 65 (1995) 1597.
12. Stringfield, T.M. and Morimoto, B.H., Biochem. Pharmacol., 53 (1997)1271.
13. Palejwala, S., Stein, D., Wojtczuk, A., Weiss, G. and Goldsmith, L.T., Endocrinology, 139 (1998) 1208.
14. Fisher, D.A., Smith, J.F., Pillar, J.S., St Denis, S.H. and Cheng, J.B., Biochem. Biophys. Res. Commun., 246 (1998) 570.

Relaxin activating effect on the adenylyl cyclase and glycogen synthase in muscles of mammals and molluscs

Ludmila Kuznetsova[1], Svetlana Plesneva[1], Eugeniya Omeljaniuk[2], Alexander Shpakov[1] and Marianna Pertseva[1]
[1]Institute of Evolutionary Physiology and Biochemistry, Russian Academy of Sciences, Thorez pr.44, 194223, St. Petersburg, Russia; [2]Institute of Obsterrics and Gynecology Russian Academy of Medical Sciences, Mendeleev line 3, 199034, St. Petersburg, Russia.

Introduction

Relaxin is one of the related peptides belonging to the insulin superfamily. The main physiological function of relaxin is relaxation effect on uterine muscles. In addition, relaxin has diverse effects (changing cAMP level, calcium current, etc.) in some other tissues [1]. We have shown for the first time [2,3] that mammalian insulin and insulin-like growth factor-1 (IGF-1) as well as an insulin-like peptide (ILP) isolated from the mollusc Anodonta cygnea are capable of activating adenylyl cyclase (AC) in the muscle tissues of mammals and mollusc. The mechanism of the AC activating effect of insulin was revealed to be via the following signalling pathway: receptor tyrosine kinase ⇒ Gi-protein (βγ) ⇒ phosphatidylinositol-3-kinase (PI3K) ⇒ protein kinase C (PKCζ) ⇒ Gs-protein ⇒ AC. According to our hypothesis [2] this AC signalling mechanism (ACSM) is involved in the mitogenic rather than metabolic action of the peptides. Until our investigations the participation of AC signalling system in the action of insulin and related peptides has not been demonstrated. Because of the evolutionary relationship between insulin superfamily peptides, we have studied the ability of relaxin to activate AC as well as the mechanism of this effect.

Results and Discussion

It has been shown that H 2 (10^{-12}-10^{-8} M) causes a dose-dependent activating effect on AC in human myometrium (with maximum +300%), in rat skeletal (+100%) and smooth foot muscles of the mollusc *Anodonta cygnea* (+56%). Mammalian insulin and IGF-1, as well as molluscan ILP have similar stimulating effects. The order of efficiency of the peptides evaluated by their maximal AC activating effect was as follows: relaxin > IGF-1 > insulin (human myometrium); IGF-1 > relaxin > insulin (rat skeletal muscle); molluscan ILP > IGF-1 > insulin > relaxin (molluscan muscle). The relaxin AC activating effect is potentiated with Gpp[NH]p giving evidence for the involvement of Gs-protein in the hormone action. Knowledge of the ACSM of insulin action ld us to study the possible participation of receptor tyrosine kinase and PI3K in relaxin AC activating action. Tyrphostin 47 (0.62-5.00 μM) decreased the relaxin AC activating effect (IC_{50} = 0.8-1.0 μM). This fact gives ground to suggest that the relaxin receptor belongs to the family of receptors with intrinsic tyrosine kinase activity. Wortmannin also blocks the realization of relaxin AC activating effect (Fig. 1).

G.W. Tregear et al. (eds.), Relaxin 2000, 317–319.
© 2001 *Kluwer Academic Publishers. Printed in the Netherlands.*

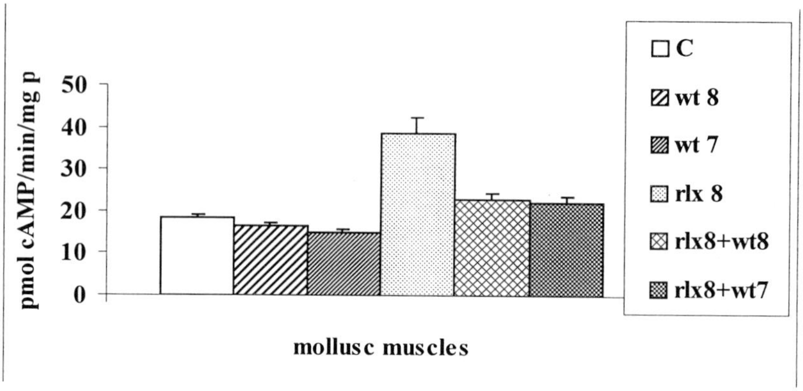

Fig. 1 Effect of wortmannin (wt, 10^{-8}-10^{-7} M) on relaxin-stimulated (rlx) AC activity in the mollusc muscle membranes.

The similarity between insulin and relaxin regulatory action is supported also by experiments showing the relaxin stimulating effect on glycogen synthase activity I (G6P-independent form) (Table 1).

Table 1. Influence of relaxin (Rlx) and insulin on glycogen synthase I activity in the muscle membranes of mollusc and rat

Glycogen synthase I activity (nmol UDP/mg of protein/10 min)		
Additions (M)	Mollusc	Rat
Relaxin		
Without	12,7±1,1	20,2±2,0
Rlx 10^{-9}	18,9±1,8	36,8±3,1
Rlx 10^{-8}	22,6±1,0	35,1±3,9
Insulin		
Without	11,2±1,1	17,8±2,2
Ins 10^{-9}	16,8±2,3	38,2±4,1
Ins 10^{-8}	19,0±1,2	57,1±2,7

The results of this study show the potential mechanism of relaxin action on AC to be the following signalling chain: receptor tyrosine kinase ⇒ PI3K ⇒ Gs-protein ⇒ AC. This is similar to the structural-functional organization of insulin ACSM. The other proposed signalling components of the relaxin mechanism need further study. Finally it should be noted that the relaxin AC effect under study manifest themselves both in the muscle tissues of mammals and mollusc. This is in a good agreement with the common principles of structural-functional organization of hormone sensitive AC signalling systems in vertebrates and invertebrates [4]. The work is supported by grant of RFBI 00-04-49465.

Acknowledgments

We express our thanks to Dr. Christian Schwabe (Dep. of Biochem. Mol. Biol., Medical University of South Caroline, USA) for a generous gift of H2 relaxin.

References

1. Bryant-Greenwood, G.D. and Schwabe, C., Endocr. Rev.,15 (1994) 5.
2. Pertseva, M.N., Plesneva, S.A., Shpakov, A.O., Rusakov, Y.I.and Kuznetsova, L.A., Comp. Biochem. Physiol., 112B (1995) 689.
3. Pertseva, M.N., Plesneva, S.A., Kuznetsova, L.A., Shpakov, A.O. and Derkach, K.V., Biochem. Pharmacol., 52 (1996) 1867.
4. Pepio A.M., Fan X. and Sossin W.S., J. Biol. Chem., 273 (1998) 19040.

The regulatory role of protein kinase C in insulin signal transduction via adenylyl cyclase signalling system

Svetlana Plesneva, Alexander Shpakov, Ludmila Kuznetsova and Marianna Pertseva

Institute of Evolutionary Physiology and Biochemistry, Russian Academy of Sciences, Thorez pr.44, 194223, St. Petersburg, Russia

Introduction

Reviewing the available data on the functional cross-talk between hormone-regulated adenylyl cyclase (AC) signalling system and protein kinase C (PKC) in the cell, the role of PKC in the initiation or modulation of transduction of insulin signal via AC signalling mechanism (ACSM) was studied. As was found for the first time in our earlier study, this mechanism is involved in the action of mammalian insulin and insulin, insulin-like growth factor-1 (IGF-1) as well as of insulin-like peptide (ILP) of the mollusc, *Anodonta cygnea* [1,2]. The identification of functional components of insulin ACSM showed the involvement of receptor tyrosine kinase, Gs-protein and AC.

The aim of the present work was to continue the deciphering of ACSM of insulin using pertussis toxin (PT) and the inhibitor of phosphatidylinositol-3-kinase (PI3K), wortmannin. A second aim was to study the influence of activators [phorbolesters (PMA), diacylglycerin (DAG)] and an inhibitor (calphostin C) of PKC on insulin signal transmission to AC. The comparative study was carried out using rat skeletal muscles and the foot smooth muscles of bivalve mollusc *Anodonta cygnea.*

Results and Discussion

PT treatment of the muscle membranes of mollusc and rat led to a small increase in the AC basal activity and to a significant decrease in the AC stimulating effect of insulin (10^{-9} M) (from 280% to 8% in mollusc; and from 71% to 2% in rat in the absence and presence of PT, respectively). We believe that it reflects the loss of ability of the G_i -protein subunits to dissociate and release the βγ-dimer. As is known, βγ-dimer mediates a number of stimulatory signals, e.g. the activating effect on PI3K. [3]. In the presence of wortmannin the AC stimulating effect of insulin in these tissues was blocked.

Concentrations of 10^{-10}-10^{-8} M PMA and DAG *in vitro* stimulated the catalytic activity of AC in the rat and mollusc muscle membranes by 34-140% over the control (Table 1). Activation of PKC by DAG and PMA led to complete inhibition of Gpp[NH]p-stimulated AC activity. This provides evidence for the impairment of functional interaction between G_s-protein and AC. In contrast, calphostin C (10^{-11}-10^{-8} M) increased AC basal activity in the mollusc and rat tissues but decreased insulin-stimulated AC activity (Table 2). Discussing PKC atypical isoforms, calphostin C

G.W. Tregear et al. (eds.), Relaxin 2000, 321–323.
© 2001 *Kluwer Academic Publishers. Printed in the Netherlands.*

inhibition of AC-mediated insulin signalling provides evidence for their direct participation in the stimulatory signal transmission, but the group of PMA-sensitive PKCs is to be excluded. Indeed, if the inhibitory effect of calphostin C was directed only to PMA-sensitive PKC isoforms, which exercise a negative influence on ACSM, then the insulin AC-activating effect would remain unchanged. Meanwhile we observed the calphostin C inhibitory action on insulin AC signalling. Consequently caphostin C inhibitory action also extends to PMA- insensitive isoforms of PKC, e.g. PKCζ which is the target of insulin regulatory action.

Table 1. Influence of DAG and PMA on the basal AC activity in the muscle membranes of rat and mollusc.

Adenylyl cyclase activity (pmol cAMP/mg of protein/min)		
Additions (M)	Mollusc	Rat
Without	16,2±1,3	50,4±4,1
DAG 10^{-9}	29,6±2,7	111,3±10,3
DAG 10^{-8}	38,4±4,1	121,3±9,4
PMA 10^{-9}	37,9±3,8	87,9±8,5
PMA 10^{-8}	35,8±2,8	71,6±5,2

Table 2. The AC-stimulating effect of insulin in the presence of different concentrations (10^{-11}-10^{-8} M) of calphostin C. Basal AC activity = 100 %.

Adenylyl cyclase activity (% over the control)		
Additions (M)	Mollusc	Rat
Insulin 10^{-8}	350±16	160±12
Calphostin 10^{-11}	157±15	67,0±2,8
Calphostin 10^{-10}	136±9	65,0±1,4
Calphostin 10^{-9}	127±4	60,2±1,3
Calphostin 10^{-8}	89±3	56,2±2,3

In our view the PKC system plays a dual role in ACSM. One role of PKC is regulatory, providing for negative regulation of AC-signaling system by PMA and DAG sensitive-PKC isoforms. The second role of PKC, we postulate to be as transductory, and in the insulin-sensitive AC system, is performed by atypical PKC isoform (evidently by PKCζ). It consists of direct participation of PKCζ in the transduction of insulin signal from the receptor to AC. On the basis of our and literature data the following insulin signal transduction cascade via AC system can be suggested: receptor tyrosine kinase ⇒ Gi-protein (βγ) ⇒ phosphatidylinositol-3-kinase (PI3K) ⇒ protein kinase C (PKCζ) ⇒ Gs-protein ⇒ AC.

References

1. Pertseva, M.N., Plesneva, S.A., Shpakov, A.O., Rusakov, Y.I. and Kuznetsova, L.A., Comp. Biochem. Physiol., 112B (1995) 689.
2. Pertseva, M.N., Plesneva, S.A., Kuznetsova, L.A., Shpakov, A.O. and Derkach, K.V., Biochem. Pharmacol., 52 (1996) 1867.
3. Houslay, M.D. and Milligan, G., Trends Biochem. Sci., 22 (1997) 217.

SECTION 11

Relaxin-like factors

The Relaxin-Like Factor: from gene to physiology

Richard Ivell[1], M Rasoul Zarreh-Hoshyari-Khah[1], Ross A D Bathgate[1,2], Marga Balvers[1], Paul J Baker[3] and Peter J O'Shaughnessy[3]

[1]*Institute for Hormone and Fertility Research, University of Hamburg, Germany;* [2]*Howard Florey Institute, University of Melbourne, Australia; and* [3]*Department of Veterinary Pre-Clinical Studies, University of Glasgow Veterinary School, Scotland.*

Introduction

The Relaxin-like Factor (RLF) is a relatively new member of the insulin-IGF-relaxin family, which now includes altogether eight members in mammals (Table 1; see also article by Jaspers, this volume), though a comparison with the diverse members of this family found in the completed *C. elegans* genome [1], suggests that more members may still be found. RLF (also known as Ley-IL) is the product of the *INSL3* gene (see Table 1) and was found independently from differential cloning projects in the testes of pigs [2] and mice [3]. Subsequently, cross-hybridization or homologous PCR analysis identified equivalent genomic or cDNA sequences from a wide range of other mammalian species (human [4,5]; cows [6]; sheep [7]; marmoset monkey [8]; rats [9], as well as other species: see article by Klonisch *et al.*, this volume). Like the closely related peptide hormone relaxin, RLF is encoded by a small gene comprising mostly only two exons (see below), with a splice junction occurring at the beginning of the region corresponding to the C-domain. Also like relaxin, RLF appears to have a three domain protein structure, with a B-domain immediately following a typical signal peptide. This is followed by the C- and A-domains respectively. Preliminary protein modelling suggests that sulfhydryl bonds link the six cysteine residues in the same configuration as for relaxin and the other members of this family.

Protein expression in the male and female.

To date we still know very little about the nature of the protein expressed *in vivo* by the *INSL3* gene. While a number of specific antibodies have been raised either against peptide domains or against recombinant RLF protein, most of these appear to function well only in an immunohistochemical context (Fig. 1). Consistent Western blots to indicate the size of the protein made *in vivo* are still lacking. The possession of a consensus signal peptide motif in a predicted protein sequence with a typical precursor structure, the detection of RLF epitopes by ELISA in male human serum [10], and the observation that recombinant RLF made in a baculovirus-infected insect cell system is preferentially secreted into the culture medium [8], all imply that RLF is made, like insulin and relaxin, as a precursor which is subsequently processed in and secreted from the cells where it is made. We still have no evidence for the form in which the RLF

G.W. Tregear et al. (eds.), Relaxin 2000, 327–335.
© 2001 *Kluwer Academic Publishers. Printed in the Netherlands.*

protein is secreted and circulates in the blood. The presence of A-, B- and C-domains, and the presence at appropriate positions of PC1/3, PC2 or furin enzyme recognition sites - as in insulin and relaxin - might presuppose cleavage into an A-B heterodimer with elimination of the connecting or C-peptide. However, in insect cells recombinant pro-RLF does not appear to be processed into a smaller form [8], also, the C-A transition domain is unlike that of insulin or relaxin, and instead of a typical furin-like cleavage signal is typified in all species so far characterized by a histidine-rich motif. Furthermore, once RLF is secreted, we have no knowledge of subsequent events such as a possible interaction with specific binding proteins or extracellular proteolytic enzymes.

name (alternative names)	abbreviations	gene name
insulin *	Ins	*INS*
insulin-like growth factor I	IGF1	*IGF1*
insulin-like growth factor II	IGF2	*IGF2*
relaxin H1 **	RLX, REL, RLN	*RELH1*
relaxin H2	do.	*RELH2*
relaxin-like factor (Leydig-insulin-like peptide)	RLF (Ley-I-L)	*INSL3*
placentin (early placental insulin-like factor)	(EPIL)	*INSL4*
INSL5 (relaxin/insulin-like factor 2)	(RIF2)	*INSL5*
INSL6 (relaxin/insulin-like factor 1)	(RIF1)	*INSL6*

*Table 1. The mammalian insulin/IGF/relaxin family of peptide hormones, indicating (in parentheses) the alternative names and abbreviations to be found in the literature. *There are two insulin genes in the rodent genome, but not other mammals. **Here the two relaxin genes in the human genome are indicated (H1 and H2). These are not present in the genomes of non-anthropoid species.*

There is a very consistent pattern where the *INSL3* gene is expressed. In most mammals the principal site of synthesis are the mature Leydig cells of the testis (Fig. 1), where the mRNA is found at very high concentrations. In the female, RLF is detected in the ovary, in particular in the theca cells of large antral follicles (Fig. 1) and in the corpora lutea of the cycle and pregnancy [6, 8, 11-13]. In addition, RLF expression, either as mRNA or as immunoreactivity, has also been detected in the brain, thyroid gland and placenta (see articles by Klonisch *et al.*, this volume; [6]), and also within the seminiferous tubules of the testis in the hamster (Balvers and Ivell, unpublished). In the majority of species the apparent levels of mRNA and protein found in the non-testicular tissues is generally very low, often only detectable by PCR or very sensitive immunohistochemistry. An exception is the ovary of cows and sheep, where both theca cells and corpus luteum are the sites of very high level expression, comparable to that in the testicular Leydig cells, and suggesting that RLF may be upregulated as a compensatory reaction to the apparent deletion of the relaxin gene in ruminants (see article by Bathgate *et al.*, this volume).

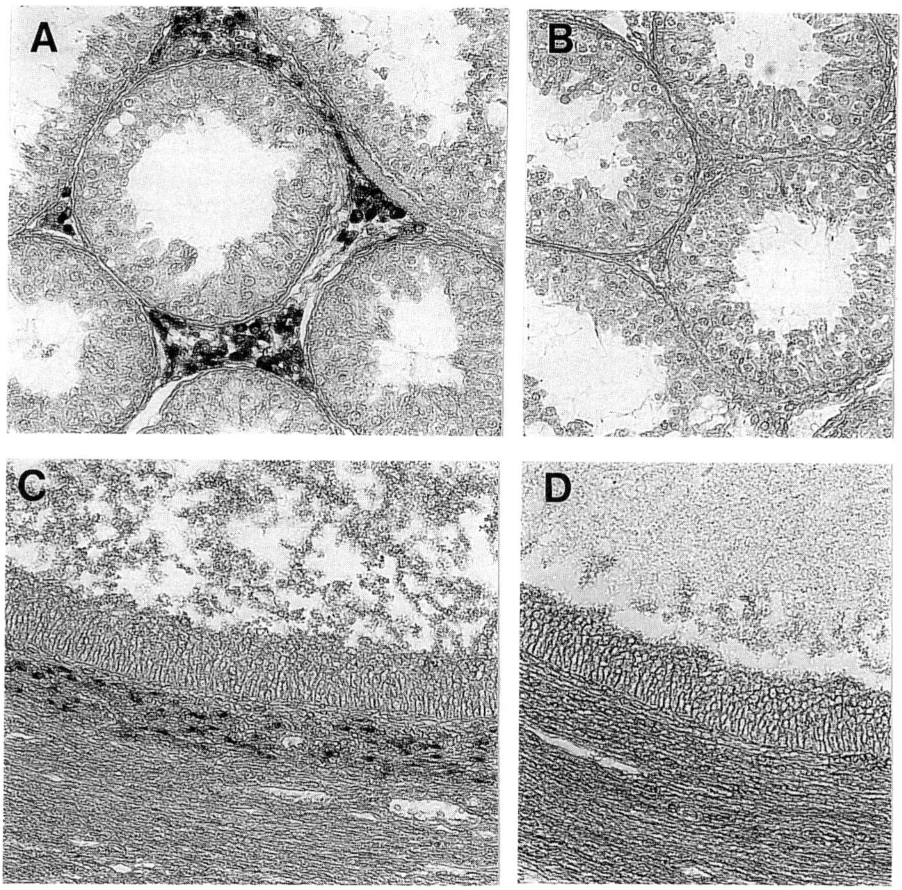

Fig. 1. Immunohistochemical staining for RLF epitopes in the bovine testis showing clear Leydig cell expression (A, B) and a large antral follicle of the ovary with expression in the theca interna cells (C, D). Negative controls using preimmune serum are shown in panels B and D.

Regulation of RLF expression

Both RLF mRNA and protein appear to be upregulated in association with achievement of a particular differentiation status in those specific cells where it is made. For example, it reaches high levels in mature fetal Leydig cells of the mouse, as well as in mature, late pubertal Leydig cells in the adult testis [12]. It is absent, however, from immature Leydig cells, or from Leydig cells which have undergone involution either postnatally in the mouse or rat [9, 12], or in hamsters as a consequence of a switch from a long-day to a short-day regime (Ivell and Balvers, unpublished). Experimental induction of Leydig cell differentiation, as in the hCG-treated *hpg* mouse, leads within several days to maturation of Leydig cells and to upregulation of the *INSL3* gene [12]. It is also down regulated in Leydig cells undergoing hypertrophy and proliferation, or which have become tumorigenic [14]. Where it is expressed, it does not appear to be acutely regulated. In both primary and tumor Leydig cell cultures, no acute effector has yet been found to influence the levels of RLF mRNA expression [12], nor are levels shown to vary in the testicular biopsies from patients with a variety of extreme endocrine pathologies [5]. Comparable results have also been obtained using primary cultures of bovine ovarian theca cells [11] (and Bathgate *et al.*, this volume), where in long-term culture levels of RLF mRNA appear to reflect the differentiation status of the cells.

One situation where specific hormonal regulation of the *INSL3* gene appears to occur is in fetal Leydig cells, where estrogen treatment *in vivo* down-regulates the gene without seeming to influence other Leydig cell parameters [15] (also Nef *et al.*, this volume).

Gene structure, regulation and alternative splicing

The two exons encoding the expressed mRNA transcripts are located within a small region of the genome (1-5 kb, depending on the species [4, 8, 9, 16] (and Bathgate *et al.*, this volume). There are two unusual features of note. Firstly, in those species where more information is available on the genomic locus (i.e. human, rat, mouse), the *INSL3* gene is shown to exist in close relation to the gene for the signal transduction molecule JAK3 [9, 16]. Indeed, in the mouse, there is an overlap of the two genes such that the *INSL3* promoter is within the intron separating exons 23A and 23B of the JAK3 gene [16], with the exon 1 of *INSL3* being also within the same intron, but exon 2 being colinear with exon 23B of that gene (Fig.2), though differently spliced. The *INSL3* promoter itself appears to be short, possibly related to its inclusion within the intron of another gene. Transfection studies indicate that as little as 700 bp is sufficient to allow high expression in Leydig cells but not Sertoli or other cell types [16]. The only elements for transcription factor interaction that have been characterized so far in the *INSL3* promoter are three putative sites for steroidogenic factor 1 (SF-1) [16, 17], and the negatively interacting factor DAX1. However, since RLF is not expressed in the adrenal gland (Ivell, Bathgate and Balvers, unpublished), where there are high concentrations of this transcription factor, and SF-1 is also expressed in immature

Leydig or theca-lutein cells when there is no RLF expression, other transcription factors either being recruited by SF-1 or interacting elsewhere within the *INSL3* promoter are probably also involved.

The second notable feature of the *INSL3* gene is its ability to undergo differential splicing (Fig. 2). In the marmoset monkey, PCR analysis indicated a second, longer transcript present wherever RLF was being expressed [8]. This was shown to be due to the splicing in of a novel exon 1A, located within the genomic region corresponding to intron 1 of other species. This splice event leads to a frame-shift in the encoded amino acid sequence and a new stop codon. The resulting protein would therefore comprise a B-domain with part of the C-domain and a novel C-terminus, although antibodies raised against the latter sequence failed to detect a protein product. It is doubtful whether this differential splice product is of functional relevance since the exon 1A sequence does not have a homologue in any other species where the sequence of intron 1 is known (i.e. mouse, rat, human, pig, cow).

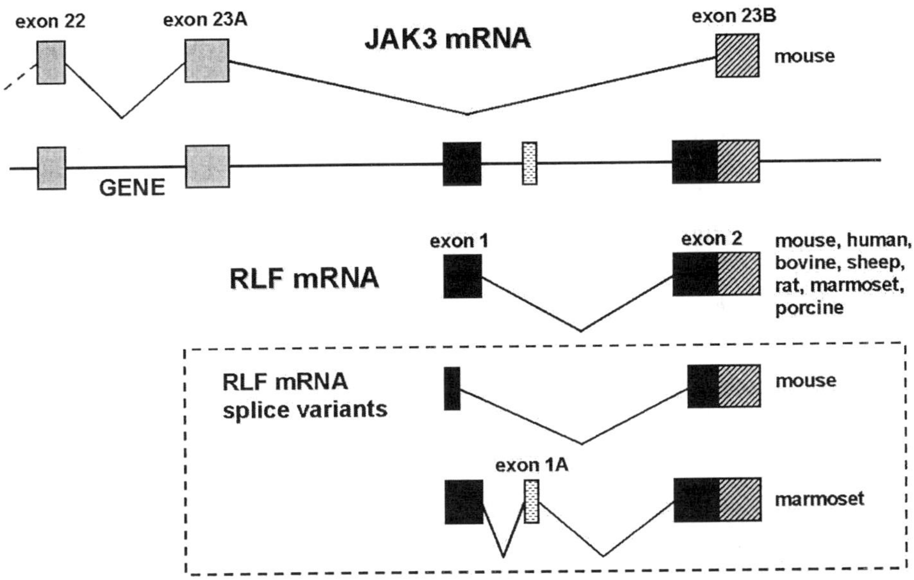

Fig. 2. Genomic organization and alternative splicing of the INSL3/JAK3 locus. The exonic composition of the functional transcripts for JAK3 and RLF are shown above and below the genomic locus, respectively. Within the boxed region are shown the alternative splice variants described for the mouse (upper schematic; P.J. Baker & P.J. O'Shaughnessy, unpublished; see text) and the marmoset monkey (lower schematic; for details, see [8].

Differential splicing has also been observed for the mouse (P.J. Baker and P.J. O'Shaughnessy, unpublished). Using oligonucleotide primers specific for the extreme 5' and 3' ends of the normal transcripts, a shorter PCR product is obtained using testicular mRNA in addition to the expected product. This novel transcript appears to be the result of a splicing event whereby the beginning of the B-domain region (encoding amino acids ALRSPQ) is linked directly to a downstream region of the C-domain (encoding amino acids LPRQ..), thus eliminating 195 nucleotides of the expected functional transcript. A protein product, if made, would therefore comprise only the first few amino acids of the B-domain connected directly to the latter part of the C-domain and the A-domain. None of these novel splice junctions are similar to those used in the expression of the *JAK3* gene (Fig. 2). It therefore appears that this region of chromatin is quite plastic and susceptible to considerable transcriptional variation. This view is supported by the occasional observation in RT-PCR reactions using testicular RNA of other mammalian species of other weaker products both longer and shorter than the principal expected PCR product (P.J. Baker & P.J. O'Shaughnessy, unpublished).

Transgenic mice and RLF physiology

The strong expression of RLF in the testicular Leydig cells (see above), and the observation of a circulating RLF-like hormone in adult men, but only at much lower levels in children and women [10], suggested an important function in the context of male reproductive physiology. However, it was only upon the production of knockout mice, where the *INSL3* gene was partially ablated, that the importance of this relaxin-like molecule became clear [18, 19]. RLF appears to be the long sought after "descendin" factor responsible for testicular descent in the perinatal period of mammals. The principal defect in the knockout mice was the failure of the testes to descend into the scrotum. Being thus cryptorchid, they are exposed to the deleterious effects of body rather than scrotal temperature and hence show also secondary defects of spermatogenesis. Testicular descent can be subdivided into three physiologically discrete though overlapping aspects (Fig. 3). The undifferentiated gonad of the early embryo becomes dorsally suspended by the cranial suspensory ligament (CSL), and ventrally by another ligament which will become the gubernaculum. In the female embryo, in the absence of testosterone and RLF, the former develops more, the latter disappears, and the resulting ovary is retained in a perirenal position. In the normal male embryo, the production of testosterone by the fetal Leydig cells leads to a dissolution of the CSL. RLF, it would appear, is partly responsible for the development of the gubernaculum and hence the transit of the male gonad to the inguinal region. This is followed by a third phase, namely the traverse by the gonad of the inguinal body wall to enter the scrotum. In the *INSL3* knockout mice the testes fail to descend, with the gubernaculum remaining relatively undeveloped. Since testosterone is nevertheless being made by the fetal Leydig cells, the CSL is regressed, and hence the cryptorchid gonad is loosely positioned in the peritoneal cavity. Only upon introducing a second gene ablation, one for the androgen receptor, is the cryptorchid testis retained closely to the kidney by a robust CSL [18]. Subsequent experiments [15, 20] (and Nef et al., this volume) have shown that the production of RLF by the fetal Leydig cells is estrogen-dependent, and hence the well-known effects of estrogens to induce cryptorchidism can now be explained. Independent experiments using fetal gubernacular explants have confirmed an effect of RLF on the growth and development of this tissue (C. Temelco, Y. Kubota, K. Smith, R. Bathgate & J. Hutson, unpublished).

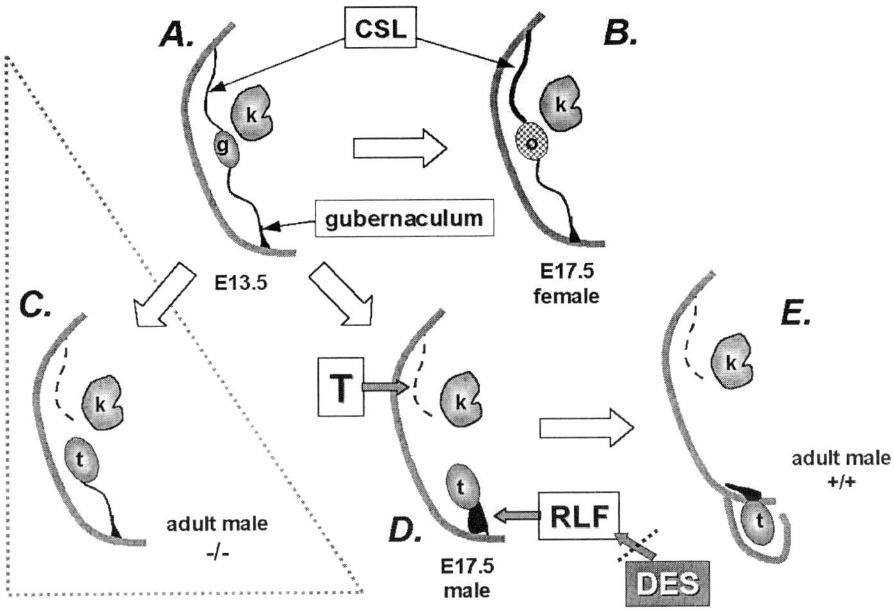

Fig. 3. Scheme to illustrate the role of RLF/INSL3 in testicular descent in the mouse. A. undifferentiated gonad (g) at embryonic day E13.5. The gonad is held loosely by an undeveloped cranial suspensory ligament (CSL) and the gubernacular ligament. B. In the female in the absence of testosterone, the CSL thickens, and the ovary (o) is retained close to the kidney (k). D. In the normal male embryo at E17.5, the CSL dispapears under androgen influence and the gubernaculum thickens and retracts the testis (t) to the inguinal region. E. In the newborn male the testis subsequently descends into the scrotum. In the adult male INSL3 knockout mouse (C), testosterone causes loss of the CSL, but the absence of RLF leads to the testis being retained in a perirenal region due to the lack of gubernacular development. DES, diethylstilbestrol, as a typical estrogen, causes cryptorchidism by disrupting the production of RLF by the fetal testis.

RLF may also be involved in the third phase of testicular descent, namely movement of the gonad into the scrotum. In an elegant set of experiments, transgenic mice were induced to produce RLF constitutively by introducing into the genome the *INSL3* gene driven by the ubiquitin-C promoter [21] (and personal communication). The resulting mice were then crossed with other transgenic mice constitutively expressing the aromatase gene, such that genetic males and females were produced expressing both RLF and estrogens constitutively from an early embryonic age. As now might be expected (see above) the young females indicated both a developed gubernaculum as well as the CSL, such that the ovaries were partially descended into the inguinal region. Most interestingly, however, both the engineered males and females showed a high preponderance to hernia of the body wall musculature in the inguinal region, suggesting that RLF may also be important for allowing the testes to traverse the body wall and enter the scrotum.

In the male, therefore, the production of RLF by the fetal Leydig cells has a clearly associated physiology in regulating testicular descent. However, we still have no information on a possible role for RLF in the adult male, where circulating levels are expected to be moderately high [10]. Surgical intervention of the young cryptorchid *INSL3* knockout mice to bring the testes into a scrotal position appeared to lead to adult males with normal spermatogenesis [18], thus effects in the adult male are likely to be of a subtle or redundant nature.

Female *INSL3* knockout mice also indicated a phenotype, though less obvious than that in males. Although homozygous knockout females were fertile, litter size appeared to be slightly reduced, and most noticeably the length of the estrous cycles was extended from the usual 6-7 days to over 15 days [19]. Given the consistent pattern of RLF expression in follicles and corpora lutea, these results suggest that RLF is probably playing an important role in follicular maturation (and hence affecting litter size), as well as in maternal recognition of pregnancy and/or luteal maintenance.

Other evidence to support a physiology for RLF is indirectly provided by the finding of specific RLF-binding sites in the uterus and brain [22, 23]. Therefore, although the knockout mice do not indicate a phenotype associated with these and other peripheral organs, there is still a likelihood for local and/or redundant effects associated with RLF in non-gonadal tissues, including the brain.

Future directions

Considerable inroads into our understanding of RLF structure, expression and physiology have been made in the last few years. What is now required is on the one hand a better selection of assays for the detection, measurement and characterization of RLF protein *in vivo* under a variety of physiological and pathological conditions. On the other hand, there is a considerable need for sufficient amounts of a biologically active peptide to test for effects of RLF both *in vivo* and *in vitro*, as well as to develop a robust pharmacology of this highly conserved and important new hormone. Preliminary studies reported in this volume (e.g. Bathgate *et al.* and Smith *et al.*, this volume), suggest that we are well on the way to achieving these goals.

Acknowledgments

We gratefully acknowledge the support of the Deutsche Forschungsgesellschaft (grant Iv7/9-1) for much of the research reported here, as well as for helpful comments and suggestions by numerous colleagues. Thanks are also due to Pasi Koskimies and Ilpo Huhtaniemi (Turku) for permission to refer to their unpublished results.

References

1. Duret, L. Guex, N., Peitsch, M.C. and Bairoch, A., Genome Research, 8 (1998), 348.
2. Adham, I.M., Burghardt, E., Benahmed, M. and Engel, W., J. Biol. Chem., 268 (1993), 26668.
3. Pusch, W., Balvers, M. and Ivell, R., Endocrinology, 137 (1996), 3009.
4. Burghardt, E., Adham. I.M., Brosig, B., Gastmann, A., Mattei. M.G. and Engel, W., Genomics, 20 (1994), 13.
5. Ivell, R., Balvers, M., Domagalski, R., Ungefroren, H., Hunt, N. and Schulze, W., Mol. Hum. Reprod., 3 (1997), 459.
6. Bathgate, R.A.D., Balvers, M., Hunt, N. and Ivell, R., Biol. Reprod., 55 (1996), 1452.
7. Roche, P.J., Butkus, E., Wintour, E.M. and Tregear, G., Mol. Cell. Endocrinol., 121 (1996), 171.
8. Zarreh-Hoshyari-Khah, M.R., Einspanier, A. and Ivell, R., Biol. Reprod., 60 (1999), 445.
9. Spiess, A.N., Balvers, M., Tena-Sempere, M., Huhtaniemi, I., Parry, L. and Ivell, R., Mol. Reprod. Dev., 54 (1999), 319.
10. Büllesbach, E.E., Rhodes, R., Rembiesa, B. and Schwabe, C., Endocrine, 10 (1999), 167.
11. Bathgate, R., Moniac, N., Bartlick, B., Schumacher, M., Fields, M. and Ivell, R., Biol. Reprod., 61 (1999) 1090.
12. Balvers, M., Spiess, A.N., Domagalski, R., Hunt, N., Kilic, E., Mukhopadhyay, A.K., Hanks, E., Charlton, H.M. and Ivell, R., Endocrinology, 139 (1998), 2960.
13. Bamberger, A.M., Ivell, R., Balvers, M., Kelp, B., Bamberger, C.M., Riethdorf, L. and Löning, T., Int. J. Gynecol. Pathol., 18 (1999), 163.
14. Klonisch, T., Ivell, R., Balvers, M., Kliesch, S., Fischer, B., Bergmann, M. and Steger, K., Mol. Hum. Reprod., 5 (1999), 104.
15. Nef, S., Shipman, T. and Parada, L.F., Dev. Biol., 224 (2000), 354.
16. Koskimies, P., Spiess, A.N., Lahti, P., Huhtaniemi, I. and Ivell, R., FEBS Letts., 419 (1997), 186.
17. Zimmermann, S., Schöttler, P., Engel, W. and Adham, I.M., Mol. Endocrinol., 12 (1998), 706.
18. Zimmermann, S., Steding, G., Emmen, J.M.A., Brinkmann, A.O., Nayernia, K., Holstein, A.F., Engel, W. and Adham, I.A., Mol. Endocrinol., 13 (1999), 681.
19. Nef, S. and Parada, L.F., Nature Genet., 22 (1999), 295.
20. Emmen, J.M., McLaskey, A., Adham, I.M., Engel, W., Verhoef-Post, M., Them, A.P., Grootegoed, J.A. and Brinkmann, A.O., Endocrinology, 141 (2000), 846.
21. Koskimies, P., Nokkala, E., Mäkelä, S., Huhtaniemi, I. and Poutanen, M., Ares-Serono Foundation Meeting on Molecular Genetics of Human Reproduction, Crete, Sep. 2000. Abstract.
22. Büllesbach, E. and Schwabe, C., J. Biol. Chem., 270 (1995), 16011.
23. Büllesbach, E.E. and Schwabe, C., J. Biol. Chem., 274 (1999), 22354.

Molecular genetics of *Insulin3*

Serge Nef, Sunita Verma-Kurvari, Tracey E Shipman and Luis F Parada

Center for Developmental Biology, University of Texas, Southwestern Medical Center 6000 Harry Hines Blvd., Dallas, Texas 75235-9133

Introduction

A critical aspect in mammalian development is the acquisition of a sexually dimorphic phenotype. The genetic sex, determined at fertilization by the inheritance of either a X or Y chromosome, directs the bipotential gonad to differentiate into either testes or ovaries. In males, expression of the Y-specific gene *Sry* in the genital ridge initiates a chain of events that induces differentiation of the uncommitted gonad into testes. Testes in turn express molecules that specify secondary traits of male sexual differentiation. At least three hormones have been shown to be important for male differentiation: a) Müllerian inhibiting substance (MIS), produced by Sertoli cells, specifies the regression of the Müllerian ducts which give rise to the uterus in the female; b) androgens, produced by Leydig cells, direct the differentiation of multiple male accessory organs including: the vas deferens, seminal vesicle, prostate, the second phase of testicular descent, and external male genitalia and finally c) Insulin 3 (*INSL3*), a hormone of the Insulin family involved in mediating the first phase of testicular descent, another specific aspect of male sexual differentiation. In this chapter, experiments addressing the role of *INSL3* in male sexual differentiation and in particular its function in testicular descent are described and discussed.

Cryptorchidism from "kryptos" (hidden) and "orchis" (testis), is a common congenital defect in humans affecting approximately 2-3% of newborn males. Complications resulting from cryptorchidism include infertility and an increased risk for development of germ-cell tumors [1]. Testicular descent occurs in two anatomically distinct, hormonally regulated phases, trans-abdominal phase and inguino-scrotal phase [1]. Two mesenchymal structures, the cranial suspensory ligament (CSL) and gubernaculum or caudal genital ligament connect the gonads, the Wolffian and Müllerian ducts to the abdominal wall. Development of both gubernaculum and CSL are sexually dimorphic [2]. During the trans-abdominal phase (15.5 dpc to 17.5 dpc in mice), the testes relocate from a high abdominal position to the bottom of the abdomen. The gubernaculum mediates the trans-abdominal descent. In males, the androgen-dependent regression of the CSL and the outgrowth and contraction of the gubernaculum anchors the testis close the base of the abdomen. In females, on the other hand, the gubernaculum remains a thin, flexible chord elongating passively with embryonic growth, whereas the development of CSL allows the ovaries to retain their original position just below the kidney. The second phase of testicular descent which involves the movement of the testis from the base of the abdomen to the scrotum occurs

G.W. Tregear et al. (eds.), Relaxin 2000, 337–345.
© 2001 *Kluwer Academic Publishers. Printed in the Netherlands.*

in the second or third postnatal week in mice. In human, both phases occur during fetal life.

Various studies have established that different factors control the two phases of testicular descent. The trans-abdominal phase does not require androgens and occurs normally in humans or animals with androgen deficiency or resistance [3,4]. The inguino-scrotal phase, on the other hand, is androgen dependant since it is absent in humans or rodents with androgen resistance [4] or androgen deficiency elicited by antiandrogens [5]. The factor promoting trans-abdominal testicular descent has been sought after for decades and has received names such as "factor X" [6], "third hormone" [7] or even "descendin"[8]. This factor must fulfill the following two criteria: it must be expressed by the testes and have the capacity to promote gubernaculum outgrowth and contraction.

Insulin-3, also known as Leydig Insulin-like hormone (Ley I-L) or relaxin-like factor (RLF), is part of the Insulin peptide hormone family that includes insulin, relaxin, and Insulin-like growth factors I and II, Insulin-4 [9], Insulin-5 [10,11] and insulin-6 [11-13]. The mature INSL3 hormone, like insulin, is composed of A and B-chains linked by disulfide bonds [14]. *INSL3* expression is sexually dimorphic and transcripts are seen primarily in the reproductive tissues [15]. The gubernacular and testicular phenotypes of *INSL3* deficient mice and that of estrogen treated male embryos lend support to the hypothesis that *INSL3*, a hormone secreted by testis mediates the first phase of testicular descent via outgrowth and contraction of the gubernaculum [16-18].

Results and Discussion

To understand the biological role of *INSL3*, we disrupted *INSL3* gene by homologous recombination in embryonic stem cells. We designed the recombination cassette such that part of exon 1 and all of exon 2 of the *INSL3* gene were deleted and replaced by the *neomycin* resistance gene. This approach insured the generation of a null allele [16]. Mice homozygous for the *INSL3* mutation were viable. In contrast to females, mutant males were infertile. Anatomical inspection revealed poorly developed scrotum and freely mobile testes in the abdominal cavity (data not shown). The intra-abdominal testes had grossly normal morphology at birth, but in adults, were considerably smaller in size than the normal or heterozygous counterparts. The mutation of *INSL3* by Zimmermann and colleagues resulted in an identical cryptorchid phenotype [18]. These independently generated results confirm that the phenotypes observed are the consequence of *INSL3* mutation and not of potential additional spurious mutations.

In male mice, *INSL3* transcripts are expressed at high levels in embryonic and adult Leydig cells. Increase of *INSL3* expression in adult Leydig cells is concomitant with sexual maturation and the initiation of spermatogenesis. In the absence of *INSL3*, the number and the morpholgy of Leydig cells is similar to wild type testes, although their density was higher in mutant testes (data not shown). Histological comparison of wild type and *INSL3-/-* testes demonstrated lesions of the seminiferous tubules including multinucleated giant cells (MGCs) and the absence of spermatids and mature sperm in the mutant testes or vas deferens. The number of primary spermatocytes was greatly reduced and a large proportion had picnotic nuclei. Sertoli cells, spermatogonia, and Leydig cells appeared similar in wild type and mutant testes [16,18].

Androgen deficiency can also lead to cryptorchidism. However, in addition to impaired testicular descent, androgen deficiency is also tightly associated with impaired virilization and differentiation of male associated genitourinal structures and passive behavioral characteristics. In contrast to the models of cryptorchidism caused by testosterone deficiency (*GnRH* deficient mice [19]) or testosterone sensitivity (Tfm mice [20]), *INSL3-/-* males exhibit virile, androgen-dependent behaviors including normal mounting and copulatory behavior (vaginal plugs, normal size of the sexual accessory organs, normal Testosterone levels [16]). Taken together, these results demonstrate that deficiency in intrabdominal descent in *INSL3-/-* males cannot be attributed to androgen-deficiency.

To examine the anatomical basis for failure of testicular descent in *INSL3* mutant mice, we investigated the developing genito-urinary system in newborn mice using scanning electron microscopy (SEM). Wild-type testes have normally separated from the original mesonephric location and are located caudal to the opening of the bladder neck in the inguinal region. In contrast, mutant neonate testes remain located adjacent to the lower pole of the kidneys. Wild-type male gubernaculae have a large bulb and a thick cord. In contrast, *INSL3* mutant male gubernaculae have a flat thin bulb with a thin elongated cord, remarkably similar to that found in wild-type females. Except for the gubernaculum, the remainder of the urogenital tract appears normal in mutant males. Similar results were obtained with embryos of earlier ages (E16.5) [16]. These data support the interpretation that *INSL3* controls normal male gubernaculum development. Since *INSL3* transcripts are present in developing testes but not in gubernacular bulb or in other neighboring tissues, the observed gubernacular phenotype is consistent with the model that embryonic testes secrete *INSL3* and that the gubernaculum is the target organ.

Estrogens and cryptorchidism

Elevated estrogen levels have been implicated as a possible contributing cause for sporadic cryptorchidism in humans. Several observations have been posited to support this theory. First, treatment of pregnant women with diethyl-stilbestrol (DES), a nonsteroidal estrogenic substance, is associated with undescended testes in male offspring [21]. One clinical study showed that mothers of cryptorchid children had higher levels of free estradiol during the first trimester versus mothers whose offspring had normally descended testes [22]. Finally, Hadziselimovic and colleagues [23] found an increased expression of estradiol in placenta of boys with cryptorchid testes compared to placenta of boys and girls with normal genitalia. Animal studies support the human correlation. In mice, *in utero* exposure to estradiol induces cryptorchidism. Indeed, the main experimental model for intra-abdominal cryptorchidism involves exposure of pregnant rodents to exogenous estrogens [24,25]. The effects of estradiol include, a reduction of gubernacular outgrowth, the induction of estrogen receptors within the Wolffian ducts, and the stabilization of the Müllerian ducts.

Since both the absence of *INSL3* and *in utero* exposure to exogenous estrogen induces bilateral cryptorchidism in newborns, a possible relationship may exist between these hormones. To determine the relationship between estrogen and *INSL3* function, we examined *INSL3* transcription in testes upon estrogen exposure [17]. Pregnant mice received a subcutaneous injection of 6 mg of 17ß-estradiol on day 13 of gestation (E13.5) corresponding to the beginning of male sexual differentiation in the gonads. The embryos from the subjected pregnant females were analyzed at either E17.5, during trans-abdominal testicular descent, or at P0, when the first phase of testicular descent is complete and the testes are located at the base of the abdominal cavity, just above the inguinal canal. SEM analysis of untreated E17.5 male embryos showed a typical, well developed gubernacular bulb and a contracted gubernacular cord. In contrast, the gubernaculum of 17ß-estradiol-treated embryos displayed a small bulb and an elongated cord. The testes of untreated neonate male mice, localized beneath the bladder neck and proximal to the inguinal region. The once swollen gubernacular bulbs have regressed and appear embedded in the peritoneal hollow. In contrast, neonates previously exposed to 17ß-estradiol show the testes located above the bladder neck or, in some cases, just below the kidney in a position similar to the ovaries. Except for the stabilization of the Müllerian structures, the urogenital anatomy of mice treated with 17ß-estradiol share striking features with *INSL3-/-* mice [16,18]. Both displayed intra-abdominal testes and feminized gubernaculae. Similar results were obtained with smaller doses of estradiol (500 μg estradiol or as little as 15 μg/g of body tissue). Although lower doses did not completely abolish *INSL3* transcription, they were sufficient to retard gubernaculum development and induce bilateral intra-abdominal cryptorchidism (data not shown,[17]).

The similar gubernacular and intra-abdominal testicular phenotype of *INSL3-/-* and 17ß-estradiol-exposed male mice prompted us to examine the possibility that fetal estrogen exposure may interfere with *INSL3* action. We therefore performed *in situ* mRNA hybridization on testis from normal and 17ß-estradiol-treated neonate (P0) mice using an *INSL3*-specific probe (Fig.1). In untreated testes, *INSL3* transcripts are abundant in Leydig cells. In contrast, we were unable to detect *INSL3* transcripts in the testes of males exposed to 17ß-estradiol *in utero* (compare Fig. 1C and D). Histological examination of the estradiol treated testes revealed normal Leydig cell numbers and morphology (Fig. 1A and B). Thus, similar to the *INSL3-/-* condition, absence of *INSL3* transcripts from embryonic Leydig cells is concomitant with impaired testicular descent. To assess whether the absence of *INSL3* transcripts was specific for the gene or whether the reduction was due to general transcription failure in Leydig cells, we performed *in situ* hybridization for the *cholesterol side-chain cleavage cytochrome P450* (*P450 scc*) mRNA, a marker specific for Leydig cells. Expression of *P450scc* was not altered in Leydig cells from treated testes (Figure 1E & F). Similarly, the expression of *Desert hedgehog* (*Dhh*), a specific marker for Sertoli cells, was normal in estradiol-exposed testes (data not shown). These data indicate that estrogen-exposed testes differentiate normally and that the general transcription in Leydig and Sertoli cells remains normal, while *INSL3* transcription is specifically reduced. We have also analyzed *INSL3* transcription during various stages of testis

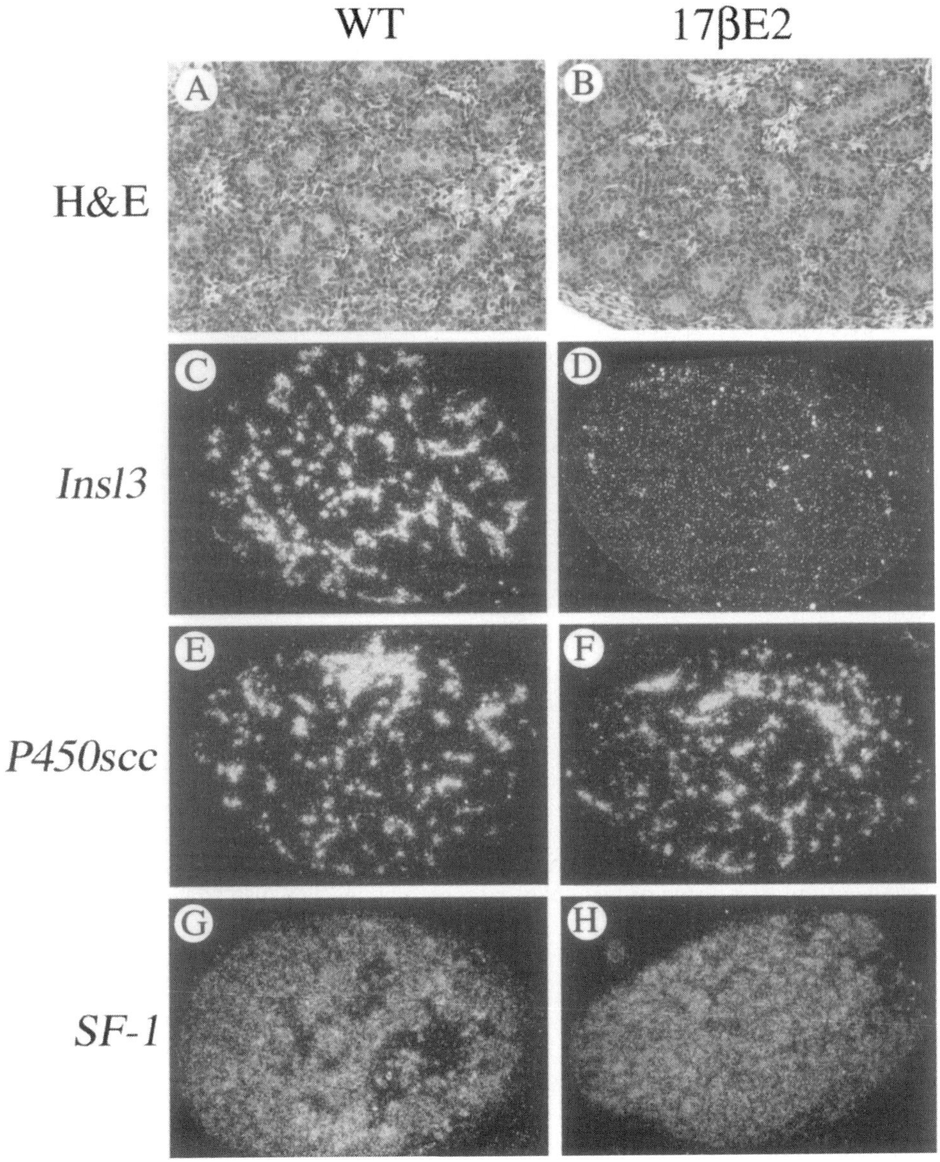

Fig.1. 17β estradiol specifically blocks INSL3 gene expression in the fetal population of Leydig cells.

development. *INSL3* transcripts disappear as early as E17.5 upon estradiol treatment but reappear at normal levels by P7 [17]. The reemergence of *INSL3* transcripts between P0 and P7 coincides with the appearance of the adult-type Leydig cell population. Thus, the data indicate that *INSL3* expression is permanently and specifically blocked in fetal-type Leydig cells by exposure to a single bolus of exogenous estradiol. Interestingly, the adult population of Leydig cells is not sensitive to estradiol treatment as injections of either one dose of 6mg estradiol or gradual administration over 6 days in the adult male has no effect on *INSL3* transcription (data not shown,[17]).

The *SF-1* gene has been reported to control the transcription of *INSL3* in a Leydig cell line (MA-10), derived from adult Leydig cells [26]. We observed normal levels of *SF-1* transcripts in estradiol exposed embryonic testes in spite of the fact that *INSL3* expression was abolished (Fig. 1G and H). These data indicate that *INSL3* expression is blocked specifically although *SF1* gene expression is not affected. Thus, exposure of male embryos to unusually high levels of estradiol resulted in specific blockade of Leydig cell *INSL3* gene expression. These seemingly contradictory data may be reconciled by the differing nature of the cells examined (embryonic versus adult cell lines).

It has been hypothesized that estrogen related molecules, including xenoestrogens, may be responsible for the increasing incidence of sporadic cryptorchidism [27,28]. To obtain experimental evidence in support of this idea, we have determined whether the estrogen stereoisomer 17α-estradiol, and the related estriol, two molecules known to have reduced estrogenic activity, affect *INSL3* gene expression and the development of the gubernaculum and testicular descent. In addition, the nonsteroidal synthetic estrogen diethylstilbestrol (DES) was tested in this assay. Preliminary results indicate that *INSL3* transcripts are down-regulated or significantly reduced in 17ß-estradiol-, estriol-, 17α-estradiol- and DES-exposed embryonic testes at P0. Thus, xenoestrogens (DES) or compounds with reduced estrogenic activity (estriol, 17α-estradiol) can promote cryptorchidism in mice through down regulation of *INSL3* gene transcription.

In summary, the above data show that *in utero* exposure to 17α-estradiol, 17ß-estradiol, estriol or DES, negatively regulate *INSL3* gene expression. The absence of *INSL3* which mediate outgrowth and contraction of the gubernaculum, result in blockade of the fisrt phase of testicular descent, a phenotype also found in the *INSL3* mutant mice.

Insulin 3 receptor identification

The first phase of testicular descent takes place between E15.5 and E17.5 in mice. Since *INSL3* is required for the transabdominal descent and for proper growth and differentiation of gubernaculum, our working hypothesis is that *INSL3* receptor must be expressed in the gubernaculum between E15.5 and E17.5. Many members of Insulin gene family (Insulin, IGF1 and IGF2) bind to a receptor tyrosine kinase (RTK) subfamily: the Insulin receptor (IR) family. The three known members of IR subfamily are Insulin receptor (IR), Insulin-like growth factor receptor 1 (IGFR1) and Insulin receptor-related receptor (IRR) share a high degree of sequence similarity both within and outside of the TK domain [29].

We have designed two PCR based strategies to search for a putative *INSL3* receptor belonging to either the IR family or the broader RTK family. First, to identify members of the IR subfamily expressed in the developing gubernaculum, we used degenerated oligonucleotides from conserved amino acid sequences both in and outside of the tyrosine kinase domain of the IR subfamily for RT-PCR. The template used was 1st strand cDNAs generated from total RNA extracted from E17.5 male mouse gubernaculum.(Forward primers 1) 5'GGGAATTCTA(T/C)GTICCIGA(C/T)GA(A/G)TGGGA3', aa: VPDEWEV or 2) 5'GGGAATTCTT(C/T)GGIATGGTITA(C/T)GA(A/G)GG3', aa FGMVYEG. Reverse primers: 1) 5' CGGGGATCC(G/T)(G/A)TA(G/A)TA(G/A)TCIGT(T/C)TC(G/A)TA 3', aa YETDYYRK 2) 5' CGGGGATCCTCCCAIA(G/A)IACICC(/G)AA 3', aa FGVVLWEI).

In the event that the putative *INSL3* receptor is not closely related to IR subfamily, but still belongs to the RTK family, the strategy using primers specifically against IR members will likely not succeed. We have therefore adopted a broader screen aimed at including additional RTKs present in gubernacular tissue. The primers employed for this experiment have been used successfully by several groups to isolate tyrosine kinases [30,31]. Similar to the above experiments, the cDNAs used for the RT-PCR were generated from total RNA from gubernaculum at E15.5- E17.5.

Using the strategies listed above we have amplified transcripts for the *IGFR-1* and *IR* genes but not for *IRR* or any novel or uncharacterized RTK genes belonging to the IR subfamily. We have identified 16 receptor tyrosine kinase and 12 cytoplasmic tyrosine kinases expressed in gubernaculum between E15.5- E17.5 (Table 1). Presently we do not know whether any of the RTKs identified in this study constitute the *INSL3* receptor. In summary, these results indicate that IR and IGFR genes are the only known members of the IR family to be expressed in embryonic gubernaculum at a stage when outgrowth and contraction occurs. It remains formally possible that a new, divergent members of the IR family, or a novel RTK may be the *INSL3*-receptor.

Table 1. Tyrosine kinases expressed in male mouse embryonic gubernaculum.

Cytoplasmic Tyrosine Kinase	Receptor Tyrosine Kinase
Jak1, Fps/Fes related, Fes, Flk1, Fak, c-Src, Arg, Tyk-2, Csk, Syk, Met, c-Abl	PDGFR α, PDGFR β, ErbB4, EGFR, FGFR1, FGFR1-IIIb, FGFR4, EphA1, EphA2, EphA3, EphA4, EphB3, EphB4, Flt3, IR, IGFR-1

The results discussed in this article demonstrate that INSL3 plays an important role in sexual differentiation by mediating the first phase of testicular descent. Intrauterine exposure to estrogens, a known high risk factor responsible of developing multiple reproductive defects such as cryptorchidism, causes specific inhibition of *INSL3* gene expression. However, despite these advances in the understanding *INSL3* genetics, very little is known about the function of INSL3 in the adult and more importantly the underlying signaling pathway. Understanding of INSL3 signaling mechanisms may provide us new tools for treatment of cryptorchidism in humans and is an area that needs to be investigated.

Acknowledgements:

We are grateful to Dr. Ross Bathgate for presenting our poster at the Relaxin 2000 meeting and for the assistance dealing with this manuscript. This work was supported by an Excellence in Education Grant to LFP.

References:

1. Hutson, J.M., Baker, M., Terada, M., Zhou, B. and Paxton, G., Reprod. Fertil. Dev., 6 (1994) 151.
2. Hutson, J.M., Hasthorpe, S. and Heyns, C.F., Endocr. Rev., 18 (1997) 259.
3. Griffin, J.E. and Wilson, J.D., N. Engl. J. Med., 302 (1980) 198.
4. Hutson, J.M., J. Pediatr. Surg., 21 (1986) 195.
5. Husmann, D.A. and McPhaul, M.J., Endocrinology, 129 (1991) 1409.
6. Habenicht, U.F. and Neumann, F., Adv. Anat. Embryol. Cell Biol., 81 (1983) 1.
7. van der Schoot, P., Vigier, B., Prepin, J., Perchellet, J.P. and Gittenberger-de, G.A., Anat. Rec., 241 (1995) 211.
8. Fentener van Vlissingen, J.M., van Zoelen, E.J., Ursem, P.J. and Wensing, C.J., Endocrinology, 123 (1988) 2868.
9. Chassin, D., Laurent, A., Janneau, J.L., Berger, R. and Bellet, D., Genomics, 29 (1995) 465.
10. Conklin, D., Lofton-Day, C.E., Haldeman, B.A., Ching, A., Whitmore, T.E., Lok, S. and Jaspers, S., Genomics, 60 (1999) 50.
11. Hsu, S.Y., Mol. Endocrinol., 13 (1999) 2163.
12. Kasik, J., Muglia, L., Stephan, D.A. and Menon, R.K., Endocrinology, 141 (2000) 458.
13. Lok, S., Johnston, D.S., Conklin, D., Lofton-Day, C.E., Adams, R.L., Jelmberg, A.C., Whitmore, T.E., Schrader, S., Griswold, M.D. and Jaspers, S.R., Biol. Reprod., 62 (2000) 1593.
14. Adham, I.M., Burkhardt, E., Benahmed, M. and Engel, W., J. Biol. Chem., 268 (1993) 26668.
15. Zimmermann, S., Schottler, P., Engel, W. and Adham, I.M., Mol. Reprod. Dev., 47 (1997) 30.
16. Nef, S. and Parada, L.F., Nat. Genet., 22 (1999) 295.
17. Nef, S., Shipman, T. and Parada, L.F., Dev. Biol., 224 (2000) 354.
18. Zimmermann, S., Steding, G., Emmen, J.M., Brinkmann, A.O., Nayernia, K., Holstein, A.F., Engel, W. and Adham, I.M., Mol. Endocrinol., 13 (1999) 681.
19. Radovick, S., Wray, S., Lee, E., Nicols, D.K., Nakayama, Y., Weintraub, B.D., Westphal, H., Cutler, G.B., Jr. and Wondisford, F.E., Proc. Natl. Acad. Sci. USA, 88 (1991) 3402.
20. Lyon, M.F. and Hawkes, S.G., Nature, 227 (1970) 1217.
21. Gill, W.B., Schumacher, G.F., Bibbo, M., Straus, F.d. and Schoenberg, H.W., J. Urol., 122 (1979) 36.
22. Bernstein, L., Pike, M.C., Depue, R.H., Ross, R.K., Moore, J.W. and Henderson, B.E., Br. J. Cancer, 58 (1988) 379.
23. Hadziselimovic, F., Geneto, R. and Emmons, L.R., J. Urol., 164 (2000) 1694.
24. Grocock, C.A., Charlton, H.M. and Pike, M.C., J. Reprod. Fertil., 83 (1988) 295.
25. Perez, M.C., Garcia, I.M., Ferreras, E.M., Bravo, M.A., Espinosa, A.J. and Escudero, D.A., J Comp Pathol 114 (1996) 407-18.

26.Zimmermann, S., Schwarzler, A., Buth, S., Engel, W. and Adham, I.M., Mol. Endocrinol., 12 (1998) 706.
27. Abney, T.O., Steroids, 64 (1999) 610.
28. Jensen, T.K., Toppari, J., Keiding, N. and Skakkebaek, N.E., Clin. Chem., 41 (1995) 1896.
29. Shier P. and Watt V.M., J. Biol. Chem., 264 (1989) 14605.
30. Lai, C. and Lemke, G., Neuron, 6 (1991) 691.
31. Sanchez, M.P., Tapley, P., Saini, S.S., He, B., Pulido, D. and Barbacid, M., Proc. Natl. Acad. Sci. USA, 91 (1994) 1819.

Testicular descent in rats: a role for relaxin?

Laura J Parry[1], David J Hornsby[2], Roman S Poterski[2], Brian C Wilson[3] and Alastair JS Summerlee[2]

[1]*Howard Florey Institute, University of Melbourne, Parkville, Victoria 3052, Australia;* [2]*Department of Biomedical Sciences, Ontario Veterinary College, University of Guelph, Guelph, Ontario N1G 2W1, Canada; and* [3]*Department of Biology, Sacred Heart University, 5151 Park Avenue, Fairfield, CT 06432-1000, USA.*

Introduction

Cryptorchidism is a relatively common congenital anomaly in humans, pigs and horses [1]. The first or transabdominal phase of testicular descent involves enlargement of the gubernaculum testis, including proliferation of the mesenchymal cells in the intraabdominal segment and an increase in extracellular matrix glycosaminoglycans and hyaluronic acid. These hydrophilic matrix proteoglycans induce water uptake by the gubernaculum which becomes gelatinous and increases in volume [2]. A failure of the gubernaculum to swell prevents the testis from descending through the inguinal canal.

In 1988, tissue extracts of fetal porcine testes, but not ovaries, were shown to induce DNA synthesis in gubernaculum cells in culture [2]. This peptide (approximately 6 kD) was named *descendin* [3] but it was not characterized. Recent evidence from the mouse suggests that *descendin* could be relaxin-like factor (RLF) [4], however, we present evidence in the rat that the factor involved in the first phase of testicular descent in rats may be relaxin.

Results and Discussion

Four experiments were carried out: (1) the expression of relaxin and/or RLF transcripts in fetal testis; (2) *in situ* hybridization for fetal testicular relaxin and RLF; (3) effect of neutralizing relaxin on gubernacular volume and testicular descent; and (4) presence of relaxin-binding sites in the gubernaculum.

Relaxin, but not RLF transcripts, were detected in fetal testes between days 14.5 and 20.5 *pc* and in the day 7-neonate (Fig. 1). The intensity of the positive signal for relaxin transcripts was highest in the fetal testis between days 14.5 and 18.5 *pc* and less in older fetuses and day 10 neonates. *In situ* studies revealed that relaxin, but not RLF, was present in testicular tissue between days 14.5 and 19.5 *pc*. In contrast RLF was present in neonatal tissue on days 2-10 *post partum* (data not shown).

Pregnant rats were immunized with a monoclonal rat relaxin antibody or treated with highly purified rat relaxin (08:00-09:00h and 20:00-21:00h) from day 15 to day 21 of pregnancy (n=16 rats per group). Treatment of pregnant dams with exogenous relaxin did not significantly ($p>0.05$: ANOVA) affect gubernacular volume compared with saline-treated controls. In contrast, relaxin-neutralization significantly ($p<0.05$: ANOVA)

G.W. Tregear et al. (eds.), Relaxin 2000, 347–348.
© 2001 *Kluwer Academic Publishers. Printed in the Netherlands.*

decreased gubernacular growth in the fetuses (Fig. 2) and testicular descent in these animals was delayed compared with controls (data not shown).

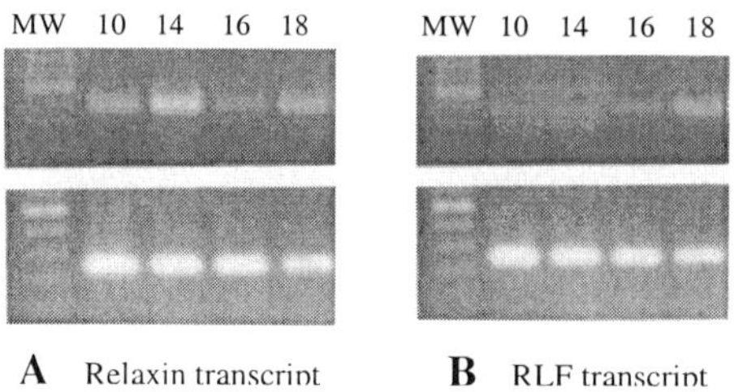

Fig. 1. RT-PCR expression of relaxin (A) and RLF (B) transcripts (upper panels) in fetal rat testicular tissue. Lower panel GAPDH controls.

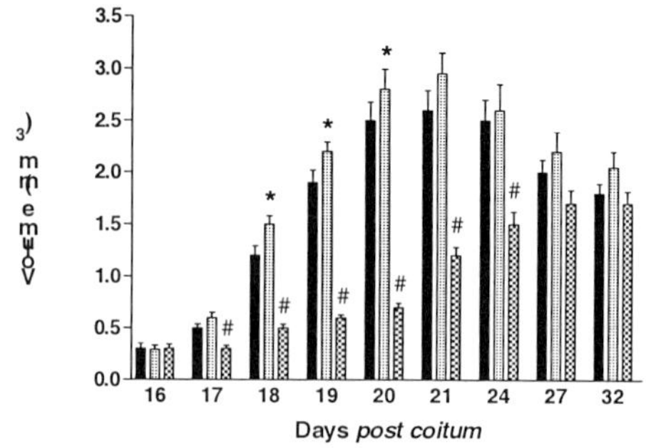

Fig. 2. Gubernacular volume in rat fetuses: intact controls; excess relaxin; and relaxin-neutralized. Note significant ($p<0.05$) increases (*) or decreases (#) in volume in fetal life.

A biologically active biotinylated-relaxin probe was used to identify specific relaxin binding sites in the caudal portion of the gubernacula. Binding was only observed between days 15 and 17 in fetal male gubernacular tissue (Fig. 3) but not in the female.

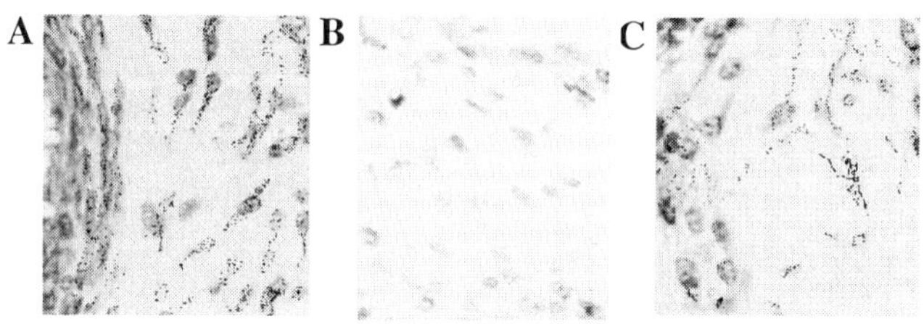

Fig. 3. Binding of biotinylated relaxin to gubernacular tissue (day15 fetus) (A). Incubation with excess unlabelled relaxin (B) displaced the binding but was unaffected by excess insulin (C).

Acknowledgements

This work was supported by NSERC (Canada).

References

1. Hutson, J.M., Hasthorpe, S. and Heyns, C.F., Endocr. Rev., 18 (1997) 259.
2. Heyns, C.F., Human, H.J., Werely, C.J. and De Klerk, D.P., J. Urol., 143 (1990) 612.
3. Fentener can Vlissingen, J.M., van Zolen, E.J.J., Ursem, P.J.F. and Wensing, C.J.G., Endocrinology, 123 (1988) 2868.
4. Nef, S. and Parada, L.F., Nat. Genet., 22 (1999) 295.

The Relaxin-Like Factor (Insulin 3) is highly expressed in the ruminant ovary: A putative ruminant relaxin?

Ross AD Bathgate[1], Nicole Moniac[2], Bettina Bartlick[2], Antonia Claasz[1], Nicola Dawson[1], Yean-Yeow Tan[1], John D Wade[1], Geoffrey W Tregear[1] & Richard Ivell[2]

[1]*Howard Florey Institute, University of Melbourne, Victoria 3010, Australia and* [2]*Institute for Hormone and Fertility Research, University of Hamburg, Hamburg, Germany.*

Introduction

The relaxin-like factor (RLF, Insl3) was initially characterized as the leydig insulin like peptide (Ley-I-L)[1] as it was highly expressed in porcine Leydig cells. More recently however, the RLF gene has also been shown to be expressed in the human ovary and trophoblasts as well as the mouse ovary [Ivell *et al.*, this volume]. This data, together with the similarity in structure of the derived protein sequence of RLF to relaxin, including a conserved, although shifted, relaxin-like receptor binding site (Fig. 1), led to the name relaxin-like factor [2]. The function of this new member of the relaxin/insulin family of peptides is discussed in detail in Ivell *et al.* (this volume).

Among mammals, domestic ruminants are unusual in that they do not appear to express a relaxin peptide; however they possess a relaxin physiology [3]. In cows and sheep typical softening and dilatation of the cervix can be achieved with injections of a porcine relaxin preparation [4]. Porcine relaxin has been reported to facilitate birth in cows [5], however other studies have not been able to confirm this result [6]. It is clear that the efficacy of these experiments is undermined by the use of heterologous relaxin peptides. There are numerous studies indicating the presence of relaxin assayable materials in the blood and tissues of ruminants, however the levels are generally 1% of those that are present in species like the pig. Relaxin-like peptides of around 6000 Daltons, with low but significant activities in relaxin bioassays, have been partially purified from cow [7] and sheep [8] reproductive tissues. Various studies using ^{32}P labelled heterologous full relaxin cDNA probes have been able to show a 1kb hybridizing signal in low stringency Northern blots in the corpus luteum (CL) of cows and goats [3], however the identity of these hybridizing bands is not known. Recently a single copy sheep genomic relaxin clone has been isolated which cannot produce a functional mRNA [9]. Therefore it would seem that ruminants do not produce a functional relaxin peptide but retain a relaxin physiology. Recently, we [10] and others [11] have shown that the CL of the ruminant does in fact express large amounts of a relaxin-like gene, namely RLF. The size of the RLF mRNA (1kb) and of the putative processed form of the peptide (6000 Daltons) correlate well with the putative ruminant relaxin mRNA and protein that were isolated previously [3, 8, 7]. Furthermore, the presence of a relaxin-like receptor binding motif (RXXXR) in RLF, especially in a shifted position, could account for the low relaxin activities and the low antibody cross

G.W. Tregear et al. (eds.), Relaxin 2000, 349–356.
© 2001 *Kluwer Academic Publishers. Printed in the Netherlands.*

reactivities of these isolated ruminant relaxins. The studies described in this paper were initiated to investigate the possibility that RLF may be substituting for relaxin in ruminants, and to further our understanding of the role of RLF in female biology.

Cloning of bovine RLF and expression in the ovary

The bovine RLF cDNA was cloned from a testis cDNA library [10]. The derived amino acid sequence conforms to the structure of the insulin-relaxin family with a B domain after the signal peptide, followed by a C- and A-domain. Using primers based on the cDNA sequence and genomic DNA we have shown that the RLF gene has an identical structure to other RLF genes as well as the relaxin gene, with a single intron interrupting the coding region. At 814bp the intron is slightly larger than that found in other species [Ivell *et al.*, this volume]. Furthermore, the full bovine RLF gene including the promoter has been isolated from a bovine genomic library [Sührbier, W., Bathgate, R., Stredonsky, K., Walther, N. and Ivell, R., unpublished]. Northern analysis with a bovine RLF probe indicates that, as expected, there was high expression of the RLF mRNA in testicular Leydig cells, however there was also high expression in the bovine CL as well as the follicular thecal cells. A partial sheep RLF cDNA sequence has also been cloned [11] and the expression pattern of the sheep RLF mRNA is very similar to the bovine. Importantly, the expression of RLF mRNA in both the ovine and bovine CL is many orders of magnitude higher than that seen in species which express a relaxin. RT-PCR analysis of other bovine tissues showed that the RLF gene is also expressed in the hypothalamus, uterus and fetal membranes, known sites of relaxin gene expression. The pattern of RLF gene expression is therefore very similar to that of relaxin in other species [13].

We have studied the temporal expression of RLF mRNA in the bovine CL [12]. During the oestrous cycle maximum RLF expression was seen in the mid to late cycle, whereas only low levels were seen in the early cycle. During pregnancy, RLF mRNA levels were maintained at mid cycle levels from day 20-80, followed by increased expression in mid pregnancy which was maintained until labour, at which time it was completely lost. The pattern of RLF gene expression in the bovine CL therefore shows many similarities to relaxin expression in the pig [13].

RLF mRNA expression and regulation in thecal cell cultures

The highest expression of RLF mRNA is in thecal cells of antral follicles. Freshly prepared bovine thecal cells show extremely high RLF mRNA expression. Furthermore, the expression is higher in cells prepared from small antral follicles compared to large or atretic follicles [12]. Combining the pattern of RLF expression with the observation that female *INSL3* knockout mice have extended oestrous cycles, suggests that RLF may be playing an important role in follicular maturation.

When thecal cells are cultured *in vitro* RLF mRNA expression is initially lost under all treatment regimes. However after 6 days in culture under serum-free conditions in the presence of IGF-1 or insulin, RLF mRNA expression is regained and increases to high levels up to 15 days in culture but, interestingly, as a transcript with a longer poly-A tail [12]. In contrast, luteinization of thecal cells resulted in low RLF mRNA expression. Interestingly, cultured granulosa cells have low RLF mRNA expression, which is also lost during culture, and is not recovered under luteinizing conditions [Bathgate, R., Bartlick, B. and Ivell, R., unpublished]. Our studies indicate that RLF mRNA expression may be an interesting and important marker for distinguishing bovine theca cell differentiation.

The RLF Protein

As has been discussed in Ivell *et al.* [this volume], the primary structure of the RLF protein *in vivo* is yet to be determined. However the predicted amino acid sequence shares the common structural elements of the relaxin preprohormone. Based on the structure of relaxin and predicted protease cleavage sites a synthetic sheep RLF peptide has been produced, the structure of which is represented in Fig. 1.

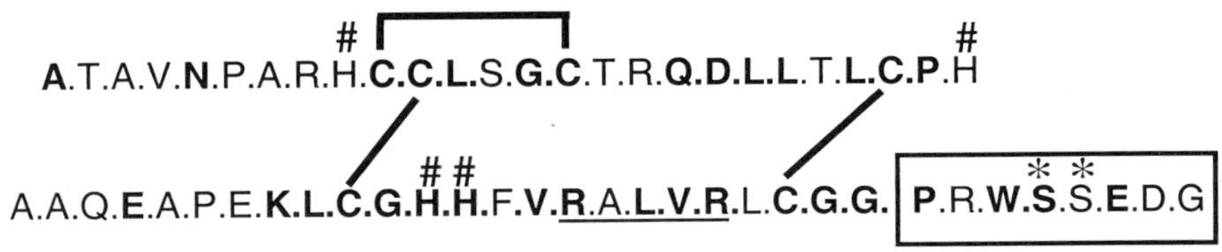

Fig. 1. Primary structure of synthetic sheep RLF (A24 top/B32 bottom). Residues conserved between RLF molecules from different species are shown in bold. The residues implicated in binding of RLF to its receptor are boxed and the displaced relaxin-like receptor binding motif is underlined. Serine residues labelled with ^{33}P (*) and histidine residues labelled with ^{125}I (#) are shown (see text for details).

This peptide at concentrations up to one micromolar has no classic relaxin activity in a rat atrial bioassay [14]. Furthermore, as described in Claasz *et al.* [this volume], generation of a sheep RLF peptide with a full relaxin binding site in the correct position only results in minimal relaxin activity. Circular dichroism (CD) spectroscopy measurements indicate a secondary structure which is less helical than relaxin [14]. Therefore, it is likely that the RLF and relaxin molecules are sufficiently different in structure, thus influencing the conformation of the binding motif, which in turn contributes to the reduced activity of the RLF analogue. The overall level of homology between RLF molecules is very high, especially in ruminants (90%). One of the highly conserved regions is at the carboxy-terminal end of the B chain and recent studies have shown that these residues are essential for RLF receptor binding [15] (Fig. 1). Hence relaxin and RLF contain very different receptor binding domains and this correlates well with the recent discovery of distinct relaxin and RLF binding sites in the mouse [2]. As RLF subserves clearly defined physiologies in mammals, quite distinct from relaxin, it would be very unlikely that it would cross-react with relaxin receptors. If RLF is assuming the role of relaxin in ruminants it is more likely to be acting through RLF receptors.

We have developed antibodies against the synthetic ovine RLF B-chain and against recombinant bovine RLF. Interestingly the sheep B chain antibody recognizes synthetic sheep RLF but not recombinant bovine RLF in dot blots, whereas the bovine antibody recognizes the recombinant protein but not the synthetic sheep RLF peptide. However both of these antibodies detect RLF protein in the bovine and ovine testis and ovary using immunocytochemistry, suggesting that RLF may be present in these tissues in both prohormone and processed forms. These antibodies do not detect native RLF peptides in Western blots of crude tissue extracts although preliminary experiments have identified immunoreactive fractions in these extracts by dot blotting. So far we have been unable to isolate RLF peptides of any form from these extracts. It would seem that the RLF protein is very elusive, which parallels the inability of researchers to isolate immunoreactive relaxin peptides from ovine [8] and bovine tissues [7, Steinetz this volume]. Future work will concentrate on the isolation of native RLF peptides from these tissues as well as the further development of specific antibodies.

Characterization of sheep RLF receptors

To understand the physiology of RLF in ruminants it will be essential to identify and localize RLF receptors. Additionally, very little is know about RLF binding sites in any species and the RLF receptor is yet to be isolated. Preliminary work has involved labelling the synthesized sheep RLF peptide with both ^{33}P and ^{125}I (see Fig. 1 for residues labelled). Although the labelling was successful in both cases, we were unable to obtain specific binding in sheep tissues with ^{33}P-labelled sheep RLF. Labelling with ^{125}I through Histidine residues resulted in two main iodinated products, both of which showed low amounts of specific binding to sheep tissues. Fig 2 shows the results of the binding of one of these ligands to sheep tissues.

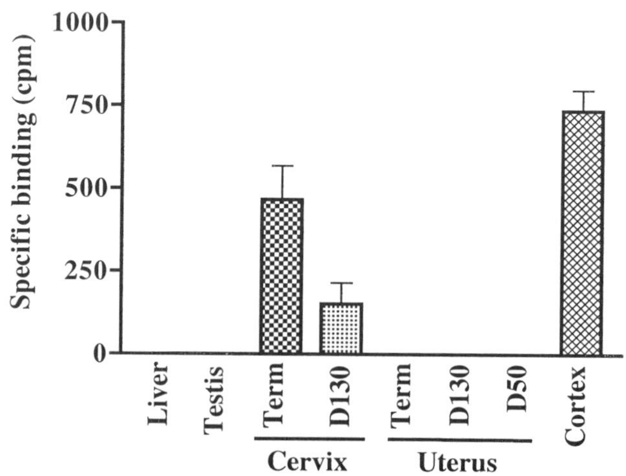

Fig. 2. Specific binding of ^{125}I-labelled sheep RLF to membrane preparations of sheep tissues. Specific binding was only observed in pregnant cervix and cerebral cortex.

Low amounts of specific binding (displaceable by a 1000 fold excess of unlabelled RLF) were found consistently in the cerebral cortex and pregnant cervix but not in the uterus or other tissues. The localizations of binding sites correlate with that of relaxin receptors in rat and human cerebral cortex [Tan et al, this volume] and a putative role for a relaxin-like peptide in cervical softening in the sheep [4]. The low amount of specific binding obtained with these ligands was explained recently by the publication of data, which showed that, unlike for relaxin, the carboxyl terminus of the RLF B-chain is essential for binding activity [15]. Both of these labelling procedures resulted in modifications of this binding region and hence dramatic decreases in binding efficiency. Based on information in the above publication and knowledge of the structure of relaxin, we are currently synthesising ligands that will be able to be ^{125}I-labelled while at the same time protecting this critical binding region. These ligands will also form the basis of strategies for the synthesis of biotin and fluorescent labelled peptides for the localization of RLF receptors.

Actions of RLF and relaxin on the sheep uterus and atria *in vitro*

In rodents and pigs relaxin has clearly been shown to inhibit spontaneous myometrial activity [13]. Experiments using crude extracts of porcine relaxin injected intravenously into ovarectomized sheep induced typical hormone specific quiescence in myometrial activity [16], whereas in another study no effect was seen when highly purified porcine relaxin was injected [17]. We have tested the ability of sheep RLF and rH2 relaxin to influence both spontaneous or KCl induced sustained myometrial contractions in isolated strips of late pregnant sheep uterus.

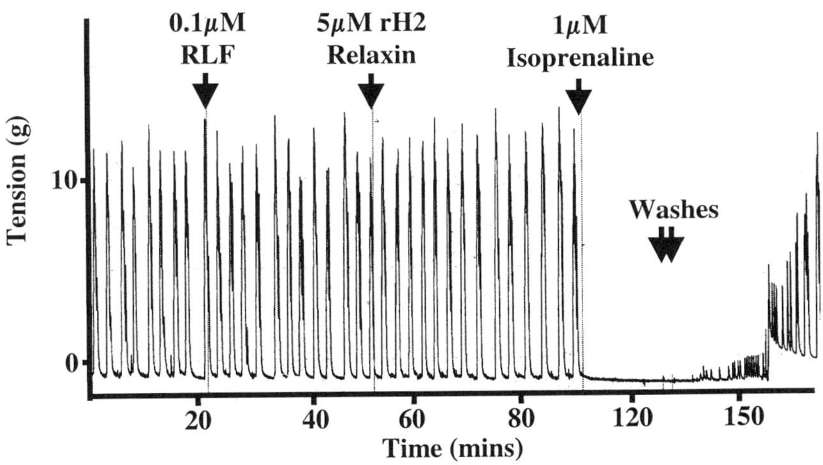

Fig. 3. Representative experiment showing that neither synthetic sheep RLF nor rH2 relaxin influence spontaneous contractions in strips of late pregnant sheep uterus in vitro.

As can be seen in Fig. 3 neither RLF nor rH2 relaxin was able to influence spontaneous myometrial activity, whereas isoprenaline completely abolished it. This was also the case for KCl induced sustained myometrial contractions (data not shown). The lack of response to RLF in these studies correlates with the absence of RLF binding sites in late pregnant sheep uterus outlined above. Furthermore it is possible that a relaxin like molecule is not involved in the regulation of myometrial contractility during pregnancy in the sheep as has been demonstrated in the human where relaxin has a minor role, if any, in inhibiting spontaneous myometrial activity [18].

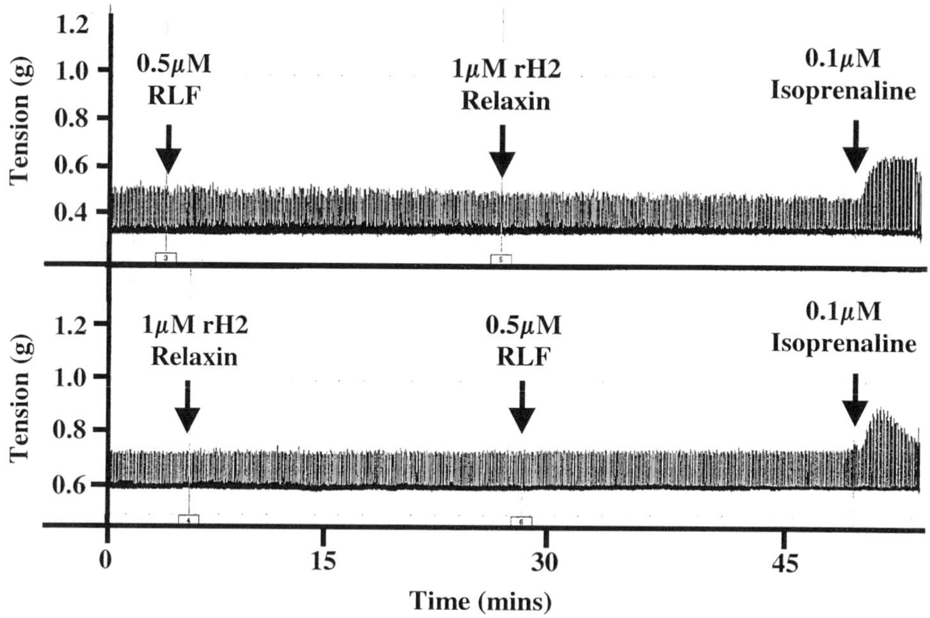

Fig. 4. Representative experiment showing that neither synthetic sheep RLF or rH2 relaxin will induce positive ionotropic responses in isolated sheep atria.

Relaxin will induce dose-dependent positive chronotrophic and inotrophic responses in isolated rat left and right atria and this is the basis for a well characterized bioassay [see Tan *et al.*, this volume]. We have tested the ability of both synthetic sheep RLF and rH2 relaxin to induce positive inotropic responses in isolated sheep atria (Fig. 4). Neither synthetic sheep RLF nor relaxin was able to induce positive ionotropic responses, whereas isoprenaline had a clear stimulatory action. ^{33}P-labelled H2 relaxin does not bind to the sheep or human atria [Tan, Y.Y., personal communication]. Furthermore, H2 relaxin at concentrations producing positive chronotropic and inotropic responses in rat atria had no effect in human atria [19]. Therefore, as for its actions in the uterus, the actions of relaxin on the atria may be species specific.

Conclusions

The studies described in this paper have been unable to provide conclusive evidence that RLF may be substituting for relaxin in ruminants. However, the fact remains that there is sufficient similarity in structure of RLF with relaxin, including the presence of a shifted relaxin binding motif, to explain the cross reactivity of relaxin antisera and the low relaxin-like activities of extracts of ruminant tissues. Furthermore, we have shown that RLF binding sites are localized in the cerebral cortex and pregnant cervix, sites which correlate with a relaxin biology. Regardless of whether RLF is determined to be the ruminant equivalent of relaxin, the high expression of RLF in the ovary of the cycle and pregnancy in the ruminant provide an ideal model to study the expression of this new peptide hormone. The elucidation of its native structure, together with the localization of binding sites, will be the first crucial steps in determining the physiology of this new hormone in the female reproductive tract.

Acknowledgments

Ross Bathgate was a recipient of an NHMRC Howard Florey Centenary Fellowship. Research was funded by grants from the Deutsche Forschungsgemeinschaft to Richard Ivell (Iv7/1-4, Iv7/9-1) and a block grant to the Howard Florey Institute (Reg Key 983001) from the NHMRC.

References

1. Adham, I.M., Burkhardt, E., Benahmed, M. and Engel, W., J. Biol. Chem., 268 (1993) 26668.
2. Büllesbach, E. and Schwabe C., J. Biol. Chem., 270, (1995) 16011.
3. Hartung, S., Kondo, S., Abend, N., Hunt, N., Rust, W., Balvers, M., Bryant-Greenwood, G. and Ivell R., In McLennan, A.H., Tregear, G. and Bryant-Greenwood, G.D. (eds.) Progress in Relaxin Research, World scientific Publishing, Singapore, 1995, p. 439
4. Musah, A.I., Schwabe, C. and Willham, R.L., Biol. Reprod., 34 (1986) 363.
5. Musah, A.I., Schwabe, C. and Willham, R.L., Endocrinology, 118 (1986) 1476.
6. Smith, D.E., Hixon, D.L., Moore, D.W., van Kirk, E.A., Alexander, B.M., Anthony, R.V. and Moss, G.E., Dom. Anim. Endo., 13 (1996) 469.
7. Fields, M.J., Fields, P.A., Castro-Hernandez, A. and Larkin, L.H., Endocrinology, 869 (1980) 869.
8. Wathes, D.C., Rees, J.M. and Porter, D.G., J. Reprod. Fertil., 84 (1988) 247.
9. Roche, P.J., Crawford, R.J. and Tregear, G.W., Mol. Cell. Endocrinol., 91 (1993) 21.
10. Bathgate, R., Moniac, N., Bartlick, B., Schumacher, M., Fields, M. and Ivell, R., Biol. Reprod., 61 (1999) 1090.
11. Roche, P.J., Butkus, E., Wintour, E.M. and Tregear, G., Mol. Cell. Endocrinol., 121 (1996) 171.
12. Bathgate, R.A.D., Balvers, M., Hunt, N. and Ivell, R., Biol. Reprod., 55 (1996) 1452.
13. Sherwood O.D., In Knobil, E. and Neill, J.D. (eds), The Physiology of Reproduction 2nd ed., Raven Press, New York, USA, 1994, p. 861.
14. Dawson, N.F., Tan, Y.Y., Macris, M., Otvos Jr, L., Summers, R.J., Tregear, G.W. and Wade, J.D., J. Peptide Res., 53 (1999), 542.

15. Büllesbach, E. and Schwabe C., Biochemistry, 38 (1999) 3073.
16. Porter, D.G., Lye, S.J., Bradshaw, J.M. and Kendall, J.Z., J. Reprod. Fertil., 61 (1981) 409.
17. Schramm, W., Einer-Jensen, N., Brown, M.B. and McCracken, J.A., Biol. Reprod., 30 (1984) 523.
18. Maclennan, A.H., Grant, P. and Bryant-Greenwood, G., J. Reprod. Med., 40 (1995) 703.
19. Castro, J.M., Eddie, L.W. & Summers, R.J., Clin. Exp. Pharmacol. Physiol. Suppl., 21 (1992) 12.

Characterization of the rat Relaxin Like Factor/JAK3 gene region

Peter J Roche, Zeyad Nasa and Ganeema Tokhi
Department of Medical Laboratory Science, RMIT University, Melbourne, Vic 3001, Australia

Introduction

Relaxin-like factor (RLF) is a recently characterized member of the insulin / IGF / relaxin hormone family [1-3]. RLF, which is also known as Leydig insulin-like peptide, is synthesized primarily in the leydig cells of the testis. RLF is also expressed in the ovary at low to high levels, depending on the particular mammalian species [1, 3]. Male mice that are deficient in RLF exhibit impaired testicular descent and abnormal spermatogenesis, while females have impaired fertility associated with deregulation of the estrus cycle [4]. Thus RLF has important reproductive roles in both males and females.

The mouse RLF gene has been recently characterized and was found to be closely associated with the JAK3 tyrosine kinase gene [5]. Exons 23A and 23B of the JAK3 gene are separated by a 2.2 kb intron, which contains the entire RLF gene sequence. Thus the promoter and coding regions of the RLF gene resides entirely within the JAK3 gene. We have investigated this unusual genomic arrangement in the rat and compared this gene region to that in the mouse.

Results and Discussion

The rat RLF / JAK3 gene region was amplified by PCR using primers to exon 23A of JAK3 and the 3' coding region of RLF. This DNA fragment contains the entire 5' flanking sequence and coding sequence of RLF. Nucleotide sequencing of the rat RLF / JAK3 gene region revealed a similar overall structure to the mouse region with 85 % nucleotide identity over the 1,120 bp 5' flanking region (Fig. 1).

The nucleotide homology was evenly distributed, except for a few regions which had areas of insertions or deletions. In comparison to the mouse sequence, insertions ($\geq$ 5 bp) are located at positions - 700 and - 210 and two distinct deletion sites are also identified at – 632 (19 bp deletion) and - 228 on the rat JAK3-RLF sequence.

The RLF / JAK3 gene region contained a number of conserved binding sequences for the transcription factors steroidogenic factor 1 (SF-1) and the sex-determining transcription factor (SRY) (Table 1). Steroidogenic factor 1 is known to mediate the regulation of several genes in steroidogenic tissues. One SF-1 binding site in the mouse gene is absent in the rat gene, however another potential SF-1 binding site is found at position – 590. Thus the rat RLF gene still has the same number of potential SF-1 sites as the mouse gene.

G.W. Tregear et al. (eds.), Relaxin 2000, 357–358.
© 2001 *Kluwer Academic Publishers. Printed in the Netherlands.*

Table. 1. The location and sequence of consensus binding sites for the transcription factors SF-1 (steroidogenic factor 1) and SRY in the rat RLF promoter.

Transcription Factor	Sequence	Rat RLF promoter
SF-1	CCAAGGCC	- 57
SF-1	CCAAGGCT	- 470
SF-1	GAGAGGTC	- 590
SRY	AAAAGGGG	- 215

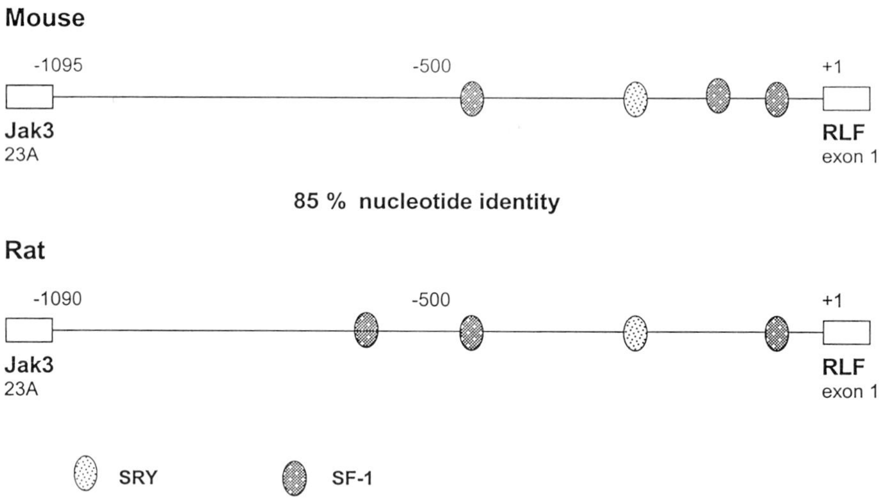

Fig. 1. Schematic comparison of the mouse and rat Jak3 / RLF promoter. The location of potential transcription factor binding sites (SF-1 and SRY) are indicated.

Acknowledgements

This work was supported by a faculty research grant from the Faculty of Life Sciences, RMIT University, Melbourne.

References

1. Adham, I.M., Burkhardt, E., Benahmed, M., and Engel, W., J. Biol.Chem., 268 (1993) 26668.
2. Burkhardt, E., Adham, I.M., Brosig, B., Gastmann, A., Mattei, M-G. and Engel, W., Genomics 20 (1994) 13.
3. Roche, P.J., Butkus, A., Wintour, E.M. and Tregear, G.W., Mol. Cell. Endocrinol., 121 (1996) 171.
4. Nef, S. and Parada, S.F., Nat. Genet., 22 (1999) 295.
5. Koskimies, P., Spiess, A., Lahti, P., Huhtaniemi, I. and Ivell, R., FEBS Letts., 419 (1997) 186.

Relaxin-Like Factor is differentially expressed in the normal and neoplastic human mammary gland

Sabine Hombach-Klonisch[1], Joerg Buchman[2], Sukhena Sarun[1], Bernd Fischer[1] and Thomas Klonisch[1]

Departments of [1]Anatomy and Cell Biology and [2]Pathology, Martin Luther University, Faculty of Medicine, Halle, Germany.

Introduction

In the female reproductive tissue, the polypeptide hormone relaxin-like factor (RLF) has been shown to be expressed in the ovarian theca interna cells [1] and in the non-pregnant and pregnant uterus [2, 3]. Within the mammary gland, the structurally closely related hormone relaxin is essential for breast and nipple development during pregnancy in various species [4-7]. Specific relaxin binding sites have been demonstrated in human and porcine breast epithelial cells and smooth muscle cells of the mammary nipples [8, 9]. In human breast cancer tissues, expression of relaxin has been reported to be up-regulated suggesting an involvement of relaxin during cancerogenesis of the human breast [10, 11]. Currently, no information is available on the expression of RLF in the human mammary gland. As a first step toward elucidating the role of RLF in the human breast, we have studied the cellular localization of RLF in the normal and neoplastic human mammary gland.

Results and Discussion

Employing RT-PCR and specific primers for human RLF, we have detected a single RLF transcripts in a cDNA library of normal human breast tissue and in cDNA preparations of human breast carcinoma tissue. When normal, benign and malignant neoplastic human mammary tissue of pre- and postmenopausal women was subjected to nonradioactive in-situ hybridization and immunohistochemistry, RLF mRNA and protein were exclusive observed in the tubulo-alveolar and ductal epithelium. Although tubulo-alveolar epithelial cells of the normal mammary tissue showed strong RLF hybridization signals, only weak RLF immunostaining was detected using a rabbit antiserum specific for RLF. In metaplastic epithelial cells of infiltrating ductal and lobular breast carcinomas and their corresponding lymph node metastases, weak expression of RLF mRNA contrasted with strong detection of immunoreactive RLF. This discrepancy between the mRNA and protein expression pattern might be suggestive of post-transcriptional mechanisms to affect RLF mRNA within the human mammary gland. In benign neoplasia, increased levels of both RLF mRNA and RLF protein were observed. Thus, upregulation of immunoreactive RLF in the human mammary gland is an early event and does not depend on malignant transformation of

G.W. Tregear et al. (eds.), Relaxin 2000, 359–360.
© 2001 *Kluwer Academic Publishers. Printed in the Netherlands.*

breast epithelial cells. In contrast to relaxin -/- mice functionally impaired mammary tissue and disturbed nipple development [6] is not a feature of RLF -/- mice [12, 13] indicating that both hormones [14] possess different, yet unknown, functional roles within the mammary gland. Given our data and the ability of RLF to interact with specific binding sites for both RLF and relaxin [15, 16], we propose an autocrine/paracrine hormonal relaxin-like network in the human mammary gland.

Acknowledgements

The authors would like to thank Christine Froehlich, Elisabeth Schlueter and Rosemarie Rappold for their excellent technical assistance. S. Hombach-Klonisch was supported by the province Sachsony-Anhalt, Germany.

References

1. Bamberger, A-M., Ivell, R., Balvers, M., Kelp, B., Bamberger, C.M., Riethdorf, L. and Löning, T., Int. J. Gynecol. Pathol., 18 (1999) 163.
2. Tashima, L.S., Hieber, A.D., Greenwood, F.C. and Bryant-Greenwood, G.D., J. Clin. Endocrinol. Metabol., 80 (1995) 707.
3. Hombach-Klonisch, S., Kauffold, J., Rautenberg, T., Steger, K., Tetens, F., Fischer, B. and Klonisch, T., Mol. Cell. Endocrinol., 159 (2000) 147.
4. Hurley, W.L., Doane, R.M., O'Day-Bowman, M.B., Winn, R.J., Mojonnier, L.E. and Sherwood, O.D., Endocrinology, 128 (1991) 1285.
5. Kuenzi, M.J. and Sherwood, O.D., Endocrinology, 131 (1992) 1841.
6. Zhao, L., Roche, P.J., Gunnersen, J.M., Tregear, G.W., Wintour, E.M. and Beck, F., Endocrinology, 140 (1999) 445.
7. Peaker, M., Taylor, E., Tashima, L., Redman, T.L., Greenwood F.C. and Bryant-Greenwood, G.D., Endocrinology, 125 (1989) 693.
8. Min, G. and Sherwood, O.D., Biol. Reprod., 55 (1996) 1243.
9. Kohsaka, T., Min, G., Lukas, G., Trupin, S., Campbell, E.T. and Sherwood, O.D., Biol. Reprod., 59 (1998) 991.
10. Mazoujian, G. and Bryant-Greenwood, G.D., Lancet, 335 (1990) 298.
11. Tashima, L.S., Mazoujian, G. and Bryant-Greenwood, G.D., J. Mol. Endocrinol., 12 (1994) 351.
12. Zimmermann, S., Steding, G., Emmen, J.M.A., Brinkmann, A.O., Nayernia, K., Holstein, A.F., Engel, W. and Adham, I.M., Mol. Endocrinol., 13 (1999) 681.
13. Nef, S. and Parada, L.F., Nat. Genet., 22 (1999) 295.
14. Büllesbach, E.E., Rhodes, R., Rembiesa, B. and Schwabe, C., Endocrine, 10, (1999) 167.
15. Büllesbach, E.E., and Schwabe, C., J. Biol. Chem., 274 (1999) 54.
16. Büllesbach, E.E., and Schwabe, C., J. Biol. Chem., 270 (1995) 16011.

Relaxin-Like Factor in the human thyroid

Thomas Klonisch[1], Sabine Hombach-Klonisch[1], Bernd Fischer[1] and Cuong Hoang-Vu[2]

Department of [1]Anatomy and Cell Biology and [2]Clinic of Surgery, Martin Luther University, Faculty of Medicine, Halle, Germany

Introduction

As a member of the insulin-like family, the relaxin-like factor (RLF) has first been cloned from testiculat tissues [1, 2] and was later shown to be a marker for postpubertal testicular Leydig cells in a number of species [3-5]. Male RLF -/- mice display impaired development of the Gubernaculum testis resulting in cryptorchidism and male infertility [6, 7]. Female RLF -/- mice show disturbances in follicular development [7]. Two RLF splice variants have been described in primates [8-10] and in the human the fusion protein between the exon 1 of RLF and JAK3 displays tissue-specific expression [8]. RLF may be involved in human carcinogenesis as we have recently shown differential expression of RLF in the normal and neoplastic human mammary gland [11]. Here we demonstrate the differential expression of RLF in histologically normal goiter tissue and epithelial carcinomas of the human thyroid gland.

Results and Discussion

Employing non-radioactive in situ hybridization and immunohistochemistry we have shown that RLF mRNA and RLF protein were exclusively expressed in follicular or papillary epithelial thyroid carcinoma. In contrast, normal human thyroid tissue was devoid of both RLF mRNA and RLF protein. RT-PCR analysis revealed variable expression of RLF transcripts in human thyroid carcinoma cell lines (Table 1).

Table 1. Expression of RLF and RLF splice variant in human thyroid tumor cell lines. Consistant expression of the RLF splice variant was exclusively observed in the undifferentiated thyroid cell line 8505C. (+) weak expression; + strong expression; - no expression.

Cell line	**FTC 133**	**FTC 236**	**FTC 238**	**BCPAP**	**8305C**	**8505C**
RLF	(+)	(+)	(+)	(+)	+	+
RLF splice variant	-	(+)	(+)	(+)	-	+

G.W. Tregear et al. (eds.), Relaxin 2000, 361–362.
© 2001 *Kluwer Academic Publishers. Printed in the Netherlands.*

Transcriptional activation of the RLF gene and the appearence of the RLF splice product were affected by the state of tumor cell differentiation. RLF gene expression was strongest in the undifferentiated thyroid carcinoma cell lines 8305C and 8505C. In addition to normal RLF, 8505C expressed a novel RLF splice variant with an unaltered B-domain and an out-of-frame insertion of a unique peptide of 32 residues at the beginning of the C-domain as demonstrated by RT-PCR cloning and Western analysis with an RLF antiserum directed against the conserved B-domain. Currently, experiments are being performed to determine the expression pattern of this novel RLF splice variant within thyroid tissues. The exclusive expression of RLF in human thyroid epithelial carcinoma but not in the normal thyroid implicates RLF to be involved in carcinogenesis in the human thyroid. In contrast to human breast carcinoma [11], two different RLF isoforms may be expressed within human thyroid carcinomas.

Acknowledgements

The authors would like to thank Christine Froehlich and Elisabeth Schlueter for their excellent technical assistance. This work was in part supported by the province Saxony-Anhalt, Germany.

References

1. Adham, I.M., Burkhardt E., Benahmed M. and Engel W., J. Biol. Chem., 268 (1993) 26668.
2. Pusch, W., Balvers, M. and Ivell, R., Endocrinology, 137 (1996) 3009.
3. Ivell, R., Balvers, M., Domagalski, R., Ungefroren, H., Hunt, N. and Schulze, W., Mol. Hum. Reprod., 3 (1997) 459.
4. Hombach-Klonisch, S., Kauffold, J., Rautenberg, T., Steger, K., Tetens, F., Fischer, B. and Klonisch, T. , Mol. Cell. Endocrinol., 159, (2000) 147.
5. Hombach-Klonisch, S., Tetens, F., Kauffold, J., Steger, K., Fischer, B., Klonisch, T., Mol. Reprod. Dev., 53 (1999) 135.
6. Zimmermann, S., Steding, G., Emmen, J.M.A., Brinkmann, A.O., Nayernia, K., Holstein, A.F., Engel, W., Adham, I.M., Mol. Endocrinol., 13 (1999) 681.
7. Nef, S., Parada, L.F., Nat. Genet., 22 (1999) 295.
8. Safford, M.G., Levenstein, M., Tsifrina, E., Amin, S., Hwakins, A.L., Griffin, C.A., Civin, C.I. and Small D., Exp. Hematol., 25 (1997) 374.
9. Koskimies, P., Spiess, A.-N., Lahti, P., Huhtaniemi, I. and Ivell, R., FEBS Letts., 419 (1997) 186.
10. Zarreh-Hoshyari-Khah, M.R., Einspanier, A. and Ivell, R., Biol. Reprod., 60 (1999) 445.
11. Hombach-Klonisch, S., Buchmann, J., Sarun, S., Fischer, B., Klonisch, T., Cancer, 89 (2000) 2161.

The genomics of Insulin 5

Stephen Jaspers, Si Lok, Catherine E Lofton-Day, Betty A Haldeman, Theodore E Whitmore, Kevin P Foley and Darrell Conklin
ZymoGenetics, Inc., Seattle, Washington, 98102 USA

Introduction

Advances in the development of sequence databases and novel bioinformatics algorithms have changed the way many new hormones are discovered. Nucleic acid sequence-based queries targeted to conserved features of the insulin/relaxin superfamily, have led to the discovery of a novel member of the relaxin family of hormones, INSL5 [1]. The discovery of a novel gene from DNA sequence permits the rapid analysis of gene expression patterns and chromosomal localization in an attempt to elucidate the normal physiological or pathological role of the hormone. However, the protein to gene discovery path practiced with all previously identified members of this hormone family benefited from the knowledge of the biologically active form of the hormone from the very onset, as well as a bioassay with which to follow the hormone. This is not the case with genomic discoveries since the signalling form of members of this hormone family are usually the product of multiple steps of post-translational modification and for many members, the structure of the mature form is not obvious. In addition, the varied biological activities and unique organ and cellular location of the current members of the insulin/relaxin family make the elucidation of the biology of its orphan members a difficult task. The first steps toward our understanding of INSL5 biology has focused on experiments necessary to provide information on the structure of the mature hormone, the identification of some of the organ systems affected by its action, and the identification of a population of cells which make this hormone.

Results and Discussion

The discovery of human INSL5 was the result of searching all available EST databases utilizing a query designed to identify sequences containing the cysteine motif common to all known members of the insulin/relaxin superfamily of hormones. A single clone containing this EST was obtained and full length cDNA sequencing revealed a polypeptide which suggested a prohormone structure in which the cysteine motif, signal sequence, B-chain, C-peptide, and A-chain structure was similar to other members of this family. Subsequent cloning of the murine orthologue of *INSL5* supported the structural motif identified in the human sequence and showed that this hormone exhibits a high degree of identity between species. Over the domains predicted to be the B-chain and A-chain regions, the mouse and human sequences are 71% identical. The primary sequences of these regions are also most similar to shark

G.W. Tregear et al. (eds.), Relaxin 2000, 363–367.
© 2001 *Kluwer Academic Publishers. Printed in the Netherlands.*

(dogfish) relaxin and human relaxin sequences compared to those found in insulin or the insulin-like growth factors. Fig. 1 illustrates the domain structure of INSL5 and the high degree of conservation between the human and murine prohormone.

```
                Signal                        B-Chain
hINSL5  MKGSIFTLFLFSVLFAISEVRS  KESVRLCGLEYIRTVIYICASSRW  RR
mINSL5  MKGPTLALFLLLVLLAVVEVRS  RQTVKLCGLDYVRTVIYICASSRW  RR
        MKG----LFL--VL-A--EVRS    V-LCGL-Y-RTVIYICASSRW  RR

                         C-Peptide
hINSL5  HLEGIPQAQQAETGNSFQLPHKREFSEENPAQNLPKVDASGEDRLWG
mINSL5  HLEGHFHSQQAETRNYLQQDRHEPSKKTLEHS-LPKTDLSGQELVRD
        HLEG----QQAET-N--Q--------------LPK-D-SG

hINSL5  GQMPTEELWKSKKHSVMSR
mINSL5  PQAPKEGLWELKKHSVVSRR
         Q-P-E-LW--KKHSV-SR

             A-Chain
hINSL5  QDLQTLCCTDGCSMTDLSALC
mINSL5   DLQALCCREGCSMKELSTLC
         DLQ-LCC--GCSM--LS-LC
```

Fig 1. Primary structure and domains of human and murine INSL5

The isolation of human and mouse cDNAs coding for INSL5 has allowed a broad based genetic analysis of the expression profile of this novel molecule. The unique tissue distribution observed for INSL5 has provided some insight into the biological systems controlled by this hormone [1,2]. Northern blot analysis has demonstrated that the *INSL5* gene is expressed in human colon, rectum, and uterus [1] and murine brain, kidney, testis and thymus [1,2]. However, the specific cell types responsible for this expression pattern is not known.

Transgenic mice have been utilized to either increase the expression of mINSL5 or to delete its function. These mice have been generated to investigate the biological consequences of these changes on embryonic development and adult physiology. Overexpression of INSL5 under a metallothionein or insulin promoter has not demonstrated an obvious effect of this gene on murine physiology however it is not known whether receptor regulatory elements may compensate for the overabundance of the ligand in this system. For example, without specific knowledge of the appropriate bioassay to perform, one may not generally be able to observe an effect of slight and chronic hyperinsulinemia.

The generation of mice which are deficient in *mINSL5* has also been performed and the animals are currently under investigation. The strategy by which the *INSL5* gene was deleted from these mice included the replacement of the deleted gene with *lacZ*. The insertion of this gene allows the staining of cells which normally express the *INLS5* gene. Preliminary results obtained from these mice have indicated that only a discrete population of cells express the *INSL5* gene. Consistent with the results obtained from Northern blot and RT-PCR analysis, the colon and uterus are two organs which exhibit significant expression of *lacZ* and specific β-galactosidase activity. While the specific localization of the uterine staining is still under investigation, the staining observed in the colon is localized to what appears to be a single dispersed cell population. The β-galactosidase staining in the colon of male and female heterozygous animals is localized to various regions of the colon and in a cell population which appears to be enteroendocrine-like cells within the crypts [3]. No *INSL5* gene dependent staining was apparent in the upper gastrointestinal tract. The stained cells are similar in shape and frequency to other endocrine hormone secreting cells found in the stomach or intestine such as those responsible for somatostatin or GLP-1 secretion [3]. The INSL5 cells are rare (<1% of the cell population) and appear to span the colonic epithelial layer, so that while the cells do appear to contact the lumen, most of the cell body is located adjacent to the basal layer. The specific identification and functional characterization of this cell type will require further study.

While progress has been made in the analysis of gene expression in vivo, the isolation of the native protein molecule and its structural characterization has remained elusive. Based on its close homology to relaxin [4] and INSL3 [5], a two-chain, disulfide linked molecule is predicted to be the physiologically relevant form of the protein. To investigate this possibility, recombinant methods have been used to generate the cDNA encoding INSL5 and multiple cell lines have been transfected to study the processing events required to convert the genetic sequence to a mature, biologically active molecule.

Expression constructs for INSL5 and INSL5-fusion proteins which contain an epitope tag, have been introduced into baby hamster kidney cells (BHK) and have allowed the determination of the signal processing site on human INSL5. Human INSL5 cDNA which also contained a C-terminal epitope tag sequence was expressed in BHK cells and the fusion-protein purified. Amino terminal sequencing was performed on the protein and the results confirmed the prediction of a signal sequence cleavage at the C-terminal side of serine-22 which produces a B-chain starting at lysine-23. Additional studies have also confirmed the susceptibility of the prohormone to processing at the dibasic site (Arg47-Arg48) of the B-chain and C-peptide junction by prohormone convertase 3 (PC3). Co-expression of hINSL5 with PC3 produces a protein which is predominantly cleaved at the dibasic site of the B-chain/C-peptide junction (data not shown). Studies have also been performed in which hINSL5 was produced in a rat insulinoma cell line (RIN 1046-38) [6] (Fig. 2). This cell line is capable of processing proinsulin and is also capable of processing INSL5 albeit at a very low efficiency. While proinsulin is known to be efficiently processed by this cell line, most of the INSL5 protein produced by this line migrates on a SDS-PAGE system at a size consistent with the prohormone. A small amount of immunoreactive protein also migrates at a molecular weight consistent with the fully processed hormone. These data suggest that RIN cells can process some of the expressed INSL5 to the mature hormone. However, due to the low proportion of mature hormone secreted by these cells, one might conclude that the prohormone convertases required to perform this function efficiently are either absent or are expressed at very low levels in this insulinoma cell line.

The information obtained from protein expression and sequencing studies have allowed the design and production of synthetic INSL5. The structure shown in Fig. 3 is based on the data generated from N-terminal sequencing of the human prohormone, analysis of coexpression studies of INSL5 with PC3, expression of INSL5 in an endocrine cell line (RIN 1046-38), a comparison of the murine and human sequences and by analyzing the common processing sites observed between species and other members of the relaxin superfamily.

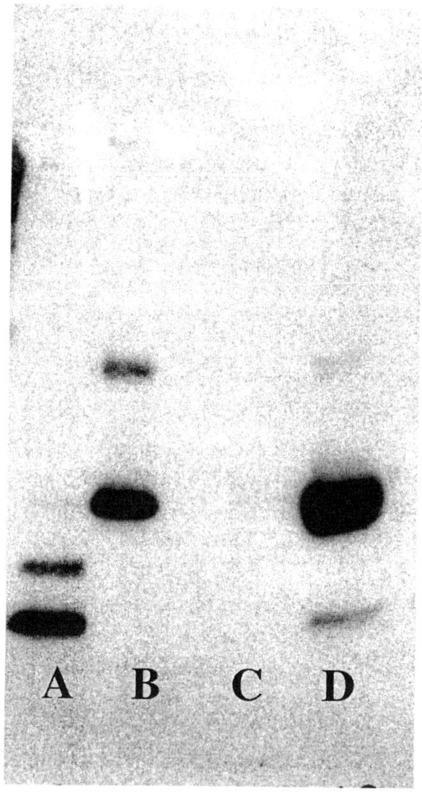

Fig. 2. Immunoblot analysis of synthetic and recombinant INSL5 with rabbit anti-proINSL5. A) synthetic INSL5, B) recombinant prohINSL5-CEE, C) RIN (wt) conditioned media, D) RIN+hINSL5 conditioned media.

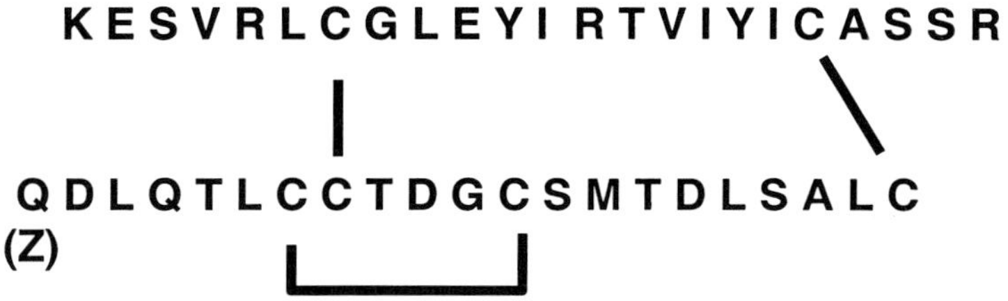

Fig. 3. Proposed structure of mature human INSL5

INSL5 is a new member of the insulin and relaxin superfamily of hormones which appears to play a unique position in mammalian physiology. Its novel cellular localization in its tissues of origin and a potential role for the protein in both gastrointestinal and reproductive physiology leads one to speculate that this hormone represents either a previously unknown biological process or one which has previously been ascribed to other hormones. Future research will be necessary to definitively ascertain the physiological or pathological role for this hormone and its signalling system.

The discovery of INSL5 by bioinformatics analysis of EST data and its subsequent biological and structural characterization has demonstrated the advantages and disadvantages of genomics based discoveries. This research has also provided a glimpse of our future in elucidating the biology of novel members of known hormone families

Acknowledgement

The authors would like to acknowledge the valuable assistance of Phil Hofstrand, Andrew Ching, Anna Jelmberg, Maria Dasovich, Mike Stamm, and Ross Hoffman. The synthesis of INSL5 was performed by the Protein Research Laboratory at the Medical University of South Carolina and the contribution of Dr. Christian Schwabe, Director of Protein Chemistry Facility is gratefully acknowledged.

References

1. Conklin, D., Lofton-Day, C., Haldeman, B.A., Ching, A., Whitmore, T., Lok, S., and Jaspers, S., Genomics, 60 (1999) 50.
2. Hsu, S.Y., Mol. Endocrinol., 13 (1999) 2163.
3. Stevens, A. and Lowe, J., Human Histology. Mosby, London, 1997.
4. Schwabe, C. and Bullesbach, E.E., Relaxin and the Fine Structure of Proteins, Springer, Berlin, 1998.
5. Adham, I.M., Burkhardt, E., Benahmed, M. and Engel, W., J. Biol. Chem., 268 (1993) 26668.
6. Clark, S.A., Quaade, C., Constandy, H., Hansen, P., Halban, P., Ferber, S., Newgard, C.B., and Norminton, K., Diabetes, 46 (1997) 958.

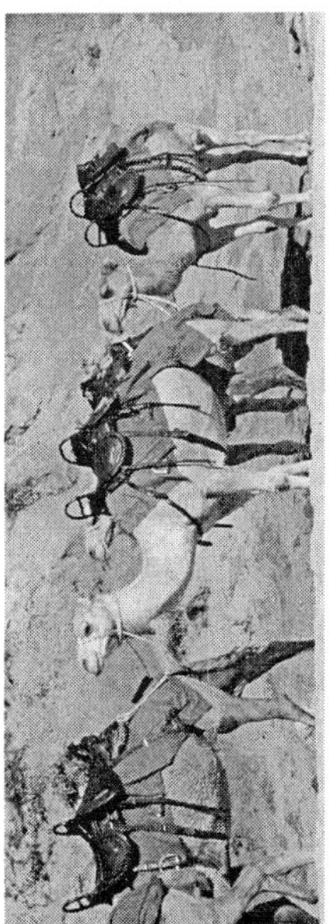

SECTION 12

Clinical aspects of relaxin I – Matrix remodelling

Relaxin clinical trials in systemic sclerosis

Marlena S Erikson and Elaine N Unemori
Connetics Corporation, Palo Alto, California, 94303, USA
[Presented at the meeting by Kristina Zsebo]

Introduction

The reproductive hormone, relaxin, has a long history as an anti-fibrotic. The name, relaxin, in fact, stems from its initially observed ability to cause relaxation of the pubic ligament in guinea pigs [1]. Relaxin-mediated modulation of connective tissue integrity in the pubic ligament, cervix, uterus, mammary gland, and uterus has been demonstrated in a number of mammalian species [2].

The hypothesis that relaxin could effect the degradation of extracellular matrix molecules in tissues outside of the reproductive system was first tested clinically in the 1950s by Casten and Boucek in a study in scleroderma [3]. Systemic sclerosis (scleroderma) is a generalized disorder of connective tissue characterized clinically by thickening and fibrosis of the skin and by distinctive forms of involvement of internal organs, including the heart, lungs, kidneys and gastrointestinal tract [4]. The etiology of scleroderma is unknown and the disease is characterized by increased accumulation of extracellular matrix in the dermis and in internal organs, prominent vascular abnormalities (renal crisis, pulmonary hypertension, and Raynaud's phenomenon), as well as immunologic abnormalities. Casten and Boucek utilized a partially purified preparation of porcine relaxin, which was available as "Releasin," from the Warner-Chilcott Laboratories, in an open label study. Twenty-three subjects with scleroderma were studied over a period, which ranged from 6 to 30 months. Seventeen of the 23 subjects were assessed to have experienced improvement in the degree of skin tightness, judged using one, and in some cases, two, semi-quantitative scoring systems. Further evaluation of relaxin in a placebo-controlled study for its beneficial effects in this disease was limited by uncertainties concerning the purity of the porcine relaxin preparation sold at that time [2], and by the removal of this pharmaceutical from the market following the Kefauver Amendment to the Food and Drug Act in the early 1960s.

More recently, recombinant human (rH2) relaxin has been tested in vivo in a series of preclinical experiments evaluating its ability to modify connective tissue in the dermis [5, 6], in the lung [7, 8], in the heart [9], and in the kidney [10]. In the skin, rH2 relaxin facilitated skin expansion in a pig model utilizing an implanted Silastic tissue expander [5]. The authors speculated that expansion caused by the mechanical expander stimulated the synthesis of extracellular matrix molecules in the dermis to allow stretching, and that rH2 relaxin modified the deposition of matrix to accommodate further expansion. rH2 relaxin has also been tested in two models of dermal fibrosis in rodents [6], as well as in a murine model of bleomycin-induced pulmonary fibrosis [7]. In the dermal models, rH2 relaxin demonstrated the ability to inhibit the rapid

G.W. Tregear et al. (eds.), Relaxin 2000, 373–381.
© 2001 *Kluwer Academic Publishers. Printed in the Netherlands.*

deposition of collagen by up to 29%, concurrent with effecting ultrastructural alterations in connective tissue consistent with degradation or altered deposition [6]. In the pulmonary model, rH2 relaxin decreased bleomycin-induced alveolar thickening by 64% and completely inhibited hydroxyproline-quantified collagen deposition in mice [7]. Hypoxia-induced pulmonary hypertension, associated with increased collagen deposition in the walls of the pulmonary arteries, was also ameliorated by rH2 relaxin treatment [8]. In the heart, rH2 relaxin inhibited collagen deposition in the infarct scar in the left ventricle following ligation of the left anterior coronary artery by 23% [9]. That relaxin may have a natural role in the regulation of collagen deposition in the left ventricle has also been suggested by results obtained in the relaxin knockout mouse [11]. Finally, in a chemically induced renal fibrosis model in the rat, 28-day treatment with rH2 relaxin decreased fibrosis score by 75% and prevented decreases in creatinine clearance [10].

A program to assess rH2 relaxin for safety in the clinical setting and for efficacy in preventing progression of disease in systemic sclerosis was initiated by Connetics Corporation (then Connective Therapeutics, Inc.) in 1994. Besides the preclinical findings on relaxin's anti-fibrotic properties listed above, a large body of work describing its fascinating pleiotropism has been compiled over the years [2]. Because rH2 relaxin was infused parenterally in the clinical studies described in this chapter, and especially in light of the species-specific reproductive physiological effects it has demonstrated, these trials offer valuable insight into the effects of rH2 relaxin in humans.

Results and Discussion

Phase I study

A phase I study (Study R9401) designed to investigate the safety and pharmacokinetics of a 28-day continuous infusion of rH2 relaxin via a microinfusion pump in subjects with systemic sclerosis with diffuse scleroderma was initiated in 1994 [12]. Thirty subjects with stable diffuse scleroderma with moderate disease received rH2 relaxin at 6, 12, 50, 100, and 200 μg/kg/day (n = 4 per group) or placebo (n = 10) in this double blind, dose escalation study. All subjects completed 28 days of study dosing. Dose-related steady state serum levels up to a mean of 55.4 ng/ml, approximately 50-fold those of normal pregnancy, were reached at the 200 μg/kg/day dose by the third day of infusion. Adverse events included local infusion site rash and pain, minor bleeding episodes, and decreased hemoglobin concentration (mean reduction 1.1 gm/dL). Standard scleroderma measurements included the Modified Rodnan Skin Score (MRSS), which measures 17 body areas for skin thickness by clinical palpitation on a 0-3 scale [13]. Other measurements included maximal oral aperture, hand extension and hand closure, modified Ritchie index (metacarpolphalangeal joints, wrists, and knees), and modified Health Assessment Questionnaire (HAQ) [14]. All measurements were unchanged, although trends toward improvement were seen in the Global Assessment and the Health Assessment Questionnaire, suggesting a clinical effect. Trends indicating decreases in serum osmolality were observed. Administration of rH2 relaxin at doses up to 200 μg/kg/day was safe and well tolerated. Patient acceptance of rH2 relaxin treatment was high for 28 days of continuous subcutaneous infusions.

Phase II study

A phase II study (Study RLXN.C.003) designed to assess the efficacy and safety, and dose-response effect of rH2 relaxin in subjects with scleroderma was completed in 1997 [15]. Sixty-eight subjects who had stable, diffuse scleroderma (moderate to severe) for less than five years received rH2 relaxin at 25 or 100 μg/kg/day or placebo in this double-blind study. rH2 relaxin was administered by subcutaneous infusion over 24 weeks. The MRSS was the primary efficacy measure. Secondary measurements included pulmonary function tests, the Health Assessment Questionnaire, measurement of maximal oral aperture, hand extension and hand closure; and the modified Ritchie index (metacarpolphalangeal joints, wrists, and knees). Subjects who received rH2 relaxin at 25 μg/kg/day had significantly lower skin scores than those who received placebo [mean change, -3.6 at 4 weeks (p=0.021), -7.5 at 12 weeks (p=<0.001), and -8.7 at 24 weeks (p=0.040)], as seen in Table 1. Similar trends were noted in other outcome measures, including forced vital capacity, measures of oral aperture and hand extension, functional status, and global assessment. Subjects who received 100 μg/kg/day did not differ from those who received placebo. The nonlinear relationship between these clinical parameters and rH2 relaxin dose was hypothesized to be attributable to the biphasic collagenolytic response to rH2 relaxin displayed by dermal fibroblasts [16]. Drug related adverse events included menorrhagia, metrorrhagia, reversible anemia, and complications of subcutaneous drug administration system (site irritation and local infection). Retrospective predicted creatinine clearance measurements, calculated using the Cockcroft and Gault method [17], showed statistically significant increases in the 100 μg/kg/day compared to changes in the placebo group in the course of rH2 relaxin infusion.

Phase II/III study

A phase II/III safety and efficacy study (Study RLXN.C.005) in systemic sclerosis was completed in the fall of 2000. Subjects were randomized to one of three dose groups (25 μg/kg/day, placebo or 10 μg/kg/day) in a 2:2:1 ratio, with a minimum of 72 subjects each in the 25 μg/kg/day and placebo groups, and 36 subjects in the 10 μg/kg/day group. A total of 231 subjects received study drug (rH2 relaxin or placebo). Subjects were required to have systemic sclerosis with diffuse scleroderma for ≤ 5 years; MRSS of ≥ 20, or ≥ 16 with truncal involvement, that was stable for at least one month; a forced vital capacity and lung diffusion capacity of > 40 % predicted. Subjects also were required to have adequate renal, cardiovascular, and hematologic status. Subjects were excluded if they had limited scleroderma, overlap syndromes with other connective tissue diseases, required > 10 mg/day of predinisone, or were on any other disease modifying drugs within four weeks of initiating study drug. The eligibility criteria were the same as for the phase II study. The primary objective of the study was to replicate a statistically significant decrease ($\alpha \leq 0.05$) of the MRSS in the 25 μg/kg/day dose group compared to the placebo dose group, as was seen in RLXN.C.003. An additional objective was to provide supportive evidence of the effect of the 25 μg/kg/day dose, compared to placebo, on additional variables relevant to assessment of this disease. These included Health Assessment Questionnaire, quality of life assessment (SF-36 Questionnaire), measurement of right and left hand extension, musculoskeletal assessment (tenderness and swelling of joints), and patient and physician global assessment by visual analog scale. Lung function measured by forced vital and diffusing lung capacity (actual and percent predicted) and the Mahler dyspnea index [18] was also assessed. Renal function by predicted creatinine clearance was also evaluated. In addition, patient assessment of Raynaud's activity by visual analog scale and enumeration of cutaneous ulcers were performed.

Table 1. Serial change in mean modified Rodnan Skin Score (Phase II Study)

		Treatment	
Visit	**Placebo** (n=19)	**rH2 relaxin 25 μg/kg/day** (n=21)	**rH2 relaxin 100 μg/kg/day** (n=24)
Screening	26.7 ± 1.4	27.5 ± 1.6	26.3 ± 1.6
p-value*		NS	NS
Week 0	26.8 ± 1.3	27.5 ± 1.6	27.5 ± 1.7
p-value*		NS	NS
Week 4	26.0 ± 1.4	23.9 ± 1.7	25.0 ± 1.8
mean change†	-0.8 ± 1.0	-3.6 ± 0.6	-2.5 ± 0.7
p-value‡		0.021	0.136
Week 12	24.4 ± 1.5	20.0 ± 1.9	24.5 ± 2.1
mean change†	-2.5 ± 0.9	-7.5 ± 0.8	-3.0 ± 0.9
p-value‡		<0.001	NS
Week 24	21.9 ± 1.8	18.8 ± 2.0	24.1 ± 2.3
mean change†	-4.9 ± 1.4	-8.7 ± 1.0	-3.4 ± 1.2
p-value‡		0.040	NS

All values are mean ± SE.

* Mean comparison of treatment groups from one-way ANOVA.

† Mean change from Week 0 (follow-up minus Week 0)

‡ ANOVA comparison of treatment group to placebo, adjusting for value at Week 0

Table 2. Patient demographics and clinical features (mean ± SE) (Phase II/III Study)

	Placebo	rH2 relaxin 25 μg/kg/day	rH2 relaxin 10 μg/kg/day
N	94	95	42
Age (yr)	46.1 ± 1.0	46.3 ± 1.2	46.4 ± 1.6
Duration SSc (yr)	2.3 ± 0.1	2.2 ± 0.2	1.9 ± 0.2
Entry MRSS	27.1 ± 0.6	28.0 ± 0.8	28.6 ± 1.1

All differences NS by one-way ANOVA

While the patient demographics (Table 2) were essentially the same as seen in the phase II study, the results of the study did not meet its primary endpoint (MRSS), as seen in Table 3. In addition, the majority of the secondary outcomes (hand extension, oral aperture, enumeration of cutaneous ulcers, pulmonary function tests, Raynaud's activity, patient and physician global assessments and functional and quality of life questionnaires) were not different between rH2 relaxin and placebo.

Table 3. Mean modified Rodnan Skin Score (Phase II/III Study)

	Placebo	rH2 relaxin 25 μg/kg/day	rH2 relaxin 10 μg/kg/day
Baseline	27.1	28.0	28.6
Week 12	24.4	24.8	24.5
Week 24	22.2	22.7	24.3
Change from Baseline (%)	-4.9 (0.7)	-5.2 (0.7)	-4.3 (1.3)

All differences NS by one-way ANOVA

One of the most striking findings in this study was the statistically significant increase from baseline in predicted creatinine clearance observed in subjects receiving rH2 relaxin (Table 4). This clinical observation is consistent with rH2 relaxin-mediated increases in renal perfusion, which are observed in rodents [19]. Moderate, statistically significant decreases in diastolic blood pressure were also observed over the course of rH2 relaxin administration in the clinical studies. These changes were not clinically meaningful.

The observations in renal perfusion and blood pressure were reversible following discontinuation of rH2 relaxin. In a small number of patients (n = 6), the withdrawal of rH2 relaxin may have led to a "rebound" hypertension. This hypertension appears to have been the result of halting rH2 relaxin infusion in patients who have a generalized vasospastic condition. Daily blood pressure monitoring, and the administration of anti-hypertensives as needed, appeared to adequately control these rebound episodes.

Table 4. Predicted creatinine clearance (mL/min) (Phase II/III Study)*

		Treatment	
Visit	**Placebo** (n=94)	**rH2 relaxin 10 μg/kg/day** (n=42)	**rH2 relaxin 25 μg/kg/day** (n=95)
Screening Mean (SE)	128.04 (4.08)	115.50 (5.93)	127.07 (4.86)
p-value[1]		NS	NS
Week 0	127.45 (3.92)	116.51 (6.25)	126.99 (4.95)
p-value[1]		NS	NS
Week 2	125.96 (4.26)	129.13 (7.13)	131.54 (4.81)
p-value[2]		0.004	0.076
Week 4	130.32 (4.49)	124.92 (7.21)	141.85 (7.41)
p-value[2]		0.309	0.107
Week 8	122.87 (4.31)	133.23 (9.45)	133.09 (5.11)
p-value[2]		0.012	0.025
Week 12	125.40 (4.26)	128.30 (7.99)	139.17 (5.92)
p-value[2]		0.022	0.001
Week 16	121.50 (4.24)	120.50 (6.85)	132.60 (5.37)
p-value[2]		0.180	0.007
Week 20	122.71 (4.30)	126.07 (7.20)	128.88 (5.40)
p-value[2]		0.026	0.013
Week 24	120.57 (4.31)	125.01 (9.25)	128.87 (5.54)
p-value[2]		0.046	0.015
Week 28	124.50 (5.27)	104.44 (8.08)	111.35 (6.52)
p-value[2]		0.072	0.072

* Computed per formula from Cockcroft and Gault [16]
[1] Mean comparison of treatment groups using the one-way ANOVA
[2] ANCOVA comparison of active treatment group to placebo, adjusting for value at Week 0.

A summary of the most frequent adverse experiences from completed Connetics-sponsored clinical studies is provided in Table 5. Adverse experiences that were found with an incidence ≥ 15% in subjects treated with rH2 relaxin (all dose groups combined, n = 239) and that occurred with statistically greater frequency in rH2 relaxin groups, compared to placebo subjects (n = 133), are listed. These experiences included menorrhagia (heavy menstrual bleeding), metrorrhagia (irregular menstrual bleeding), anemia and abdominal pain. Menorrhagia and metrorrhagia are believed to reflect rH2 relaxin's effects on endometrial blood supply [20]. Anemia (decreases in hemoglobin) occurred without alterations reflecting either erythrocyte underproduction or destruction, and may instead reflect an increase in intravascular volume [21]. These events were reversible (anemia) or subsided (menorrhagia, metrorrhagia, abdominal pain) following discontinuation of study drug. "Abdominal pain" included upper and lower quadrant tenderness, abdominal pain/tenderness, epigastric pain/tenderness, and abdominal menstrual cramping, and therefore, the significance of each of these reports is unclear at present. Local skin reactions, including pain, inflammation and rash, were common at infusion sites but were approximately balanced among groups.

Table 5. All adverse events during treatment with an incidence ≥ 15% and statistically different from placebo

	Combined Placebo (n = 133)	Total Relaxin (n = 239)
Menorrhagia	8%	33%
Metrorrhagia	11%	21%
Anemia	9%	22%
Abdominal Pain	7%	17%

The three placebo-controlled clinical trials described in this section were augmented by three additional open-label studies in patients with scleroderma, in which safety information on extended rH2 relaxin administration (up to one year) was collected. Although analysis on the last of these studies was not completed at the time these proceedings are in press, results of the two completed studies are consistent with observations made in the placebo-controlled studies.

Conclusion

Although the pivotal study in systemic sclerosis did not meet its primary efficacy endpoint, the conduct of these well-controlled clinical trials allows the definitive conclusion that relaxin is biologically active in humans. Increases in predicted creatinine clearance observed in these studies are consistent with recent findings of rH2 relaxin-mediated increases in renal vasodilation in rodents. Together these results have implications regarding physiological changes in renal perfusion that occur in the menstrual cycle [22] and early pregnancy [23]. Menorrhagia and metrorrhagia may reflect increased perfusion of the uterine lining, mediated perhaps both by increased angiogenesis and vasodilation that occur during the luteal phase of the menstrual cycle and pregnancy. Menorrhagia and metrorrhagia are consistent with histological changes in the endometrium that have been observed following systemic administration of porcine relaxin to monkeys [24]. Anemia may be the result of rH2 relaxin-mediated increases in intravascular volume, an effect previously described in animals [21] and which may have implications in early pregnancy in humans [25]. These clinical observations support the preclinical work of many not mentioned here, and lay the foundation for future clinical trials designed to test rH2 relaxin's potential in other clinical indications.

The failure of the Phase II/III study to meet its primary efficacy endpoint in scleroderma was a disappointment to those who study relaxin, but it was devastating to the patients afflicted with this disease and to those who care for them. We at Connetics hope that our trials will help other companies to meet the challenge of treating or curing this unrelentingly cruel disease.

Acknowledgements

The authors would like to acknowledge the following individuals, who worked tirelessly in the Connetics-sponsored clinical trials in the belief that rH2 relaxin held the potential to help patients with systemic sclerosis: Dr. Martin E. Sanders, Dr. James R. Seibold, Dr. Philip Clements, Dr. Daniel E. Furst, Dr. David Collier, Dr. Mary Ellen Csuka, Dr. Michael Ellman, Dr. Vivien Hsu, Dr. Arthur Kavanaugh, Dr. Youn Kim, Dr. Joseph Korn, Dr. Maureen Mayes, Dr. Thomas Medsger, Dr. Peter Merkel, Dr. Larry Moreland, Dr. Naomi Rothfield, Dr. Richard Silver, Dr. Robert Simms, Dr. Edwin Smith, Dr. Virginia Steen, Dr. Fredrick Wigley, Susan E. Rocco, Dr. Edward P. Amento, Mark E. Erikson, and Beverly Grove.

References

1. Hisaw, F.L., Proc. Soc. Exp. Biol. Med., 23 (1926) 661.
2. Sherwood, O.D., In Knobil, E., Neill, J.D. (eds.), Relaxin, The Physiology of Reproduction, 2nd Ed., Raven Press, New York, 1994, p. 861.
2. Casten, G.G. and Boucek, R.J., J.A.M.A., 166 (1958) 319.
4. Seibold, J.R. Scleroderma. In Kelley, W.N., Harris, E.D., Ruddy, S., Sledge C.B. (eds.) Textbook of Rheumatology. 5th ed. Philadelphia, WB Saunders, 1997, p. 1133.
5. Kibblewhite, D., Larrabee, W.F., and Sutton, D., Arch. Otolaryngol. Head Neck Surg., 118 (1992) 153.
6. Unemori, E.N., Beck, L.S., Lee, W.P., Xu, Y., Siegel, M., Keller, G., Liggitt, H.D., Bauer, E.A., and Amento, E.P., J. Biol. Chem., 101 (1993) 280.
7. Unemori, E.N., Pickford, L.B., Salles, A.L., Piercy, C.E., Grove, B.H., Erikson, M.E., and Amento, E.P., J. Clin. Invest., 98 (1996) 2739.
8. Tozzi, C.A., Poiani, G.J, McHugh, N.A., Riley, D.J., and Unemori, E.N., Am. J. Resp. Crit. Care Med.,161 (2000) A641.
9. Pickford, L.B., Guzman, D., Karimimanesh, N., Salles, A.L., Frey, L., Emlen, W., Lewis, M., Deshpande, U., Grove, B.H., Erikson, M., and Unemori, E.N., FASEB J., 13 (1999) A488.
10. Garber, S.L., Mirochnik, Y., Brecklin, C., Unemori, E.N., Singh, A.K., Slobodskoy, L., Grove, B.H., Arruda, J.A.L., and Dunea, G., Kidney Int., 59 (2001) 876
11. Du, X.J., Zhao, L., Goao, X.M., and Tregear, G., these proceedings.
12. Seibold, J.R., Clements, P.J., Furst, D.E., Mayes, M.D., McCloskey, D.A., Moreland, L.W., White, B., Wigley, F.M., Rocco, S., Erikson, M., Hannigan, J.F., Sanders, M.E., and Amento, E.P., J. Rheumatol., 25 (1998) 302.

13. Clements, P., Lachenbruch, P., Seibold, J., White, B., Weiner, S., Weinstein, M.R., Weisman, M., Mayes, M., et al., J. Rheumatol. 22 (1995) 1281.
14. Pope, J.E., Baron, M.E., Bellamy, N., Campbell, J., Carette, S., Chalmers, I., Dales, P., Hanly, J., Kaminska, E.A., Lee, P. et al., J. Rheumatol., 22 (1995) 1271.
15. Seibold, J.R., Korn, J.H., Simms, R., Clements, P.J., Moreland, L.W., Mayes, M.D., Furst, D.E., Rothfield, N., Steen, V., Weisman, M., Collier, D., Wigley, F.M., Merkel, P.A., Csuka, M.E., Hsu, V., Rocco, S., Erikson, M., Hannigan, J., Harkonen, W.S. and Sanders, M.E., Ann. Int. Med., 132 (2000) 871.
16. Unemori, E.N., and Amento, E.P., J. Biol. Chem., 265 (1990) 10681.
17. Cockcroft, D.W. and Gault, M.H., Nephron, 16 (1976) 31.
18. Mahler, D.A., Weinberg, D.H., Wells, C.K., and Feinstein, A.R., Chest, 85 (1984) 751.
19. Danielson, L.A., Sherwood, O.D., and Conrad, K.P., J. Clin. Invest., 103 (1999) 525.
20. Unemori, E.N., Erikson, M.E., Rocco, S.E., Sutherland, K.M., Parsell, D.A., Mak, J., Grove, B.H., Hum. Reprod., 14 (1999) 800.
21. Zhao, S., Malmgren, C.H., Shanks, R.D., and Sherwood, O.D., Endocrinology, 136 (1995) 1892.
22. Chapman, A.B., Zamudio, S., Woodmansee, W., Merouani, A., Osorio, F., Johnson, A., Moore, L., Dahms, T., Coffin, C., Abraham, W.T. and Schrier, W.R., Am. J. Physiol., 273 (1997) F777.
23. Davison, J.M., and Noble, M.C.B., Br. J. Obstet. Gynecol., 88 (1981) 10.
24. Dallenbach–Hellweg, G., Dawson, A.B. and Hisaw, F.L., Am. J. Anat., 119 (1966) 61.
25. Linheimer, M.D. and Davison, J.M., Eur. J. Endocrinol., 132 (1995) 133.

Relaxin decreases renal interstitial fibrosis: A novel treatment to slow the progression of kidney disease?

Sandra L Garber[1,2], Yelena Mirochnik[3], Carolyn Brecklin[2], Leonid Slobodskoy[3], Elaine N Unemori[5], Beverly Grove[5], Jose A L Arruda[1,2,4] and George Dunea[1,2,3]

[1]Section of Nephrology, Cook County Hospital, Chicago, IL 60612, USA; [2]Section of Nephrology, University of Illinois at Chicago, Chicago, IL 60612, USA; [3]The Hektoen Institute for Medical Research, Chicago, IL 60612, USA; [c]Section of Nephrology, Veterans Administration West Side Medical Center, Chicago, IL 60612, USA; and [5]Connetics Corp, Palo Alto, CA 94303, USA

Introduction

Interstitial fibrosis is a characteristic finding in numerous pathologies. In the kidney this condition is manifested by a disturbance in tubular architecture such that there is tubular atrophy accompanied by tubular dilation. The normally sparse intercellular space is expanded and filled with extracellular matrix composed mainly of collagen and fibronectin as well as mononuclear cells. If left untreated this may progress to encompass the glomeruli and renal insufficiency may result.

In the rat renal papillary necrosis and interstitial fibrosis may be reproducibly induced by a single intraperitoneal injection of bromoethylamine (BEA) [1]. We have studied this model extensively and have shown that at one month after injection there is complete papillary necrosis accompanied by interstitial fibrosis mainly in the corticomedullary area. Bowman's capsule is also thickened in the affected areas of the cortex. At this time point the lesion encompasses approximately 15-20% of the kidney [2]. By six months there is extensive glomerular involvement, and by nine months the animals show changes typically associated with focal segmental glomerulosclerosis [3]. These histological changes are accompanied by functional changes consistent with the nephrotic syndrome. There is a rise in serum creatinine and a fall in the glomerular filtration rate (GFR) and a profound albuminuria.

Relaxin, a hormone of the insulin growth factor family has been shown to decrease fibrosis in several models [4] including two models of renal disease [5]. Here we report the effect of a continuous infusion of relaxin on the BEA model of papillary necrosis and interstitial fibrosis.

Male, Sprague Dawley rats were injected *ip* with 200mg/kg body weight with BEA. One week later, at a time when papillary necrosis had already been initiated, osmotic pumps delivering 2 μg/hr of relaxin were implanted subcutaneously. Treatment continued for one month at which time the rats were evaluated for renal functional and structural changes.

G.W. Tregear et al. (eds.), Relaxin 2000, 383–388.
© 2001 *Kluwer Academic Publishers. Printed in the Netherlands.*

Results and Discussion

Renal balance studies were done one week after BEA injection and showed that all animals has a concentrating defect as typically seen in this model. One month of relaxin treatment failed to correct this defect (Table 1). Tail cuff blood pressures were taken prior to and during the experimental period. ANOVA failed to show any significant differences between or among the experimental groups (Table 1).

Table 1. Urine specific gravities (sp.gv.) and mean arterial blood pressures (mm Hg)

Treatment		sp.gv. pre relaxin	sp.gv. post relaxin	MAP pre relaxin	MAP post relaxin
None	N=8	1.026 ± .005	1.032 ± .003	116 ± 2	117 ± 1
Relaxin	N=6	1.031 ± .005	1.037 ± .008	119 ± 3	123 ± 3
BEA	N=11	1.017 ± .001	1.017 ± .002	109 ± 6	114 ± 4
BEA + Relaxin	N=15	1.018 ± .001	1.019 ± .002	118 ± 4	110 ± 7

Animals receiving only BEA had an increased serum creatinine and decreased glomerular filtration rate (GFR) whether measured by inulin or creatinine clearance. Relaxin decreased serum creatinine to normal levels, with a corresponding increase in the GFR (table 2).

Table 2. Serum creatinine and creatinine clearances after one month of relaxin treatment
** = p <0 .05 compared to control ** = p <0 .05 compared to BEA alone*

Treatment	Serum creatinine mg/dl	Creatinine clearance ml/min/100g BW
None	0.58 ± .04	0.58 ± .06
Relaxin	0.50 ± .03	0.49 ± .08
BEA	0.83 ± .04*	0.38 ± .03*
BEA + Relaxin	0.61 ± .03**	0.49 ± .04**

These functional changes were accompanied by a striking improvement in the histology of the lesion. In BEA animals, 100% of the papillary area was necrotic whereas in the relaxin treated animals this number ranged from 3-21%. There was also a general preservation of the normal papillary structure in the relaxin group. (Fig. 1)

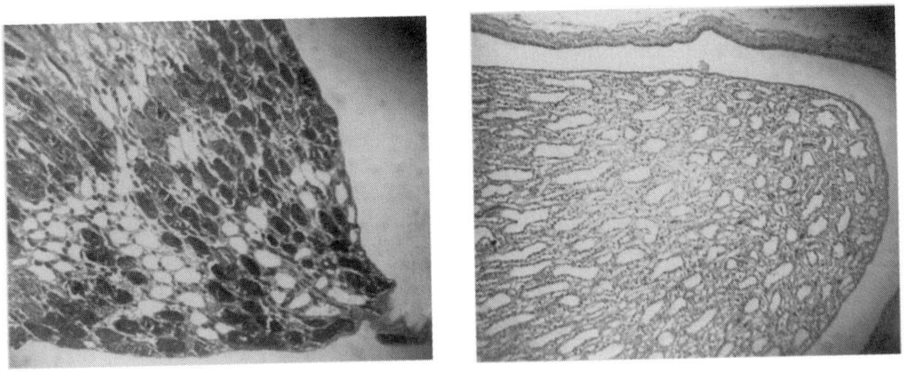

BEA BEA + Relaxin

Fig 1. Rat papilla one month after the injection of bromoethylamine (BEA)

There was also an improvement in the extent of the interstitial fibrosis as may be seen below.

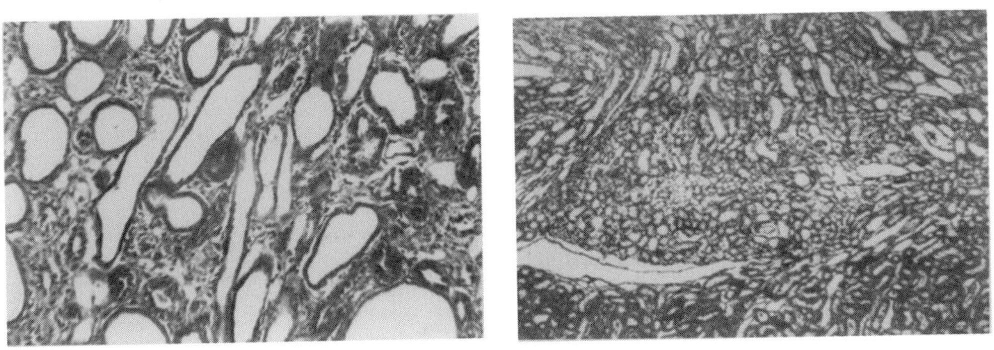

BEA BEA + Relaxin

Fig 2. Fibrosis at the corticomedullary junction. In the rats receiving only BEA there is tubular dilation, a mononuclear cell infiltrate and an increase in intercellular matrix. In the relaxin treated animals the normal renal architecture is preserved.

Area measurements performed on digitized section of the corticomedullary area showed that in the BEA animals 19% of the tissue was fibrotic whereas in the relaxin treated group this number fell to 5%. This is the equivalent of a 75% decrease in the extent of fibrosis.

BEA alone caused a 6-fold increase in the number of infiltrating macrophages at the cortico-medullary junction as evidenced by the increase in the number of ED-1 positive cells. In the relaxin-treated BEA animals there was a decrease of over 50% in the number of these cells (BEA 2336 ± 616 vs. BEA + Relaxin 1043 ± 236 cells/mm^2, $p < 0.05$). The number of ED-1 positive cells was increased in those animals receiving only relaxin (Control 355 ± 80 vs. Relaxin 704 ± 120 cells/mm^2, $p < 0.05$).

TGFβ is a factor frequently associated with increased matrix production [6]. We have previously shown that BEA administration is accompanied by an increase in this factor [7]. In the normal kidney there is a basal level of TGF staining, primarily in the tubules. After BEA there is an increase in TGFβ staining primarily in the tubules adjacent to the fibrosis (Fig.3, left). The fibrotic area itself is negative. In the treated animals staining appeared to be only slightly above control levels (Fig. 3, right). The patchiness of the lesion made it difficult to do whole tissue ELISA quantification; therefore we digitized antibody-stained sections and measured the staining intensity in a defined area of fibrosis and in adjacent normal tubules. The normal area provided the level of background staining intensity and the increase in staining was compared to this level. BEA treated animals had a 77% increase in TGFβ staining as compared to the controls. In BEA animals receiving relaxin, the intensity of TGFβ staining was not different than the controls and significantly lower than that of BEA treated animals (Cont 1.56 ± 0.34 vs. BEA 2.77 ± 0.83 vs. BEA + Relaxin 1.60 ± 0.41 $p = 0.026$ BEA vs. BEA+ Relaxin).

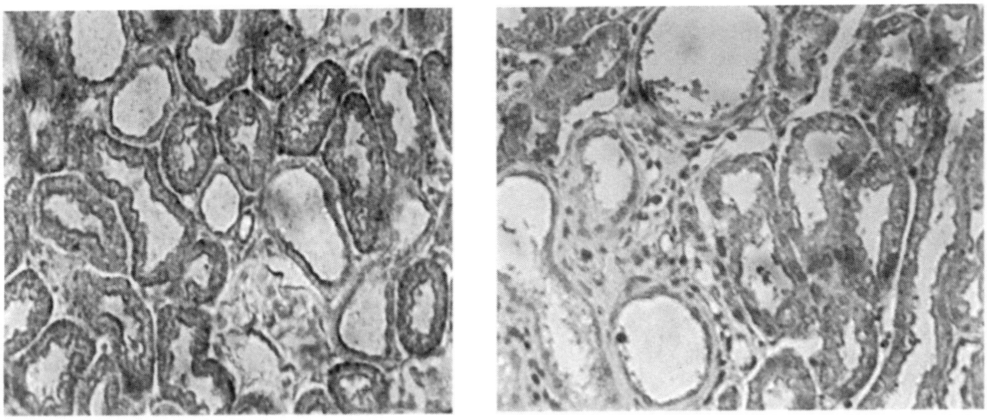

BEA   BEA + Relaxin

Fig. 3 Immunohistochemical staining of TGFβ at one month.

In conclusion, in this model of an existing progressive interstitial fibrosis, relaxin administration prevented the progression of fibrosis as measured both functionally and histologically. This is evidenced by an improvement in GFR, a decrease in the number of ED-1 positive cells, a decrease in immunohistochemical staining of TGFβ, and a decrease in the amount of fibrosis in the kidney. The ability of relaxin to prevent the progression of fibrosis makes it tempting to suggest that relaxin has promise as a therapeutic agent for the treatment of a multitude of renal diseases having interstitial fibrosis as a part of their pathologic picture.

Acknowledgements

This study was funded in part by a grant from the National Kidney Foundation of Illinois to SLG. JAL Arruda was supported by the Merit Review program from the Department of Veteran Affairs. Portions of this work were presented at the American Society of Nephrology Meeting (1999) and have been published in abstract form (J. Am. Soc. Nephrol. 10, (1999) A659).

References

1. Murray, G., Wyllie, R.G., Hill G.S., Ramsden, P.W. and Heptinstall, R.H., Am. J. Pathol., 67 (1972) 285.
2. Garber, S.L., Mirochnik, Y., Desai, S.S., Arruda, J.A.L. and Dunea, G., J. Am. Soc. Nephrol., 9 (1998) 1052.

3. Garber, S.L., Mirochnik, Y., Arruda, J.A.L. and Dunea, G., Am. J. Kidney Dis., 33 (1999) 1033.
4. Unemori E.N. and Amento, E.P., J. Biol. Chem., 265 (1990) 10681.
5. Unemori E.N., Pickford, L.B., Salles, A.L., Piercy, C.E., Grove, B.H., Erikson, M.E. and Amento, E.P., J. Clin. Invest., 98 (1996) 2739.
6. Yamamoto, T., Nobel, N.A., Miller, D.E. and Border, W.A., Kidney Int., 45 (1994) 916.
7. Mo, W., Brecklin, C., Garber, S.L., Song, R.H., Peroraro, A.A., Au, J., Arruda, J.A.L., Dunea, G. and Singh, A.K., Kidney Int., 56 (1999) 145.

Relaxin modulates fibrillin-2, but not fibrillin-1, gene expression by human dermal fibroblasts

Chrishan S Samuel[1,2], Lynn Y Sakai[4] and Edward P Amento[1,2,3]
[1]*Molecular Medicine Research Institute, Mountain View, CA 94043, Departments of* [2]*Dermatology and* [3]*Medicine, Stanford University School of Medicine, Stanford, CA 94305 and* [4]*Research Department, Shriners Hospital for Crippled Children, Portland, OR 97201, USA.*

Introduction

Fibrillins-1 and -2 are the main constituents of extracellular elastic microfibrils and contribute significantly to tissue integrity and elasticity [1]. Fibrillin-1 (FBN-1) was first discovered from the medium of cultured human dermal fibroblasts [2] and was identified as part of elastic-associated and non-elastic tissue microfibrils [3,4]. Mutations in the gene for FBN-1 have been linked to Marfans Syndrome (MS) [5,6], a disease characterized by defects in the cardiovascular, skeletal and ocular systems. Fibrillin-2 (FBN-2) was discovered and partially characterized later [5,7] and was found to be highly homologous and as widely expressed in many tissues as FBN-1, although FBN-2 was found to be highly expressed in embryonic tissues, while FBN-1 was increasingly expressed in maturing tissues. Furthermore, alterations to the FBN-2 gene caused the overlapping phenotype (to MS), of congenital contractural arachnodactyly [8]. While structurally similar, FBN-1 and –2 are thought to have diverse roles in elastic tissues, while their physiological role in non-elastic tissue is yet to be defined. It has been hypothesized that FBN-1 provides the force bearing structural support to elastic-associated tissues, while FBN-2 plays more of a role in the regulation of elastogenesis [1].

Relaxin is a 6kDa polypeptide hormone, best known for its potential roles in mammalian pregnancy, which include an ability to stimulate a remodeling of the connective tissues, in particular, collagen within the pelvic girdle and female reproductive tract in preparation for parturition [9]. Outside the female reproductive tract, relaxin is a potent modulator of collagen expression by human dermal fibroblasts in vitro [10] and in vivo [11,12], establishing relaxin as a mediator of extracellular matrix turnover. Studies using male pigs have demonstrated the ability of relaxin to enhance skin expansion [13]. Based on this finding, we hypothesized that relaxin may influence extracellular matrix proteins other than collagen that contribute to the elasticity of tissues. Therefore, the purpose of this study was to determine the effects of relaxin on elastic microfibril expression, in particular, on fibrillin mRNA and protein expression.

G.W. Tregear et al. (eds.), Relaxin 2000, 389–392.
© 2001 *Kluwer Academic Publishers. Printed in the Netherlands.*

Results and Discussion

The effects of relaxin on fibrillin mRNA expression

Dose-response studies

Normal human dermal fibroblasts (provided by Dr. E. Unemori, Connectics Corporation, Palo Alto, CA) were plated as described previously [14] and grown for three days to confluence and for a further five days past confluence (hyperconfluent culture). At each time point (day 3, day 8) the cells were treated with increasing concentrations (0, 0.1, 1, 10, 100ng/ml) of human recombinant (rH2) relaxin for 72 hours before RNA was extracted and analyzed by RT-PCR, using primers to FBN-1 [15] and FBN-2 [8]. β2-microglobulin and GAPDH were used as house-keeping genes, to ensure equal loading of the cDNA samples.

FBN-1 and FBN-2 were detected in both confluent and hyperconfluent cultures. rH2 relaxin had no effect on FBN-1 mRNA levels when applied to confluent and postconfluent cultures. Similarly, rH2 relaxin administration to confluent cultures did not effect FBN-2 mRNA levels, further supporting previous observations that the peptide does not regulate the ECM under basal conditions. When incubated with postconfluent cultures, rH2 relaxin (0.1-100ng/ml) caused a dose-dependent decrease in FBN-2 mRNA levels.

Time studies

We designed studies to examine the effects of rH2 relaxin on FBN-2 expression, when incubated as a bolus or with continuous culture. Cells were plated at postconfluent densities and grown for up to 6 days before being treated with rH2 relaxin (30ng/ml) for 18hrs or 72hrs, for 18hrs or 72 hrs after 3 days in postconfluent culture or continuously for 6 days in culture.

rH2 Relaxin had no effect on FBN-1 mRNA levels at any of the time points examined. Furthermore, rH2 relaxin had no effect on FBN-2 gene expression when applied to cultures over an 18-72 hr time period. The addition of rH2 relaxin for 72 hrs to day 3 hyperconfluent cultures, however, decreased FBN-2 mRNA expression by 15% of control (untreated) samples, while continuous exposure of the cells to rH2 relaxin (30ng/ml) for 6 days caused a 40% decrease in FBN-2 mRNA, compared to controls. These findings confirmed that rH2 relaxin induced a specific down-regulation of FBN-2 mRNA levels and that the effects of rH2 relaxin were more evident when present continuously in culture.

Relaxin inhibits TGF-β-induced increase in FBN-2

The addition of TGF-β (1ng/ml) to HDF was shown previously to be a potent modulator of collagen expression, up-regulating collagen synthesis to 9-fold of basal levels [10]. The effects of TGF-β on fibrillin expression, however, are unknown.

TGF-β induced a marked increase in FBN-1 and FBN-2 mRNA expression when applied to postconfluent HDF. The addition of rH2 relaxin and TGF-β to HDF had no effect on the TGF-β-mediated upregulation of FBN-1, but decreased FBN-2 mRNA levels such that they resembled those of control values.

The effects of relaxin on fibrillin protein expression

We next determined the effects of rH2 relaxin on FBN protein expression. Cells were plated at postconfluent densities and treated with or without rH2 relaxin for 6 continuous days in culture, the first 5 days in the presence of serum and the final 24 hours in serum-free conditions. HDF were also incubated in the presence of the collagen cross-link inhibitor, β-aminopropionitrile (BAPN) for the 6 days of culture. Aliquots of medium samples were run on 4.5% acrylamide gels under nonreducing conditions and analyzed by Western blot analysis, using the monoclonal antibodies to FBN-1 and FBN-2.

Fibrillins-1 and -2 were detected by Western blot analysis, with FBN-production greater than FBN-2. rH2 relaxin had no effect on FBN-1 protein expression, while FBN-2 secretion was decreased, consistent with our mRNA findings.

These studies further extend the number of components of the extracellular matrix, mediated by relaxin. In this study, we show that rH2 relaxin regulates specifically the expression of FBN-2, but not FBN-1. rH2 Relaxin decreased FBN-2 mRNA levels as well as FBN-2 protein production. These findings may be consistent with a potential role for relaxin in regulating the elastic component of tissues via regulation of FBN-2, without adversely affecting the structural integrity of tissues, which may be dependent on FBN-1.

References

1. Ramirez, F. and Pereira, L., Inter. J. Biochem. Cell. Biol., 31 (1999) 255.
2. Sakai, L.Y., Keene, D.R. and Engvall, E., J. Cell. Biol., 103 (1986) 2499.
3. Maddox, B.K., Sakai, L.Y., Keene, D.R. and Glanville, D.R., J. Biol. Chem., 264 (1989) 21381.
4. Sakai, L.Y., Keene, D.R., Glanville, R.W. and Bachinger, H.P., J. Biol. Chem., 266 (1991) 14763.
5. Lee, B., Godfrey, M., Vitale, E., Hori, H., Mattei, M.G., Sarfarazi, M., Tsipouras, P. and Ramirez, F., Nature, 352 (1991) 330.
6. Maslen, C.L., Corson, G.M., Maddox, B.K., Glanville, R.W. and Sakai, L.Y., Nature, 352 (1991) 334.
7. Zhang, H., Apfelroth, S.D., Hu, W., Davis, E.C., Sanguineti, C., Bonadio, J., Mecham, R.P. and Ramirez, F., J. Cell. Biol., 124 (1994) 855.
8. Maslen, C., Babcock, D., Ragunath, M. and Steinmann, B., Am. J. Hum. Genet., 60 (1997) 1389.
9. Sherwood, O.D. In Knobil, E. and Niell, J. (eds.), The Physiology of Reproduction. 2nd ed., Raven Press, New York, USA, p861.

10. Unemori, E.N. and Amento, E.P., J. Biol. Chem., 18 (1990) 10681.
11. Unemori, E.N., Beck, L.S., Lee, W.P., Xu, Y., Siegel, M., Keller, G., Liggitt, H.D., Bauer, E.A. and Amento, E.P., J. Invest. Dermatol., 101 (1993) 280.
12. Unemori, E.N., Pickford, L.B., Salles, A.L., Piercy, C.E., Grove, B.H., Erikson, M.E. and Amento, E.P., J. Clin. Invest., 98 (1996) 2739.
13. Kibblewhite, D., Larrabee, W.F. and Sutton, D., Acta. Otolaryngol. Head. Neck. Surg., 118 (1992) 153.
14. Brenn, T., Aoyama, T., Francke, U. and Furthmayr., H., Lab. Invest., 75 (1996) 389.
15. Ljubimov, A.V., Saghizadeh, M., Spirin, C.S., Mecham, R.P., Sakai, L.Y. and Kenney, M.C., Cornea, 17 (1998) 309.

Effect of relaxin on normal and impaired wound healing in rodents

Xinfan Huang, Gretchen Arnold, Martyn Lewis, Lorenzo Guzman, Beverly H. Grove, Elaine N. Unemori & Krisztina Zsebo
Connetics Corporation, 3400 West Bayshore Road, Palo Alto, CA 94303 USA

Introduction

Wound healing is a well orchestrated complex series of events and can be divided into three phases: inflammation, granulation tissue formation and re-epithelialization, and final tissue remodeling [1]. The common element that governs all three phases is local vascular supply [2]. One of the few axioms of surgery is that damaged tissue will heal and resist infection in direct proportion to its local perfusion. Feet heal more slowly and become infected more commonly than chests, necks, and heads in that order. Impaired local or regional perfusion seems to be the major factor in the failure to heal [2]. Common vascular conditions that lead to chronic wounds are diabetes, venous stasis, arterio-sclerosis, pressure sores, steroid hormone use, radionecrosis and ischemic ulcers.

Diabetics exhibit delayed wound healing due to skin ischemia, probably because of changes in vascular basement membranes and extracellular matrix that delay the entry of cells into wounds [3]. Loss of nutrient capillaries in the skin begins within a few years of the onset of diabetes and can precede the manifestation of major vascular disease. Contributors may include mechanical compression of capillaries from tissue edema and the basement membrane thickening of capillaries. As arterial insufficiency progresses, tissue loses its ability to resist infection and to repair itself after injury. Thus, the slightest injury can become significant, and anaerobic gangrenous infections become a hazard. This type of ulcer usually is deep and often malodorous. Granulation tissue rarely appears except after treatment.

Full thickness skin excisions in the db/db mouse provide a reproducible model of impaired wound healing and have been widely used to demonstrate wound healing defects and possible mechanisms. A complete explanation of the causes of the delay in healing in these animals is not yet available but several factors probably contribute. An important component is the delayed entry of inflammatory cells and fibroblasts into these wounds [4]. Some of the delay in cell entry may be due to the poor blood supply that results from pathological change in vasculature. The delay in cell entry may also reflect reduced or abnormal function of the inflammatory cells in these mice. The obesity of the db/db mice may further contribute to the impairment of healing by limiting the contribution of wound contraction to wound closure. Wounds in the db/db mice heal primarily by filling with granulation tissue and re-epithelialization. A third factor that may affect healing in these mice is their diabetes and the metabolic changes that accompany it. In addition to severe derangement in glucose and insulin levels, other metabolic systems may be changed. However, because of the impairment in the vascular

G.W. Tregear et al. (eds.), Relaxin 2000, 393–397.
© 2001 *Kluwer Academic Publishers. Printed in the Netherlands.*

component of healing in this model, full thickness skin wounds in the db/db mouse provide a useful model to explore the potential of various agents to stimulate wound healing. In this chapter, we describe the effects of systemic administration of rH2 relaxin on wound healing in an ischemic setting in the diabetic mouse.

Results and Discussions

To assess the effect of rH2 relaxin treatment on granulation tissue formation in normal and ischemic wounds, Hunt-Schilling wound chambers were implanted subcutaneously in either the shoulder or the more hypoxic hip region of female Sprague Dawley rats (Charles River Laboratories, MA). The wound chambers, fine stainless steel mesh cylinders sealed with medical grade silicone, provide a defined dead space for granulation tissue formation. rH2 relaxin or vehicle was administered continuously via an Alzet pump implanted subcutaneously. On day 18 after surgery, the granulation tissue contained within the chambers was harvested. The mean wet weight of granulation tissue (n=6) contained within the wound chambers from the hip region was lower than that in the wound chambers from shoulder region, presumably reflecting relatively delayed healing in the more ischemic posterior site. Systemic administration of rH2 relaxin was associated with a significant increase in the weight of granulation tissue formation in the wound chambers at the hip region ($p<0.05$), but not at the shoulder (Fig. 1).

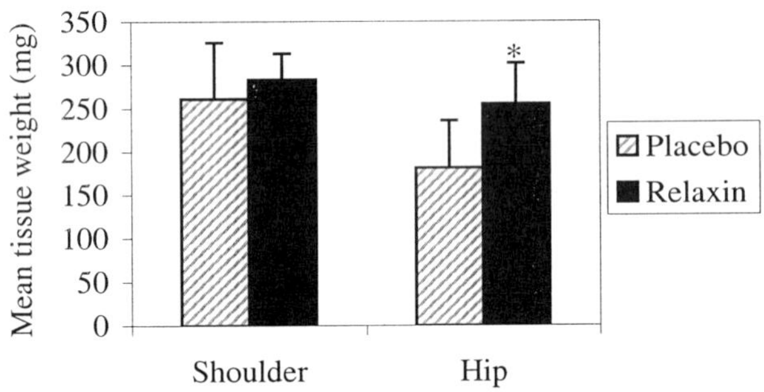

*Fig. 1. Systemic administration of rH2 relaxin on the weight of granulation tissue contained within wound chambers implanted at either the shoulder or the hip region. At 18 days after wound chamber implantation and rH2 relaxin treatment, the granulation tissue contained within the chambers was harvested. Relaxin treatment was associated with a significant increase in granulation tissue weight in the hip region, but not in the shoulder. Bar represent means ± SEM, n=6 animals/group; * $p<0.05$ vs vehicle-treated hip group by Student-Newman-Kuels test.*

The increase in granulation tissue in the chambers in the hip region was accompanied by an increase in bFGF and VEGF expression in wounds cells and an increase in new blood vessel formation [5]. These results suggest that rH2 relaxin treatment selectively enhances granulation tissue formation in ischemic wounds by stimulating angiogenesis via the induction of VEGF and bFGF in wound cells.

To determine whether rH2 relaxin would improve healing in the db/db mouse model of impaired healing, standardized full thickness (1.5 cm X 1.5 cm) wounds were made on the back of mice (Jackson Laboratories, Bar Harbor, ME). Mice were treated continuously for 21 days with rH2 relaxin via a subcutaneously implanted Alzet pump at a dose of either 0.1 mg/kg/day or 0.2 mg/kg/day. Wound areas were measured in the course of the study. Healing is presented as wound area on day 14 relative to the original size of the wound on day 0 and is calculated as: % of original wound area = [Area on day 14/area on day 0] x 100. Systemic administration of rH2 relaxin at a dose of 0.2 mg/kg/day significantly improved wound closure compared to that of animals treated with vehicle at 14 days (Fig. 2). The differences in the extent of wound healing were small among vehicle control and low dose of rH2 relaxin treatment.

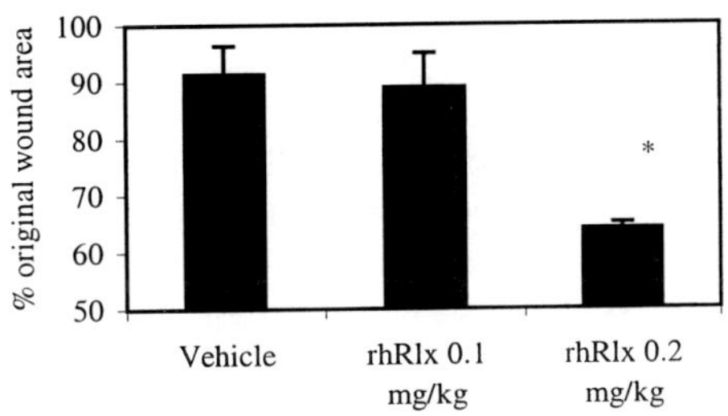

*Fig. 2. Wound areas on day 14 in diabetic mice treated with vehicle, and two doses of rH2 relaxin (rhRlx). Wound closure is presented as "% of original wound area" calculated as described above. Values are mean ± SEM, n=8-20 animals/group; * p<0.05 vs vehicle-treated group by Student-Newman-Kuels test.*

At day 21-post surgery, the wounds were also excised and fixed in formalin prior to paraffin embedding. Sections from the middle of the wound were stained with Masson's trichrome and evaluated by two investigators. Wounds treated with rH2 relaxin were covered by a thick, cellular and vascular granulation tissue (Fig. 3). Wound treated with vehicle demonstrated a very thin layer of granulation tissue and little angiogenesis. Inflammatory cell accumulation was present at the wound edges.

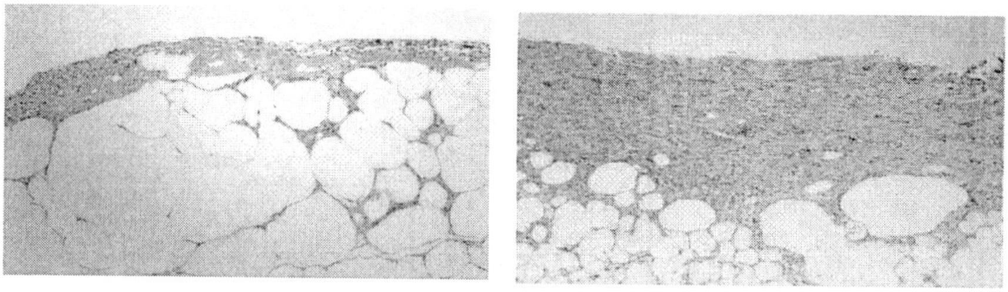

Fig. 3. Photomicrographs of a histological section taken midway through full thickness wounds stained with Masson's Trichrome at day 21. A vehicle-treated wound has modest cellular infiltration of inflammatory cells, fibroblasts, and macrophages. B. A rH2 relaxin (0.2 mg/kg) treated wound has a thick, cellular, vascular granulation tissue in the wound bed. Collagen deposition is in regular arrays.

Neovascularization is a key component of granulation tissue formation. New blood vessel formation is stimulated by a variety of growth factors including bFGF, and VEGF. Recently, we demonstrated relaxin's ability to induce both VEGF and bFGF in THP-1, a cell line of monocyte lineage, in vitro and in ischemic wound sites in vivo [5]. Immunohistochemical staining of the midsection of full thickness wounds using a polyclonal antibody to Von Willebrand factor VIII-related antigen was used to identify new blood vessels in the wounds (Fig. 4). Administration of rH2 relaxin was associated with stronger positive staining in the wounds from rH2 relaxin treated mice compared to the wounds of vehicle treated animals.

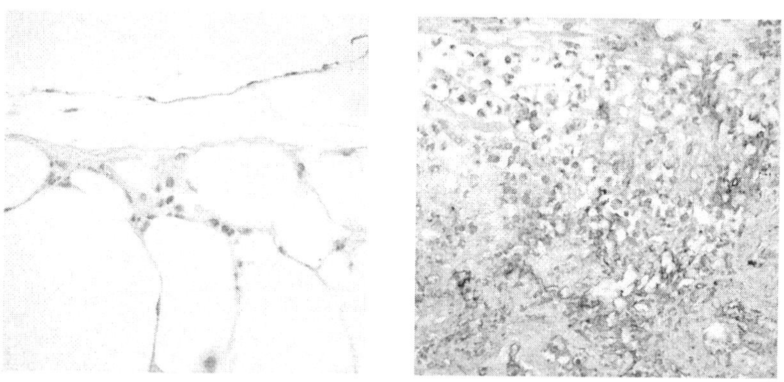

Fig. 4. Photomicrographs of immunohistochemical staining with Von Willebrand Factor VIII-related antigen (Dako Corporation, Carpinteria, CA) in vehicle (left panel) and rH2 relaxin (right panel) treated animals.

The observations from this study are consistent with the clinical observation made almost 30 years ago on porcine relaxin's effect in patients with scleroderma and peripheral vascular disease [6,7]. In these clinical studies, evidence of local vascular supply, including a rise in skin temperature, improved plethysmographic segmental and digital blood flow, and improved ulcer healing, were reported.

Since the healing potential of ischemic wounds is dependent upon the local blood supply and rH2 relaxin administration may increase vascular perfusion to wounds, relaxin may offer a new class of pharmacological therapy to enhance healing of ischemic wounds. Systemic administration of rH2 relaxin can enhance granulation tissue formation and speed wound closure by improving the vascular supply selectively in the ischemic wound sites. This biological activity of relaxin appears to be mediated via the dilation of local blood vessel and induction of angiogenesis. Its vasodilitory effects on local vessels is believed to be the result of blunting angiotensin II induced vasoconstriction [9], activating endothelin type-B receptors [10] and increasing local nitric oxide production [11]. The pro-angiogenic activity is mediated via the induction of angiogenic cytokines, such as VEGF and bFGF, in specific targets cells, likely include macrophages. This induction results in local enhancement of the concentration of these potent stimulators, and new blood vessel growth results.

In summary, the findings described here demonstrate that systemic administration of rH2 relaxin can stimulate earlier and more mature granulation tissue formation and speed the closure of ischemic wounds significantly in diabetic mice. Therefore, rH2 relaxin may offer a safe and convenient way to significant enhance clinical revascularization and healing of chronic wound. Recently, Connetics has initiated a clinical study to examine the effect of rH2 relaxin on ischemic wound healing.

References

1. Clark, R., J. Am. Acad. Derm., 13 (1985) 701.
2. Hunt, T. In Barbul, A., Caldwell, M., Eaglstein, H., Hunt, T., Marshall, D., Pines, E. (eds.) Clinical and experimental approaches to dermal and epidermal repair. Normal and chronic wounds, Wiley-liss New York, USA, 1991, p1.
3. Harris, M., Hadden, W.C., Knowles, W.C. and Bennet, P.H., Diabetes, 36 (1987) 523.
4. Greenhalgh, D., Srugel, K., Murray, M. and Ross, R., Am. J. Pathol., 136 (1990) 1235.
5. Unemori, E., Lewis, M., Constant, J., Arnold, G., Brove, G., Normand, J., Desphade, U., Salles, A., Pickford, L., Erikson, M., Hunt, T., and Huang, X., Wound Repair and Regeneration, 8 (2000) 361.
6. Casten, G., and Boucek, R., J.A.M.A., 166 (1958) 319.
7. Casten, G., Gilmore, H., Houghton, F. and Samuel, S., Angiology, 11 (1960) 408.
8. Seibold, J., Korn, J., Simms, R., Clements, P., Moreland, L., Mayes, M., Furst, D., Rothfield, N. and Steen, V., Ann. Int. Med., 132 (2000) 871.
9. Danielson, L., Sherwood, D. and Conrad, K., J. Clin. Invest., 103 (1999) 525.
10. Bani-Sacchi, T., Bigazzi, M., Bani, M., Mannaioni, P. and Masini, E., Br. J. Pharmacol., 116 (1995) 1589.
11. Toth, M., Taskinen, P. and Ruskoaho, H., J. Endocrinol., 150 (1996) 487.

The formation of amyloid fibrils by relaxin

Cait E MacPhee[1], Christopher M Dobson[1], and John D Wade[2]
[1]*Oxford Centre for Molecular Sciences, University of Oxford, New Chemistry Laboratory, South Parks Rd, Oxford OX1 3QT, UK; and* [2]*Howard Florey Institute, University of Melbourne, Parkville, Victoria 3052, Australia.*

Introduction

The amyloid diseases involve the deposition of normally soluble proteins in the form of stable insoluble fibrils in a variety of tissues. Amyloid formation is associated with a number of conditions including Alzheimer's disease, Creutzfeldt-Jakob Disease (CJD) and late-onset diabetes [1,2]. In addition to the amyloidogenic proteins associated with disease, it is now becoming clear that the formation of fibrils is a fundamental property of the polypeptide backbone [3,4,5,6]. A wide variety of structurally and functionally unrelated proteins are now known to self-associate into amyloid structures [7,8], yet the fibrils formed by different proteins share a common fundamental architecture. Amyloid is characterised by a typical 'cross-β' X-ray diffraction pattern indicative of a stacked β-sheet structure in which the strands are arranged perpendicular to the fibril axis; by a long, unbranched appearance under electron microscopy; and by staining with the dye Congo Red, which exhibits a green birefringence when viewed under polarised light [9].

The gelation of insulin is well-known and has been investigated for decades [10], although the realization that insulin assembles into structures that resemble the amyloid fibrils implicated in human disease is more recent [11]. The aggregation of insulin in the form of amyloid fibrils has posed an obstacle to the development of implantable devices for the long-term delivery of insulin to diabetics [12,13], and has also been implicated in the development of an immunological response to insulin [14]. More disturbingly, insulin amyloid fibril deposits have been identified at the site of injection in some diabetics [15]. Relaxin is a member of the insulin superfamily and is structurally closely related to insulin, consisting of two chains covalently linked by three highly conserved disulphide bonds. Other than these cysteine residues however, the sequences of insulin and relaxin show little identity. Given the recent interest in possible therapeutic uses of relaxin [for review, see 16], it is of interest to investigate the amyloid fibril forming potential of the recombinant human protein (rH2 relaxin).

Results and Discussion

The formation of fibrils by rH2 relaxin was initially examined under the conditions employed for insulin fibril formation. Insulin fibrils can be formed by heating a concentrated (2.5 mM) stock solution of protein at pH 2 over a period of 5 h at 70°C. Under these conditions, the protein loses α-helical structure (Fig.1) acquiring a partially unfolded conformation as indicated by a blue-shift in the CD spectrum and a decrease

G.W. Tregear et al. (eds.), Relaxin 2000, 399–404.
© 2001 *Kluwer Academic Publishers. Printed in the Netherlands.*

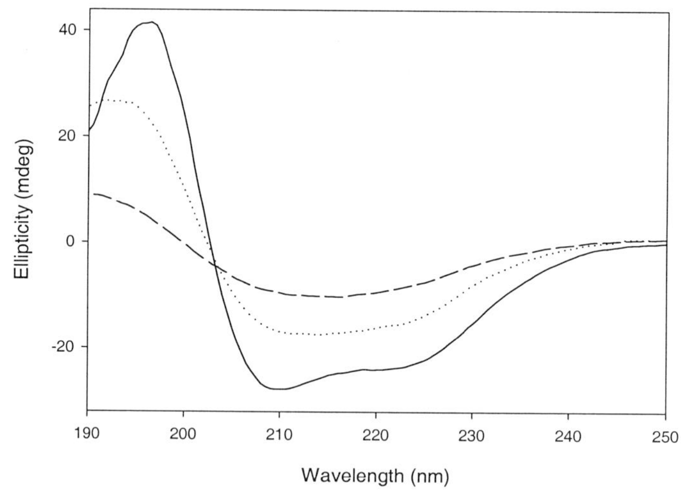

Fig. 1. Far-UV circular dichroism spectra illustrating the transition from the soluble, α-helical form of insulin (solid line), to a partially folded form after 1 h incubation at 70°C (dotted line) to β-sheet insulin amyloid fibrils after 5 h at 70°C (dashed line).

the observed ellipticity. After heating for 5 h the CD spectrum shows a single minimum at 216 nm indicative of the adoption of a β-sheet conformation, which is not reversible on return to room temperature. Fibrillar aggregates are observed by transmission electron microscopy (EM; Fig. 2).

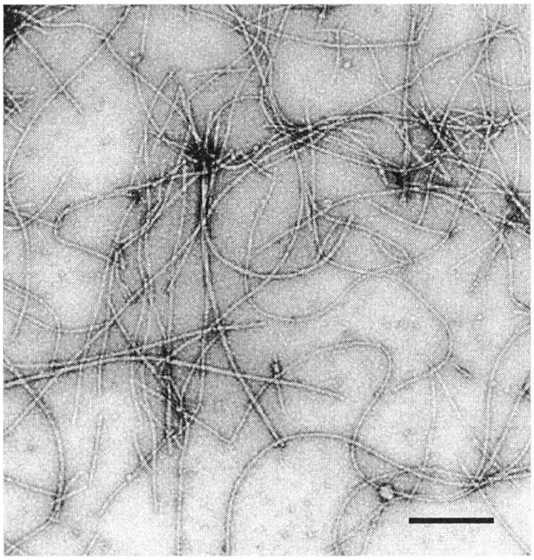

Fig. 2. A transmission electron micrograph of amyloid fibrils formed after incubation of a 2.5 mM stock solution of insulin for 5 h at 70° and pH 2.0. Scale bar: 200 nm.

Prolonged heating of a 3 mM stock solution of rH2 relaxin at 70°C also induces a partial unfolding of the protein (Fig. 3) similar to that described earlier for insulin, however the protein refolds to a predominantly α-helical conformation on return to room temperature. rH2 relaxin is therefore substantially less susceptible than insulin to

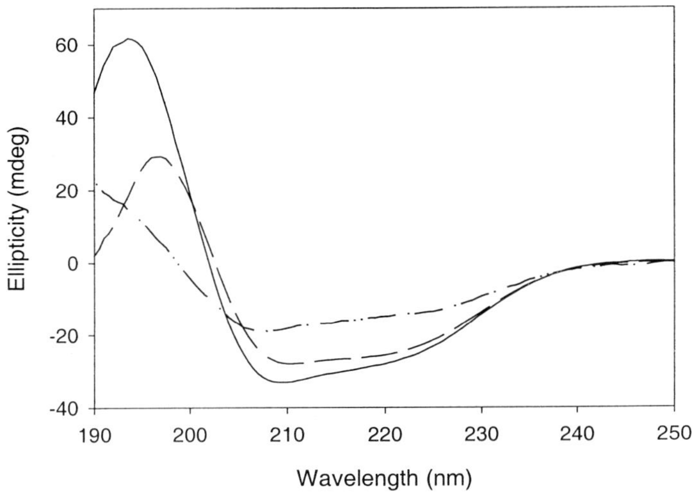

Fig. 3. CD spectra of relaxin at pH 2 at room temperature (solid line), 70°C (dash-dotted line) and on return to room temperature after heating at 70°C for 5 days (dashed line).

irreversible aggregation under conditions of high temperature and low pH. The change in the CD spectrum on cooling is consistent with the development of a small amount of random coil conformation, and EM indicates the formation of aggregates (Fig. 4a). The low concentration of these aggregates does not, however, permit characterization of their nature.

In contrast to native rH2 relaxin in which the three disulphide bonds are intact, reduced rH2 relaxin readily and rapidly forms fibrils at low pH. Addition of 1 mM of the reducing agent tris-carboxyethylphosphine (TCEP) rapidly induces the gelation of a 3 mM stock solution of rH2 relaxin, and ordered fibrils are visible by EM within 24 h at room temperature (Fig. 4b). The CD spectra indicate the formation of β-sheet structure (*data not shown*). The B chain of rH2 relaxin is notoriously difficult to study because it rapidly aggregates and precipitates after synthesis, making formation of the native heterodimeric complex with the rH2 relaxin A chain problematic [17]. It is therefore possible that only one of the two rH2 relaxin peptides is aggregating to form the fibrils observed in Fig 4b. There is no evidence from CD spectroscopy, however, for the presence of a significant proportion of random coil conformation in the solution, suggesting that both chains are incorporated into β-sheet aggregates. It is not as yet known whether the two chains segregate into separate populations of aggregates, or coprecipitate within mixed fibrils as has been observed in other systems [6].

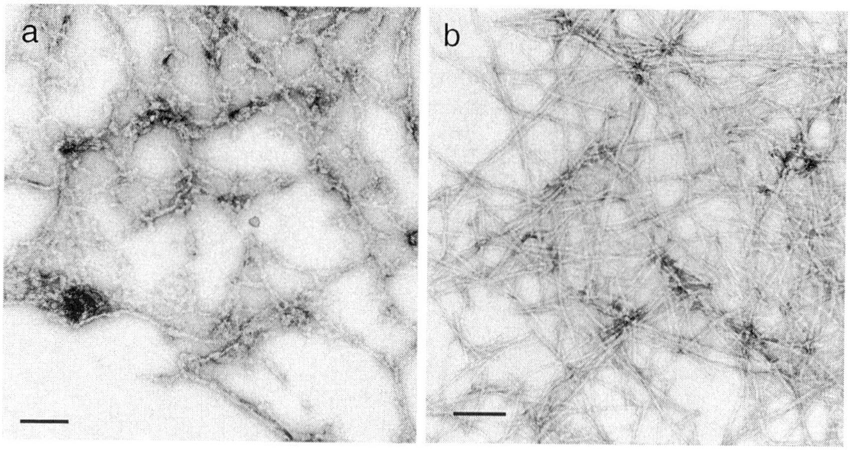

Fig. 4. Transmission electron micrographs of aggregates formed by 3 mM relaxin (a) after 5 days incubation at 70 C, pH 2.0 and (b) after addition of 1 mM TCEP pH 2.0 and incubation for 24 h at room temperature. Scale bar: 200 nm.

In further experiments we examined the influence of pH on the potential of rH2 relaxin to form amyloid fibrils. All relaxins have higher pI values than insulins due to a number of conserved arginine residues in the native sequence. The predicted pI of rH2 relaxin is above 10, and all known relaxins have predicted pI values above 9. Human insulin in contrast exhibits a pI of 5.6, and all known insulins have pI values below 6. We incubated rH2 relaxin at 50°C for 24 h at pH 8.0 at the relatively low concentration of 50 μM. The solution pH was adjusted by the addition of NaOH. Over this period, the CD spectrum indicates the formation of predominantly random coil structure (Fig. 5a),

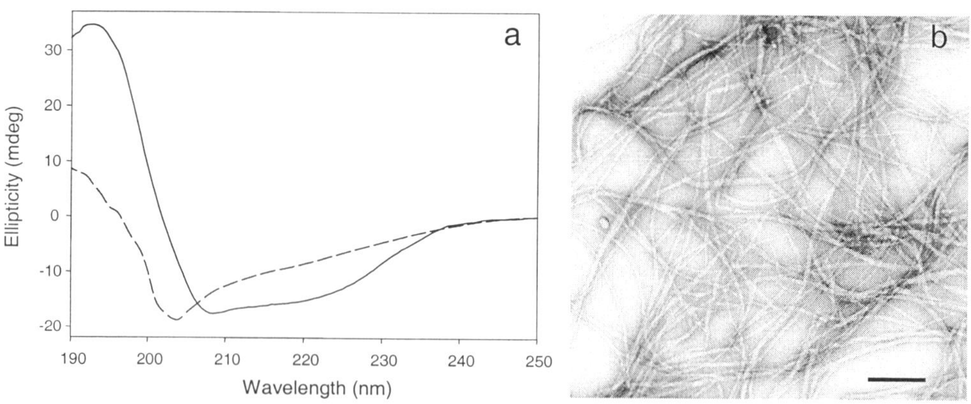

Fig. 5 (a) The formation of fibrils from native relaxin monitored by CD spectroscopy following incubation of a 50 μM stock solution at pH 8.0 for 24 h at 50°C (dashed line). The CD spectrum of relaxin at pH 2.0 is plotted for comparison (solid line); (b) TEM shows the formation of fibrillar aggregates.

which did not refold to the native state on return to room temperature. This result confirms that the folded state of rH2 relaxin is unstable under conditions of high pH. Examination of the solution by EM indicates the formation of fibrillar aggregates (Fig. 5b) suggestive of amyloid fibril formation. The solution of rH2 relaxin does not convert entirely to β-sheet structure, as was observed at pH 2.0 both for insulin (Fig. 1) and for reduced rH2 relaxin. The CD spectrum in Fig 5b may contain contributions from β-sheet structure; β-sheet conformations, however, give rise to a dichroic band of significantly lower magnitude than that of unfolded polypeptides. Given the difficulty in determining absolute peptide concentrations in a mixture of aggregated and soluble material it is not possible to deconvolute the spectrum with confidence. Further work is underway to characterize the nature of the aggregate.

This study has demonstrated that rH2 relaxin is relatively stable to aggregation under conditions that induce fibril formation by insulin. However, even at relatively low concentrations and moderate temperature at high pH, rH2 relaxin self-associates into fibrillar aggregates. The basic pH can be rationalized by the higher pI of rH2 relaxin. These studies were performed in the absence of salts, however, and ionic contributions may stabilize the secondary structure sufficiently to prevent aggregation at basic pH. However the results suggest that rH2 relaxin may share with the structurally-related insulin the potential to readily form amyloid superstructures. Such phenomena need to be studied in detail from the point of view of storage and safe therapeutic administration of this small peptide hormone.

Acknowledgements

The Oxford Centre for Molecular Sciences is supported by the U.K. BBSRC, the EPSRC, and the MRC. C.E.M is a Royal Society Dorothy Hodgkin Research Fellow, and was supported in part by a McIlrath Research Fellowship from St Hilda's College, Oxford. The research of C.M.D. is supported in part by the Wellcome Trust.

References

1. Lansbury, P.T., Proc. Natl. Acad. Sci. USA, 96 (1999) 3342.
2. Perutz, M.F., Trends Biochem. Sci., 24 (1999) 58.
3. Guijarro, J.I., Sunde, M., Jones J.A., Campbell, I.D. and Dobson, C.M., Proc. Natl. Acad. Sci. USA, 95 (1998) 4224.
4. Chiti, F., Webster, P., Taddei, N., Clark, A., Stefani, M., Ramponi, G. and Dobson, C.M. Proc. Natl. Acad. Sci. USA, 96 (1999) 3590.
5. Dobson, C.M., Trends Biochem. Sci., 24 (2000) 329.
6. MacPhee, C.E. and Dobson, C.M., J. Am. Chem. Soc., (2000) in press.
7. Sunde, M. and Blake, C.C., Q. Rev. Biophys., 31 (1998) 1.
8. Kelly, J.W., Curr. Opin. Struct. Biol., 8 (1998) 101.
9. Sunde, M., Serpell, L.C., Bartlam, M., Fraser, P.E., Pepys, M.B. and Blake, C.F., J. Mol. Biol., 273 (1997) 729.
10. Waugh, D.F., J. Am. Chem. Soc., 68 (1946) 247.

11.Glenner, G.G., Eanes, E.D., Bladen, H.A., Linke, R.P. and Termine, J.D., J. Histochem. Cytochem., 22 (1974) 1141.
12.Hutchison, K.G. J. Pharm. Pharmacol., 37 (1985) 528.
13. Sluzky, V., Tamada, J.A., Klibanov, A.M. and Langer, R., Proc. Natl. Acad. Sci. USA, 88 (1991) 9377.
14. Brange, J., Andersen, L., Laursen, E.D., Meyn, G. and Rasmussen, E., J. Pharm. Sci., 86 (1997) 517.
15.Dische, F.E., Wernstedt, C., Westermark, G.T., Westermark, P., Pepys, M.B., Rennie, J.A., Gilbey, S.G. and Watkins, P.J., Diabetologia, 31 (1988) 158.
16. Bani, D., Gen. Pharmacol., 28 (1997) 13.
17. Wade, J.D. and Tregear, G.W., Meth. Enzymol., 289 (1997) 637.

SECTION 13

Clinical aspects of relaxin II – Joints, breast cancer & delivery

Pelvic girdle relaxation, developmental dysplasia of the hip and the hormone relaxin – are they loosely connected?

Alastair H MacLennan
Department of Obstetrics and Gynaecology, Adelaide University, South Australia, 5000

Introduction

A decade ago Dr Richard Vandlen from Genentech Inc. and I were invited to a conference on symptom-giving pelvic girdle relaxation of pregnancy and postnatal pelvic joint syndrome in Tromsø, Norway, to give talks on the hormone relaxin and its possible association with pelvic changes before and after human pregnancy. I was surprised to find alongside this medical conference in the Artic Circle another conference where the delegates were women severely disabled with pelvic joint pain. Nearly 100 women had gathered on the floor of a neighbouring gymnasium and moved with the aid of wheel chairs, walking frames and walking sticks. I was informed that these were members of the Norwegian Association for Women with Pelvic Girdle Relaxation and that there were 1,609 members. I was informed that this problem, during and after pregnancy, was relatively common in countries bordering the west coast of Northern Europe where other similar national societies existed. No comprehensive survey of this clinical problem had been conducted at that time and so, in collaboration with the Norwegian Association, an epidemiological survey of its members was conducted and is described below.

Excessive pelvic ligament relaxation during pregnancy causing considerable pain and/or pelvic instability, which is sufficient to impair normal daily function, has been termed *symptom-giving pelvic girdle relaxation* [1]. Similar pains, following pregnancy, in relation to one or more of the pelvic joints, have been recommended to be called *pelvic joint syndrome* [1]. Although pelvic and low back pain can be due to other causes, and no unambiguous definition of pelvic girdle relaxation or pelvic joint syndrome exists, the classic woman presents during or after pregnancy with typical walking dysfunction, low back and pelvic joint pain with reproduction of the pain by sacroiliac provocation tests [2]. The world-wide prevalence of this condition has not been clearly defined but there are wide variations in different populations from less than 1% to 9% [3]. Northern European countries, particularly Norway, have reported the highest incidence of these conditions [4]. In Norwegian there is a specific word for this condition - *bekkenlosning*. A family history of this disorder has been reported in patients with these conditions, suggesting the possibility of a genetic inheritance [4]. This gene or genes may be more common in Scandinavian populations.

Symptom-giving pelvic girdle relaxation and pelvic joint syndrome have been inadequately studied and many patients with these conditions complain about the lack of medical understanding or acceptance of their condition and the lack of effective treatments offered. The etiology of the pelvic relaxation and pain is unknown but a susceptibility to the hormones of pregnancy, particularly the hormone relaxin, has been

G.W. Tregear et al. (eds.), Relaxin 2000, 407–413.
© 2001 *Kluwer Academic Publishers. Printed in the Netherlands.*

postulated [5]. Developmental dysplasia (congenital dislocation) of the hip is another pathology that has been postulated as possibly due to the susceptibility of the female fetus to the hormone relaxin [6] and there is also a high incidence of this condition in the Norwegian population [7].

In 1988 large numbers of Norwegian women purporting to suffer from these problems joined together to form an association, "Landsforeningen for Kvinner med Bekkenløsningsplager" (LKB) to promote better care for women in similar situations. This has allowed an international collaboration of Australian researchers with a background of research in the hormone relaxin, and the LKB, to conduct the largest survey to date of women who have suffered from symptom-giving pelvic girdle relaxation and pelvic joint syndrome. Our objective was to gather greater descriptive and epidemiological information on these problems and to determine the incidence of congenital hip disorders in the children of women with symptom-giving pelvic girdle relaxation during pregnancy.

Patients and Methods

Questionnaires were posted to all 1,609 members of the Norwegian Association for women with pelvic girdle relaxation who had previously indicated that they had experienced either symptom-giving pelvic girdle relaxation syndrome or postnatal pelvic joint syndrome. Members were recontacted once if there was no initial response. 1,269 members replied (79% response rate). Amongst these responses there where 38 members who did not wish to complete the questionnaire and 17 members who said that they had inappropriately registered as having experienced this type of pelvic problem. Thus, 1,214 questionnaires were available for analysis.

The questionnaire initially established whether the respondents' pelvic symptoms had begun in pregnancy or immediately after a pregnancy and whether the diagnosis of pelvic girdle relaxation of pregnancy or pelvic joint syndrome had been confirmed by a doctor, physiotherapist, midwife or qualified manual therapist. 99 respondents had never had their diagnosis confirmed by one of these health professionals. The main data analysis was performed on the remaining 1,115 respondents whose diagnosis had been confirmed by a health professional. The data was entered and analysed on a DBASE IV database software package.

The circulated questionnaire was in Norwegian and the data derived from this was coded and entered into the database by Norwegian research staff. The 56 questions identified baseline demographic characteristics such as age, parity, employment and current and past smoking status. Specific questions concerning the pelvic symptoms ascertained the pregnancy of onset, the duration of symptoms before and after delivery, sites of pain, extent of incapacity and the influence of subsequent menstrual cycles on the symptoms. Obstetric data included numbers of pregnancies and pregnancy outcomes including mode of delivery, gestation at delivery and method of infant feeding. The women were asked whether they perceived the therapies they received to be effective or ineffective and this included the effect of hormonal contraception. Specific questions were asked about whether the respondent had delivered any child who was subsequently diagnosed by X-ray or ultrasound examinations as having a congenital dislocation (developmental dysplasia) of the hip or a presumed diagnosis of this condition which required a major therapeutic intervention such as splints, plaster casts or surgery. A family history (first or second degree relatives) of pelvic girdle relaxation and/or developmental dysplasia of the hip was also sought. Severity of current and past pelvic joint pain was estimated using a visual analogue scale of 0 (no pain) to 5 (extreme).

Results

Analysis of the data was performed on the 1,115 respondents whose diagnosis had been confirmed by a health professional. Mean parity of the respondents was 1.98 (SD 1.15) and 34.9% had previously been smokers. Parturition details were generally unremarkable with a 9.2% prematurity rate, 4.4% cesarean section rate and a 4.4% instrumental delivery rate. A history of back pain prior to the first pregnancy was reported by 481 (23.2%) of the respondents. For 95.2% of respondents the pelvic pains had begun during pregnancy; most having begun in the first trimester of pregnancy. The problems began in the first pregnancy for 74% of respondents. Of the 317 women who had two pregnancies 81.4% responded that their symptoms were worse in their second pregnancy. 230 women had three pregnancies and 60% of these said the symptoms in the third pregnancy were worst. The main length of pelvic problems during and after pregnancy was 6.25 years (SD 5.68) with a range of 1 to 420 months. Some respondents still had continuing pain.

The most common sites of pelvic pain recorded on an anatomical chart by the respondents were sacroiliac (50.0%) and symphysis pubis (31.7%). However, 64% of respondents complained of pains at non-pelvic sites including neck (59.8%), shoulders (58.6%) and thighs (51.1%). Nearly all respondents graded their pains as very severe (5 on the 0 to 5 analogue scale) when they were most incapacitated. The problems, when at their most severe, had required the use of crutches (64.4%), a wheel chair (13.8%), cushions (8.3%), a walking frame (4.8%) or a walking stick (2.8%). 84.5% of employed women required sick leave specifically due to this problem.

All respondents nominated that their pelvic problems had a negative effect on at least one aspect of their lives. The activities which were worst affected were, in ranking order: housework, exercise, activities with their children, employment, leisure/hobbies and their personal relationships or married life.

A wide variety of non-pharmaceutical treatments had been tried by the respondents with improvement or temporary relief in most. No therapy gave clear long term relief. Bed rest and physical support were ranked more highly than alternative therapies. (Table 1). Analgesics were used by 49.4% of respondents, for their pelvic pain. A hormonal oral contraceptive had been used following the onset of pelvic joint syndrome in 31% of women and 75% of these thought that such therapy made no difference to their symptoms. Equal numbers of the remainder thought that oral contraception made the symptoms better or worse. However, 87% of women not on oral contraception thought that their symptoms fluctuated during their menstrual cycles, being worse mostly in the second half of the cycle. Cessation of breast feeding had no effect in 69%, improvement in 9% and exacerbated the symptoms in 22%.

Table 1. Respondents evaluation of therapies for pelvic joint syndrome

TREATMENT	NUMBER TREATED	TEMPORARILY IMPROVED %	TEMPORARILY WORSE %
BED REST	803	94	2
RELAXATION	337	94	<1
CRUTCHES	652	92	<1
MANUAL THERAPY	114	88	6
PHYSIOTHERAPY	440	78	11
PELVIC BELT	574	76	11
TAPING	144	72	8
ACUPUNCTURE	52	64	6
HOMEOPATHY	47	56	4
CHIROPRACTIC	132	27	18
OPERATION	5	20	20

Only 933 (83.7%) of the respondents answered the questions about whether they had children with possible developmental dysplasia (congenital dislocation) of the hip. These respondents had a total of 87 children with this diagnosis, giving an incidence of 4.7% in the children delivered to this group. To further refine the respondent's diagnosis, 84 (4.5%) of the children delivered had been diagnosed by X-ray or ultrasound or had required a major therapeutic intervention eg splinting. The sex ratio of the affected children was 2:1, female:male. 13.5% of the babies born with a diagnosis of developmental dysplasia of the hip were in a breech presentation at delivery.

4.6% of the respondents thought they themselves had been born with a hip dysplasia, 13.5% reported that other members of their family (first and second degree relatives) had a history of developmental dysplasia of the hip and 29.0% reported a family history of symptom-giving pelvic girdle relaxation in pregnancy and pelvic joint syndrome.

Discussion

A postal survey of self-selected women who have a presumptive diagnosis of pelvic girdle relaxation and pelvic joint syndrome has clear methodological limitations and cannot, for instance, allow calculation of prevalence of the syndrome or allow prospective correlation of other possible etiological factors such as serum relaxin or tissue receptor levels. However, this is the largest published survey of women with these poorly understood but debilitating conditions describing the clinical factors surrounding this problem. It may also allow generation and testing of hypotheses for its etiology and treatment. The survey findings are in broad agreement with a previously published survey of 153 Norwegian women with pelvic pain and joint instability [8].

The results of this survey clearly show that the respondents found this condition to be severely debilitating and that the problem often continues for many years after its initiation during pregnancy. In general, subsequent pregnancies are even more debilitating; having major repercussions on a woman's physical capability and social interaction. Although these women had tried many therapies which had often brought temporary relief, no one therapy was associated with permanent relief. The cessation of breast feeding and the return of menstruation sometimes exacerbated the pains premenstrually but the avoidance of ovulation by the use of combined hormonal oral contraception was not generally associated with an improvement in symptoms. Despite the reported severity of the pain, only half of the respondents regularly used analgesics over long periods. Most found rest to be the best temporary remedy. Many women expressed that there was inadequate community support for people who have pelvic joint syndrome. In countries where this problem is less common these women feel very lonely and misunderstood with many physicians unaware of the nature of the problem and therefore unsure how to explain the problem or treat the patient [5].

The survey suggests that there is a strong family history of pelvic joint syndrome and this may correlate with the high prevalence of this problem in Norway [4]. This supports the hypothesis that these women may have a genetic susceptibility to the hormones of pregnancy, and possibly in particular, the hormone relaxin. During their pregnancies the women surveyed also experienced pain in other major joints eg neck and shoulder suggesting a general arthropathy rather than just a local pelvic effect. A significant increase in peripheral joint laxity during pregnancy has been reported previously [9]. Excessive laxity of the pelvic ligaments, together with other effects of pregnancy, including increased weight and postural changes, may make the pelvic joints vulnerable to unusual trauma or stress. Whether it is joint trauma sustained during the joint laxity of pregnancy that takes years to recover, or whether there is a continuing post-natal pathology is not clear. Radiological examination of 121 women with pelvic joint syndrome by Walde [10] revealed a high incidence (31%) of osteosclerotic changes in the symphysis, sacroiliac joints and lumbar discs, the origins of which the author related to prior pregnancies with "pelvic relaxation".

Two studies have shown statistically significant higher serum relaxin levels in many women with severe pelvic or back pain in pregnancy [11, 12]. Two other smaller studies have not confirmed a relationship between serum relaxin level levels and pelvic pain during or after pregnancy [13, 14]. Serum relaxin levels are almost undetectable outside pregnancy except in the mid-luteal phase, but are at their highest in early to mid pregnancy [15]. In this survey the onset of most patients' symptoms was at this gestation. Relaxin is both an endocrine and a paracrine hormone and therefore serum levels may not reflect their local influence [16]. In the human, serum relaxin levels generally reflect ovarian production but later in pregnancy the decidua and placenta produce relaxin without contributing to serum levels [16, 17]. A local effect within the pelvis and uterus on myometrial relaxation, cervical ripening and other connective tissue softening is one of the major postulates for this ubiquitous hormone which has many actions [5]. It is unlikely that relaxin would have a continuing effect on joint laxity after delivery of the placenta. Therefore, slow or inadequate postpartum resolution of the joint pathology and trauma sustained in pregnancy would have to be postulated if relaxin was to be incriminated as the cause of persisting pelvic joint syndrome.

Relaxin has also long been postulated as a possible agent in the pathology of developmental dysplasia of the hip [6]. In this survey of Norwegian women who had experienced pelvic girdle relaxation and pelvic joint syndrome the incidence of confirmed and treated development dysplasia of the hip was 45/1,000 which is five times the reported incidence for the Norwegian population (9.0/1,000) [7] and about six times greater than many other countries such as Australia (7.5/1,000) [18]. Some of the increased incidence of hip dysplasia reported in children of the women surveyed may be associated with the high incidence (13.5%) of their self reported family history of this disorder. There was also a strong family history of pelvic joint syndrome in the women surveyed. We have previously reported a study of serum relaxin levels in the cord blood of 24 babies who were later confirmed as having congenital dislocation of the hip (now termed developmental dysplasia of the hip) matched against the levels in the cord blood of babies without hip pathology [16]. Mean cord relaxin levels were higher in the babies with hip dysplasia (28.7 versus 19.8 pg/ml) but this did not reach statistical significance (p=0.086). Until the relaxin receptor has been identified, isolated and can be measured in target tissues it is not possible to know if some mothers with pelvic pathology and some fetuses with hip dysplasia have an increased susceptibility to the hormone relaxin. On the data, to date, the possibility of a link between the hormone relaxin, pelvic girdle relaxation of pregnancy and developmental dysplasia of the hip remains tantalising but unproven. This qualitative survey underscores the long term debilitating effects of pelvic joint syndrome, the lack of effective treatment for this condition and the need for greater understanding of its aetiology. Identification of relaxin receptor changes in affected joints is a hypothesis requiring testing before selective relaxin receptor modulators could be designed as possible therapeutic agents in affected pregnancies. The potential to effectively treat these pelvic or hip pathologies following pregnancy seems remote.

Acknowledgments

This survey was published in detail as MacLennan, A.H., MacLennan, S.C. and L.K.B. Acta Obstet Gynecol Scand, 76 (1997) 760.

References

1. Dietrichs, E., Kogstad, O., Scand. J. Rheumatol. Suppl., 88 (1991) 3.
2. Kogstad, O. Scand. J. Rheumatol. Suppl., 88 (1991) 21.
3. Ostergaard, J., Bonde, B. and Thomsen, B.S., UGESKR-LAEGER, 154 (1992) 3568.
4. Kogstad, O. and Biornstad, N., TIDSSKR-NOR-LAEGEFOREN, 110 (1990) 2209.
5. MacLennan, A.H., Scand. J. Rheumatol. Suppl., 88 (1991) 7.
6. Wilkinson, J.A., J. Bone Jt. Surg., 45B (1963) 268.
7. Hinderaker, T., Daltveit, A.K., Irgens, L.M., Uden, A. and Reikerås, O., Acta. Orthop. Scand., 65 (1994) 239.
8. Saugstad, L.F., Eur. J. Obstet. Gynec. Reprod. Biol., 41 (1991) 197.
9. Calguneri, M., Bird, H.A. and Wright, V., Ann. Rheum. Dis., 41 (1992) 126.
10. Walde, J., Acta. Obstetrica. Scan., XLIL (1962) 11.
11. MacLennan, A.H., Nicholson, R., Green, R.C. and Bath, M., Lancet, ii (1986) 243.
12. Kristiansson, P., Svärdsudd, K. and von Schoultz, B., Am. J. Obstet. Gynecol., 175 (1996) 1342.
13. Hansen, A., Jensen, D.V., Larsen, E., Wilken-Jensen, C. and Petersen, L.K., Acta. Obstet. Gynecol. Scand., 75 (1996) 245.
14. Schauberger, C.W., Rooney, B.L., Goldsmith, L., Shenton, D., Silva, P.D. and Schaper, A., Am. J. Obstet. Gynecol., 174 (1996) 667.
15. MacLennan, A.H., Nicolson and R., Green, R.C., Lancet, ii (1986) 41.
16. Borthwick, G.M., Borthwick, A.C. and Grant, P., In MacLennan, A.H., Tregear, G.W. and Bryant-Greenwood, G.D. (eds) Progress in Relaxin Research, World Scientific Publishing Co., Singapore, 1995, p. 251.
17. Weiss, G., In MacLennan, A.H., Tregear, G.W. and Bryant-Greenwood, G.D. (eds) Progress in Relaxin Research, World Scientific Publishing Co., Singapore, 1995, p. 199.
18. Annual Report of the South Australian Birth Defects Register, Medical Genetics and Epidemiology, Women's & Children's Hospital, Adelaide, 1994, p. 9.

Back pain in IVF-induced and spontaneous pregnancies

Per Kristiansson[1,2], Lena Nilsson-Wikmar[3], Bo von Schoultz[1], Kurt Svärdsudd[2] and Håkan Wramsby[1]

[1]Karolinska Institute, Department of Women and Child Health, Karolinska Hospital, S-171 76 Stockholm, Sweden; [2]Uppsala University, Department of Public Health and Caring Sciences, S-751 85 Uppsala, Sweden; and [3]Karolinska Institute, Department of Physical Therapy, S-141 57 Huddinge, Sweden

Introduction

It is well known that the pelvis undergoes changes during pregnancy [1,2] and that pelvic pain is common among pregnant women [3,4]. MacLennan *et al.* [5] were the first to show an association between reported severe incapacitating pain located in the symphysis in pregnant women and elevated levels of circulating relaxin.

Relaxin is a polypeptide hormone of ovarian origin of the insulin-like growth factor family. Relaxin is known to remodel pelvic connective tissue in several mammalian species during pregnancy [6]. We have observed that high mean serum relaxin levels were correlated to low back pain or symphyseal pain in late pregnancy [7]. In another study, we found that the serum relaxin levels were about ten times higher among women pregnant after IVF treatment than in spontaneously pregnant women [8]. Moreover, in IVF pregnant women the relaxin level during pregnancy was closely correlated to the number of follicles developed during ovarian stimulation as well as the number of oocytes recovered at ovum pick-up in the treatment cycle, i.e. direct effects of the ovarian stimulation regimen.

If relaxin is essential for the development of pregnancy-related pelvic pain, including the sacral area, one would expect a higher incidence of back pain among IVF pregnant women than among spontaneously pregnant women. The present study was designed to evaluate back pain during IVF pregnancies prospectively and compare it with that in spontaneous pregnancies and we hypothesised that reported and provoked back pain would be more frequent in the IVF pregnant women.

Results and Discussion

The prevalence rate of back pain with onset during pregnancy among the 31 women pregnant after IVF treatment increased from 41% at the first visit to 67% and 76% at the second and third visit, respectively. The corresponding rates for the spontaneously pregnant women were 19%, 47% and 49%, respectively. The difference in prevalence rate at the first and third visit was statistically significant ($p=0.01$). These differences in back pain reported were mainly confined to the lumbar, lumbosacral and sacral area. The prevalence rate of sacral pain among the IVF women at the third visit was 60%,

G.W. Tregear et al. (eds.), Relaxin 2000, 415–420.
© 2001 *Kluwer Academic Publishers. Printed in the Netherlands.*

which was 2.3 times higher than the rate among the spontaneously pregnant women (p=0.0001).

There was no significant difference between the two groups in the proportion of earlier back pain problems, either in connetion with pregnancy or not. Among the IVF-pregnant women 65% reported earlier back pain problems unrelated to pregnancy and 18% related to pregnancy. The corresponding proportions among the spontaneously pregnant women were 58% and 40%, respectively.

Of the thirteen singleton pregnant IVF treated women, nine (69%) reported back pain and seven (54%) sacral pain at the last visit. Among the twelve multiple pregnant IVF-treated women, ten (83%) reported back pain and eight (67%) sacral pain at the last visit. There was thus a non-significant tendency towards a higher prevalence of pain among multiple pregnant IVF-treated women than among singleton pregnant women. When the IVF-treated women and spontaneously pregnant women were analysed together, the 14 multiple pregnant women reported a significantly higher prevalence of back pain (p=0.01), as well as sacral pain (p=0.02), at the last visit, than singleton pregnant women.

The prevalences of positive pelvic pain provocation tests in late pregnancy among the IVF pregnant women and among the spontaneously pregnant women are shown in Table 1. The IVF pregnant women had a significantly higher prevalence rate of a positive femoral compression test result, as well as a tender sacrospinous ligament test result, than the spontaneously pregnant women.

Table 1. Occurrence of a tender sacrospinous ligament test result and a painful femoral compression test result in late pregnancy among IVF pregnant and spontaneously pregnant women. The p-values refer to the difference between the groups.

	IVF pregnant					Spontaneously pregnant					
		moderately positive		strongly positive			moderately positive		strongly positive		
Test	n	n	%	n	%	n	n	%	n	%	p<
Tender sacrospinous ligament test	19	3	16	9	47	154	19	12	36	23	0.02
Painful femoral compression test	22	5	23	15	68	175	33	19	53	30	0.0001

The relaxin levels at the 16th gestational week were related to the result of the tender sacrospinous ligament test and the femoral compression test performed in late pregnancy in the women pregnant after IVF treatment. Women with a negative test result had the lowest relaxin levels (mean ± SEM), 5779 ±1133 and 6550 ± 1150 ng/l, respectively, those with a strongly positive test result had the highest, 11428 ± 2329 and 9840 ± 1506 ng/l, respectively. Those with a moderately positive test result had relaxin levels in between regarding the tender sacrospinous ligament test, 8833 ± 612 ng/l, but low regarding the femoral compression test, 4625 ± 1726 ng/l. However, for both tests there was a significant ($p<0.05$) relationship between serum relaxin level and test result.

Among the thirty-one IVF pregnant women, the average mean relaxin level was higher in women reporting sacral pain with onset during pregnancy at the third visit, 8921 ± 1546 ng/l, than in women with no such pain, 5406 ± 846 ng/l, but the difference was not statistically significant. However, when the IVF pregnant women and spontaneously pregnant women were analysed together as one group the women reporting sacral pain with onset during pregnancy at the third visit had higher relaxin values at all five sampling visits, all differences being highly significant ($p<0.0001$). The mean relaxin value across all sampling visits for women reporting sacral pain was 2786 ± 597 ng/l as compared to 1016 ± 121 ng/l for those not reporting sacral pain. For other back pain locations there were no similar consistent differences of relaxin values.

The variables study population group (IVF pregnant or spontaneously pregnant), number of foetuses and serum relaxin level were thus all correlated to the prevalence of back pain during pregnancy. Since these variables in addition were intercorrelated (multiples were more common in IVF pregnancies, relaxin was higher in IVF pregnancies etc.) an attempt was made to estimate the influence of each factor on pain prevalence with the influence of the other factors taken into account in a set of multivariate analyses. Age was not related to back pain prevalence but was included in the analyses to control for possible confounding. When all four variables were included no one reached significance, but after stepwise backward elimination of the least significant variables, relaxin level was the only one that remained significantly associated with pelvic pain prevalence ($p=0.0008$).

The size of the effects of these variables on reported sacral pain at the third visit is summarised in table 2.

Table 2. Effect of age, number of foetuses, conception status and mean serum relaxin level on reported sacral pain in late pregnancy after univariate and multivariate logistic regression analyses. O.R.=Odds ratio. C.I.=95% confidence interval for O.R.

		Reported sacral pain					
		Univariate analysis				Multivariate analysis	
	n	n	%	O.R.	C.I.	O.R.	C.I.
Age, 18-23 years	27	8	29.6	1.00	-	1.00	-
Age, 24-34 years	151	44	29.1	0.98	0.40-2-40	0.81	0.32-2.06
Age, 35-42 years	22	8	36.4	1.36	0.41-4.50	0.74	0.19-2.85
Spontaneous pregnancy	175	45	25.7	1.00	-	1.0	-
IVF-induced pregnancy	25	15	60.0	4.33	1.82-10.33	0.97	0.06.14.75
Singleton pregnancy	186	52	28.0	1.00	-	1.0	-
Multiple pregnancy	14	8	57.1	3.44	1.14-10.38	0.80	0.19-3.44
Mean relaxin 186-382	29	5	17.2	1.00	-	1.0	-
Mean relaxin 383-1665	144	38	26.4	1.43	0.54-3.78	1.48	0.55-3.94
Mean relaxin 1666-16800	26	16	61.5	6.40	1.94-21.11	6.12	0.48-77.90

In the univariate analyses, shown in the left hand part of table 2, women in the oldest age group had 1.36 times higher odds of reporting pain at the third visit than the youngest women. IVF pregnant women had 4.33 times higher odds than spontaneously pregnant women, women with multiple pregnancies had an odds ratio of 3.44 compared to women with singleton pregnancies, and women with the highest mean relaxin values had an odds ratio of 6.40 compared to those with the lowest values. The effects adjusted in multivariate analyses are shown in the right hand part of table 2. Now the odds ratio declined somewhat by age, the odds ratios for IVF-induced pregnancy and multiple pregnancy were reduced to almost unity, whereas the odds ratios for the relaxin levels remained almost unchanged.

In this study, a high prevalence rate of back pain among women pregnant after IVF, as compared to spontaneously pregnant women, was found. The two times higher chance of reporting sacral pain might be due to the supraphysiological concentrations of relaxin among women pregnant after IVF. In fact, the serum relaxin level was the factor that best explained differences in sacral pain prevalence. When the influence of serum relaxin level on back pain prevalence was taken into account women carrying multiple pregnancies had no more pain than women carrying singletons, and IVF pregnant women had no more pain than spontaneously pregnant women. Thus, the ovarian stimulation would affect the prevalence rate of low back pain since the number of follicles developed was significantly correlated to relaxin concentrations during subsequent pregnancy [8]. Furthermore, the data suggest that milder ovarian stimulation might lower the pain prevalence rate. The IVF pregnant women had many days of sickness benefit, to which back pain may have contributed. This fact would have lowered the difference in prevalence of reported pain between spontaneously and IVF pregnant women. Interestingly, in a multivariate logistic regression analysis the influence of multiple pregnancy did not remain significant but the serum relaxin was.

Data were collected prospectively and in a standardised way in the two groups of women. In the cohort of spontaneously pregnant women our observations were certainly representative of pregnant women in the area and in the cohort of IVF pregnant women they were sampled consecutively. The IVF group was relatively small. Still, a highly significant correlation between prevalence of sacral pain and relaxin levels was observed. In addition, no confounding factor that would affect the association between relaxin levels and sacral pain or the pain provocation tests has been found, which makes the results even more valid. Moreover, a sensitive and reliable ELISA measuring human relaxin was used [9].

This is, to our knowledge, the first study presenting data on back pain in women pregnant after IVF. Earlier studies in spontaneously pregnant women have reported a relationship between relaxin levels and pain in the pelvis including the sacral region [5,7]. Petersen *et al.* [10] could not present data supporting relaxin´s having this effect. However, the present study gave further evidence for the hypothesis that relaxin may be related to back pain with onset during pregnancy.

A proposed effect of relaxin in women is remodelling of the connective tissue of the pelvis in preparation for parturition. A side-effect of this remodelling might be a predisposition to pain in the pelvic area. Results from several studies indicate that pregnancy affects ligaments and bones in the pelvic region. For example, in anatomical studies, variably deep coalesced bony pits and craters have been observed at the attachments of the ligaments of the pelvis in women who had been pregnant, but not in women who had never been pregnant or in men [1]. The pain has been proposed to be triggered from the insertions of the ligaments of the pelvis [11]. One of the possible locations for pain release is the sacrospinous ligament, where the loading power of the back is concentrated [12]. Provoked pain in this region was also related to elevated relaxin levels in the present study. This view is also supported by the finding that pain in the lower back could only occasionally be induced from the internal genital organs [13].

The effect of human relaxin on the pelvis remains speculative. Human relaxin decreased synthesis and secretion of collagen in normal human skin fibroblasts [14] and reduced collagen accumulation in rodents [15]. Also, human relaxin reduced the total collagen content in nonpregnant rat pubic symphyses [16]. The reduced soft tissue collagen synthesis may predispose developing pain when loading the pelvis.

In conclusion, we found that hyperrelaxinaemia was closely correlated to a high prevalence of sacral pain in late pregnancy among IVF pregnant women. This is evidence supporting the theory that relaxin is involved in the generation of pelvic pain in pregnant women.

Acknowledgements

This study was supported by grants from the Swedish Medical Research Council (project B91-17X-05982-15B) and Uppsala University. We thank Genentech Inc., San Francisco, California, USA for generously providing the reagents for the relaxin ELISA.

References

1. Houghton, P., Bull. N. Y. Acad. Med., 51 (1975) 655.
2. Putschar, W.G.J., Am. J. Phys. Anthropol., 45 (1976) 589.
3. Kristiansson, P., Svärdsudd, K. and von Schoultz, B., Spine, 21 (1996) 702.
4. Östgaard, H.C., Andersson, G.B.J. and Karlsson, K., Spine, 16 (1991) 549.
5. MacLennan, A.H., Nicolson, R., Green, R.C. and Bath, M., Lancet, ii (1986) 243.
6. Sherwood, O.D., In Knobil, E. and Neill, J.D. (eds.) The Physiology of Reproduction. 2nd ed., Raven Press, New York, USA, p. 861.
7. Kristiansson, P., Svärdsudd, K. and von Schoultz, B., Am. J. Obst. Gyn., 175 (1996) 1342.
8. Kristiansson, P., Svärdsudd, K., von Schoultz, B. and Wramsby, H., Hum. Reprod., 11 (1996) 2036.
9. Lucas, C., Bald, L.N., Martin, M.C., Jaffe, R.B., Drolet, D.W., Mora-Worms, M., Bennett, G., Chen, A.B. and Johnston, P.D., J. Endocrinol., 120 (1989) 449.
10. Petersen, L.K., Hvidman, L. and Uldbjerg, N., Gynecol. Obstet. Invest., 38 (1994) 21.
11. Kristiansson, P. and Svärdsudd, K., Spine, 21 (1996) 2337.
12. Irvin, R.E. In Vleeming A, Mooney V, Dorman T, Snijders C, Stoeckart R (eds.) Movement, stability and low back pain, Churchill Livingstone, 1997, p 133.
13. Koninckx, P.R. and Renaer, M., Hum. Reprod., 12 (1997) 1785.
14. Unemori, E.N. and Amento, E.P., J. Biol. Chem., 265 (1990) 10681.
15. Unemori, E.N., Beck, L.S., Lee, W.P., Xu, Y., Siegel, M., Keller, G., Liggitt, H.D., Bauer, E.A. and Amento, E.P., J. Invest. Dermatol., 101 (1993) 280.
16. Samuel, C.S., Butkus, A., Coghlan, J.P. and Bateman, J.F., Endocrinology, 137 (1996) 3884.

Immunohistochemical detection of relaxin binding to ligaments of the thumb and knee

Stephanie Galey[1], John Lubahn[1], David Ivance[1], Elisa Konieczko[2], Christopher Arnold[3] and Timothy Cooney[1]
[1]Hamot Medical Center, Erie, PA 16550, USA; [2]Gannon University, Erie, PA 16541, USA; and [3]The Arnold Knee Clinic, Fayetteville, AK 72701, USA

Introduction

Osteoarthritis of the thumb basal joint and rupture of the anterior cruciate ligament of the knee are particularly vexing issues for women. Prevalence rates for these pathologies range from two to more than ten times that of age- and activity-matched males [1,2]. The reason for this gender discrepancy is unclear, but hormonal-mediated changes in joint stability are a plausible explanation. Joint laxity has been proposed as a risk factor for joint morbidity [2-6]. Furthermore, estrogen release during menstruation has been linked to ACL injury [7]. Other studies have revealed *in vivo* and *in vitro* perturbations in ACL ligament physiology as a function of estrogen dosing [8,9]. Relaxin may be a less obvious but equally culpable hormonal agent of ligamentous changes. Previous studies have demonstrated its ability to alter collagen and collagenase production in cultured fibroblasts [10-13]. Gravidity studies have also linked relaxin to increased laxity in knee and finger joints [14-16]. Given its systemic distribution [17], we hypothesized that relaxin binds to the ACL and palmar oblique ligaments, prime stabilizers of the knee and thumb, respectively. Since neither receptor antibodies nor cDNA probes are available, detection was based on an indirect immunohistochemical method.

Methods

Fourteen ACL specimens and eight palmar oblique ('beak') ligaments were obtained intraopertively from reconstruction or arthroplasty surgeries. Both male and female patients (average age: 44 years) provided ACL specimens while thumb ligaments were derived from perimenopausal women. Additionally, uterus and cervix, obtained from biopsy or hysterectomy, and meniscus, derived from knee arthroplasty, were used as positive and negative staining controls, respectively. Use of these controls was based on the literature or empirically. Specimens were snap-frozen and stored at -80^{0}C. Serial sections (5 μm) were cut in a cryostat, rinsed with TBS (Tris-buffered saline, pH 7.4) and then flooded with a chelating buffer solution (10mM Tris, 0.2% Tween 20, 5mM EDTA) for 30 minutes. This was followed by a rinse with TBS supplemented with 5.4 mM KCl, 1.3 mM CaCl, and 0.5 mM MgCl to restore physiologic ionic balance. Two final rinses with unadulterated TBS followed. As a procedural control, the chelating wash was omitted for randomly selected samples and replaced by a TBS

G.W. Tregear et al. (eds.), Relaxin 2000, 421–424.
© 2001 *Kluwer Academic Publishers. Printed in the Netherlands.*

rinse. Sections were then incubated overnight with rh-relaxin (Immundiagnostik, Bensheim, Germany) (1 μg/ml in TBS/1% bovine serum albumin [BSA]), in a humidified chamber. They were then fixed with 4% formalin in TBS for thirty minutes, rinsed twice in TBS, followed by a final rinse in TBS containing 1% BSA.. Nonspecific antibody binding was blocked via a two-hour incubation with Protein Blocking Agent (Immunotech, Marseille, France). Afterwards, excess blocking agent was decanted and the sections were incubated for one hour with rabbit polyclonal anti-relaxin IgG (Immundiagnostik, Bensheim, Germany) (5μg/ml in TBS/1% BSA). This was followed by triplicate rinses in TBS. All incubation steps occurred at room temperature. Ultratech HRP Streptavidin Biotin Universal Detection System and Ultratech AEC Chromogen kits (Immunotech, Marseille, France) were then used per manufacturer instructions to impart a red stain to areas with ligand-antibody complexes. Hematoxylin was used as a blue counterstain. Stained slides were covered with aqueous mounting medium (Crystal Mount, Biomedia, Foster City, California) and stored at room temperature until examined. Most sections were reviewed within 24 hours of staining. Positive and negative control tissues were processed in an identical manner as ACL specimens. Additionally, ligand-negative and primary antibody-negative controls were run on select uterine and meniscus sections using TBS/1% BSA as the substitute. Finally, mouse anti-insulin antibodies (Sigma Chemical, St. Louis) were used to assess the binding specificity of the anti-relaxin antibodies. A concentration of 1:1000 in TBS/1% BSA was used. Tissue treated in this manner included ACL, uterine (relaxin-positive control), and rat pancreatic (insulin-positive control) sections.

Results and Discussion

Staining was apparent in all uterine, ACL, and beak ligament specimens (Figs. 1-3, respectively). Age and gender produced qualitative differences in staining intensity for ACL specimens, with young females evidencing the most abundant hormonal binding. Little or no staining occurred in the negative tissue control (Fig.4). Additional controls (not shown) further verified staining specificity. Primary antibody-negative sections did not stain while ligand-negative sections revealed substantially reduced staining. Pretreatment with chelating buffer completely eradicated this endogenous hormone background. This was reversible, as addition of ligand fully restored staining competence. Finally, substituting anti-insulin antibodies for the primary antibody similarly failed to evoke a staining reaction.

Figure 1: Uterine Staining
(demarcated)

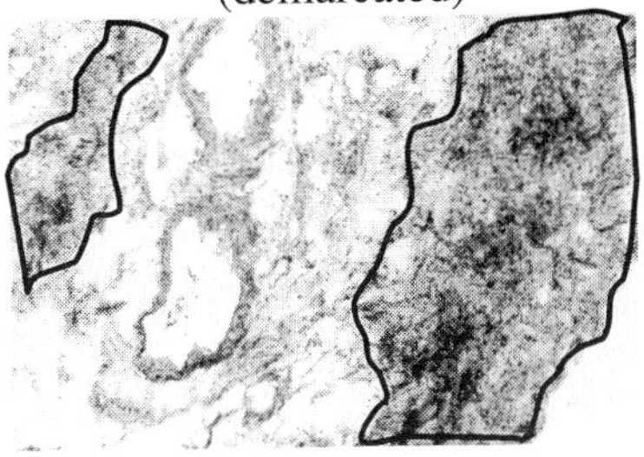

Figure 2: ACL Staining
(demarcated)

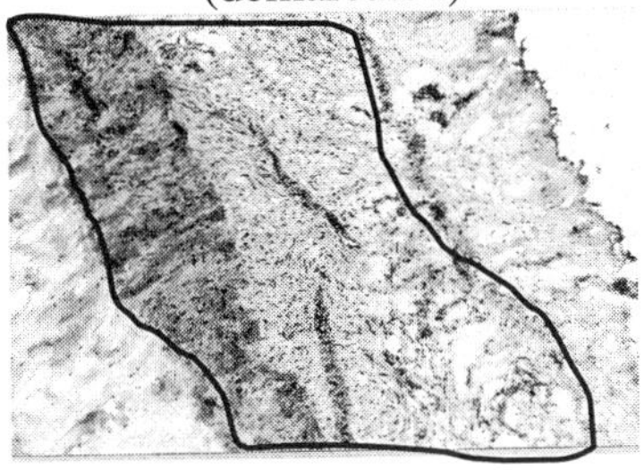

Figure 3: Beak Ligament Staining
(demarcated)

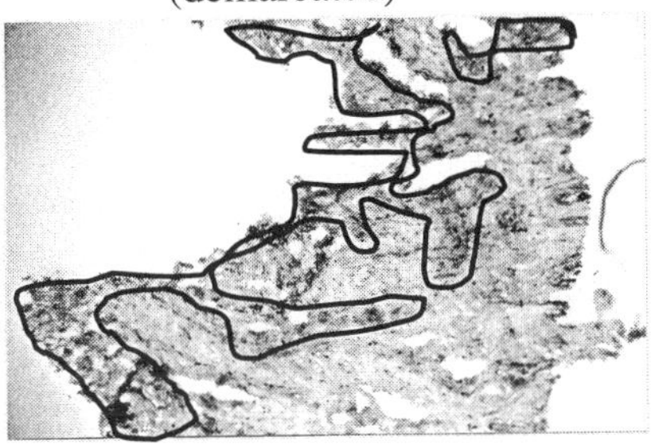

Figure 4: Negative Control
(staining absent in meniscus)

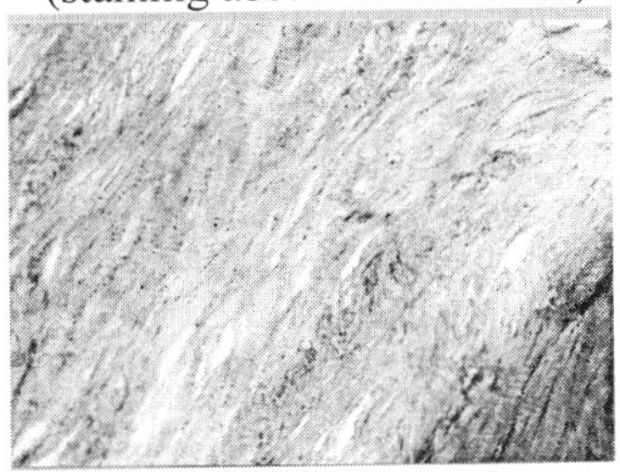

These data suggest that relaxin binds to the ACL of the knee and the beak ligament of the thumb. Relaxin binding was specific as both ligand and detection antibody were required for reactivity. That binding is tenable coincides with the observable changes in joint laxity associated with relaxin during pregnancy [14-16], as well as reports of *in vitro* collagen remodeling effects [10-13]. In this context, we propose that relaxin binding to the ligaments under study reflects areas of cellular or extracellular matrix receptors, facilitating the aforementioned effects. Previous autoradiographic determination of relaxin receptors in other areas of the body underscores the ubiquitous nature of the tissues affected and further lends some credibility to our assertion [18,19]. Speculatively, relaxin may be involved in normal ligament physiology, antagonizing collagen anabolism and contributing to ligament elasticity and compliance. The indirect method used herein, however, necessitates more supporting data to arrive at definitive conclusions. Furthermore, there is no evidence at the present time to suggest a causal relationship between joint morbidity or pathology and relaxin. Limitations in this study included the lack of a comprehensive set of samples from age-matched males, the use of polyclonal detection antibodies, and the absence of histomorphometry-based measurements. The lack of representative samples reflects the female bias associated with these morbidities, resulting in sporadic sample collection from males. While the use of polyclonal antibodies enhances epitope recognition, lot-to-lot variation may reduce or completely eliminate detection of ligand binding. This argues for further refinement of our method to identify suitable replacement monoclonal antibodies, as well as adjunctive studies to show that binding is receptor-mediated.

Acknowledgements

The authors wish to thank the Hamot Second Century Fund for underwriting research costs, to the Hamot Histology Department, for processing assistance, to the Hamot Research Center for editorial help, and especially, Dr. Danielle Bani for her invaluable advice.

References

1. Huston, L.J., Greenfield, M.L. and Wojtys, E.M., Clin. Orthop., 372 (2000) 50.
2. Newport, M.L., Clin. Orthop., 372 (2000) 85.
3. Malcolm, L.L., Daniel, D.M., Stone, M.L. and Sachs, R., Clin. Orthop., 196 (1985) 35.
4. Pelligrini, V.D., Orthop. Clin. North Am., 23 (1992) 83.
5. Doerschuk, S.H., Hicks, D.G., Chinchilli, V.M. and Pellegrini, V.D. Jr., J. Hand Surg., 24 (1999) 496.
6. Eaton, R.G. and Glickel, S.Z., Hand Clinics, 3 (1987) 455.
7. Wojtys, E.M., Huston, L.J., Lindenfeld, T.N., Hewett, T.E. and Greenfield, M.L., Am. J. Sports Med., 26 (1998) 614.
8. Slaughterbeck, J., Clevenger, C., Lundberg, W. and Burchield, D.M., J. Bone Joint Surg., 81 (1999) 54.
9. Yu, W.D., Liu, S.H., Hatch, J.D., Panossian, V. and Finerman, G.A.M., Clin. Orthop., 366 (1999) 229.
10. Unemori, E.N., Beck, L.S., Lee, W.P., Xu, Y., Siegel, M., Keller, G., Liggitt, H.D., Bauer, E.A. and Amento, E.P., J. Invest. Dermatol., 101 (1993) 280.
11. Unemori, E.N. and Amento, E.P., J. Biol. Chem., 265 (1990) 10681.
12. Unemori, E.N., Pickford, L.B., Salles, A.L., Piercy, C.E., Grove, B.H., Erickson, M.E. and Amento, E.P., J. Clin. Invest., 98 (1996) 2739.
13. Samuel, C.S., Butkus, A., Coghlan, J.P. and Bateman, J.F., Endocrinology, 137 (1996) 3884.
14. Calguneri, M., Bird, H.A. and Wright, V., Ann. Rheum. Dis. 41 (1982) 126.
15. Klein, K.K., Am. Correct Ther. J., 26 (1972) 43.
16. Blecher, A.M. and Richmond, J.C., J. Arthroscop. Rel. Surg., 14 (1998) 77.
17. Bani, D., Gen. Pharmacol., 28 (1997) 13.
18. Yang, S., Rembiesa. B., Büllesbach, E.E. and Schwabe, C., Endocrinology, 130 (1992) 179.
19. Mercado-Simmen, R.C., Byrant-Greenwood, G.D. and Greenwood, F.C., Endocrinology, 110 (1982) 220.

Preterm delivery predicted by serum relaxin

Ida Vogel[1], Marianne Glavind-Kristensen[1], Jannie Dalby Salvig[1], Poul Thorsen[2], Puk Sandager[1] and Niels Uldbjerg [1]

[1]*Department of Obstetrics and Gynecology, Aarhus University Hospital, Skejby, Denmark; and* [2]*Danish Epidemiology Sciences Center, Aarhus University, Denmark and Developmental Disabilities Branch, Centers for Disease Control and Prevention, Atlanta, USA.*

Introduction

Preterm delivery is by definition delivery before 37 weeks' gestation. Preterm delivery contributes 50-70% of neonatal deaths [1]. Preterm delivery is associated with cerebral palsy, but it has also been related to disabilities such as minimal brain damage, decreased intelligence, and reading difficulties. An American study on the cognitive outcome of infants born at 24, 25, and 26 weeks' gestation showed a normal cognitive outcome in 28, 47, and 71%, respectively, of the survivors at a 36-month follow-up [2].

Measures against preterm delivery include certain forms of treatment during pregnancy, e.g. cervical cerclage, and prevention schemes, e.g. cessation of smoking or sick leave from physically demanding jobs. Other forms of treatment aim to limit the mortality and morbidity in the newborns, e.g. cortisone treatments. However, the use of such measures is rational only when a limited risk group can be identified.

Results and Discussion

Risk factors and biomarkers predicting preterm delivery

Many risk factors for preterm delivery have been described, but the prediction of preterm delivery from single risk factors has very limited value. Risk scoring systems, combining anamnestic information, are often very laborious and have up to 80% false positive tests. The numbers of potential biomarkers increase rapidly, but again with reduced sensitivity at an acceptable false positive rate. Concerning both risk factors and biomarkers we must conclude that it is not presently possible to identify a limited high-risk population early in pregnancy in low-risk pregnancies. However, the use of a combined score system, including both risk factors and several biomarkers, remains to be examined.

Relaxin and preterm delivery in a low risk population

The first report of serum relaxin and preterm delivery described continuously raised serum relaxin levels through late pregnancy in 5 women who delivered preterm [3]. Later, increased serum relaxin in the 30th gestational week was shown to be associated with preterm delivery, and markedly so in women developing PPROM [4]. Among women subjected to follicle-stimulating treatment for in vitro fertilization purposes,

G.W. Tregear et al. (eds.), Relaxin 2000, 425–428.
© 2001 *Kluwer Academic Publishers. Printed in the Netherlands.*

there is a correlation between raised serum relaxin in 6-12th weeks of gestation and increased risk of preterm labor [5].

In a nested case control study we examined the association between serum relaxin in the 18th gestational week and preterm delivery [6]. In order to evaluate serum relaxin's potential as a predictor, serum relaxin was dichotomized into raised (≥1648 pg/ml) or normal levels with the use of ROC curves. This test would have identified 27% of the very preterm deliveries, but with a false positive rate of 3%, and raised serum relaxin during the 18th gestational week was associated with increased risk for very preterm delivery [OR=11.3, 95% Confidence Interval (95% CI): 2.1-59.1].

Serum relaxin is not presently a clinically useful screening test for preterm delivery, though this study gave a likelihood ratio (sensitivity/1-specificity) of 6.8. Other predictors gave following likelihood ratios in low-risk populations: fetal fibronectin: 2.7 [7], cervical length: 7.7 [8] and corticotropin releasing hormone: 4.9 [9]. However this study is very limited, and larger investigations are to be awaited. We have a study in process using a combination of risk factors and biomarkers (serum relaxin, infectious biomarkers, cervical length measured by ultrasound, anamnestic information). This study may show whether these biomarkers are overlapping or whether a better risk profile can be obtained by the combined use of biomarkers.

Relaxin and threatening preterm delivery

In a prospective study, we have collected serum samples from 93 women with spontaneous preterm labor, PPROM or preterm progression of cervical ripening before 34 weeks gestation. When women with threatening preterm delivery are admitted, it is important to identify those who progress into preterm delivery, in order to intensify treatment among the high-risk pregnancies, and minimize treatment and stress on the lower-risk pregnancies.

We found that serum relaxin was significantly higher in the women with PPROM, when compared to women in labor, or with ripe cervices.

Approximately 3 out of 4 of the women with raised serum relaxin levels deliver preterm, compared with one out of three of the women with low serum relaxin levels. Raised serum relaxin gave a 5 times increased risk of preterm delivery when used as a single predictor of preterm delivery. Additionally, sensitivity, specificity, positive predictive value, and negative predictive value for high serum relaxin achieved similar results in comparison with fetal fibronectin [10]. Also, fetal fibronectin can only be estimated in women without PPROM, without pelvic inspection, and no intercourse within the last 24 hours, which indicates that serum relaxin may have certain advantages as a useful predictor at the department of labor and delivery.

Results from this study suggest that serum relaxin may be a potential predictor of very preterm delivery in women with threatening preterm delivery, and especially in women with preterm labor. The association between serum relaxin and cervical fetal fibronectin awaits analysis in future studies, where better predictive skills may be achievable by combinations of tests.

The course of serum relaxin throughout pregnancy among women delivering preterm

Thirdly, in a prospective study, serum samples were obtained in early pregnancy, at preterm delivery with gestationally matched controls, and from a random sample of women delivering at term. The study consists of 84 women delivering preterm, 347 delivering at term and 53 women delivering post term. We found that women, who deliver preterm, do not just have upward shifted s-relaxin levels throughout pregnancy, but also have an abnormal s-relaxin curve with initial low, but stable, s-relaxin levels resulting in elevated relaxin levels in second and third trimester. Additionally, it was possible to establish a longitudinal model for the prediction of preterm delivery based on the change in serum relaxin over time as well as anamnestic information.

Conclusions

Women, who deliver preterm have an abnormal s-relaxin curve with initial low, but stable, s-relaxin levels resulting in elevated relaxin levels in second and third trimester. Raised serum relaxin in second and third trimester is a potential predictor of preterm delivery in low- and high-risk population. Further, the change in serum relaxin over time may be the most sensitive predictor preterm delivery, as women delivering preterm have a lesser decrease per week compared with term deliveries. However, at present serum relaxin seems closest to clinical usefulness in high-risk populations.

Among women with threatening preterm delivery, serum relaxin was highest in women with PPROM. This is in line with relaxin inducing a weakening of the biomechanical strength of the human fetal membranes [11]. These results suggest that relaxin is also part of the physiology and pathology of human delivery.

Acknowledgments

This study was supported by: Institute for Experimental Clinical Research, Aarhus University, Lægekredsforeningen Forskningsfond, Kong Christian den Tiendes Fond, Kornings foundation, and Snedkermester Sophus Jacobsens og Hustru Astrid Jacobsens Fond. The staff at Skejby Hospitals Delivery Wards and especially Anne Mouritzen are thanked for their help in collecting the samples. The contributions to the statistical analysis by Assistant Professor Heidi Holmager Hundborg, Institute for Biostatistics, Aarhus University, are gratefully acknowledged.

References

1. Rush, R.W., Keirse, M.J., Howat, P., Baum, J.D., Anderson, A.B. and Turnbull, A.C., Br. Med. J., 2 (1976) 965.
2. Piecuch, R.E., Leonard, C.H., Cooper, B.A., Kilpatrick, S.J., Schlueter M.A. and Sola, A., Obstet. Gynecol., 90 (1997) 809.
3. Bell, R.J., Eddie, L.W., Lester, A.R., Wood, E.C., Johnston, P.D. and Niall, H.D., Br. J. Obstet. Gynaecol., 95 (1988) 1264.

4. Petersen, L.K., Skajaa, K. and Uldbjerg, N., Br. J. Obstet. Gynaecol., 99 (1992) 292.
5. Weiss, G., Goldsmith, L.T., Sachdev, R., Von-Hagen, S. and Lederer, K., Obstet.Gynecol., 82 (1993) 821.
6. Vogel, I., Salvig, J.D., Secher, N.J. and Uldbjerg, N., Am. J. Obstet. Gynecol., 184 (2001) 390.
7. Leitich, H., Egarter, C., Kaider, A., Hohlagschwandtner, M., Berghammer, P. and Husslein, P., Am. J. Obstet. Gynecol., 180 (1999) 1169.
8. Iams, J.D., Goldenberg, R.L., Meis, P.J., Mercer, B.M., Moawad, A., Das, A., Thom, E., McNellis, D., Copper, R.L., Johnson, F. and Roberts, J.M., N. Engl. J. Med., 334 (1996) 567.
9. McLean, M., Bisits, A., Davies, J., Walters, W., Hackshaw, A., De Voss, K. and Smith, R., Am. J. Obstet. Gynecol., 181 (1999) 207.
10. Revah, A., Hannah, M.E. and Sue, A.Q.A., Am. J. Perinatol., 15 (1998) 613.
11. Petersen, L.K., Helmig, R., Oxlund, H., Vogel, I. and Uldbjerg, N., Eur. J. Obstet. Gynecol. Reprod. Biol., 57 (1994) 123.

High serum concentrations of relaxin correlate with dissemination of breast cancer

Claudia Binder[1], Lutz Binder[2], Lüder Gurlit[2] and Almuth Einspanier[3]
Departments of [1]Hematology/Oncology and [2]Clin. Chemistry, Georg-August-University, Robert-Koch-Street 40, 37075 Göttingen; [3]Dep. of Reproductive Biology, German Primate Center, Kellnerweg 4, 37077 Göttingen, Germany.

Introduction

Until recently, relaxin has mainly been known as a hormone of the reproductive system [1-4]. However, identification of relaxin mRNA transcripts and specific binding sites in many non-reproductive tissues [5] suggests a physiological role for relaxin apart from these functions. Of particular interest is the ability of relaxin to induce modifications of stromal tissue components. Relaxin decreases the synthesis and deposition of interstitial collagens in human normal and scleroderma fibroblasts [6]. Expression and catalytic activity of collagenase-1 (MMP-1) as well as stromelysin-1 (MMP-3), members of the matrix metalloprotease family (MMPs), is increased by relaxin [7-9].

MMPs, on the other hand, play an important role for tumor progression, as they facilitate invasion and metastasis through dissolution of the basement membrane and degradation of the extracellular matrix. Upregulation of MMP expression and activity has been found in many tumor tissues [10-12] and is associated with an invasive phenotype [13]. In breast cancer, high expression levels of MMP-1, MMP-2, MMP-9 and MMP-14 are positively correlated with invasiveness and shortened survival [14-16]. MMP-2 and -9 upregulation has been demonstrated particularly in tumors overexpressing her2/c-erb-B2, another predictor of early dissemination and unfavourable clinical outcome [17, 18].

As relaxin is involved in the regulation of MMP expression and MMPs, for their part, are involved in tumor progression, it is tempting to speculate upon a role of relaxin and its tissue remodeling ability not only for benign tissues, but also for invasiveness and metastatic potential of malignant cells. The assumption is particularly attractive for breast cancer, as relaxin immunostaining has been demonstrated in epithelial and myoepithelial cells of normal mammary ducts as well as in breast cancer cells [19].

Results and Discussion

To test the hypothesis, relaxin concentrations were determined with a sensitive enzyme immunoassay (EIA) in the sera of 160 patients with histologically confirmed breast cancer. Details of the applied EIA as well as its evaluation regarding sensitivity and specificity have been described elsewhere [20]. Sera from 27 healthy blood donors (20 females at different stages of the menstrual cycle, 7 males, age 18 - 39 years) were used

G.W. Tregear et al. (eds.), Relaxin 2000, 429–435.
© 2001 *Kluwer Academic Publishers. Printed in the Netherlands.*

as controls. Additionally, relaxin concentration was determined in the sera of 27 patients of both sexes (age 24 - 83 years) with various diseases: heart failure, hypertension, pneumonia, hyperthyroidism, lymphomas and acute leukemias, metastatic germ cell tumors, non-metastatic carcinoma of the pancreas.

The breast cancer patients had been enrolled in a follow-up program during the years 1991-95 after removal of the primary tumor (Table 1). As surgery in some cases had been performed in other hospitals, data about histological subtype and exact tumor staging according to the International Tumor-Node-Metastasis (TNM) Classification were not available for all of the 160 patients. The median follow-up time was 3 years (range 0.5 - 16 years). During the observation period, none of the patients received treatment with estrogens or derivatives.

Table 1. Patients' characteristics at surgery of the primary tumor (T1= <2cm, T2= 2-5cm, T3= >5cm, T4= any tumor size + involvement of skin/thoracic wall; N≥1=presence, N0= absence of lymph node involvement; M1=presence, M0=absence of distant metastases).

Total number	160 patients (pts)
Age	58.9 ± 10.9 years (mean ± SD, range 32 - 82 years)
Premenopausal	35/160 pts
Histology	ductal invasive 85/98 pts, tubular 2/98 pts, medullary 3/98 pts, inflammatory 2/98 pts, cystosarcoma phylloides 1/98 pts, DCIS (ductal carcinoma in situ) 7/98 pts
Size of the primary tumor	T1: 39/128 pts, T2: 62/128 pts, T3: 4/128 pts, T4: 16/128 pts, Tis (carcinoma in situ): 7/128 pts
Lymph node involvement	N≥1: 74/148 pts, N -: 74/148 pts
Distant metastases	M1: 13/160 pts, M0: 147/160 pts

12/160 patients had a history of secondary carcinoma, diagnosed either previously or during the course of follow-up. Six of these malignancies were of gynaecological origin (uterus and ovary), the rest of them were stomach (1), bladder (1) and kidney (2) carcinomas as well as 1 non-Hodgkin's lymphoma and 1 carcinoid tumor. Except of the carcinoid tumor, none of them was metastatic.

At the time of relaxin determination, 94 of the 160 patients (58.8%) were free of disease. 66 patients (41.2%) showed evidence of active metastatic disease. The diagnosis was made when metastases or locoregional relapses were present at the time of relaxin measurement and up to 6 months later. Patients with locoregional relapses which had been irradiated or removed by surgery and had shown no sign of recurrence for at least 1 year were not included in the metastatic group.

In comparison to the group of healthy blood donors, serum relaxin concentrations were higher in breast cancer patients (Fig. 1). Median values were 0.3 ng/ml (0.28 - 0.33 ng/ml 95% confidence interval (CI)) for the former and 0.475 ng/ml (0.42 - 0,55 ng/ml 95% CI) for the latter ($p<0.0001$, Mann-Whitney test). Equally, the group of patients with various other diseases had lower relaxin levels than the breast cancer patients (median value 0.33 ng/ml, 0.25 - 0.4 ng/ml 95% CI, p=0.0006).

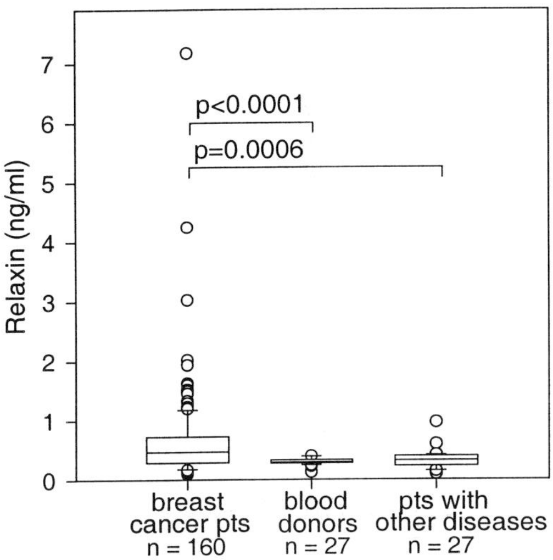

Fig. 1. Serum relaxin concentrations in breast cancer patients, healty blood donors and patients with various other diseases. Data are given as box plots, the box encompassing the median value as well as the 25th through 75th percentile. Values outside the 10th and the 90th percentile (bars), respectively, are shown as circles.

As relaxin concentrations in blood donors and patients with other illnesses were not significantly different, both groups were combined and used as a control cohort for the comparison of patients in complete remission with those who had evidence for metastases. A highly significant positive correlation between metastatic disease and relaxin values could be demonstrated (Fig. 2). Relaxin concentrations were higher in metastatic (0.625 ng/ml, 0.53 - 0.77 ng/ml 95% CI) than in disease-free patients (0.385 ng/ml, 0.34 - 0.44 ng/ml 95% CI, p<0.0001) as well as in controls (0.3 ng/ml, 0.28 - 0.34 ng/ml 95% CI, p<0.0001).

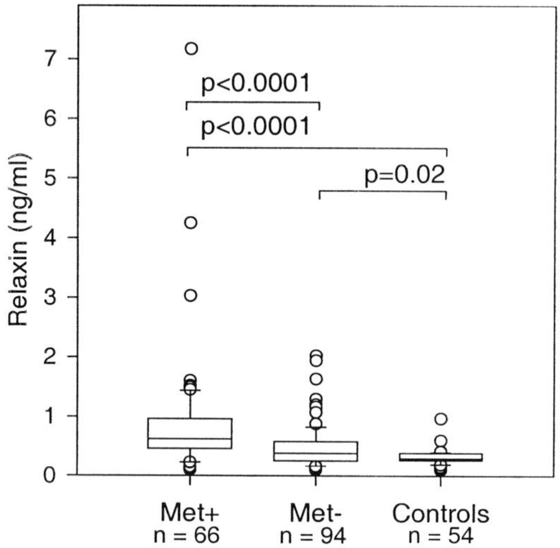

Fig. 2. Serum relaxin concentrations in breast cancer patients with presence (Met+) or absence (Met-) of metastatic disease and control patients.

The results clearly point to a significant role of relaxin in breast cancer dissemination, although the molecular basis of this effect is still unclear. Up to now, data about involvement of relaxin in mammary gland pathology are rare. Relaxin has been reported to influence the growth of experimental mammary carcinomas and breast cancer cell lines in either a positive or negative way [21, 22]. Immunohistochemistry demonstrated augmented relaxin staining in neoplastic breast tissues as compared to their normal counterparts [23].

However, the most probable explanation for the putative effect of relaxin on tumor progression might come from its ability to upregulate MMPs. In parallel to the activation of MMPs during extracellular rearrangement of the uterine cervix [24], relaxin has been shown to upregulate MMPs in other non-malignant tissues and to favour degradation and remodeling of connective tissues [7]. As stromal matrix degradation is essential for tumor cell invasion [25], relaxin may prepare the ground for invasion of malignant cells into the surrounding stromal compartment or other tissues by induction of MMPs. Additionally, relaxin has been described to upregulate expression of vascular endothelial growth factor [26] and to induce vasodilation by stimulation of nitric oxide-production [27]. As a consequence, tissue microcirculation is enhanced which may further contribute to the relaxin effect.

There was still a considerable part of patients who did not show increased relaxin levels despite evidence for active metastases as well as patients with augmented relaxin concentrations without any sign of dissemination. Part of the discordant results may be explained by the retrospective design of the study. Serum samples for relaxin determination had been obtained during routine presentation of the patients with no particular respect to the clinical course. Metastatic disease had been diagnosed either before, at or after serum collection, part of the patients receiving some kind of therapy, others none. Assuming that relaxin as a hormone may exert its auto- or paracrine effect on tissue degrading enzymes in close temporal association with the ongoing metastatic process, prospective investigations with optimised serum harvest may yield more conclusive results.

Additional factors, such as presence or absence of cardiovascular disturbances, may also have influenced relaxin levels, although control cohort and breast cancer patients were comparable regarding age and underlying diseases. A variety of relaxin effects on the cardiocirculatory system have been described [28, 29], however, clinical data about relaxin concentrations in patients with heart failure or hypertension are still lacking. In the present study, patients with cardiovascular problems in the control group did not show elevated relaxin levels, except one patient with non-severe myocardial insufficiency.

Among the non-metastatic patients with "false positive"relaxin concentrations there were two with inflammatory carcinomas and one with a cystosarcoma phylloides. Both entities are characterised by high local invasiveness and might not be comparable with other histologic subtypes. For the rest of the group, follow-up may have been too short, and some of the patients may have developed metastases later on. The surprising observation of significantly higher relaxin concentrations in the non-metastatic group than in the control cohort (Fig. 2) also suggests the existence of hidden metastatic cases within the controls which may have become detectable during longer follow-up.

To further elucidate the putative relation between relaxin and tumor spreading, CA 15-3, a so-called tumor marker, was measured in the same samples. CA 15-3 is a tumor cell-associated epithelial mucin, representing a very sensitive marker for metastatic breast cancer [30] which is widely used during follow-up programs. To allow comparative investigation of relaxin and CA 15-3, patients were divided in a relaxin - positive (serum concentration >0.7 ng/ml) and -negative group. The cut-off was chosen, as values >0.7 ng/ml had never been detectable in the 27 healthy blood donors and only once in a control patient with heart failure (see above).

CA 15-3 was elevated (>30 kU/l) in 40 of 160 patients (25%), 31/40 (77.5%) having metastatic disease. Relaxin was positive in 47 patients (29.3%), 27 of them (57.5%) showing evidence of dissemination. In the non-metastatic group (20/47, 42.5%), there were four patients with concomitant secondary malignancies as well as two with inflammatory carcinomas, which might have contributed to the "false positive"relaxin levels. Surprisingly, there was no correlation between elevated relaxin levels and CA 15-3-positivity (p=0.368, Fisher's exact-test). However, all 9 patients with pathological concentrations of both relaxin and CA 15-3 suffered from aggressive disease with particularly short survival. It is tempting to speculate that, although these two parameters are obviously connected with malignant progression, they may be linked to different aspects of the pathology of tumor dissemination. Elevated relaxin levels may reflect the actual process of tissue invasion by metastatic cells, thus representing rather a functional marker than a structural quality of the tumor cells, such as CA 15-3.

In conclusion, there is considerable evidence that relaxin, apart from its reproductive function, plays a role in tissue degradation and invasion of breast cancer cells. However, as retrospective analyses have to be interpreted with due care, several open questions remain to be answered. It is still unclear, how and to what extent relaxin participates in the metastatic process and whether this phenomenon is restricted to tumors of the mammary gland in female patients. Although preliminary experiments corroborate the hypothesis that relaxin enhances in vitro invasiveness of tumor cells by upregulation of a variety of MMPs [31], further investigations as well as prospective studies are needed to confirm these results and to elucidate the molecular effects of relaxin.

Acknowledgements

This work was supported by grants Ei 333/8-1 (Deutsche Forschungsgemeinschaft) and 10/1270-Bi I (Deutsche Krebshilfe). We thank A. Marten and M. Schulz for technical assistance. We are also grateful to Prof. O.D. Sherwood for generous provision of porcine relaxin and the respective antibody as well as to Connetics Corporation for human recombinant relaxin.

References

1. MacLennan A.H., In MacLennan, A.H., Tregear, G.W., Bryant-Greenwood, G. (eds.) Progress in relaxin research, World Scientific Publishing, Singapore, 1994, p 263.
2. Luque, E.H., Munoz de Toro, M.M., Ramos, J.G., Rodriguez H.A. and Sherwood, O.D., Biol. Reprod., 15 (1998) 169-175.
3. Sherwood, O.D., In MacLennan, A.H., Tregear, G.W., Bryant-Greenwood, G. (eds.) Progress in relaxin research, World Scientific Publishing, Singapore, 1994, p 529.
4. Zhao, L., Roche, P.J., Gunnersen, J.M., Hammond, V.E., Tregear, G.W., Wintour, E.M. and Beck, F., Endocrinology, 140 (1999) 445.
5. Osheroff, P.L., Ling, V.T., Vandlen, R.L., Cronin, M.J. and Lofgren, J.A., J. Biol. Chem., 265 (1990) 9396.
6. Unemori, E.N. and Amento, E.P., In MacLennan, A.H., Tregear, G.W., Bryant-Greenwood, G. (eds.) Progress in relaxin research, World Scientific Publishing, Singapore, 1994, p 481.
7. Unemori, E.N., Pickford, L.B., Salles, A.L., Piercy, C.E., Grove, B.H., Erikson, M.E. and Amento, E.P., J. Clin. Invest., 98 (1996) 2739.
8. Qin, X., Chua, P.K., Ohira R.H. and Bryant-Greenwood, G.D., Biol. Reprod., 56 (1997) 812-820.
9. Kapila, S. and Xie, Y., Lab. Invest., 78 (1998) 925.
10. Garbett, E.A., Reed, M.W. and Brwon, N.J., Br. J. Cancer, 81 (1999) 287.
11. Brummer, O., Athar, S., Riethdorf, L. Riethdorf, L., Loning, T. and Herbst, H., Virchows Arch., 435 (1999) 566.
12. Kugler, A,. Hemmerlein, B., Thelen, P., Kallerhoff, M., Radzun, H.J., Ringert, R.H., J. Urol., 160 (1998) 1914.

13. Ramos DeSimone, N., Hahn-Dantona, E., Sipley, J., Nagase, H., French D.L. and Quigley, J.P., J. Biol. Chem., 274 (1999) 13066.
14. Ginestra, A., La Placa, M.D., Saladino, F., Romano, F.M. and Vittorelli, M.L., Anticancer Res., 18 (1998) 433.
15. Talvensaari-Mattila, A., Paakko, P., Hoyhtya, M. Blanco-Sequeiros, G., Turpeenniemi-Hujanen, T., Cancer, 83 (1998) 1153.
16. Jones, J. L., Walker, G. and Walker, R., J. Pathol., 189 (2000) 161.
17. Roetger, A., Merschjann, A., Dittmar, T. Jackisch, C., Barnekow, A. and Brandt, B., Am. J. Pathol., 153 (1998) 1797.
18. Bitran, J.D., Samuels, B., Trujillo, Y. et al., Clin. Cancer Res. 2 (1996) 1509.
19. Tashima, L. and Bryant-Greenwood, G., In MacLennan, A.H., Tregear, G.W., Bryant-Greenwood, G. (eds.) Progress in Relaxin Research. World Scientific Publishing, Singapore, 1994, p 552.
20. Einspanier, A., Nubbemeyer, R., Schlote, S. Schumacher, M., Ivell, R., Fuhrmann, K. and Marten A., Biol. Reprod., 61 (1999) 512.
21. Bigazzi M., Bani Sacchi, T., Bani, G., In MacLennan, A.H., Tregear, G.W., Bryant-Greenwood, G. (eds.) Progress in Relaxin Research, World Scientific Publishing, Singapore, 1994, p 541.
22. Bani, D., Flagiello, D., Poupon, M.F. Nistri, S., Poirson-Bichat, F., Bigazzi, M. and Bani Sacchi, T., Virchows Arch., 435 (1999) 509.
23. Mazoujian, G. and Bryant-Greenwood, G., Lancet, 335, (1990) 298.
24. Leppert, P.C., In MacLennan, A.H., Tregear, G.W., Bryant-Greenwood, G. (eds.) Progress in relaxin research, World Scientific Publishing, Singapore, 1994, p 341.
25. Chenard, M.P., Lutz, Y., Mechine-Neuville, A. Stoll, I., Bellocq, J.P., Rio, M.C. and Basset, P., Int. J. Cancer, 82 (1999) 208.
26. Unemori, E.N., Erikson, M.E., Rocco, S.E. Sutherland, K.M., Parsell, D.A., Mak, J. and Grove, B.H., Hum. Reprod., 14 (1999) 800.
27. Bani, D., Failli, P., Bello, M.G. Thiemermann, C., Bani Sacchi, T., Bigazzi, M. and Masini, E., Hypertension, 31 (1998) 1240.
28. Toth, M., Taskinen , P. and Ruskuaho, H., J. Endocrinol., 150 (1996) 487.
29. Danielson, L.A., Sherwood, O.D. and Conrad K.P., J. Clin. Invest., 103 (1999) 525.
30. Stearns, V., Yamouchi, H. and Hayes, D.F., Breast Cancer Res. Treat., 52 (1998) 239.
31. Hagemann, Th., Einspanier, A., Schulz, M., Simon, A., Binder, C., Onkologie, 23 (suppl. 7), (2000) 154.

Relationship of serum relaxin levels to knee joint laxity in female athletes

Christopher Arnold[1], Craig Van Bell[2], Vincent Rogers[3], and Timothy Cooney[3]

[1]Arnold Knee Clinic, Fayetteville, AK 72701, USA, [2]Edinboro University of Pennsylvania, Edinboro, PA 16444, USA; and [3]Hamot Medical Center, Erie, PA 16550, USA

Introduction

Female athletes have a substantially greater risk of an anterior cruciate ligament (ACL) rupture than their age- and activity-matched male counterparts [1-10]. Anatomical, neuromuscular, and hormonal theories have attempted to explain this discrepancy. Hormonal changes to ligament compliance could result in excessive joint laxity, and subsequently injury. Recently, estrogen has been linked to changes in the collagen content and strength of the ACL [11-13]. Relaxin is equally culpable. Dosing studies in cultured fibroblasts evidenced altered collagen and collagenase production [14-18]. Studies of pregnancy have intimated an association between elevated relaxin levels and joint hypermobility [19,20]. With respect to the latter, we hypothesized a similar relationship in eumenorrheic female athletes.

To study this, 41 collegiate female athletes and 21 age-matched controls (including males, females, and previously injured athletes), were assessed for serum relaxin levels and knee joint translation weekly for four consecutive weeks. RIA was used for the former, a standard knee arthrometer for the latter. Oral contraceptive (OC) use was noted. Data were analyzed by a repeat measures ANOVA controlling for OC use. Bivariate regression was employed to quantify the strength of the relationship between laxity and hormone level.

Results and Discussion

Our data revealed significant weekly fluctuation in mean relaxin levels in all females, from 382 ± 91to 474 ±1 42 pg./ml (p=0.035). Neither group, gender, nor OC use exerted an impact ($p \geq 0.139$). Knee translation in males averaged 5.4 ±1.8 mm; in females, 7.1 ± 1.2 mm (p=0.063). Weekly variation was insignificant (p=0.912). Injured females evidenced an average of two mm higher translation than nonathletes (p=0.111). No significant correlation was obtained between relaxin level and knee translation for any time period ($r^2 \leq 0.001$).

The data show: (1) the absence of a relationship between serum relaxin levels and knee joint laxity; (2) no significant change in knee joint laxity over the course of menstruation; (3) weekly fluctuation in serum relaxin levels in females; (4) an heretofore unreported similarity between male and nonpregnant female relaxin levels;

G.W. Tregear et al. (eds.), Relaxin 2000, 437–438.
© 2001 *Kluwer Academic Publishers. Printed in the Netherlands.*

and, (5) a trend towards significantly less anterior translation in males. These observations argue against a relaxin-based etiology for ACL injury in female athletes.

Acknowledgement

The authors wish to acknowledge the Foundation for Sports Medicine Education and Research (Rosemont, ILL, USA) and the Hamot Second Century (Erie, PA, USA) for underwriting the cost of this study.

References

1. Albohm, M., Phys. Sports Med., 4 (1976) 46.
2. Arendt, E.A., Clin. Sports Med., 13 (1994) 483.
3. Arendt, E. and Randall, D., Am. J. Sports Med., 23 (1995) 694.
4. Chandy, T.A. and Grana, W.A., Phys. Sports Med., 13 (1985) 106.
5. Emerson, R.J., Clin. Sports Med., 12 (1993) 317.
6. Ferretti, A., Papndrea, P., Conteuca, F. and Mariani, P.P., Am. J. Sports Med., 20 (1992) 203.
7. Gerberich, S.G., Luhmann, S., Finki, C., Priest, J.D. and Beard, B.J., Phys. Sports Med., 15 (1987) 75.
8. Lindenfeld, T.N., Schmitt, D.J., Hendy, M.P., Mangine, R.E. and Noyes, F.R., Am. J. Sports Med., 22 (1994) 364.
9. Whiteside, P.A., Phys. Sports Med., 8 (1980) 130.
10. Zelisko, J.A., Noble, H.B. and Porter, M., Am. J. Sports Med., 19 (1982) 297.
11. Liu, S.H., al-Shaikh, R., Panossian, V., Yang, R.S., Nelson, S.D., Soleiman, N., Finerman, G.A. and Lane, J.M., J. Orthop. Research, 14 (1996) 526.
12. Slaughterbeck, J., Clevenger, C., Lundberg, W. and Burchield, D.M., J. Bone Joint Surg., 81 (1999) 54.
13. Yu, W.D., Liu, S.H., Hatch, J.D., Panossian, V. and Finerman, G.A.M., Clin. Orthop., 366 (1999) 229.
14. Unemori, E.N., Beck, L.S., Lee, W.P., Xu, Y., Siegel, M., Keller, G., Liggitt, H.D., Bauer, E.A. and Amento, E.P., J. Invest. Dermatol., 101 (1993) 280.
15. Unemori, E.N. and Amento, E.P., J. Biol. Chem., 265 (1990) 10681.
16. Unemori, E.N., Bauer, E.A. and Amento, E.P., J. Invest. Dertmatol., 99 (1992) 337.
17. Unemori, E.N., Pickford, L.B., Salles, A.L., Piercy, C.E., Grove, B.H., Erickson, M.E. and Amento, E.P., J. Clin. Invest., 98 (1996) 2739.
18. Samuel, C.S., Butkus, A., Coghlan, J.P. and Bateman, J.F., Endocrinol., 137 (1996) 3884.
19. Calguneri, M., Bird, H.A. and Wright, V., Ann. Rheum. Dis., 41 (1982) 126.
20. Klein, K.K., Am. Correct. Ther. J., 26 (1972) 43.

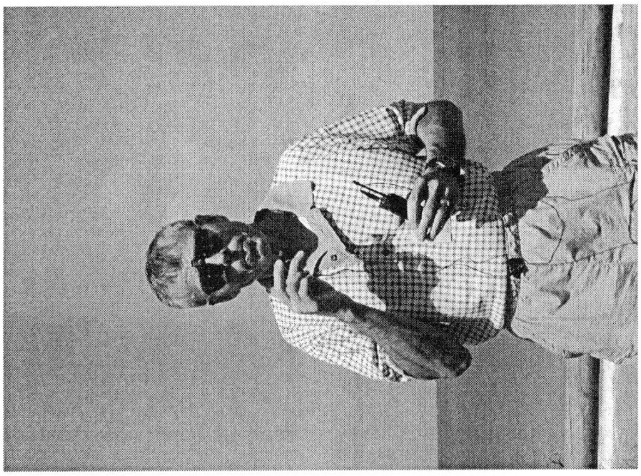

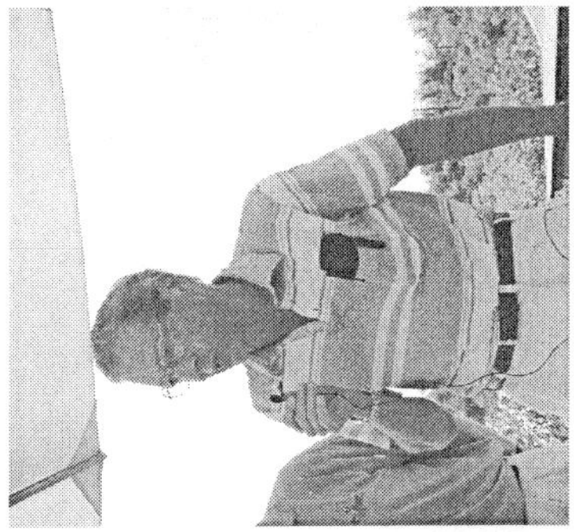

Relaxin in the new Millennium – a round table discussion

The final session of the Relaxin 2000 conference was held with the aim of involving as many delegates as possible to provide insights into the possible directions of future relaxin research. It was co-chaired by Alastair MacLennan and Geoffrey Tregear.

Alastair MacLennan:

I invite you all to help sum up the Conference and give us your current overview of relaxin and its future.

Gerson Weiss:

I have been involved in relaxin research since 1976 when Bernie (Steinetz) and I proved that relaxin is a human product of the corpus luteum. Working with relaxin for this long is a statement of the triumph of hope over reality. I have enjoyed the work, the collegiality and what I have learned. Research is the spice of clinical practice. Advances have been slow but relaxinologists know how to have a good time!

There have been several lessons. One is that there are no stand alone hormones. One needs only appreciate the vast diversity and complexity of biology and the richness of multiple overlapping and complementary systems. Looking for relaxin as the only cause of an effect is usually doomed to failure.

We should always differentiate between pharmacological and physiological studies and not kid ourselves that we are learning physiology from the pharmacological studies. Finally let us hope that we advance knowledge as we develop products.

Mario Bigazzi:

At this meeting, we heard some new topics that were very interesting and very promising. Still, the general impression is that we are still going around and around in search of a concrete core on which we could finally organize a concrete understanding of and use for relaxin. Probably we should stop thinking that some isolated (even if deep) results from the lab can sole the problem of understanding relaxin and its role. I shall return home from this conference more and more convinced that we should take a step backwards. I shall try to explain my thoughts. I am really convinced that relaxin probably does not do anything directly but that it acts mostly as a controller of other hormone actions.

Both seriously and in jest, let us for a moment think of relaxin as a "capitalistic hormone" or a "supervisor hormone". In a sense it assures that the work done by many others will reach the wished goal. It is similar to a "manager" of a farm or a boss who superintends many employees and workers. He has just to control the work they do. This is not useless, and in the post-communist era, is a well-recognized and more and more appreciated role.

G.W. Tregear et al. (eds.), Relaxin 2000, 441–451.
© 2001 *Kluwer Academic Publishers. Printed in the Netherlands.*

Now if you had followed me, and although I am only partly joking though, I hope you can understand the way relaxin -/- animals can be fertile, can be pregnant, can deliver, can even have some milk in their mammary ducts, but are useless in assuring the survival of their species because their puppies cannot eat just because a stupid nipple has not grown!! They did not have a "manager" who made the final business plan! That is the reason why I am always more convinced that we should go back to a more "philosophical" consideration of the role of relaxin. It is also the reason why I did the work on the cardiovascular system looking for a kind of supervisory action of relaxin to achieve the improved cardiac output, better blood/gas exchanges in capillaries, and faster growth of new vessels. I guess I am right!!

My love to all of you. Good work, and many thanks for your endurance and your friendship.

Alastair MacLennan:

That was a nice simile comparing relaxin to a manager of other hormones and playing a part as a conductor of the orchestra. I'm going to be the conductor of the orchestra now, and ask you to play your tune. We would love to have a summary of how you see the future of relaxin after five days of hearing diverse, interesting and multi-disciplinary presentations. Where do you think relaxin research should be directed in the next four years before our next meeting in North America in 2004.

Edouard Nice:

I think some times it's very nice to come to a meeting where you are not a specialist in that area, and actually look at it from the side lines. I came to this meeting as a non-relaxinologist. I like relaxing and I like partying, but the other side of it was very very new to me. I have really enjoyed the conference, and was very surprised that there were lots and lots of overlaps into what is my area of interest, which is the EGF-receptor system. Obviously some of the things we are seeing at the moment in relaxin is a family of ligands and a family of receptors. We that we are seeing with very slight differences, the on rates and off rates of some of the receptors, or some of the different cells where they are behaving very slightly differently. I think that might be giving us some clues and may be this is the area we are going to go. So for me, the next few years are going to be very interesting in trying to chase receptors and binding partners. I hope again, as it has with the EGF-receptor system, that this ultimately will give us the insight in where relaxin's going to go and let us develop both agonists and antagonists.

Barbara Sanborn:

I would like to continue on from that last comment and reflect a bit about signalling in the context of what we've heard over the last week. I suggest that we have to look at signalling pathways and how we work these out in different tissues from two points of view. Wherever the system is intact and functioning, one can look at the basic molecular mechanisms that are involved. However, it is important that we then place these mechanisms back into the context of physiology. I think that was the comment that was made by Mario (Bigazzi). It is an important one to remember. After we work out the principal pathways, we have to go back and determine how relevant a given pathway is in a given tissue. We also need to think about how the receptor may influence different signalling pathways in the context of cell-specific responses. It may well be that the responses to relaxin vary in different tissues due to hormonal control of regulatory effectors that influence the intracellular signalling pathways initiated. We may also have a situation where a small number of receptors in one cell may not couple to as many different pathways as in a cell that expresses a larger number of receptors. So there will probably be a wide variety of responses. Even if we define specific signalling pathways, we may find in different tissues that the receptor tends to use one or another pathway. Hormone-regulated changes in responses, for example, in pregnancy versus nonpregnancy, and other factors may influence the signalling pathways. So I think it is going to be an exciting four years. I think we will know a lot more about the receptor, a lot more about these signalling pathways, and a lot more about the complexity of these mechanisms over that period of time.

Alastair Summerlee:

I want to make personal comments about relaxin and where we're going. For me, I think, picking on the great phrase that Gerson provided has stimulated my thoughts. Working with relaxin is more about hope than reality. This is a field that I think has always embodied both art and science. That for me is one of the reasons I have enjoyed remaining in the field of relaxin. So I want to hook together the ideas of hope and reality and art and science. Probably most of you have noted how simply and how elegantly Bernie (Steinetz), and David, ask penetrating questions. Usually they are of less than one sentence in length, and they completely put you back on your heels. Whereas, I am struggling to form a long question usually with a preamble, and by the time I've formulated my thoughts they've asked their question and taken the discussion in a completely different direction. So for me, the hope is that the field of relaxin will be like the life path of somebody who has been involved in the field for a long time. That is we'll go through a period of struggling with all sorts of complex and unintelligible pieces of information and then somewhere something will happen that will make this all seem very simple and very elegant. We'll be able to look back and understand what it was that relaxin was doing from that elegant perspective. Meanwhile, we are stuck with struggling, over more than the next four years, with all of the things we have been talking about this week. For those of you who don't know, I don't sleep a great deal, and one of the things I was particularly struck about was the talk from Kris (Szebo) and the problems of finding a use for relaxin and the really disappointing news about scleroderma. I don't know how many of you have read about the Scarlet Pimpernel and the French Revolution, but there was a very famous short poem that the author wrote about the hero, and I thought that applied brilliantly to relaxin. So I would like to share with you a poem as my hope for the future: –

"We seek it here, we seek it there,
we seek a clinical use for it everywhere.
Is it in the kidney and the skin as well,
that damned elusive chemical."

David Sherwood:

First, I would like to express my gratitude to Geoff Tregear and all of his colleagues in Melbourne for organizing this superb conference. It has been incredible, Geoff, and we appreciate it.

I go away from this meeting with a sense of optimism. Much has been accomplished in the last few years. Precisely where relaxin research will take us is hard to predict. Six years ago at the Third International Relaxin Conference in Adelaide, characterization of the relaxin receptor was an attractive goal that was often mentioned. That objective remains to be accomplished, and description of the relaxin receptor would be a superb contribution at the Fourth International Relaxin Conference. Additionally, we have learned that relaxin has previously unidentified activities in organs such as the kidney and liver. There are so many diverse aspects of relaxin that are likely going to be investigated over the next few years that the next international relaxin conference will surely reveal many additional novel findings.

Finally, one of the important things I'm taking away from this conference is new friendships and a heightened sense of camaraderie that seems as important as the acquisition of new knowledge. I think that the openness and spirit of sharing that we have experienced at this meeting is potentially the finest thing we have attained. If we continue to collaborate and share with one another, not only will progress with relaxin research benefit but also participation in the process will be more enjoyable.

Bernie Steinetz:

I would like to echo Dave's sentiments about the organization and conduct of this meeting and the selection of the venue. It's just incredible, and on a scale of 1 to 10, I would certainly give this meeting an 11! Regarding my thoughts, Alastair (Summerlee) mentioned that I speak in single sentences. That's all I can remember at any one time! Incidentally, that problem was exacerbated by Alastair's good work last night. I happened to be at his table, and in addition to his renowned scientific abilities, he is an incredible wine expert. He managed to juggle four bottles of wine at once without spilling a drop, and of course, I had to try them all!

I think what I get from this wonderful conference is that actually the relaxin world is expanding with these previously elusive factors that are being newly described every day. What it illustrates is the very basic importance of this structural type, this whole family of substances that we call relaxin. (I'm prejudiced, so it's "relaxin", but I will settle for "insulaxin".) During evolution, this structure has probably been as important as the steroids in promoting life and diversity of life, and when you stop to think that small changes in either the A- or B-chain can convert the substance from having one biological activity to an entirely different type, it's pretty mind-boggling!

One thing that is definitely needed, particularly for the in vitro work, is a control substance that binds to either the A- or B-chain or to the whole relaxin molecule, without activating post-receptor pathways. Incidentally, heating alone is not sufficient to inactivate relaxin; it can be autoclaved and still retain its activity! What I would envision would be a substance that would have a relaxin receptor binding domain that would bind to the relaxin receptor with high affinity and block test substances (with putative relaxin activity) from binding to the receptor. Something like that is really needed to prove the specificity in many of the activities that are being looked at. There are many activities that have been described that may or may not be post receptor events; there is really no way now of knowing which comes first. Right now, where Geoff is involved, I wonder whatever happened to histidine decarboxylase. It would seem to me that a lot of the actions of relaxin could be explained on the basis of histamine release, but that's not very popular anymore. There are examples in invertebrates such as insects, where it's been shown that insulin-like compounds do activate histidine decarboxylase. That might be worth looking at in mammals.

Regarding the clinical trials, I hope that pharmaceutical companies will not continue to have premature fixations with what the hormone should do, and then design their studies on that basis; otherwise they'll be going up the primrose path one more time. I have seen it happen *five* times now and unless there is really a change in attitude, I think it's going to happen again. We should not give up on relaxin's use in the softening of the cervix. Dr. Vogel's work clearly showed there were marked effects on parturition and also shortening of labour, and so, I think if it were investigated in a proper way, elucidation of the mechanism for this could prove that relaxin is a clinically useful substance.

The scleroderma trial may also have not been optimally designed. The original observations were made with porcine relaxin, which may have properties which are somewhat different from those of human relaxin II. I think that another careful look at that should compare activities of porcine and human relaxins in vivo.

Finally, with respect to the mammary gland, there is a subtle difference, at least in the rat, between what Dave Sherwood sees with his anti-relaxin NCA 1 monoclonal antibody treatment of pregnant rats and what you see when you ovariectomize rats and maintain them with estrogen and progesterone. When you add relaxin in the latter case, it seems to have a much greater effect on the mammary gland development, whereas the NCA 1 seems to block primarily nipple growth in pregnant rats. I wonder if the difference isn't due to the production of relaxin by the gland itself; the NCA I may not neutralize the locally produced relaxin which may, itself exert autocrine actions on the gland. Anyway, with those thoughts, I'll say thanks again, it's been a real experience seeing all my old friends here, and meeting so many enthusiastic new young investigators. Most of all, despite all of the setbacks, I am delighted that there are more and more people getting hooked on relaxin. Thank you.

Peter Ryan:

I'm not an expert in the field of relaxin, and I'm not really going to speak on the science, as I regard myself as a neophyte, but I'd like to share something with you all that I remember distinctly from Dr. Fred Greenwood in 1992 when the Porter Lab and the Bryant-Greenwood-Greenwood Lab met at the Vancouver meeting of the Society of Reproduction and Fertility. One of the things he said to us as a group was that as relaxinologists we really have to look out for each other, we have to stand behind each other when we submit our manuscripts for review, at the peer review level, and when it comes to consideration with out grant applications. This sort of sentiment has been echoed here today. The relaxin field is larger than the individual group or the individual researcher. It's a collective endeavour, and the field is only going to move forward if we continue in the collective endeavour. I really am heartened by the degree of collaboration that we have witnessed here this week. We owe a great degree of gratitude to the senior scientists in the field of relaxin, like David Sherwood and Bernie Steinetz, and it's heartening to see such a large group of young scientists who are willing to crack this very difficult, and as Alastair eloquently put it, elusive hormone. So with that I would like to encourage all of us to go away not disheartened by some of the events that we have heard this week, but invigorated, and look forward to an exciting and forward moving exercise in the year 2004 somewhere in the United States.

Richard Ivell:

I came here, of course, with the real hope that we would have another indication for relaxin in the context of hair loss, but obviously that indication hasn't been mentioned yet, but I'm ever hopeful!

Two aspects I want to underscore. The first is, we haven't got the receptor. That's right, but I've come away with a much better image, even though it's seen nebulously through a veil of what that receptor probably looks like. Having said that I'm going to put my money on the table and say, we are probably going to have a type of receptor we don't know about yet. It will be a tyrosine kinase yet but it will probably belong to a sub-family of which we probably don't have any members yet. It's out there somewhere and that's why we're having such difficulty trying to understand and trying to trace it. I would bet that the next time we're here together it will all be so clear. Some one made the comment a couple of days ago that, this all sounds like insulin was ten – fifteen years ago until they found the receptor. I think that's very true indeed. We are in that very similar situation, and I think it could all become so easy and so clear once we have that molecule.

The second point which really struck home to me was the plethora of different clinical indications we have. We've seen some very exciting work from being just suppositions a few years ago. We are now seeing some hard data on real clinical consequences or clinical effects. Linked to that is another little point – these effects are often quite subtle and we don't see them first time round. This suddenly came also out of the work with the knock-out mice. When they were first described we all thought

OK let's phenotype this one, we listed two or three different symptoms and that was the knock-out. Now, a few years down the road, whether we're talking about the RLF knock-out or the relaxin knock-out, we are seeing now subtle effects here, and subtle effects in the male tract, may be something else happening there. I'm seeing that not only as interesting signs but also as a warning that we still really don't know enough about relaxin in the whole animal context, or, of course, in a chronic context. So my bottom line here is that I still think before we go rashly through all these clinical trials, we have to see some more good whole animal physiology in model systems with human relevance. I think we could all benefit from that immensely. I was extremely impressed with the talk yesterday about the breast cancer situation where, just by looking at the wide picture, in this case, the dissemination of metastases, we might suddenly see something which we've been missing all this time, when we've only been looking at the origins of breast cancer. So I think we've got to be more subtle and think a few steps more ahead. That's basically the things I've wanted to say, and I'm hoping the next time round when we meet it will all be so much easier.

Thanks again Geoff, and to all the Melbourne team who put so much into this. I think it was absolutely super. Thank you.

Elaine Unemori:

The most enjoyable thing about this meeting is that it gives people like us from the biotech industry the opportunity to appreciate the depth and breadth of relaxin biology. We tend to be focussed on human disease, human indications and often times that makes us myopic. So it's really nice to be able to come, particularly to a setting like this, to be able to understand what relaxin does in other animal species, which is equally important to the overall understanding of relaxin.

I would like to say something about the scleroderma clinical trial because, obviously, it has a great impact on a lot of people, if not everyone, here. Of course, we were deeply disappointed by the results of the clinical trial, but I think one of the things that we are satisfied with is that we conducted the best possible clinical trial in which to test relaxin in this indication. This indication was not entered into lightly. There was a great deal of preclinical work that went into it, including animal models that are as close as we could get to the human system without actually going into the human. Of course, when you're talking about human disease, that's ultimately the experiment that you have to do. The first clinical trial in scleroderma with relaxin was planned in 1993, started in 1994, and subsequently, there have been three clinical trials in this area. We gained a lot of experience about relaxin in those first studies and we consulted with the world's best experts in the area to try to design a study that would give relaxin its best shot. A couple of people have mentioned that part of the problem was that we took patients that had progressed so far in their disease that it would not give relaxin a fair chance. But scleroderma is an incredibly complicated, disease of multiple etiologies that nobody quite understands. What happens in the first few years of the disease is that the mortality rate is extremely high, and so, given this disease course, it was decided that this would not be the appropriate interval in which to test relaxin's efficacy. So the inclusion criteria were designed to include patients that were far enough along in the disease to avoid this complication, but not so far along that the disease would be refractile to any sort of treatment. Anyway, having said that, of course, the results are extremely disappointing. We don't feel that this necessarily means that relaxin does not have potential in other fibrotic diseases. Of course, the trial results have given us lots of data, which we are still collecting. There is still a lot to analyse, and we hope that a lot more interesting information on relaxin in the human will come out of this data. Kris (Zsebo) mentioned some of the things that we've learnt so far from the trial. My personal opinion is that you can learn a great deal about physiology by studying pharmacology. For example, the results of the clinical trial with regard to the haemodynamic effects of relaxin, in conjunction with the work of Lee Danielson and Kirk Conrad have shed a lot of light on the physiology of early pregnancy.

Kris (Zsebo) also gave you an overview of our development program, and I realize that some of you had questions regarding the pace at which we are moving forward in the clinic in specific indications. Just to reassure you, we also have plans for studies whose sole purposes are to understand relaxin's biology from a human safety standpoint, as well as from a physiological standpoint. We have a red cell mass study that is planned, and the sole purpose of that study is to understand what relaxin's effect is on the human bone marrow. We've seen from our clinical trials that you get changes in haematocrit and haemoglobin, and have hypothesized that this can be ascribed to increases in intravascular volume and not haemolysis. The purpose of that study is to give us, and hopefully all of you, a better understanding of what relaxin does in the human. The other study is designed to look at the haemodynamics of relaxin in the human. We are going to be looking at things like cardiac output, peripheral vascular resistance, blood pressure, ejection fraction - to try to understand, as best as we can, what relaxin's effect on the cardiovascular system in the human is. These studies should give all of us some interesting results. If the NIH feels that relaxin needs some validation in human data, those trials should offer that information without the risk of having a failure in a clinical indication.

I hope I am reassuring all of you that we're trying to do the right thing. We're trying to do it the right way, and certainly if any of you have questions about the scleroderma trials, there several of us here who would be happy to answer whatever questions you have as completely as we can.

Ernst Rinderknecht:

I have been working 20 years for a biotechnology company, actually two of them, and my motivation has been doing great science and, hopefully, at the end being able to produce something that helps patients. In the course of that, I have, of course, observed that clinical development is and has always been very difficult. There have been failures in the past and there will probably be more failures in the future. We hope that at some point we can find something that actually will benefit patients and the research field. I want to follow up on Elaine, and I want to express our thanks to all of you here. I want to pay tribute to your research in relaxin. We know that we cannot do this alone.

We need input. This is a two-way street, and we hopefully at some point can give something back to you the research community. We very much appreciate your efforts in the field of relaxin. Thank you.

Laura Goldsmith:

I think there are people who like to make things more focussed and specific and there are those of us who like to see things in more of a global picture. You know, over the years I have been studying relaxin I received a lot of comments such as relaxin doesn't do anything. Study section members may be prejudiced that relaxin may not be that important. I think that we all have to recognize that relaxin is really the embodiment of several biological and specifically endocrinological principles, and one of the most important is biological redundancy. The problem with the knock-outs in general is that when you knock-out one gene the systems are very complex. In order to ensure survival of the species, you get redundancy and overkill from other systems that will replace those that you've knocked-out. I think one of the problems with the knock-out mice for relaxin is that there has probably been quite a bit of that. The other biological principle is species specificity. Relaxin is probably more unique than many other substances in terms of this true species specificity. What relaxin does and what it does not in the context of a hormone is really very species specific. It is also specific in terms of cell type. For example, what was shocking to us is that we can take an endometrium and see relaxin inhibit matrix metalloproteinases and yet take cervical tissue and see it increase these matrix metalloproteinases. So, presumably it's using the same receptor. The ligand is the same, the receptor is likely to be the same, but the downstream signalling pathways are going to be modulated by other physiological things that that cell has been exposed to, and tissue specific differences. So I think that when we all work with relaxin we really have to keep that in mind. This meeting I think has been incredibly excellent by giving us an additional whole novel breadth of things that relaxin does. I think in the next few years there also has to be a lot more depth. I think that's true for what everyone is doing, and that's difficult without a larger critical mass of investigators. This meeting has shown us all that we can encourage new young investigators to be interested. In the United States, part of the problem is a funding one. It's difficult to get relaxin grants funded. There is a certain amount of difficulty to have people who review grants think that relaxin is important. So I look forward to more people being involved so that more of us can sit on study sections and help the cause of additional work being done in relaxin. I believe that relaxin is a modulator of what other hormones do, but a critical modulator such that there may be elements of redundancy about how it works in the normal situation. That's the reason why the pathological is not always a good way to study the normal.

I would like to thank Geoff and all of you for organizing this just fabulous meeting. Thank you.

Carol Bagnell:

I want to echo the words of those who have come before me, and add a few other comments. Relaxin is a very old hormone, as we all know. There have been some newly discovered hormones, like leptin, and it is amazing to me that research on these newer hormones has moved forward so quickly, while we continue to search for basic information on the mechanism of relaxin action and the relaxin receptor. I am still very encouraged, especially after coming to this meeting, about the work on relaxin because it's incredible to me to hear about the breadth and depth of functions of this hormone. It's also encouraging to hear about the new related insulin-like factors. New developments in this area will likely help us with finding the receptor, which will be very important in moving the field forward. If we can characterize the receptor, which we were talking about at the last meeting, that will provide new clues to relaxin's functions and clinical uses for the hormone, which is what's also going to drive the research. I also agree with Laura and Peter that since we're such a small field we really have to help each other and encourage and support new investigators entering the field. I look forward to seeing you in North America next time.

Alastair Summerlee:

I just wanted to follow up on the things that Elaine was saying. I think it's very important for us, as well as having reservations about moving forward with the clinical work, to be very clear that we as a small scientific community cannot do this, without the continued support of industry. Even though all of us are profoundly disappointed about the scleroderma trials, I'm actually amazingly excited about the fact that Connetics is going to continue the work on this from the clinical aspect as it was from the basic side. I hope that you as a group have not gone away with any sense of reservation on our part. This is an amazingly important partnership for us, not only as scientists but as human beings because we need a high level of support among all of the people who are interested in working on relaxin. So for me, listening to Elaine, I felt guilty that we had probably erred to much on the side of saying let's not rush into things clinically, because there is no doubt that the work that has gone on over the last ten years has really made a significant contribution to the basic biology of relaxin. On behalf of everyone here I would like to thank Connetics very much for all that they've done.

Alastair MacLennan:

It is my duty to wrap up this final discussion. Those who only read the transcript of this session will find it hard to visualise the spectacular and unique setting in which this academic debate has taken place. Just to give some feeling for the occasion, delegates are sitting having a tropical breakfast under parasols on the sand dunes overlooking an almost empty 30 kilometre idyllic beach with a very blue Indian Ocean gently rolling onto the shores of Cable Beach, Western Australia. We are 200 kilometres from any metropolis and yet each day we have reached the heights of modern sophistication in our scientific sessions. Fred Greenwood would have been more than pleased to see the bonding, the unity, the friendship and the collaboration amongst the relaxinologists of 2000. He always tried to encourage such an ethic. I am sure this unofficial family and its clear support for each other will greatly facilitate progress in relaxin research in years to come.

As an obstetrician and gynaecologist, I have greatly enjoyed this meeting with its integration of the clinical and basic sciences. Although it is a commercial necessity to look for early applications of human relaxin to alleviate disease, I encourage future researchers to look also at areas where the effects of relaxin in the human may not always be advantageous, such as premature labour and, in particular, the long-term effects of connective tissue remodelling in pregnancy. Over 50% of older women have significant pelvic floor dysfunction as exhibited by urinary and rectal incontinence and genital prolapse. These problems are related to the hormonal changes of pregnancy rather than vaginal delivery. Evolutionary mechanisms that favour reproduction and easier childbirth may not be an advantage after menopause. Longevity may be soured by morbidity contributed to by relaxin and its effect on connective tissue relaxation in pregnancy. Therefore, in time we must look at mechanisms that are the opposite of relaxin which encourage connective tissue recovery.

To encourage recovery from this week of work I invite you now to dive into the beautiful sea in front of you and take advantage of this wonderful venue, but firstly, I bestow gold medals to Geoff Tregear and the Howard Florey team and congratulate them on your behalf for making this such a successful meeting.

Geoff Tregear:

I think this is an appropriate time to wrap up this session. If I could just thank Alastair MacLennan for all his work in chairing this session, and indeed, in helping us throughout the organization of this meeting. If you had as much fun as the Florey team had in getting this meeting ready and running it, I would be pleased because it's been a ball for us and we have enjoyed every minute of it. I just wish to thank you all for coming. We realize it's a long way and it was expensive but I do appreciate you coming here. I would now just like to wish you a safe journey home, and look forward to seeing you in four years time. Thank you.

RELAXIN'
on the Beach

Author index

Index